T0270950

Dynamics of Multiphase Flows

Understand multiphase flows using multidisciplinary knowledge in physical principles, modeling theories, and engineering practices. This essential text methodically introduces the important concepts, governing mechanisms, and state-of-art theories, using numerous real-world applications, examples, and problems. It covers all major types of multiphase flows, including gas–solid, gas–liquid (sprays or bubbling), liquid–solid, and gas–solid–liquid flows. It introduces the volume–time-averaged transport theorems and associated Lagrangian-trajectory modeling and Eulerian–Eulerian multifluid modeling. It explains typical computational techniques, measurement methods, and four representative subjects of multiphase flow systems. Suitable as a reference for engineering students, researchers, and practitioners, this text explores and applies fundamental theories to the analysis of system performance using a case-based approach.

Chao Zhu is a Professor in the Department of Mechanical and Industrial Engineering at the New Jersey Institute of Technology. He co-authored *Principles of Gas–Solid Flows* (Cambridge, 1998) and is a Fellow of the American Society of Mechanical Engineers.

Liang-Shih Fan is a Distinguished University Professor and The C. John Easton Professor of Engineering in the Department of Chemical and Biomolecular Engineering at The Ohio State University. He is the Editor-in-Chief of *Powder Technology* and a Member of the U. S. National Academy of Engineering.

Zhao Yu was a postdoctoral researcher at The Ohio State University. He is now an Associate Senior Consultant Engineer in Bioproduct Research and Development at Eli Lilly and Company.

Cambridge Series in Chemical Engineering

Series Editor
Arvind Varma, *Purdue University*

Editorial Board
Juan de Pablo, *University of Chicago*
Michael Doherty, *University of California-Santa Barbara*
Ignacio Grossmann, *Carnegie Mellon University*
Jim Yang Lee, *National University of Singapore*
Antonios Mikos, *Rice University*

Books in the Series
Baldea and Daoutidis, *Dynamics and Nonlinear Control of Integrated Process Systems*
Chamberlin, *Radioactive Aerosols*
Chau, *Process Control: A First Course with* MATLAB
Cussler, *Diffusion: Mass Transfer in Fluid Systems, Third Edition*
Cussler and Moggridge, *Chemical Product Design, Second Edition*
De Pablo and Schieber, *Molecular Engineering Thermodynamics*
Deen, *Introduction to Chemical Engineering Fluid Mechanics*
Denn, *Chemical Engineering: An Introduction*
Denn, *Polymer Melt Processing: Foundations in Fluid Mechanics and Heat Transfer*
Dorfman and Daoutidis, *Numerical Methods with Chemical Engineering Applications*
Duncan and Reimer, *Chemical Engineering Design and Analysis: An Introduction 2E*
Fan, *Chemical Looping Partial Oxidation Gasification, Reforming, and Chemical Syntheses*
Fan and Zhu, *Principles of Gas–Solid Flows*
Fox, *Computational Models for Turbulent Reacting Flows*
Franses, *Thermodynamics with Chemical Engineering Applications*
Grossmann, *Advanced Optimization for Process Systems Engineering*
Leal, *Advanced Transport Phenomena: Fluid Mechanics and Convective Transport Processes*
Lim and Shin, *Fed-Batch Cultures: Principles and Applications of Semi-Batch Bioreactors*
Litster, *Design and Processing of Particulate Products*
Maravelias, *Chemical Production Scheduling*
Marchisio and Fox, *Computational Models for Polydisperse Particulate and Multiphase Systems*
Mewis and Wagner, *Colloidal Suspension Rheology*
Morbidelli, Gavriilidis, and Varma, *Catalyst Design: Optimal Distribution of Catalyst in Pellets, Reactors, and Membranes*

Dynamics of Multiphase Flows

CHAO ZHU

New Jersey Institute of Technology

LIANG-SHIH FAN

Ohio State University

ZHAO YU

Ohio State University

CAMBRIDGE
UNIVERSITY PRESS

University Printing House, Cambridge CB2 8BS, United Kingdom

One Liberty Plaza, 20th Floor, New York, NY 10006, USA

477 Williamstown Road, Port Melbourne, VIC 3207, Australia

314–321, 3rd Floor, Plot 3, Splendor Forum, Jasola District Centre, New Delhi – 110025, India

79 Anson Road, #06–04/06, Singapore 079906

Cambridge University Press is part of the University of Cambridge.

It furthers the University's mission by disseminating knowledge in the pursuit of
education, learning, and research at the highest international levels of excellence.

www.cambridge.org
Information on this title: www.cambridge.org/9781108473743
DOI: 10.1017/9781108679039

First published 2021

A catalogue record for this publication is available from the British Library.

ISBN 978-1-108-47374-3 Hardback

Dedicated to

Anqi Zhou,

Cheng-Yuan Rao Fan,

and

Li An Bao and Ning Liu

Contents

Preface

Multiphase flows refer to the flows with distinctively different dynamic responses of each phase in the transport and phase interactions that impact the transport phenomena. Knowledge of multiphase flows is important for researchers, scientists, and engineers in various disciplines. Our motivation for writing this book was propelled by the rapidly expanding interest in the optimized design, operation, and applications in the field of multiphase flows and the ever-increasing needs of higher education in this field. Understanding multiphase flows requires both the multidisciplinary knowledge in inherent flow concepts and theories, and their associated engineering practices. There is indeed a considerable literature that is already available for students, researchers, and other readers who are interested in multiphase flows. This literature includes research articles on fundamental concepts or problems of an applied nature, edited books on the progress of specific topics of interest, monographs on focused subjects, special collections of an individual researcher's publications, and handbooks contributed by various topic experts that outline basic principles and design criteria or formulae. However, these monographs and edited books on multiphase flows tend to have a more narrowly focused scope of subject matter and are often concerned with advanced topics, making them a desirable reference source for researchers rather than a foundational, comprehensive, and systematic introductory guide to the fundamental theories of the field, particularly for beginners. While there are a number of books that do address some fundamental theories, they either lack exercise examples and homework problems that are essential to deep learning for students, or they are not accompanied by solution manuals for homework problems. Their pedagogical effectiveness is therefore limited as they cannot be conveniently adapted by instructors as textbooks for use in the classroom.

A unified textbook in the field that can methodically encompass a broad range of the important elements of knowledge in the multiphase flow field, along with sufficient theoretical and applied details in a manner suitable to both introductory and advanced level learning in an instructional setting would be a welcome addition. This book is conceived for just this purpose, serving as an educational, learning-oriented text that introduces multiphase flows to engineering students, advanced researchers, and other readers. This book may also be regarded, to some degree, as an expanded sequel to an earlier book published in 1998 by two of the current co-authors (Fan and

Zhu), titled *Principles of Gas–Solids Flows* (Cambridge University Press). This earlier book was limited only to the gas–solid flows and did not include the other topics of multiphase flows such as gas–liquid flows (sprays or bubbling flows), liquid–solid flows (slurry flows), and gas–solid–liquid flows (sprays in gas–catalysts flows or catalysts in bubbly liquid reactors). Thus, this book seeks to more fully articulate the coherent linkages among, for example, hydrodynamics, phase changes (reactions, mass, and heat transfer), and electrostatic charges in multiphase flow applications, as well as provide an introduction to computational multiphase flow theories and basic measurement methods.

The book contains twelve chapters. The first eight chapters cover fundamentals, and the last four chapters are dedicated to the introduction of representative applications. Specifically, this book has the following characteristics:

– It provides a systematic introduction and review of multidisciplinary fundamentals critical to understanding multiphase flows (Part I, Chapters 2–6). Specifically, Chapter 2 offers a summarized review of various single-phase flows, including the turbulent modeling of Newtonian fluid flows, viscous flow through porous media, and inertia-dominated granular flows. Chapter 3 focuses on the introduction of phase transfer between an isolated object and a surrounding fluid flow, including various physical mechanisms and their representative formulations for momentum, heat, and mass transfer across the interface between the object and fluid with a relative motion. Chapter 4 discusses the impact of neighboring objects and their dynamic behaviors on the phase transfer mechanisms between the object of concern and its surrounding fluid flow. Such an impact can be affected by the noncontact interobject interactions via the intervening common fluid, by the direct contact (collisions) of the interacting objects, by the noncontact field interactions (such as electrostatic or electromagnetic fields), or by all three of the aforementioned scenarios. Chapter 5 and Chapter 6 introduce the two most popular methods in the modeling of multiphase flows. Chapter 5 is primarily focused on Eulerian–Lagrangian modeling where the Lagrangian equations are used to account for motions or trajectory tracking of individual objects and the Eulerian equations are used to describe the Eulerian field behavior of the fluid flow surrounding those objects. Chapter 6 presents Eulerian–Eulerian modeling (also known as continuum modeling) to yield field descriptions of individual component phases that constitute a multiphase flow. The key conceptual elements in continuum modeling include the constitution of each continuum phase that co-occupies a physical space, the continuum description of phase interactions or coupling, and the in-phase transport due to phase nonuniformity (hence the definition or formulation of transport coefficients).

– Part II of the book is concerned with the characteristics of four selected subjects of multiphase flow systems. They are introduced in the last four chapters. This part of the book is designed not only for readers who wish to explore important engineering applications involving multiphase flows, but also for those readers who wish to apply the fundamental knowledge gained from Part I in an integrated manner to the analysis of the performance of the systems with practical engineering relevance. It should be noted that the real multiphase flow systems can be complicated by such factors as irregular vessel geometry, varied flow regimes with complex interphase transport and phase change properties, and polydispersity of the size of the discrete phases. Such complexity is well beyond what the theoretical modeling based on the multiphase flow principles described in Part I can fully account for. However, it is expected that by following the flow principles presented in Part I while, over time, augmenting them with additional knowledge on model formulation and closure relationship, and the computing capabilities, the gap of the comparison between the model prediction and experimental data can significantly narrow down. Specifically, Chapter 9 introduces typical multiphase flow systems of phase separation, with highlighted fundamentals of separations by inertia, filtration, and electrostatic precipitation. Chapter 10 describes the basic principles of fluidization and typical fluidization systems including dense-phase fluidized beds and circulating fluidized beds. Fluidization is a major industrial practice applied not only for phase mixing and reactions but also for phase transport. Fluidization in vertical columns or pipes is also discussed in this chapter. Chapter 11 focuses on phase transport by multiphase flows through horizontal pipes. Chapter 12 explores some complex multiphase flow systems with strong coupling to physical phase changes and chemical reactions.

- The book also contains two chapters dedicated to the numerical methods and experimental techniques of multiphase flows. These chapters can be found between Part I and Part II. Chapter 7 provides a brief introduction to common numerical methods that can be used for solving coupled differential equations with boundary conditions in various types of multiphase flow models. Highlights include the methods for numerical treatments of interfacial phase transfer, the algorithm coupling equations of individual phases, and the multiscale behavior among various phases or transport properties. Chapter 8 summarizes some common measurement technologies in the experimental studies of multiphase flows. The sections in this chapter are organized according to the specific purposes of the measurements such as particle size or volumetric concentrations.

- Most established theories of multiphase flow are based on simple multiphase flows characterized by, for example, dispersed monosized rigid spheres of solids being transported in the laminar flows of Newtonian fluids. It is natural, therefore, that the core of the book is oriented toward the introduction of basic theories that are formulated for such idealized multiphase flows. Nevertheless, in order for the theories to be more applicable to practical situations, this book also aims to include more complex flows to which a simple theoretical treatment can be reasonably extended. Such treatments and situations can be exemplified by the mechanistic modifications associated with deformable particles (such as droplets and bubbles), porous or permeable particles, phase changes (such as vaporization and reactions), complex interfacial transfers (such as collision-induced phase transfers), turbulence, non-Newtonian fluid flow, and coupling with external fields (such as electric, electromagnetic, or acoustic fields).

- Throughout the book, the structure of each chapter is organized based on the following framework: (1) basic concept and formulation, and (2) examples or case studies with detailed explanation and analysis.

- Homework problems are included in each chapter. These problems are designed to provide practice opportunities for readers to gauge and assess their progress. The companion solution manual for the end-of-chapter problems is available upon request for instructors from the publisher at cambridge.org/9781108473743. Some problems are open ended and are posed only with necessary conditions rather than the usual "sufficient and necessary conditions," which yield unique answers to the problems. Thus, for these open-ended problems, various possible solutions can result from the reasonable assumptions of additional conditions required to ensure problem closure or the uniqueness of the solution. Some homework problems are also designed to further explore relevant theories or applications, with the aid of references cited in the problems.

- References for each chapter are given at the end of each chapter. Throughout the book, most of the references adapted are classical in the field, which have endured years of validation. Some more recent publications are also referenced. They are associated more with case studies and advanced homework problems, typically in modeling and simulations, for readers who wish to keep pace with related advances, understandings, and emerging subjects.

This book should by no means be used to the exclusion of other published books devoted to multiphase flows. Rather, readers are encouraged to adopt an integrative approach, as a number of these books complement each other. Thus, using this book in conjunction with others will likely maximize the development of the reader's knowledge and understanding. Such books include, but are not limited to, the following:

- *Bubbles, Drops and Particles*, R. Clift, J. R. Grace, and M. E. Weber, Dover Publications, 1978.
- *Multiphase Flow in Porous Media: Mechanics, Mathematics, and Numerics*, M. B. Allen III, G. A. Behie, and J. A. Trangenstein, Springer, 1988.
- *Particulate and Continuum-Multiphase Fluid Dynamics*, S. L. Soo, CRC Press, 1989.
- *Multiphase Fluid Dynamics*, S. L. Soo, Science Press, 1990.
- *Boiling Heat Transfer and Two-Phase Flow*, L. S. Tong and Y. S. Tang, 2nd edition, CRC Press, 1997.
- *Principles of Gas-Solid Flows* , L.-S. Fan and C. Zhu, Cambridge University Press, 1998.
- *Dynamics of Bubbles, Droplets and Rigid Particles*, Z. Zapryanov and S. Tabakova, Springer, 1998.
- *Fluid Dynamics and Transport of Droplets and Sprays*, W. A. Sirignano, Cambridge University Press, 1999.
- *The Dynamics of Fluidized Particles*, R. Jackson, Cambridge University Press, 2000.
- *Two-Phase Flow: Theory and Application*, C. Kleinstruer, CRC Press, 2003.
- *Computational Models for Turbulent Reacting Flows*, R. O. Fox, Cambridge University Press, 2003.

- *Fundamentals of Multiphase Flow*, C. E. Brennen, Cambridge University Press, 2005.
- *Multiphase Flows with Droplets and Particles*, C. T. Crowe, J. D. Schwarzkopf, M. Sommerfeld, and Y. Tsuji, 2nd edition, CRC Press, 2011.
- *Theory and Modeling of Dispersed Multiphase Turbulent Reacting Flows*, L. Zhou, Butterworth-Heinemann, 2018.
- *Essentials of Fluidization Technology*, Ed. J. R. Grace, X. T. Bi, and N. Ellis, Wiley-VCH, 2020.

Like our 1998 book, *Principles of Gas–Solid Flows*, this book can be used at the advanced undergraduate and graduate levels in courses related to mechanical and power engineering, chemical reaction engineering, pharmaceutical engineering, environmental engineering, and process system engineering. The pre-requisite subject knowledge for this book includes basic thermodynamics, fluid mechanics, and heat and mass transfer. To use this book in its entirety for teaching, a two-semester course sequence is ideal. For a 14-week or one-semester tech-elective course offered to undergraduate students, the following selective subjects taken from this book can be covered:

1. Introduction to Multiphase Flows (Chapter 1)
   - Basic concepts and definitions
   - Examples of multiphase flows in engineering applications
2. Single-Phase Flows (laminar flow of Newtonian fluids) (Chapter 2)
   - Flow over a sphere
   - Pipe flow
   - Jet flow
3. Phase Interaction with a Sphere (Chapters 3 and 4)
   - Drag force and other particle-fluid forces
   - Modification in presence of neighboring spheres
   - Momentum transfer by collisions
   - Heat and mass transfer
4. Lagrangian Trajectory Modeling (Chapter 5)
   - BBO equation
   - Deterministic Lagrangian trajectory modeling
   - Stochastic Lagrangian trajectory modeling
   - Collision-dominated trajectory modeling
5. Eulerian Continuum Modeling (Chapter 6)
   - Phase averages and averaging theorems
   - Volume-fraction-based equations
   - Time-fraction-based equations
6. Systems for Phase Separation (Chapter 9)
   - Sedimentation chamber
   - Cyclone
   - Impactor and scrubber
   - Granular and fibrous filters
   - Electrostatic precipitator

7. Systems for Phase Mixing (Chapter 10)
   - Fluidized beds
   - Jet mixer
   - Rotational blenders
8. Systems for Phase Transport (Chapter 11)
   - Pneumatic conveyers
   - Risers
   - Sprayers

For a 14-week or one-semester course offered to graduate students, the following selective subjects taken from this book can be covered:

1. Introduction to Multiphase Flows (Chapter 1)
   - Basic concepts and definitions
   - Examples of multiphase flows in engineering applications
2. Modeling of single-phase flows (Chapter 2)
   - Classification of viscous fluids
   - Laminar flows of Newtonian fluids
   - Turbulent flows of Newtonian fluids: RANS and turbulence modeling
   - Flow through porous media
   - Granular flows: kinetic theory modeling
3. Particle–Fluid Interactions (Chapters 3 and 4)
   - Momentum transfer: drag and other particle–fluid interaction forces
   - Heat transfer by conduction and convection
   - Mass transfer by diffusion and phase change
4. Collision of Solid Particles (Chapters 4 and 5)
   - Hard-sphere model
   - Normal collision characteristics of elastic objects
   - Oblique collision with tangential friction and torsional traction
   - Inelastic collisions and plastic deformation
   - Soft-sphere model
5. Formation and Interactions of Fluid Particles (Chapters 4, 6, 10, and 12)
   - Bubble formation and bubble column
   - Droplet formation and spray atomization
   - Breakup and coalescence of fluid particles
   - Population balance model
6. Continuum-Discrete Tracking Modeling of Multiphase Flows (Chapter 5)
   - Lagrangian Trajectory Modeling
   - Transport Coupling in Eulerian–Lagrangian Modeling
7. Continuum Modeling of Multiphase Flows (Chapter 6)
   - Phase averages and averaging theorems
   - Volume-averaged equations
   - Volume–time-averaged equations
   - Turbulence modulation

8. Multiphase Flow Applications with Modeling Analysis (Chapters 9–12)
   - Cascade impactors
   - Cyclone separation
   - Electrostatic precipitation
   - Fibrous filtration
   - Pneumatic pipeline transport
   - Fluidization and fluidized beds
   - Spray drying
   - Bubble columns
   - FCC riser reactor
9. Projects: Presentation of a Research Paper (from published references)
   - CFD modeling and simulation of a multiphase flow (Chapter 7); or
   - Experimental method or system of a multiphase flow, with data analysis (Chapter 8)

Such coverage could naturally be supplemented with material from the reference books given above and some state-of-the-art publications. The projects indicated above, preferably presented and discussed in class, can be extraordinarily beneficial to the students since they represent the more recent advances and practical applications of modeling, simulation, and/or measurements of multiphase flows. The draft of this book had been used effectively in one- and/or two-semester formats for mechanical engineering and chemical engineering students, respectively, at the New Jersey Institute of Technology (NJIT) and the Ohio State University (OSU).

The feedback provided by students who took the multiphase flow course or related courses such as particle technology, particulates flows, and multiphase reaction engineering from both NJIT and OSU for the past two decades has been extraordinarily helpful in shaping the contents and the structure of the book in its present form. The authors are deeply thankful for this constructive feedback and the valuable insights of the students. Several colleagues, Professor Alissa Park, Dr. Teh C. Ho, Dr. Chao-Hsin Lin, and Professor Jonathan Fan have been very generous with their time in reading the book manuscript in full or in part during its preparation. They have provided invaluable comments that have been incorporated in this book, and the authors are very appreciative of their engagement. A final stage in the revision of part of the book manuscript was undertaken by Dr. Liang-Shih Fan during the spring semester 2019 when he took a sabbatical leave at ETH, Switzerland hosted by Professor Sotiris Pratsinis. Professor Pratsinis's thoughtful and supportive facilitation of LSF's work at this stage was invaluable, and the authors are indebted to him for his assistance. Gratitude is also extended to Dr. Dawei Wang, Dr. Pengfei He, Dr. Bo Zhang, and Mr. Soohwan Hwang for dedicated help on solution manual preparation; to Mr. Guangyu Guo and Ms. Hongling Deng for excellent figure drawings; and to Ms. Phoebe Del Boccio and Mr. Nicholas Almerini for outstanding editorial assistance. The financial support provided by the C. John Easton Professorship funds of the Ohio State University is gratefully acknowledged.

Part I

Principles

1 Introduction to Multiphase Flows

1.1 Multiphase Flow Phenomena

Multiphase flows are found in abundance in both natural phenomena and industrial processes. Natural phenomena sources include such pathways as migration of deserts by sand storm and distributions of rain/snow/hail by wind. Industrial sources include pneumatic transport of solids, spray operations for surface coating and fire quenching as well as pollution control, solid-waste incineration by fluidized bed combustion, and gas–liquid bubble column reactors. In this section, several examples of multiphase flows will be examined in order to demonstrate the distinctly different transport patterns of each phase in the illustrated multiphase flow case. It is also intended to show the naturally caused or intentionally designed consequences resulting from such a difference in phase transport or flow pattern.

1.1.1 Sedimentation in a Particulate Flow

Typically initiated and maintained by a surrounding flow, sedimentation is a phenomenon describing the settling of particulate matter by gravity from the original state of suspension. For example, consider the migration of sand dunes by a strong wind as an illustration of this process. As shown in Figure 1.1-1, the blowing wind mobilizes and even lifts the sand particles from the wind front edge of the sand dune. The airborne sand is carried by wind until gravity intervenes to settle the sands some distance downstream of the wind. Some of the sand in the windward (wake region) may also be disturbed and mobilized by the wake flow. As the shaded layer of wind front resettles into the windward region, the sand dune proceeds to migrate downstream of the wind. Bear in mind that the flow pattern, or transport, of airborne sand is distinctively different from the moving air. Similar patterns may also be observed in sand sedimentation in a strong sand storm. In a strong turbulent wind, the sand stays airborne until the gravity force overpowers the wind's lifting force; this occurs when local wind strength is weakened by flow dispersion (or redistribution) by surface obstructions such as forest or city buildings.

Sand beach is another possible consequence of sand sedimentation dispersal; this phenomenon occurs when sand and fine gravel are carried and deposited by ocean waves brushing the bank, as illustrated by Figure 1.1-2. The waves work to help carry sand toward the bank, and the waves also help carry sand back out to sea when they

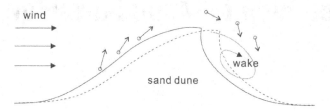

Figure 1.1-1 Migration of sand dune by wind.

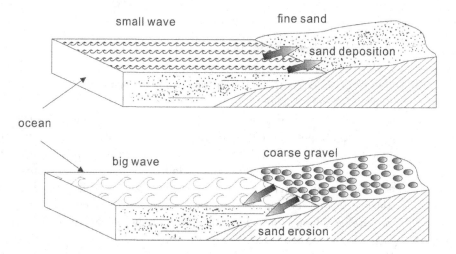

Figure 1.1-2 Sedimentation and erosion on beach by ocean waves.

recede. For that reason, large waves are constantly causing the erosion of the coastal bank and morphological changes of the coastline sand, especially during an ocean storm or hurricane. Thus, the transport (or sediment) of sand by waves depends upon the nature of wave flows as well as their interactions with sands, irrespective of the changes caused by gravity.

Sedimentation principles are also often applied to the design and applications of engineering devices. Figure 1.1-3 shows a particle size selector demonstrating gravitational settling. The mixture of various-sized particles is fed through a hopper and then propelled into the air by a powerful air blower. As the particles are projected into the air, initial momentum is obtained via the air jet. As the air jet flow is quickly dispersed, the particles begin to settle through gravity, with different inertia and particle-flow interactions that are mostly size based, assuming that all of the particles are composed of the same materials. The coarse particles settle nearest to the injection site, and the fine particles settle the farthest away. The ultrafine particles may remain airborne and may be carried away out of the collection zone by the ambient wind or dispersed air flow.

Sedimentation can also be achieved by employing other body forces besides gravity, such as centrifugal forces and electrostatic forces. A tangential-inlet cyclone represents the type of devices used for centrifugal separation, or sedimentation of

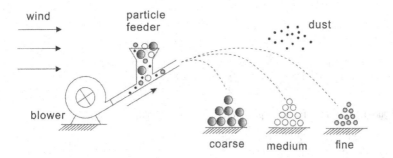

Figure 1.1-3 Particle sizing device by gravitational settling.

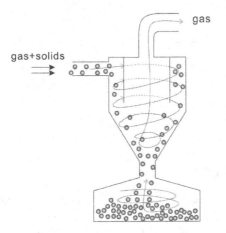

Figure 1.1-4 Cyclone separation of solids.

solid particles, from their carrying air flow. As depicted in Figure 1.1-4, the air flow first forms a tornado-like rotating core inside the cyclone and then leaves the cyclone through the exit. Under the action of the rotating core, the solid particles are subjected to centrifugal force. The centrifugal force then drives the radially migrating particles toward the cyclone wall, while the drag force tends to carry the particles along with the exiting air flow. Since the centrifugal force is proportional to the mass, whereas the drag force is proportional to the surface area, the centrifugal effect of separation is more significant for larger particles than smaller ones. Thus, larger-size particles tend to more easily reach the cyclone wall during their time within the cyclone. When the particle–wall collisions occur, most of those particles lose their momentum and are then trapped in the boundary layer of the wall where air velocity is too low to provide drag forces against the gravity. Hence, the particles on the wall will begin descending along the wall to the collection bunker located at the bottom of the cyclone. The very small and fine particles are unlikely to reach the wall surface during their time in the cyclone, and thus become uncollected and remain airborne in the exiting air flow.

Figure 1.1-5 Wind effect on spray watering from sprinkler.

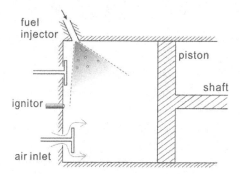

Figure 1.1-6 Fuel injection in engine cylinder.

1.1.2   Dispersion by Sprays or Multiphase Jets

Spray is a common technique for the quick dispersion of droplets or particulate matters. The distribution of sprayed particles is dependent not only upon the properties of injection but also upon the surrounding air flow. Figure 1.1-5 gives an example of the wind effect on sprinkler sprays used for irrigation. Despite the symmetric nature of the liquid spray injection, the droplet deposition on the vegetation becomes asymmetric due to the convective blowing of the wind. In order to obtain a better coverage symmetry, one technically plausible remedy is to adjust the injection angle by tilting the sprinkler toward the wind. Another solution is to adjust the mass flow asymmetrically, with more flow on the wind front side and less on the windward side. In either case, effective adjustment requires knowledge of dynamic interactions between sprayed droplets and wind.

In the form of an atomized spray jet, fuel injection is designed for quick dispersion and vaporization of liquid fuel in many combustion processes, such as hydrocarbon-fuel combustion in automobile engines, as illustrated in Figure 1.1-6. To avoid carbon deposits on the cylinder wall, it may not be desirable to have a direct impingement of liquid fuel spray onto the wall surface of the engine. Therefore, the atomization and injection of the spray needs to be designed with considerations of the hydrodynamic behaviors of droplets and air flow as well as the nature of vaporization and combustion within the engine.

A cooling tower is a device employed in many coal-fired and nuclear power plants and chemical plants, commonly used for thermal energy removal via evaporative water cooling or air cooling of a flow stream in a process. As shown in Figure 1.1-7, a

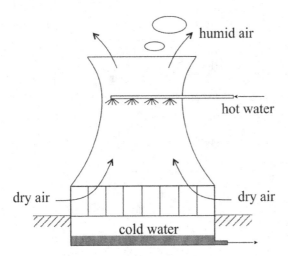

Figure 1.1-7 Sprays in cooling tower.

cooling tower is used to lower the temperature of hot water. A relatively dry and cold air flow passes through the cooling tower from bottom to the top, whereas the atomized droplets from sprays settle downward in a counterflow mode, reacting against the upward-moving gas flow. The droplet vaporization increases the air humidity while lowering the temperatures of the remaining droplets via the route of latent heat transfer. Cooling towers are also used for the pollution control of airborne particulate and hazardous gases via a wet scrubbing process. The absorption of hazardous gases by the atomized droplets consequently reduces the pollution level of exhaust gas. In addition, the airborne particulates are washed away by attaching to the droplets through particle–droplet collisions, thus forming a slurry flow at the bottom of the cooling tower, in a process known as wet scrubbing.

Spray is also commonly used to spread a thin coating layer (typically a polymeric thin film) onto a targeted surface, as shown in Figure 1.1-8. The polymeric liquid (or slurry) is atomized into fine droplets and injected into the air by aerated spray jets. While the gaseous component of the jets is quickly dispersed and merged into ambient air, the droplets, with most of the initial momentum preserved, adhere to the target surface. The coating is formed by the surface drying of the spread droplets attached to the surface. Note that a spray set at an incorrectly high jetting velocity may cause some adverse effects to the smoothness of the coating, such as a rebound of droplets from the target surface or an uneven surface coating.

A sand blaster has characteristics of multiphase flow similar to those of spray coating. Instead of the droplet attachment seen in surface coating, a sand blaster is designed for surface cleaning, in which a spray jet of abrasive sands with high momentum makes a direct impingement onto a cleaning surface, as shown in Figure 1.1-9. The sand particles gain their momentum from the assisting air jet. While the air jet is quickly dispersed and loses its momentum, the sands, with much higher inertia, continue their journey with most of the initial momentum preserved until the

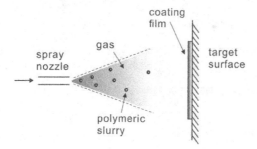

Figure 1.1-8 Spray coating.

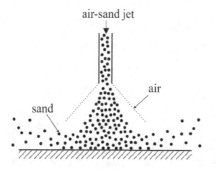

Figure 1.1-9 Sand blasting.

impingement target is reached. After the collisions of abrasive sands and targeted surface, both sand residues and chipped-off bits of surface material may become airborne and dispersed again by the air jets as they reach the surface.

1.1.3   Mixing and Material Processing

The mixing of solids and liquids is an important aspect of applications involving multiphase flow. Figure 1.1-10 shows an industrial device for concrete mixing. The mixture of cement, gravel, and water is combined by rotational stirring within a cylindrical chamber. In this case, the aggregate of each phase may initially exhibit a different dynamic response to the rotational agitation. As the mixing operation progresses, the chemical bonding formation among the cement and water makes the mixture and gravel appear as a paste-like slurry in which the difference in each component's flow dynamic responses eventually disappear.

Mixing can also be achieved via rotational agitation by applying rotating blades in a stirred tank, as shown in Figure 1.1-11. In the case of slurry mixing, the blade's rotational movement agitates the liquid to generate a vortex core that engulfs the surrounding particles into spiral motions, along with the rotating fluid. In the case of dry mixing or blending, the rotating blades cause the movement of dry particles by direct collisions between the blade and particles as well as among the particles themselves. The air in the stirred tank of dry mixing also begins flowing as a result

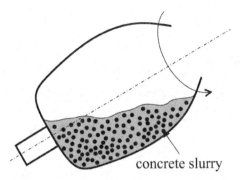

Figure 1.1-10 Concrete mixing.

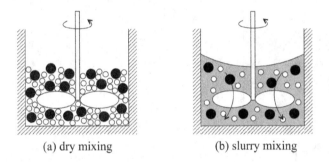

(a) dry mixing   (b) slurry mixing

Figure 1.1-11 Mixing in blade-stirring tanks.

of the blade's agitation and particle movement. Consequently, in terms of uniformity in particle size and concentration distributions, the mixing quality in a stirred
tank depends upon the particle-related coupling among the dynamic motions of fluid,
blades, and particles.

Mixing of granular particles can also be achieved by gas fluidization, as shown in
Figure 1.1-12. The fluidizing gas moves upward to suspend and mobilize the particles. The gas bubbles, or voids, are typically distinguishable in the dense suspension,
as their wobbly movement through the bed further agitates the suspended particles.
When the gas passes through the bed, the particles (suspended at a moderate gas
velocity) remain inside the bed, with in-bed recirculation and many collisions among
and between the particles and the bed wall. Typical examples of the types of applications for this process include fluidized bed combustors for coal-fired boilers, and
catalytic fluidized bed reactors for chemical synthesis.

Figure 1.1-13 illustrates a three-phase fluidized bed used as a chemical reactor
whereby the gaseous reactants, in the form of bubbles rising through the bed, react
with the liquid reactants in the presence of solid particles, which function as catalysts.
The rising bubbles carry a certain amount of liquids or liquid–solid in the wake
regions. The transport phenomena in the bed are strongly affected by the wake and
wake shedding behavior, which establishes a circulatory liquid or liquid–solid flow
pattern in the bed. When the particle size is small, the reactor is known as a slurry

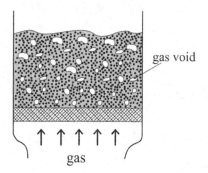

Figure 1.1-12 Gas–solid fluidized bed.

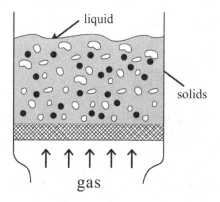

Figure 1.1-13 Gas–liquid–solid three-phase fluidized bed.

bubble column for which the liquid–solid mixture or the slurry remains within the reactor and gas flows in and out of the reactor.

The storage of granular food, such as beans and wheat grains, constantly requires drying to remove the excessive moisture. A spouting bed is a typical type of device for such a drying operation, as illustrated in Figure 1.1-14. Most granular food particles are just too big and heavy to be fully or uniformly fluidized; rather, they are merely lifted by the drying gas jet and then quickly settle down by gravity once the jet disperses to form a spouting among the surrounding particle cloud. The jet entrainment at the bed bottom introduces new particles, and the completely dried particles may be flushed out of the bed because they become lighter through a loss of moisture. The magnitude of the drying effect depends upon the multiphase flow characteristics, which includes the rate of solids recirculation between the spout and annulus regions as well as the contact time between the gas and food particles.

Jet milling is a process used to break large-sized particles or particle agglomerates into smaller fragments via the head-on collision of particles, in a process that is assisted by a pair of opposite particle-laden jets, as shown in Figure 1.1-15. The damaged fragments may be further broken down by an additional attrition method, such as grinding, and the attrited bits are removed along with the exit gas streams.

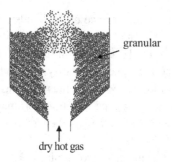

Figure 1.1-14 Spouting-bed dryer.

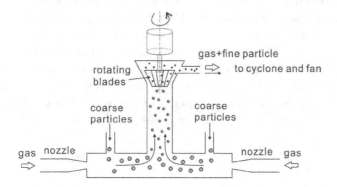

Figure 1.1-15 Jet milling.

1.1.4 Pipeline Transport

Pneumatic conveying is a common means of solids transport, where solid particles are aerated and transported in a pipe or channel from one location to another through gas flowing by using either blowing or suction action. In a horizontal transport, the suspended solids are subjected to both horizontal carrying force (drag force) and gravitational settling force, as shown in Figure 1.1-16. This results in a transport flow pattern that has a sliding layer of dense concentrated solids near the pipe bottom and fast moving particles of a dilute concentration near the top part of the pipe section. The transport flow can become unsteady and wavelike with the increase of solid loading to the flow, and sedimentation of particles and collisions may even form moving dunes.

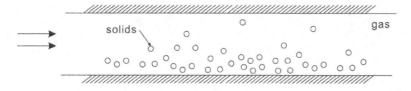

Figure 1.1-16 Pneumatic conveying of solids.

Similar to the way in which the pneumatic conveyance of dry solids is done via gas pipe flows, the solid particles can also be transported by liquid flows through pipe lines such as sewage drainage pipes or offshore crude oil transport pipelines, as shown in Figure 1.1-17. In a slurry pipe flow, the slurry may not fill the entire pipe cross-section all the time, leaving a void space of gas, thus creating an interface between the gas and slurry. Therefore, a slurry pipe flow transport may involve motions of gas, solids, and liquids, with a wave-free surface between the gas and slurry.

There are many types of multiphase flows through vertical pipes. Riser flows in a reactor system represent typical vertical pipe flows in which both fluid and particles are transported upward against the force of gravity, as shown in Figure 1.1-18. One example of riser flow application is the fluidized catalytic cracking (FCC) process seen in petroleum oil refineries, from which most gasoline, diesel, and heating oils are produced. In a FCC process, the feed stock vapor interacts with the catalytic solids alongside their transport in the riser. The subsequent catalytic reaction cracks the heavier molecular feed into low-molecular-weight fuels resulting in the booming of gas velocity with an increased number of moles and volume. This response changes not only the transport flow pattern of nonreaction but also the cracking itself,

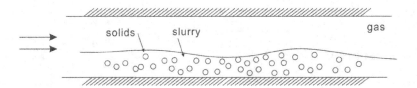

Figure 1.1-17 Slurry pipe flow.

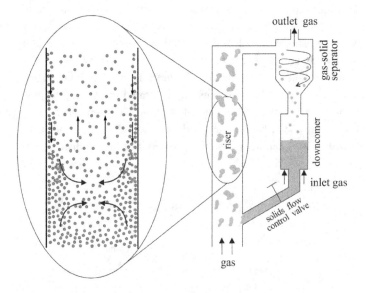

Figure 1.1-18 Riser flow in a circulating fluidized system.

due to the changes in contact time and temperature between feed vapor and catalysts, as well as the change in the local volumetric concentration of catalysts.

1.1.5   Flows with Charged Particles

In flows of dry granular materials, the solid particles become electrostatically charged via either collisions or by electric fields. Laser printing is a typical example of the controlled deposition of particles (toner) in an electric field, as shown in Figure 1.1-19. The printing cycle starts from charging the rotating photosensitive drum and exposing it with image information from laser beam. The particles are fed by a roller under a funnel-shape toner, and picked up by the charged drum. Via another corona transfer, these well-patterned particles then deposit onto the paper. The particle flow is mainly subjected to gravity force in the hopper, while dominated by electric force only during the remainder of the process.

Another example of this type of process is that of electrostatic precipitation for aerosol removal from aerosol-laden gas flows, as shown in Figure 1.1-20. In these cases, the particles suspended in a gas stream are first charged by a corona charger. Next, the charged particles are separated from the gas stream by an external electric

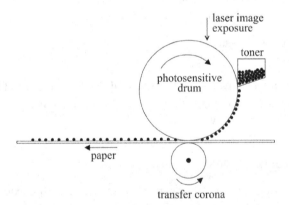

Figure 1.1-19 Laser printing with dry toner.

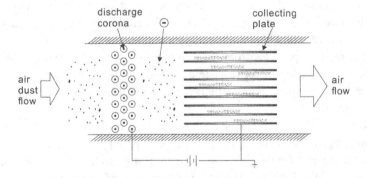

Figure 1.1-20 Electrostatic precipitator.

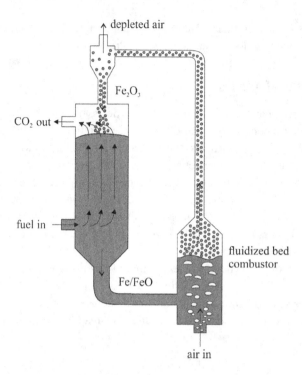

Figure 1.1-21 Chemical looping reactors.

field in which electric forces drive the particles toward the collectors. In addition to the electric force, the particle motion is also affected by the hydrodynamic force as well as other forces, such as gravity.

1.1.6   Flows with Chemical Reactions

In the flows of metal oxide oxygen carrier particles represented by iron oxides in the chemical looping reactor system as given in Figure 1.1-21, the fully oxidized iron oxide particles regenerated from a fluidized bed combustor are transported through a riser to a moving bed reducer after a cyclone separation of the oxygen-depleted air from the riser flow stream (Fan, 2010; 2017). The fully oxidized iron oxide particles are then reduced after reacting with the carbonaceous fuels introduced to the moving bed. The reduced particles from the moving bed are then transported through an L-valve to the combustor for regeneration. The reduction–oxidation reactions take place at high temperatures (*e.g.*, higher than 900°C) and can be at high pressures (*e.g.*, about 10 atm).

Several flow regimes are involved in this particle circulatory loop system including bubbling or turbulent fluidization regime for the combustor operation, fast fluidization regime for the riser flow, moving bed flows for the reducer, and vertical moving bed and dense horizontal transport for the L-valve. Gas aeration to the L-valve is used to control the particle flow rate in the nonmechanical valve setting.

Fines can be generated from attrition due to the particle stresses induced by both the flow and the chemical reactions. The cyclone is designed also to separate the attrited particles.

1.2   Definition of Multiphase Flow

The main characteristics of a multiphase flow include distinctively different dynamic responses of each phase and strong phase interactions that impact the transport phenomena. Hence it is essential to understand the differences between a multiphase flow and a multicomponent single-phase flow, as well as the differences between a dense-phase multiphase flow and a dilute-phase multiphase flow.

1.2.1   Multiphase Flows versus Multicomponent Flows

Multiphase flow is a term commonly used to refer to a type of flow that involves more than one physical state of the medium (*i.e.*, gas, liquid, or solid phase). Multiphase flow can also describe a flow where the flow media present in the flow field exhibits different dynamic behaviors when subjected to a particular driving force (Soo, 1965). To illustrate the dynamic phases involved in these processes, consider a situation in which the wind blows over a pile of fine dust, granular sand, and large rock. The fine dust will be easily airborne or suspended in the wind; the granular sand may be partially blown off and quickly resettled by gravity, and the large rocks typically remain stationary. In this particular situation, just using the term "solid phase" to describe the three types of distinctively different dynamic response behaviors of solid particles from the blowing wind will be inadequate. This is especially true when considering a consequent analysis of the transport mechanisms and mathematical modeling involved in a multiphase flow system. Therefore, in this book, a phase will be defined based on the medium dynamic response characteristics that are present, rather than just the physical state of the medium.

With the different dynamic responses of the phases, the interactions among the different phases will occur through their common interface. Such interactions may include various forms of momentum transfer (such as drag force and collision force), energy transfer (such as conductive and convective heat transfer), charge transfer (such as collision-induced electrification), and mass transfer (such as evaporation, reaction, and attrition). It is noted that a dynamic phase may contain multiple components or chemical species. Interactions can also take place between the various components through either diffusion or chemical reactions. However, bear in mind that there is no distinct boundary between the components. For example, for an evaporating fuel spray of bidispersed droplets that are injected into an air-filled combustor, there are three distinctive dynamic phases involved in the transport process: gas mixture phase, droplet phase of small droplets, and droplet phase of large droplets. The gas mixture phase is a multicomponent phase that consists of such gaseous components as fuel vapor, oxygen, and nitrogen.

Many multiphase flows have a continuous spectrum in their dynamic responses, which theoretically would require identification of an infinite number of phase groups to fully describe the multiphase transport properties. One example is that of a sand storm, in which the size distribution of sand particles is continuous over a wide range, varying in size from submicron to a few millimeters. In this instance, a finite number of lumped phases can be introduced to represent groups of particles with similar dynamic responses, as characterized by their averaged property. The exact number of the lumped phases depends on the particular tolerance level that is acceptable to the applications of interest.

1.2.2   Dilute Phase versus Dense Phase

In a multiphase flow, at least one phase is a continuum fluid, such as gas or liquid, which typically functions as a carrying medium for the phase transport. The other phases are usually composed of various rigid particles, such as solid particulates, or deformable particles such as droplets and bubbles. These particles interact with each other by either modifying the flow field, thereby indirectly influencing the hydrodynamic forces, or directly affecting their dynamics through collision or by endured contact. The particles may also interact through other mechanisms, such as electrostatic forces or colloidal forces. Depending on the nature and effective range of the particle–particle interactions that are of interest, the particle phase can be categorized into dilute and dense phases. A dilute phase refers to a particle phase in which the interactions among those constitutive particles are so insignificant (compared to the interaction between the carrying fluid and particles) that they can be ignored in the mechanism analysis and modeling of the phase transport. In contrast, a dense phase refers to a particle phase where the interactions among the constitutive particles play an important role in the phase transport. Particle–particle interactions in a dense phase can be divided into two interaction modes: noncontact interaction, by which the particles affect each other indirectly via the fluid–particle interactions or via noncontact mechanisms such as charge interactions, and contact interaction, in which interactions take place directly through the contact interface between particles.

The demarcation between the dilute and dense phases is usually based on the consideration of hydrodynamic interactions. In this case, the demarcation can be conveniently estimated based on the relative interparticle distance, which is the ratio of the distance to particle size, or it can be equivalently based on the particle volumetric fraction of that phase. For instance, a particle phase in a typical turbulent flow transport can be regarded as a dilute phase when the particle volume fraction is less than 10^{-6}. A particle phase in a laminar flow transport may be regarded as a dilute phase when the particle volume fraction is less than 10^{-4}. When the particle volume fraction is higher than 10^{-2}, the contact interaction becomes more important. It should be noted that when a long-range interaction is involved, such as the electrostatic interactions between charged particles in a suspension flow or the thermal irradiation among molten metallic particles in a reactor, the demarcation between dilute and dense phases can be quite different from that defined only from the hydrodynamics perspective.

1.3   Modeling Approaches

In the modeling theory of multiphase flows, the basic approaches can be roughly divided into two categories. The first category, known as the Eulerian–Lagrangian modeling, looks at fluid motion and the dynamics of particle phases in a particle trajectory-tracking frame. The second category, known as the Eulerian–Eulerian modeling, describes a fixed space domain filled with a pseudo-continuum of the particles. In both approaches, the dynamics of the actual continuum fluid phase is typically described as an Eulerian model similar to that for a single-phase fluid dynamics but modified by interactions with particles.

1.3.1   Eulerian–Lagrangian Modeling

Tracking the dynamic behaviors of each individual particle in a multiphase flow is an intuitive choice among various modeling options. This is due to the merits of the convenient revelation of actual phase transport and the direct application of physical laws that govern the particle transport. The Lagrangian trajectory modeling also poses a significant advantage in using the mathematical formulation of governing equations. The transport equations of each particle are expressed by a set of ordinary differential equations in their Lagrangian coordinates and hence can be readily solved using numerical methods such as the Runge–Kutta algorithms.

Typical Lagrangian approaches to be used include the deterministic trajectory method and the stochastic trajectory method; other methods include the probability-density-function (PDF) based trajectory method and the discrete element method (DEM). The deterministic trajectory method disregards all randomness involved in the particle transport, particularly the random nature of the turbulent flow to which the particle is exposed. The stochastic trajectory method takes into account the randomness that is in the turbulent flow, along with other random effects, such as the initial conditions of the particle's motion. To obtain the statistic average over such randomness, statistical computations covering a large number of possibilities for the same particle need to be performed, by the use of various techniques, such as the Monte Carlo method. Alternatively, a probability-density-function method can be introduced to tailor the randomness involved in the action. Such an approach would require the predetermination of the random nature of the stochastic process, along with the coupling with the deterministic transport equations. The DEM model, also known as the soft-sphere model, handles the type of cases where the detailed contact transfer among interparticle collisions is important. The collision dynamics in a DEM model is based on the analogy to the spring-dashpot-slider modes for normal impact, tangential, and rotation contact, with or without sliding. All of the above mentioned trajectory models for particle motion are coupled with models for the continuum fluid flow in which the particles are exposed or suspended.

The development of Lagrangian models has been limited mainly by the inherent need for the large computing capacity capable of tracking each particle in the system, to conduct statistic averaging over the randomness, and the need to compute various interactions with other phases, system boundaries, and particles of the same phase. In practice, the applications of Eulerian–Lagrangian modeling are focused on the dilute transport cases, where multi-continuum approaches may not be appropriate, or in cases in which the temporal tracking of particles is important, such as in spray drying of slurry droplets in a furnace or the tracking of radioactive particles in a particulate flow.

1.3.2   Eulerian–Eulerian Modeling

The performance of a multiphase flow system constantly calls for the microscopic "field" descriptions of dynamics of each phase. To this end, the Eulerian–Eulerian description of phase transport is the most convenient option among available modeling choices, as it is a direct extension of the mathematical formulation of fluid dynamics from a continuum single phase fluid, to a multi-continuum multiphase "fluid medium." There are three key issues in this extension: one is the construction of pseudo-continuum of the particle phase(s) that "co-share" the common space of multiple phases or multi-continuum at the same moment; the second is the interphase interactions and transfer among the multi-continuum phases; the final is the transport properties, such as viscosity and diffusivity of each pseudo-continuum, that need to be determined via constitutive equations or correlations of pseudo-continuum variables.

The pseudo-continuum of a particle phase can be established by applying the volume averaging theorems over a control volume. To account for the turbulence effect, the time averaging should be adopted after the volume averaging (Soo, 1989). Since the pseudo-continuum is based on the volume averaging, the pseudo-continuum phase should be regarded as compressible, due to the nonuniform distribution of the phase volume fraction, even if the materials that constitute the phase are incompressible.

In the collision-influenced transport of a dense phase, the modeling of the collision stress of the pseudo-continuum particle phase becomes important. The kinetic theory of granular flows has been developed to account for this collision stress by using the analogy to the molecular kinetic theory of dense gas transport.

1.4   Case Studies: Peculiarities of Multiphase Flows

The following case studies further examine some simple yet interesting phenomena of multiphase flows whose transport mechanisms are hard to be intuitively understood or difficult to be modeled. These discussions are mainly aimed to stimulate readers' interests in learning the multiphase flow theories in the following chapters, rather than to provide a detailed modeling analysis or solution.

1.4.1 Bubble Acceleration

Problem Statement: Consider a small spherical bubble initially at rest in a stagnant liquid column. What is the initial acceleration of the bubble?

Analysis: Considering the steady-state motion of a particle in a fluid, the typical forces acting on the particle include the gravitational force, the buoyancy force, and the drag force. Directly applying these forces to the transient motion of a bubble in a fluid based on the Newton's second law of motion, and if there is a relative motion between the bubble and the fluid, the equation becomes:

$$\rho_b V_b \frac{dU_b}{dt} = (\rho_l - \rho_b)V_b g - C_{Db} A_b (U_b - U)^2. \tag{1.4-1}$$

From Eq. (1.4-1), the initial acceleration can be expressed by:

$$\left.\frac{dU_b}{dt}\right|_{t=0} = \left(\frac{\rho_l}{\rho_b} - 1\right) g. \tag{1.4-2}$$

The acceleration calculated by Eq. (1.4-2) gives rise to about $1,000g$ (assuming the liquid density is about 1,000 times more than the bubble density). This result, however, is contrary to experimental measurements, which suggests about $2g$! What is wrong with the purported solution?

The answer is that additional forces are applied to a bubble when it undergoes an acceleration, deceleration, or transient state of motion including the carried mass force and Basset force (see Section 3.2.2). In this case, with the historical integral in Basset force neglected, the force balance equation of Eq. (1.4-1) is corrected by including the carried mass force as:

$$(\rho_b V_b + C_A \rho_l V_b)\frac{dU_b}{dt} = (\rho_l - \rho_b)V_b g - C_{Db} A_b (U_b - U)^2, \tag{1.4-3}$$

where C_A is a shape factor of carried mass (0.5 for a sphere). Based on Eq. (1.4-3), the bubble's initial acceleration is given by:

$$\left.\frac{dU_b}{dt}\right|_{t=0} = \left(\frac{\rho_l - \rho_b}{C_A \rho_l + \rho_b}\right) g, \tag{1.4-4}$$

which is about $2g$ since $\rho_l \gg \rho_b$.

Comments: This case analysis shows that, in the modeling of a transient velocity of the disperse phase, one should consider all the major forces acting on the disperse phase, including the carried-mass force in this case, in order to yield the correct velocity prediction.

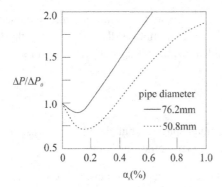

Figure 1.4-1 Pressure drop as a function of solids volume fraction (10μ zinc particle; Pipe Re=53,000, after Shimizu *et al.*, 1978).

1.4.2   Pressure Drop Reduction in Pneumatic Transport

For a fully developed particle-free single-phase turbulent pipe flow, the pressure drop per unit pipe length can be determined by the force balance against the wall friction by:

$$\frac{dP}{dz}\bigg|_0 = \frac{4}{D}\tau_{w0} = \frac{4}{D}\mu_{T0}\frac{\partial U}{\partial r}\bigg|_{w0}. \tag{1.4-5}$$

For a pneumatic pipeline transport of dilute particle suspension, the pressure drop per unit pipe length for a fully developed pipe transport flow can be expressed by:

$$\frac{dP}{dz} = \frac{4}{D}(\alpha_w\tau_w + \alpha_{sw}\tau_{sw}) = \frac{4}{D}\alpha_w\mu_T\frac{\partial U}{\partial r}\bigg|_w + \alpha_{sw}\mu_T\frac{\partial U_s}{\partial r}\bigg|_w. \tag{1.4-6}$$

In a range of the solids volume fraction at 0.1–0.5%, it is found that the pressure drop can be even smaller than that of particle-free flows under the same flow conditions (by a factor up to 30%), as shown in Figure 1.4-1. This has a clear benefit of energy savings for the pneumatic transport of granular materials.

Problem Statement: It has been experimentally shown that the radial profiles of the gas velocity (with and without dilute particle suspensions) are almost the same, and the solids volume fraction is very small (so the gas volume fraction is almost the same as the particle-free case). However, since the wall friction of solids cannot be negative (based on the physics and factual observations), how does one explain this multiphase flow transport phenomenon?

Analysis: Since the equation stated above represents a scientific fact, there must be an underlying physical mechanism to cause the action. By comparing Eq. (1.4-6) to Eq. (1.4-5), in the transport regime where pressure drop reduction occurs, the following is obtained:

$$\frac{4}{D}\alpha_w\mu_T\frac{\partial U}{\partial r}\bigg|_w + \alpha_{sw}\mu_T\frac{\partial U_s}{\partial r}\bigg|_w < \frac{4}{D}\mu_{T0}\frac{\partial U}{\partial r}\bigg|_{w0}. \tag{1.4-7}$$

Since the solids wall friction must be positive, Eq. (1.4-7) reduces to

$$\frac{4}{D}\alpha_w\mu_T \left.\frac{\partial U}{\partial r}\right|_w \ll \frac{4}{D}\mu_{T0} \left.\frac{\partial U}{\partial r}\right|_{w0}. \tag{1.4-8}$$

Based on experimental findings showing that radial profiles of gas velocity (both with and without dilute particle suspensions) are almost the same and that the gas volume fraction is also almost the same as the particle-free case, the following relationship must be valid:

$$\mu_T \ll \mu_{T0}. \tag{1.4-9}$$

Indeed, the above equation is true due to the turbulence damping effect by the presence of suspended particles leading to a low viscosity (an important phase interaction to be discussed directly in Section 6.7.3).

Comments: This case study shows the analysis that helps direct the discovery of one of the major mechanisms (turbulence modulation) in multiphase flows, through logical progression.

1.4.3   Acceleration of Solids in a Dense Gas–Solid Riser

One puzzling observation in the circulating fluidized bed systems is the acceleration length of solids in a dense gas–solid riser. According to the momentum equation of the particle phase over the cross-section of the riser, it gives:

$$\alpha_s\rho_s U_s\frac{dU_s}{dz} = -\alpha_s\frac{dP}{dz} + F_D - \alpha_s\rho_s g - \frac{4}{D}\tau_{sw}. \tag{1.4-10}$$

The drag force may be correlated to the local multiphase flow properties by the Richardson–Zaki equation as

$$F_D = \frac{18\mu}{d_s^2}\frac{\alpha_s}{(1-\alpha_s)^n}(U - U_s). \tag{1.4-11}$$

Combining the above equations with the continuity equations of solids and governing equations of the gas phase, the velocity distribution of solids along the riser can be solved, as schematically described in Figure 1.4-2.

Problem Statement: Why is the acceleration length of solids from the model prediction always much shorter than the measurements, typically by a factor of two orders of magnitude? What could be wrong with the above modeling?

Analysis: The discrepancy between the model prediction and measurements clearly indicates at least one important governing mechanism is missing, or the quantification of governing equations above is ill-formulated, or possibly both.

The analysis is started by examining the term-by-term comparisons in the major governing equation (*i.e.*, the momentum equation of solids). There are a few important experimental facts that are very useful for this comparison: (1) the pressure drop per unit riser length is much higher than the suspended solids weight; (2) the wall friction is relatively much smaller than other terms, except for risers of very small

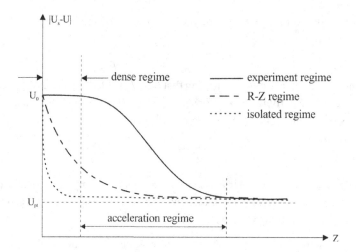

Figure 1.4-2 Solid velocities in a riser: predictions based on a model of isolated particle or Richardson–Zaki equation against experimental results.

column diameters; (3) the gas velocity is much higher than the solids velocity, typically by one to two orders of magnitude; and (4) solids accelerate quite slowly in the dense phase transport region.

Based on the truism of fact three (3), the wall friction effect may be ignored for riser transport of large column diameters (which is true for most industrial riser applications). Fact four (4), indicates that the summation of various momentum contributions on the right-hand side of Eq. (1.4-10) should be small, or basically balanced for the dense phase transport. However, both fact one (1) and fact two (2) show that the combined term would be much higher than the suspended solids weight; hence, it is impossible to balance these three remaining terms. Since the drag force cannot be negative (according to fact two (2)), there has to be another resistant force to counterbalance the rest. This additional resistance to the solids acceleration is supposedly due to the nonequilibrium collisions along the solids transport (Zhu and Wang, 2010).

Comments: This case analysis not only shows the logical path that leads to the discovery of the missing governing mechanisms (nonequilibrium interparticle collision force) in dense gas–solid flows, but also shows the misinterpretations, and hence the great need in the development of multiphase flow theories for engineering applications.

1.4.4   Cluster Formation and Instability

In particulate suspension flows, clusters are constantly formed. Clusters are different from agglomerates. A cluster refers to a loose pack of particles that will be soon dispersed or reorganized, as observed in a riser flow shown in Figure 1.4-3. The

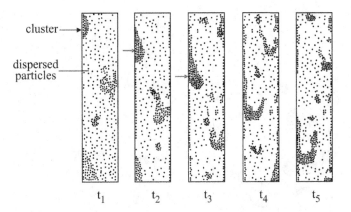

cluster

dispersed
particles

t_1 t_2 t_3 t_4 t_5

Figure 1.4-3 Schematic diagram of clusters and dispersed primary particles.

cluster is a transient group of interactive particles, with a short lifetime of the group stability.

Problem Statement: What causes the formation of a cluster and the cluster's transient transport behavior?

Analysis: In fast fluidization, particle collisions and particle-turbulence interactions yield high concentrations of solids in the low-velocity wall region. As the solids concentration reaches a certain level, the solids accumulate at various locations on the wall in the form of a cluster. The cluster, identified with arrows in Figure 1.4-3, is of a wavy shape. As it builds up, more particles are moving into the wall region. It moves downward as a result of the net effect of gravitational and drag forces imposed by the flow stream in the core region. When the cluster becomes highly wavy, and the frontal face (leading edge of the layer) becomes bluff; particles in this region are swept away from the wall region in much the same way as turbulent bursts occurring in the boundary layer (Praturi and Brodkey, 1978). The bursting process is quite sudden compared to the growing process. Considerable amounts of solids are entrained from the wall region to the core region during this process. A complete cycle of the evolution of such a cluster is illustrated in Figure 1.4-3. The measurement of local voidage suggests that the large cluster is fragmented into relatively small clusters during the ejection process as a result of large eddies. The bursting causes a surge in the local solids concentration, which in turn enhances the local turbulence. Thus, the bursting phenomenon evolves into a large-scale structure characteristic of the core region, while the flow in the core region affects the frequency of bursting. The existence of the cluster and the event of bursting can be detected from the measurements of the instantaneous local solids concentration variation (Jiang *et al.*, 1993), visualization, or computation (Gidaspow *et al.*, 1989).

It is noted that a cluster as referred to here is a lump of solid particles over which flow properties such as voidage do not vary substantially. It is formed mainly as a result of hydrodynamic effects. The mechanism of particle clustering is different

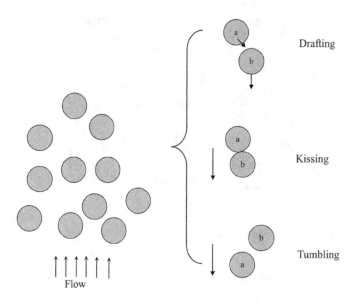

Figure 1.4-4 Transient characteristics of settling particles in a liquid.

from that of agglomeration, in which particles adhere to one another mainly by surface attraction (*i.e.*, van der Waals force and electrostatic forces) and mechanical or chemical interaction (Horio and Clift, 1992).

Comments: This case study illustrates the formation of typical clusters in a riser. It is noted that the formation of clusters alters the large-scale motion in the gas–solid flow, which in turn affects the cluster size and motion.

1.4.5   Wake-Induced Phenomena

Problem Statement: In multiphase suspensions, it is commonly seen that falling drops or rising bubbles align with the stream. This also happens to particles when they are settling in a liquid. What causes the alignment; can it be understood through a simple mechanistic model?

Analysis: In a multiphase flow, due to the relative velocity between the fluid and particles, there is a wake region of each individually suspended particle. The random fluctuation in the particle motions causes a situation where some particles fall into the wake regions of other particles. When this occurs, the drag force on the trailing particle is suddenly reduced due to the low pressure in the wake, leading to an accelerated approach ("drafting") and eventual collision ("kissing") of the pair. The vertical in-line configuration of the pair may become unstable and encourage tumbling, causing the prior trailing particle then to take the lead. The new trailing particle, originally the leading one, can be attracted to the wake of the new leading particle and the cycle repeats. This drafting-kissing-tumbling phenomenon of settling particles (Fortes *et al.*, 1987; Feng *et al.*, 1994) is illustrated in Figure 1.4-4.

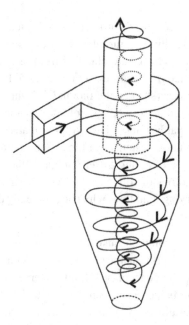

Figure 1.4-5 Trajectories of fluid tracers in a tangential inlet cyclone.

Rising bubbles and falling drops also undergo drafting, aligning, colliding, and even rotating about each other (Fan and Tsuchiya, 1990; Joseph and Renardy, 2013). In most cases, they tend to stay in line with an equilibrium distance at which the wake-induced attractive force and potential repulsive force balance (Yuan and Prosperetti, 1994).

Comments: This case study illustrates the mechanistic effects due to wakes on particle or bubble motion. Note, that with the governing mechanism identified in a given multiphase flow (*i.e.*, the wake phenomena in the particle or bubble flow of this case) a simple model that includes this governing mechanism can reasonably be used to characterize this flow phenomenon.

1.4.6   Particle Trajectories in a Cyclone Separator

Problem Statement: In a numerical simulation of gas–solid flow in a cyclone, the flow field is calculated using a commercial CFD code. In order to study the collection efficiency, trajectories of fluid tracers are plotted, as schematically exemplified in Figure 1.4-5. However, it appears that all tracer trajectories are leading to the cyclone exit, which means no collection of particles. What is wrong with the modeling or simulation?

Analysis: The velocity and pressure fields of fluid phase (air in this case) are solved using the Eulerian equations of mass and momentum conservations. Since the flow is turbulent, the transport properties or Reynolds stresses need to be solved by using a turbulence model. For simplicity, the standard k–ε model was employed here;

although, for better predictions, a more advanced turbulence model that accounts for the effects of buoyancy, wall, and multiphase turbulence modulation could be substituted. The trajectories of fluid tracers hence are obtained from integrating the associated fluid velocities, as demonstrated by path lines in Figure 1.4-5. It is important to understand that an ideal fluid tracer (such as the one in CFD simulation) is a conceptual fluid element having the fluid density but no volume. That is, the fluid tracer cannot represent a solid particle that has not only a different material density but also a finite volume. Hence, the dynamic response and transport characteristics of a solid particle are different from those of a fluid tracer. To obtain the collection efficiency of solids in a cyclone, trajectories of solids have to be solved separately from (and maybe coupled with) the carrying fluid flow, which are typically described by Lagrangian trajectory models of particles.

Comments: This case study illustrates a common misunderstanding between flow patterns (marked by trajectories of fluid tracers) and the trajectories of actual particles in multiphase flow applications. While the concept of a fluid tracer is simple, the actual use of it in real applications can be confusing since the actual fluid tracers for flow visualization, used primarily for illustrative purposes only, are real particles (such as droplets, bubbles, and solid spheres) whose dynamic transport would be different from that of fluid phase. The key in selecting those "tracer particles" in a multiphase flow application is ensuring that the dynamic response is nearly identical to that of fluid phase, yet far away from those of concerned particle phases in the application.

1.5   Summary

This chapter provides an overview of the concepts and exemplified applications of multiphase flows. It illustrates the distinctly different transport patterns or phenomena of individual phases in a multiphase flow, which either have naturally caused or intentionally designed consequences.

The chapter conveys the basic definitions of a multiphase flow, the phase interactions, and the associated modeling approaches, which include:

- the difference between a multiphase flow and a multicomponent single-phase flow, and hence identification of each phase and component involved;
- the difference between a dilute-phase multiphase flow and a dense-phase multiphase flow, and hence determination of the inclusion or exclusion of interparticle interactions in the multiphase flow of interest;
- the difference between a continuum phase and a discrete phase in describing the flow regimes;
- the difference in Eulerian–Lagrangian modeling and Eulerian–Eulerian modeling, and the intents of each approach.

In the case studies, the unique phenomena of multiphase flows are discussed that involve:

- hydrodynamic forces governing the motion of a bubble in liquid;
- pressure drop reduction in a particulate-laden pipe flow;
- clusters and their transport characteristics in a particulate-fluid flow;
- wake-induced particle motion and collision;
- motions of ideal fluid tracers and real particles in a cyclone flow simulation.

Nomenclature

A	Cross-section area
C_A	Shape factor of carried mass
C_D	Drag coefficient
D	Pipe diameter
d	Particle diameter
F_D	Drag force
g	Gravitational acceleration
n	Richard-Zaki index
P	Pressure
r	Radial coordinate in a cylindrical coordinate system
t	Time
U	Velocity
V	Volume
z	Axial coordinate in a cylindrical coordinate system

Greek symbols

α	Volume fraction
μ	Dynamic viscosity
ρ	Density
τ	Shear stress

Subscripts

0	Particle-free
b	Bubble
l	Liquid
s	Solid
T	Turbulence
w	Wall

Problems

P1.1 What is the major difference between a multiphase flow and a multicomponent flow? Give an example for each flow.

P1.2 List the major advantages and disadvantages between a Lagrangian–Eulerian model and an Eulerian–Eulerian model for modeling a bubble column (gas–liquid) bioreactor.

P1.3 Consider a mono-droplet generator that relies on an atomized nozzle spray of liquid with a cross-flow air blowing. How many dynamic phases should be considered in this application?

P1.4 Identify if the following flow phenomena or applications belong to the category of single-phase flows or multiphase flows. For a multicomponent phase, identify the multiple components:
(a) cold nitrogen jet into a hot steam stream;
(b) mist flow generated by ultrasonic humidifier;
(c) wind and cloud movements;
(d) liquid fuel spray in engine; and
(e) underground water flow through porous layer of rocks.

P1.5 Identify if the following flow systems or applications belong to the category of single-phase flows or multiphase flows. For a multicomponent phase, identify the multiple components:
(a) crude oil-natural gas flow through porous layer of rocks;
(b) liquid coolant flow with boiling in porous layer of solids surround a nuclear reactor;
(c) mono-sized solid particle movement in a vacuum container on a vibrator; and
(d) rotating hybridizer of granular particles (with a loop structure of centrifugal fan and pipe bend).

P1.6 Identify if the following flow systems or applications belong to the category of single-phase flows or multiphase flows. For a multicomponent phase, identify the multiple components:
(a) ball miller (in vacuum or inert gas container);
(b) dry coating of fine particle on coarse particle in a vibrating vacuum container;
(c) segregation of solids in dry powder mixture;
(d) pulverized coal combustion.

P1.7 Scaling two-phase flows to Mars and Moon gravity conditions: The interest in the liquid-vapor two-phase flows for space applications requires the understating of such flows under different gravity conditions. An experiment was conducted to investigate the relation between pressure drop in a pipe and the gravity. It is found that in the annular flow regime, their relationship can be expressed by $Eu = 12.7Fr^{0.41}$ (Hurlbert *et al.*, 2004). The gravity is 1.622 m/s^2 on moon, and 3.71 m/s^2 on Mars. If a pressure drop of 1,000 Pa is measured during an experiment on Earth, what will be the pressure drop in the same pipe if the same experiment is carried out on Mars and on the Moon with the same fluid and velocities of the two phases?

P1.8 Porous filters are often used for filtration and purification of fluid with impurities such as dust particles, oil drops, or bacteria. While the filter size often ranges from centimeters to meters, the impurities and the pores are of only micrometer or nanometer size. Describe the multiphase flow phenomena on different scales and interaction between scales.

PI.9 Pneumatic conveying is a common method to transport dispersed dry powders via gas pipe flows. Consider a pneumatic conveying of fine powder dispersed in an air flow through a pipe of 50 mm in diameter. The transport is undertaken at the atmosphere pressure and room temperature, with an average air velocity of 30 m/s and a mass flow rate of 0.1 kg/s of fine powders. The average size of powder is 100 μm, with a density of 1,500 kg/m^3. Is the pipe flow turbulent or laminar? Do you expect any pressure drop reduction in this pneumatic transport?

PI.10 Droplets injected from a spray nozzle are typically polydispersed (*i.e.*, of different sizes). The sprayed droplets from a lawn sprinker (such as the one in Figure 1.1-5) consist of three different sizes: mist (*e.g.*, $d_d < 100\mu$ m), fine (*e.g.*, 100μ m$d_d < 1$ mm), and coarse (*e.g.*, 1 mm $< d_d < 10$ mm). If the injection velocity and angle are the same for all droplets, do the different-sized groups of droplets have the same watering coverage area if there is no wind? If there is a strong cross wind, how does the wind alter the watering coverage area? Can mist cover further than the coarse droplets do under a strong crosswind?

PI.11 In the hydraulic fracking process of crude oil and natural gas production, a well is drilled vertically beyond a mile under surface to extract the fluid that is under enormously high pressure. The extracted fluid is a mixture of natural gas, oil, sand, salts, and water that was originally in a supercritical state (*i.e.*, at pressures higher than the critical pressure of the mixture). As the fluid is extracted to the surface, the pressure is drastically reduced down to the ambient pressure (such as the atmospheric pressure). How many different dynamic phases are there in the pipe flow during the fluid extraction from the shale (where the fluid is originated) to the surface?

PI.12 In the vapor-compression air-conditioning system, the refrigerant recirculates in a closed piping loop that is connected through an evaporator, a compressor, a condenser, and a throttling valve. Analyze the dynamic phase (*i.e.*, single-phase or two-phase) of refrigerant flow when it goes though each of these components.

References

Fan, L.-S., and Tsuchiya, K. (1990). *Bubble Wake Dynamics in Liquids and Liquid-Solid Suspensions*. Butterworth-Heinemann.

Fan, L.-S. (2010). *Chemical Looping Systems for Fossil Energy Conversions*. Wiley/AIChE.

Fan, L.-S. (2017). *Chemical Looping Partial Oxidation: Gasification, Reforming and Chemical Syntheses*. Cambridge University Press.

Feng J., Hu, H. H., and Joseph, D. D. (1994). Direct simulation of initial value problems for the motion of solid bodies in a Newtonian fluid Part 1. Sedimentation. *J. of Fluid Mech.*, **261**, 95–134.

Fortes, F., Joseph, D. D., and Lundgren, T. S. (1987). Nonlinear mechanics of fluidization of beds of spherical particles. *J. of Fluid Mech.*, **177**, 467–483.

Gidaspow, D., Tsuo, Y. P., and Luo, K. M. (1989). Computed and experimental cluster formation and velocity profiles in circulating fluidized beds. In *Fluidization VI*. Ed. Grace, Shemilt, and Bergougnou. New York: Engineering Foundation.

Horio, M., and Clift, R. (1992). A note on terminology: clusters and agglomerates. *Powder Technol.*, **70**(3), 196.

Hurlbert, K. M., Witte, L. C., Best, F. R., and Kurwitz, C. (2004). Scaling two-phase flows to Mars and Moon gravity conditions. *Int. J. Multiphase Flow*, **30**, 351–368.

Jiang, P. J., Cai, P., and Fan, L.-S. (1993). Transient flow behavior in fast fluidization. In *Circulating Fluidized Bed Technology IV*. Ed. A. A. Avidian. New York: AIChE Publications.

Joseph, D. D., and Renardy, Y. (2013). *Fundamental of Two-Fluid Dynamics: Part I: Mathematical Theory and Applications*. Springer Science & Business Media.

Praturi, A., and Brodkey, R. S. (1978). A stereoscopic visual study of coherent structures in turbulent shear flow. *J. of Fluid Mechanics*. **89**, 251–272.

Shimizu, A., Echigo, R., Hasegawa, S., and Hishida, M. (1978). Experimental study on the pressure drop and the entry length of the gas-solid suspension flow in a circular tube. *Int. J. Multiphase Flow*, **4**, 53.

Soo, S. L. (1965). Dynamics of Multiphase Flow System. *I&EC Fundamentals*, **4**(4), 426–433.

Soo, S. L. (1989). *Particulate and Continuum: Multiphase Fluid Dynamics*. New York: Hemisphere.

Yuan, H., and Prosperetti, A. (1994). On the in-line motion of two spherical bubbles in a viscous fluid. *J. of Fluid Mechanics*, **278**, 325–349.

Zhu, C., and Wang, D. (2010). Resistant effect of non-equilibrium inter-particle collisions in dense gas-solids riser flow. *Particuology*, **8**(6), 544–548.

2 Continuum Modeling of Single-Phase Flows

2.1 Introduction

The most important component of multiphase flow modeling is continuum modeling of single-phase flows. In this context, the term single-phase flows refers to the flows that have dynamic movement of only one phase, which is typically treated as a continuum. When other types of phases are present, if their presence has negligible effect on the flow dynamics of the system, or if their motion is negligible, the flow can be approximated by a single-phase flow. An example where the presence of one phase is negligible is treating gas as a vacuum in granular flows. An example where the movement of one phase is negligible is treating solids as stationary rigid bodies in modeling of flows through porous media. The four sections of this chapter discuss the basic governing and constitutive equations of single-phase flows that are closely related to the study of multiphase flows. These respective single-phase flows include:

1. laminar flows of viscous fluids;
2. turbulent flows of viscous fluids;
3. flows of viscous fluids in porous media; and
4. inertia-dominated granular flows.

The most fundamental theory of fluid mechanics is the continuum modeling of laminar flows of viscous fluids, whose rheological property may be either Newtonian or non-Newtonian. In Section 2.2, the general transport theorem in continuum mechanics is introduced and its derivation is explained. The governing equations for mass, momentum, and energy transport are readily obtained by applying the general transport theorem to Newtonian fluids.

In a turbulent flow, transport is enhanced by turbulent eddies (in contrast to the characteristic molecular diffusion of laminar transport). Various turbulence models have been developed for the turbulent momentum transfer (*i.e.*, the Reynolds turbulence stresses). Turbulent transport of other quantities is typically modeled by a ratio to the turbulence momentum transport, which is formulated based on an analogy to such ratios in laminar flows. Section 2.3 describes the modeling of turbulent flow using the Reynolds averaged Navier–Stokes (RANS) approach, along with three commonly used turbulence models (the mixing length model, the k–ε

model, and the Reynolds stress model), as well as the large eddy simulation (LES) approach.

Flow of Newtonian fluids through porous media can be described, in principle, by directly solving the Navier–Stokes equations at the scale of micropores. However, such scaled modeling would demand a computational capacity far beyond what is practically feasible today. In addition, there is a requirement for information on detailed micropore structures of the entire porous medium. In practice, the modeling of viscous flows through porous media is based upon a semiempirical equation, known as Darcy's law. This equation correlates the pressure gradient across a homogeneous porous medium to the superficial flow velocity, fluid viscosity, and permeability of the porous medium. There are two useful extensions of Darcy's law. One is the Ergun equation that incorporates both the viscous and inertial effects at high flow velocities (within the laminar flow regimes) in an unbounded porous medium of interlinked particles. The second is the Brinkman equation that incorporates the boundary effects at interfaces between a porous medium and an open flow medium. In Section 2.4, the basic equations of Darcy's law, the Ergun equation, and the Brinkman equations are introduced. The turbulent effects are assumed to be insignificant for flows in porous media. This may be partially justified by considering the micropore scale at the same order of the Kolmogorov scale, the smallest length scale of energetic turbulent eddies, so that any possible turbulence may be subdued by the viscous damping from the solid boundaries in micropores.

Inertia-dominated granular flow represents another interesting branch of single-phase flows in which the effect of fluids (such as gas) is negligibly small in comparison to the interparticle collisions. Hence, by postulating an analogy between the dynamic movements of granular materials and those of gas molecules, the kinetic theory of gases is extended to the modeling of granular flows. In Section 2.5, the theoretical basis for kinetic theory modeling is introduced, which includes the transport theorem, the governing equations, and the constitutive relations. The kinetic theory model is not only used to describe the transport of granular flows, but it is also extensively used to describe the collisional stress of solids in general multiphase flows that are associated with solid phases.

2.2   Flow of a Viscous Fluid

This section is focused on the general transport theorem in continuum mechanics, which formulates the most basic equation governing the single-phase viscous fluid flow. The classification of various viscous fluids, including Newtonian fluid, is introduced. The individual governing equations for mass, momentum, and energy transport are obtained from applying the general transport theorem to Newtonian fluids. The interfacial conditions and boundary conditions are presented. Finally, the applicability due to theory simplification and validations are discussed.

2.2.1 Constitutive Relation of a Viscous Fluid

Fluids are materials that continuously deform or yield whenever they are subjected to a shear stress. The total stress tensor $\mathbf{T}$ in a viscous fluid is given by:

$$\mathbf{T} = -p\mathbf{I} + \tau. \tag{2.2-1}$$

When the fluid is at rest, the total stress in the fluid is equal to the negative of the thermodynamic pressure p. For a viscous fluid, the viscous stress tensor τ includes the contribution both from the shear stress and from the bulk volume change that is due to compressibility. Thus, the relationship between the viscous stress and the rate of strain can be written as:

$$\tau = 2\mu\mathbf{S} + \left(\mu' - \frac{2}{3}\mu\right)(\nabla \cdot \mathbf{U})\mathbf{I} = \mu\left[\nabla\mathbf{U} + (\nabla\mathbf{U})^{\mathrm{T}}\right] + \left(\mu' - \frac{2}{3}\mu\right)(\nabla \cdot \mathbf{U})\mathbf{I}, \tag{2.2-2}$$

where the spatial deformation rate of fluid, $\mathbf{S}$, also known as the strain rate, is a tensor defined as:

$$\mathbf{S} = \frac{1}{2}\left[\nabla\mathbf{U} + (\nabla\mathbf{U})^{\mathrm{T}}\right]. \tag{2.2-3}$$

$\mathbf{I}$ is the identity tensor, and μ is the dynamic viscosity of the fluid. μ' is the second viscosity or bulk viscosity, which reflects the deviation of the averaged pressure from the pressure at equilibrium due to the nonuniformity in the local velocity distribution. Except for the strong nonequilibrium regions such as the shock wave region, it is assumed that $\mu' = 0$. Equation (2.2-1) is regarded as the general constitutive relation between stress and strain rate of a viscous fluid.

The dynamic viscosity, μ, is a material property. It can be regarded as the transport coefficient of momentum transfer. Most gases under normal temperature and pressure, and common liquids, such as water and ethanol, have linear relationships between shear stresses and rates of strain; thus, viscosity is independent of strain rate. A fluid whose viscosity is independent of strain rate is classified as a Newtonian fluid. In contrast, a non-Newtonian fluid's viscosity is dependent on strain rate.

Non-Newtonian fluids include large molecule liquids, such as polymeric fluids, or mixtures in the pseudo-homogeneous regime, such as solids suspensions in liquid, which is considered an extreme case of multiphase transport and described in more detail in Chapter 11. Typical examples of non-Newtonian fluids include the shear thinning fluids, the shear thickening fluids, and viscoplastic fluids. Shear thinning fluids are those whose viscosity decreases with an increase in shear rate; milk and blood are illustrative examples. Shear thickening fluids are those whose viscosity increases with an increase in shear rate, such as a suspension of corn starch in water. Viscoplastic fluids only begin to yield after the shear stress exceeds a threshold, such as mud and plastic melts. The stress–strain rate relationships of these non-Newtonian fluids are schematically illustrated in Figure 2.2-1. In addition to the three types of non-Newtonian fluids mentioned above, some non-Newtonian fluids exhibit a time-dependent stress–strain rate relationship. For example, when exposed to a shear strain

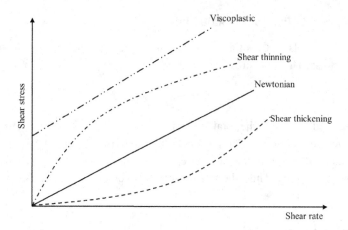

Figure 2.2-1 Classification of Newtonian and typical non-Newtonian fluids.

or extension, viscoelastic fluids, such as a slurry of pulverized coal at high concentration, glasses, rubbers, tar, or nylon, initially respond as elastic solids, then exhibit continuous deformation characteristic of fluids.

2.2.2   General Transport Theorem

Equations for the transport of any quantity, such as mass, momentum, energy, and species, can be derived from the general transport theorem. Consider a quantity $f(\mathbf{r}, t)$, which is a function of time t and position vector $\mathbf{r}$. If f can be integrated in the r-space, $\mathbf{r} = (r_1, r_2, r_3)$, a variable $F(t)$ may be defined as the volume integral over an arbitrary volume $v(t)$ in the r-space so that:

$$F(t) = \int_{v(t)} f(\mathbf{r}, t) dv. \tag{2.2-4}$$

To calculate the rate of change of F, it is necessary to introduce a ξ-space where the volume is fixed with respect to time so that an interchange of differentiation and integration is possible for the derivative of F. A transformation of the volume from r-space to ξ-space can be expressed as:

$$dv = dr_1 dr_2 dr_3 = J d\xi_1 d\xi_2 d\xi_3 = J dv_0, \tag{2.2-4a}$$

where J is the Jacobian determinant, defined as:

$$J = \frac{\partial(r_1, r_2, r_3)}{\partial(\xi_1, \xi_2, \xi_3)}. \tag{2.2-4b}$$

It can be shown that:

$$\frac{dJ}{dt} = (\nabla \cdot \mathbf{U}) J, \tag{2.2-4c}$$

where the velocity $\mathbf{U}$ is defined as:

$$\mathbf{U} = \frac{d\mathbf{r}}{dt}. \tag{2.2-4d}$$

Therefore, it gives:

$$\frac{dF}{dt} = \frac{d}{dt} \int_{v(t)} f(\mathbf{r}, t) dv = \frac{d}{dt} \int_{v_0} f(\boldsymbol{\xi}, t) J dv_0$$

$$= \int_{v_0} \left(\frac{df}{dt} J + f \frac{dJ}{dt} \right) dv_0 = \int_{v(t)} \left(\frac{df}{dt} + f \nabla \cdot \mathbf{U} \right) dv. \tag{2.2-5}$$

Note that df/dt is the substantial derivative, which can be expressed by:

$$\frac{df}{dt} = \frac{\partial f}{\partial t} + \mathbf{U} \cdot \nabla f. \tag{2.2-5a}$$

The equation describing the general transport theorem is obtained by substituting Eq. (2.2-5a) into Eq. (2.2-5). It gives:

$$\frac{d}{dt} \int_{v(t)} f(\mathbf{r}, t) dv = \int_{v(t)} \left(\frac{\partial f}{\partial t} + \nabla \cdot f \mathbf{U} \right) dv. \tag{2.2-6}$$

The rate of change of F can be found from the net flow of f across the closed surface of $v(t)$ and the generation of f inside $v(t)$. ψ is denoted as the flux of f and Φ as the rate of generation of f per unit volume. Thus:

$$\frac{d}{dt} \int_v f dv = - \int_A \mathbf{n} \cdot \psi dA + \int_v \Phi dv, \tag{2.2-7}$$

where $\mathbf{n}$ is a unit normal vector directed outwardly such that the first term on the right-hand side represents the net inflow of f across the closed surface A. Note that, according to the Gauss theorem, a surface integral over a closed area can be transformed into a volume integral, which yields:

$$\int_A \mathbf{n} \cdot \psi dA = \int_v \nabla \cdot \psi dv. \tag{2.2-7a}$$

Combining Eqs. (2.2-6), (2.2-7), and (2.2-7a) leads to:

$$\int_v \left(\frac{\partial f}{\partial t} + \nabla \cdot f \mathbf{U} + \nabla \cdot \psi - \Phi \right) dv = 0. \tag{2.2-8}$$

Since the volume, v, is arbitrarily selected, the general conservation equation in a single phase of fluid can be written in the form:

$$\frac{\partial f}{\partial t} + \nabla \cdot f \mathbf{U} + \nabla \cdot \psi - \Phi = 0. \tag{2.2-9}$$

2.2.3 Governing Equations of Viscous Flows

The conservation equations of mass, momentum, and energy of a single-phase flow can be obtained from the above general conservation equations. For a multicomponent fluid, the species equation of each component accounts for the mutual diffusion among components. The equation of state correlates the local thermodynamic properties such as pressure, temperature, and density of the fluid.

2.2.3.1   Continuity Equation and Equation of Species

For the total mass conservation of a single-phase fluid without phase change, the quantity f in Eq. (2.2-9) represents the fluid density ρ. ψ represents the diffusive flux of total mass, which is zero. For systems without mass sources, $\Phi = 0$. Therefore, from Eq. (2.2-9), the continuity equation becomes:

$$\frac{\partial \rho}{\partial t} + \nabla \cdot \rho\mathbf{U} = 0. \tag{2.2-10}$$

For a multicomponent fluid, the mass balance for individual species i considering convective and diffusive mass transfer and reactions can be expressed by:

$$\frac{\partial \rho Y_i}{\partial t} + \nabla \cdot (\rho\mathbf{U}Y_i) = \nabla \cdot (\rho D_i \nabla Y_i) - \gamma_i, \tag{2.2-11}$$

where Y_i is the mass fraction of species i, and γ_i is the rate of reaction for species i. The conservation of mass fractions for all species requires:

$$\sum_i Y_i = 1. \tag{2.2-11a}$$

2.2.3.2   Momentum (Navier–Stokes) Equation

For the momentum conservation of a single-phase fluid, the momentum per unit volume f is equal to $\rho\mathbf{U}$. The outgoing momentum flux is expressed by the negative of the stress tensor $\psi = -\mathbf{T} = (p\mathbf{I} - \tau)$, as defined by Eq. (2.2-1). Since $\Phi = -\rho\mathbf{f}$ where $\mathbf{f}$ is the field force per unit mass, Eq. (2.2-9) gives rise to the momentum equation as:

$$\frac{\partial \rho\mathbf{U}}{\partial t} + \nabla \cdot (\rho\mathbf{U}\mathbf{U}) = -\nabla p + \nabla \cdot \tau + \rho\mathbf{f}, \tag{2.2-12}$$

where τ is defined in Eq. (2.2-2).

2.2.3.3   Energy Equation

For total energy conservation, f represents the total energy per unit volume ρE. The total energy per unit volume for the flux and generation rate has three sources: the heat flux vector, $\mathbf{J}_q$; the rate of heat generation per unit volume J_e that includes Joule's heating and thermal radiation; and the rate of work done by the surface force and the field force. Thus, the energy equation can be given as:

$$\frac{\partial \rho E}{\partial t} + \nabla \cdot (\rho\mathbf{U}E) = -\nabla \cdot \mathbf{U}p + \nabla \cdot (\mathbf{U} \cdot \tau) + \rho\mathbf{U} \cdot \mathbf{f} - \nabla \cdot \mathbf{J}_q + J_e. \tag{2.2-13}$$

In the above equation, $E = e + 1/2||\mathbf{U}||^2$ and e is the internal energy per unit mass.

An alternate form of the energy equation can be expressed in terms of internal energy, e:

$$\frac{\partial \rho e}{\partial t} + \nabla \cdot (\rho\mathbf{U}e) = -p\nabla \cdot \mathbf{U} + \varphi - \nabla \cdot \mathbf{J}_q + J_e. \tag{2.2-14}$$

In Eq. (2.2-14), φ is the dissipation function defined by:

$$\varphi = \nabla \cdot (\mathbf{U} \cdot \tau) - \mathbf{U} \cdot (\nabla \cdot \tau) = \tau : \nabla\mathbf{U}. \tag{2.2-14a}$$

This equation represents the dissipation rate of energy per unit volume due to viscous effects. The energy is dissipated in the form of heat. For Newtonian fluids, φ can be proven to always be positive.

The energy equation expressed in terms of temperature is convenient for evaluating heat fluxes. For an ideal gas, $e = c_v T$. In this equation, c_v is the specific heat at constant volume and T is the absolute temperature of the fluid. Assuming the heat flux J_q obeys Fourier's law, Eq. (2.2-14) becomes:

$$\frac{\partial}{\partial t}(\rho c_v T) + \nabla \cdot (\rho \mathbf{U} c_v T) = -p\nabla \cdot \mathbf{U} + \phi + \nabla \cdot (K\nabla T) + J_e. \qquad (2.2\text{-}15)$$

In the above equation, K is the thermal conductivity of the fluid.

The energy equation can also be expressed in terms of enthalpy, $h \; (= e + p/\rho)$, as:

$$\frac{\partial}{\partial t}(\rho h) + \nabla \cdot (\rho \mathbf{U} h) = \varphi - \nabla \cdot \mathbf{J}_q + J_e + \frac{dp}{dt}. \qquad (2.2\text{-}16)$$

For an ideal gas, $h = c_p T$. The variable c_p represents specific heat at constant pressure. Again, assuming the heat flux J_q obeys Fourier's law, it gives:

$$\frac{\partial}{\partial t}(\rho c_p T) + \nabla \cdot (\rho \mathbf{U} c_p T) = \varphi + \nabla \cdot (K\nabla T) + J_e + \frac{dp}{dt}. \qquad (2.2\text{-}17)$$

2.2.3.4   Equation of State

The equation of state is the thermodynamic correlation of the pressure, temperature, and density of a fluid, which can be expressed in a general form by:

$$f(p, \rho, T) = 0. \qquad (2.2\text{-}18)$$

For example, the equation of state of an ideal gas is given by:

$$\rho = \frac{Mp}{RT}, \qquad (2.2\text{-}18a)$$

where M and R are the molar mass of the gas and the universal gas constant, respectively.

2.2.4   Interfacial Phenomena and Boundary Conditions

While transport equations are always applied within a flow domain, boundary conditions must be applied on the domain boundaries. A domain boundary is identified as an open boundary if the fluid can flow across it freely, into or out of the domain. Other boundaries are considered closed if the fluid is confined by the interface with another medium. For most applications, the boundary region is infinitesimally small compared to the domain size; that skewed ratio allows the same fluid properties in the bulk to be extended up to the boundary.

2.2.4.1   Boundary Conditions at Fluid–Solid Interface

Given an impermeable solid surface and the absence of any phase change, the mass conservation of the fluid requires that the mass flux across the Fluid–Solid interface be zero. Therefore, the normal velocity of the fluid and the solid surface must be equal so that the fluid does not penetrate into the solid. Thus, it gives:

$$\mathbf{U} \cdot \mathbf{n} = \mathbf{U}_s \cdot \mathbf{n}. \tag{2.2-19}$$

For a viscous fluid at a solid surface, the most common condition for the tangential velocity component is the no-slip condition, which requires the tangential velocity of the fluid to be identical to that of the solid surface. This condition is given by:

$$\mathbf{U} \cdot \mathbf{t} = \mathbf{U}_s \cdot \mathbf{t}. \tag{2.2-20}$$

For a conjugated heat transfer problem, requiring a solution to the energy equations in both fluid and solid domains, the temperature and heat flux on each side of the interface must be continuous, leading to:

$$T = T_s \tag{2.2-21}$$

$$J_q \cdot \mathbf{n} = \sum_i q_{si}. \tag{2.2-22}$$

When considering just one side of the interface, the choice of the temperature condition, the heat flux condition, or a mixed condition is case dependent.

2.2.4.2   Surface Tension and Wetting

The stress balance for the interface between two immiscible fluids usually needs to include the effect of surface tension. From an energy standpoint, surface tension can be defined as the amount of free energy required to create a unit of new interfacial area. This process is important to understand, since the transport of molecules from the bulk of the fluid to the interface requires doing work against the intermolecular forces. On the other hand, surface tension can also be correctly defined as the force acting on a unit length of interfacial section line. The two definitions are equivalent, as illustrated in Figure 2.2-2. At equilibrium, an infinitesimal movement, dx, would create a new surface area of $2ldx$. The increase in free energy must equal the work performed by the force F, thus:

$$2\sigma ldx = Fdx, \tag{2.2-23}$$

where σ is the surface tension with the dimension of energy per area. The above equation can also be written as:

$$\sigma = F/2l. \tag{2.2-23a}$$

Equation (2.2-23a) reflects the mechanical definition of surface tension, which has the dimension of force per length. It can be readily shown that those two dimensions are equivalent.

The magnitude of surface tension is determined by the two fluid media on two sides of the interface, and the value typically decreases as temperatures increase. For

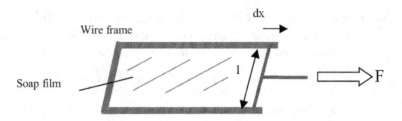

Figure 2.2-2 Force required to stretch a soap film on a wire frame.

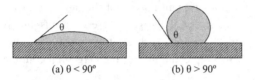

(a) $\theta < 90°$ (b) $\theta > 90°$

Figure 2.2-3 Contact angle of a liquid droplet on a solid surface.

a curved interface, the surface tension induces a pressure difference across the interface, and the fluid on the convex side has a higher pressure than that on the concave side. The pressure difference can be calculated by the Laplace–Young equation as:

$$\Delta p = p_1 - p_2 = \sigma \left(\frac{1}{R_1} + \frac{1}{R_2} \right), \qquad (2.2\text{-}24)$$

where R_1 and R_2 are the two principal radii of curvature.

When three media are in contact, the balance of surface energy determines the angle formed between the interfaces. For example, Figure 2.2-3 shows the contact angle of a liquid at the gas–liquid–solid contact line. For contact angle $0° < \theta < 90°$, the liquid is referred to as a wetting liquid on that particular solid surface, while for $90° < \theta < 180°$ the liquid is referred to as a nonwetting liquid on that particular solid surface. If the liquid is water, a wettable solid surface is known as hydrophilic, and a nonwettable surface is known as hydrophobic.

The contact angle is a function of the material properties of the three contacting media, and it is also affected by other factors. The contact angle measured under equilibrium conditions on a clean, smooth surface is referred to as the equilibrium contact angle, θ_e. When the contact line is in motion, as in the case of a liquid droplet sliding on a solid surface, different values of contact angles can be observed depending on the direction of the contact line motion. Advancing contact angle θ_a is observed when the contact line motion results in an increased liquid–solid contact area, while receding contact angle θ_r is observed with a reduced liquid–solid contact area. Typically, $\theta_r < \theta_e < \theta_a$. This effect is known as contact angle hysteresis, and it is often related to the roughness and contamination of the solid surface.

2.2.4.3   Boundary Conditions at Interface between Immiscible Fluids

If no phase change is present, the interface between two immiscible fluids must also satisfy the so-called no-penetration condition for the normal velocity, represented by:

$$\mathbf{U}_1 \cdot \mathbf{n} = \mathbf{U}_2 \cdot \mathbf{n}. \tag{2.2-25}$$

Continuity of the momentum flux across the interface requires that the stresses of the two fluids are balanced by the interfacial tension. Therefore it gives:

$$\mathbf{T}_1 - \mathbf{T}_2 = \kappa_1 \sigma \mathbf{n}_1, \tag{2.2-26}$$

where κ_1 and $\mathbf{n}_1$ are the curvature and unit normal of the interface measured from fluid 1. Since the viscous shear stress of the gas phase on a gas–liquid interface is negligible, the corresponding liquid-phase shear stress must also diminish. This special case gives rise to the free-slip condition for the tangential component of the liquid velocity at the interface.

2.2.4.4   Balances at a Phase Boundary

In general, on an interface moving with velocity $\mathbf{U}_s$, the rate of mass transported across the interface from fluid phase 1 is given by:

$$\dot{m}_1 = \rho_1 (\mathbf{U}_1 - \mathbf{U}_s) \cdot \mathbf{n}_1. \tag{2.2-27}$$

The local mass balance requires that:

$$\rho_1 (\mathbf{U}_1 - \mathbf{U}_s) \cdot \mathbf{n}_1 + \rho_2 (\mathbf{U}_2 - \mathbf{U}_s) \cdot \mathbf{n}_2 = 0. \tag{2.2-28}$$

The balance of local momentum flux across the interface is given by:

$$\sum_{k=1,2} (\dot{m}_k \mathbf{U}_s - \mathbf{T}_k) \cdot \mathbf{n}_k = \nabla_s \sigma - 2\kappa_k \sigma \mathbf{n}_k, \tag{2.2-29}$$

where $\mathbf{T}_k$ is the stress tensor of phase k, ∇_s is the surface gradient, and κ_k is the mean curvature along the direction normal $\mathbf{n}_k$.

With the capillary energy neglected, the total energy balance at the interface is given by:

$$\sum_{k=1,2} \left(\rho_k E_k (\mathbf{U}_k - \mathbf{U}_s) \cdot \mathbf{n}_k + \mathbf{J}_{qk} \cdot \mathbf{n}_k - \mathbf{U}_k - \mathbf{T}_{qk} \cdot \mathbf{n}_k \right) = 0. \tag{2.2-30}$$

The energy balance can also be given in the form of internal energy as:

$$\sum_{k=1,2} \left(\rho_k e_k (\mathbf{U}_k - \mathbf{U}_s) \cdot \mathbf{n}_k + \mathbf{J}_{qk} \cdot \mathbf{n}_k - p_k (\mathbf{U}_k - \mathbf{U}_s) \cdot \mathbf{n}_k \right) = 0 \tag{2.2-30a}$$

or in the form of enthalpy as:

$$\sum_{k=1,2} \left(\rho_k e_k (\mathbf{U}_k - \mathbf{U}_s) \cdot \mathbf{n}_k + \mathbf{J}_{qk} \cdot \mathbf{n}_k - p_k (\mathbf{U}_k - \mathbf{U}_s) \cdot \mathbf{n}_k \right) = 0 \tag{2.2-30b}$$

2.2.5 Theory Simplifications and Limitations

It is important to understand the applicability of the general transport theory, which may be limited to the validity of the assumption in the theory derivation. The following is focused on the minimum scale of a fluid element that constitutes a continuum of the fluid, the effect of fluid compressibility, the kinetic energy dissipation from viscous shearing, and limiting conditions of viscous effect for further simplification of the theory.

2.2.5.1 Minimum Scale of Continuity

This leads us to the question: to what minimum scale do the continuum fluid mechanics still apply? The basic validation criterion of the fluid continuum is that the *molecular mean free path << characteristic length of a fluid element*. From a statistics point of view, the basic fluid element should contain "enough" molecules so that the statistical error margin is less than "engineering tolerance"; that is to say:

$$\frac{1}{\sqrt{N}} < \delta. \tag{2.2-31}$$

The smallest characteristic length of a fluid element with a tolerance of δ can be estimated via the molecular concentration, C, as:

$$l_{min} > (\delta^2 C)^{-\frac{1}{3}}. \tag{2.2-32}$$

In computational fluid dynamics, the characteristic length of a fluid element can be regarded as the finest grid length of simulation. The size of the flow system needs to be much larger than the length of the fluid elements calculated above in order to have enough resolution for the model. Typically, the continuum assumption is considered applicable for a flow system with the Knudsen number (Kn) satisfying the following condition:

$$Kn = \frac{l_{mfp}}{L} < 0.01 \tag{2.2-33}$$

where l_{mfp} is the mean free path of the fluid molecules, and L is the characteristic length of the flow system. For a gas flow under standard ambient conditions, the continuum fluid mechanics modeling can be applied down to a micron-sized system (about 10 μm), but it is not applicable for a nano-sized system. An example will be analyzed in Section 2.6.2.

2.2.5.2 Flow Compressibility

Flow compressibility is different from thermodynamic compressibility of fluid. The former is only related to the relative density change by the flow velocity, whereas the latter is due to the changes of thermodynamic conditions, such as thermodynamic pressure or temperature. The basic criterion of flow compressibility is demarcated by the Mach number, defined as the ratio of the flow characteristic velocity to the speed of sound in the given fluid media:

$$\text{Ma} = \frac{U}{a} = \frac{U}{\sqrt{\left(\frac{\partial \rho}{\partial p}\right)_s}} = \begin{cases} < 0.2 & \text{incompressible} \\ > 0.3 & \text{compressible} \end{cases} \tag{2.2-34}$$

The speed of sound in a gas or liquid medium can be estimated as follows:

$$a = \begin{cases} \sqrt{\gamma R T} & \text{ideal gas} \\ \sqrt{\frac{E}{\rho}} & \text{liquid or elastic solid} \end{cases} \tag{2.2-35}$$

With cases where Ma > 1, the flow becomes supersonic and "discontinuities" of the flow in terms of pressure, density, and temperature occur due to the formation and transport of shock waves. These discontinuities are only mathematical phenomena, and they are due to the simplification of the "infinitely thin" thickness of the shock waves in most modeling approaches. The shock wave intensity becomes extremely pronounced when Ma > 3; hence, the flow is classified as hypersonic.

2.2.5.3 Viscous Dissipation of Kinetic Energy

In the absence of external heat transfer or internal reaction, a given flow is often assumed to be isothermal; hence, the energy equation is typically ignored. Since all real fluids are viscous, such a simplification may not be valid; accuracy depends on the characteristic of the conversion of kinetic energy to thermal energy in terms of viscous dissipation. One quick metric to assess the approximation is based on the maximum possible temperature rise caused by the flow velocity, which may be estimated as:

$$\Delta T = \frac{1}{2} \frac{U^2}{c_p}. \tag{2.2-36}$$

However, for flow velocity of less than 100 m/s, the increase in temperature is only a few degrees.

A more rigorous assessment should be based on the evaluation of the dissipation function, as defined by Eq. (2.2-13a). It can be shown that significant viscous dissipation takes place, typically from strong boundary-layer shearing or from normal compression in cases of shock waves. Such viscous heating, however, is most likely to be restrained to the local boundary layer or the shock wave region, rather than to the entire bulk flow region.

2.2.5.4 Limiting Flows: Creeping Flow and Potential Flow

For a steady, incompressible, and isothermal flow, Eq. (2.2-12) can take a dimensionless form and can be expressed by:

$$\frac{\partial \mathbf{U}^*}{\partial t^*} + \nabla^* \cdot (\mathbf{U}^* \mathbf{U}^*) = -\nabla^* p^* + \nabla \cdot \left(\frac{1}{\text{Re}} \nabla^* \mathbf{U}^*\right) + \mathbf{f}^*, \tag{2.2-37}$$

where the dimensionless parameters are defined using a characteristic length, L, and a characteristic speed, U_∞, as:

$$\mathbf{U}^* = \frac{\mathbf{U}}{U_\infty}; \quad t^* = \frac{U_\infty}{L}t; \quad \nabla^* = L\nabla; \quad p^* = \frac{p}{\rho U_\infty^2}; \quad \mathbf{f}^* = \frac{\mathbf{g}L}{U_\infty^2}; \quad \mathrm{Re} = \frac{\rho L U_\infty}{\mu}.$$

$$(2.2\text{-}37a)$$

The creeping flow refers to a viscous-dominated flow regime where Re $\ll$ 1. In this limiting flow regime, the convection-induced momentum transfer can be ignored. Hence Eq. (2.2-37) reduces to a linear equation of the form:

$$\frac{\partial \mathbf{U}^*}{\partial t^*} + \nabla^* p^* = \nabla \cdot \left(\frac{1}{\mathrm{Re}}\nabla^* \mathbf{U}^*\right) + \mathbf{f}^*. \qquad (2.2\text{-}37b)$$

Many phase interactions of momentum transfer in multiphase flow theory are predicated on the solutions of creeping flows, such as Stokes drag, Basset force, Saffman force, and so on.

On the other hand, the potential flow appears in limiting cases where the convection overdominates the flow viscous diffusion, with Re $\rightarrow \infty$ (but neglecting turbulence effects). In this case, Eq. (2.2-37) becomes:

$$\frac{\partial \mathbf{U}^*}{\partial t^*} + \nabla^* \cdot \left(\mathbf{U}^*\mathbf{U}^*\right) = -\nabla^* p^* + \mathbf{f}^*. \qquad (2.2\text{-}37c)$$

The solution to the above equation leads to the celebrated Bernoulli equation. It also provides the solutions for the external flow and pressure distribution for various boundary-layer theories.

2.3   Turbulence

When the velocity of a viscous fluid flow exceeds a certain threshold, typically characterized by the critical Reynolds number that corresponds to the specific type of fluid transport involved, the flow becomes turbulent. The turbulent flow transport is enhanced by turbulent eddies, with the self-sustained continuous eddy generation and dissipation. This section first introduces the typical origins and length scales of turbulent flows. Then the introduction goes to various commonly used turbulence models of Newtonian fluid flows, including the mixing length model, the k–ε model, the Reynolds stress model, and the large eddy simulation (LES) approach.

2.3.1   Turbulent Flows

Turbulent flow is characterized by the random fluctuations, eddy transport, the cascade process, and self-sustainability. The random fluctuations occur in all flow transport parameters, such as $\mathbf{U}$, p, T, and Y_i. The transport is caused by eddy movements, which is much more vigorous when compared to the molecular transport characteristic of laminar flows. When self-sustaining new eddies are generated to replace those destroyed by viscous dissipation, the turbulent eddies undergo a cascade process, shifting from a large energetic eddy to a completely dissipated fluid

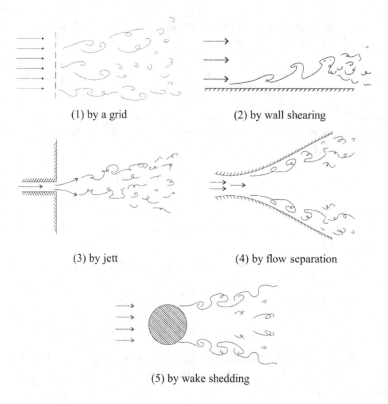

(1) by a grid   (2) by wall shearing

(3) by jett   (4) by flow separation

(5) by wake shedding

Figure 2.3-1 Typical origins of turbulent flows.

element. The size of the smallest eddy that is still active in the eddy transport can be characterized by the Kolmogorov scale.

The true mechanistic origins of turbulence are currently not fully understood. Phenomenological causes of the eddy generation include the flow separation, wall shearing, jetting, or rotation, as illustrated in Figure 2.3-1. The associated eddy generation mechanisms can be directly attributed to the reverse pressure gradients and boundary-layer or shearing instability. The reverse pressure gradients lead to eddy generation in flow systems with diffusers, sudden expansion sections, bluff bodies, and mesh screens or grids. Flow instability, such as boundary-layer instability in cases of wall shearing and continuous shearing instability in jetting and rotating flows, also causes eddy generation, shedding, and stochastic transport in turbulent flows.

With the increase of Reynolds number (Re) beyond a critical value, the molecular transport of a laminar flow becomes dominated by the eddy transport phenomenon of turbulent flow. The critical transition Reynolds number, Re_{cr}, describing the transition from laminar to turbulent flow, depends on many flow parameters including the types of flow and the external disturbance. In general, the turbulent intensities increase with Re and the characteristic eddy sizes decrease with Re.

2.3.2   Length Scales in Turbulence

In a turbulent flow, fluid transport properties such as mass, momentum, and energy are greatly enhanced by the vigorous movement of eddies. Conceptually, large eddies are generated from the inherent flow instability (*i.e.*, that triggers the flow transition from laminar regime to turbulent regime) if the Reynolds number is increased over a critical value. The turbulence kinetic energy, defined later by Eq. (2.3-15), dissipates as larger eddies break down into smaller ones in cascade. For homogeneous turbulence flows, the size of the smallest eddies is of the Kolmogorov scale and given by (Lumley, 1970):

$$l_k = \left(\frac{\mu^3}{\rho^3 \varepsilon} \right)^{1/4},$$ (2.3-1)

where ε is the dissipation rate of turbulence kinetic energy, defined by Eq. (2.3-16). Once an eddy cascades down to the size below the Kolmogorov scale, its turbulence kinetic energy is completely dissipated to heat.

Based on the scale analysis, the dissipation rate can be related to the turbulence kinetic energy, k, defined by Eq. (2.3-15), and the average size of the energy-containing eddies, l_e, as:

$$\varepsilon \propto \frac{k^{3/2}}{l_e},$$ (2.3-2)

where the proportional coefficient is on the order of unity. l_e is on the order of the characteristic length of flow system (such as pipe diameter), L, and the turbulence kinetic energy is on the order of the kinetic energy at the averaged flow velocity. Combining these two approximations, along with Eq. (2.3-1) and Eq. (2.3-2), the ratio of the flow characteristic length to the Kolmogorov scale can be estimated as

$$\frac{L}{l_k} \approx \left(\frac{\rho U L}{\mu} \right)^{3/4} = \mathrm{Re}^{3/4}.$$ (2.3-3)

Figure 2.3-2 represents the energy spectrum $E(\kappa)$, which shows the allocation of turbulence kinetic energy contained in eddies of different sizes. The wave number, κ, is defined by $2\pi/l$, where l is the eddy size. Depending on the slope of the curve, the spectrum can be divided into three regions. The energy-containing range contains most of the turbulence kinetic energy of the flow, and it is characterized by large-scale anisotropic eddies. The demarcation between the energy-containing range and the inertial subrange is determined by the length scale $l_{EL} \approx l_e/6$. Small-scale eddies with size less than l_{EI} are isotropic. The inertial subrange and dissipation range, in which the eddy motion is dominated by inertia and viscous dissipation respectively, are separated by the length scale $l_{DI} \approx 60 l_k$. In a turbulent flow, the large eddies in the energy-containing range are unstable and break up into successively smaller eddies. The turbulence kinetic energy is thus transferred to smaller scales until it is dissipated by viscosity in the smallest scale.

Comprehensive modeling of turbulent flow would require that the computational cell size be small enough to match the Kolmogorov scale. In addition, the time-step

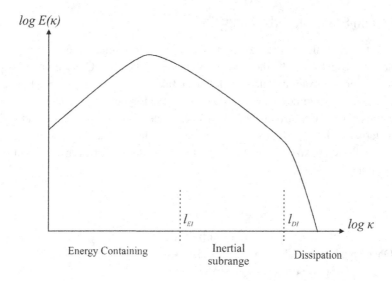

$log\ E(\kappa)$

l_{EI}

l_{DI}

$log\ \kappa$

Energy Containing

Inertial
subrange

Dissipation

Figure 2.3-2 Power spectrum of turbulent flow.

scale needs to be proportional to the Kolmogorov scale. Hence, to directly simulate a real-time, three-dimensional turbulent flow, known as Direct Numerical Simulation, or DNS, the total number of grid points in all coordinates, including the time coordinate, must be proportional to the fourth power of the ratio of the flow characteristic length to the Kolmogorov scale. Thus:

$$\text{Number of Grids} \propto \left(\frac{L}{l_k}\right)^4 \approx \text{Re}^3. \qquad (2.3\text{-}4)$$

For typical turbulent flows, the Reynolds number is within a range from 10^5 to 10^8. For a complete simulation of such a turbulent flow, the total number of grids is expected to be at least on the order of 10^{15} to 10^{24}, which is far beyond the computing capability of even the most powerful computers available today. As a compromise, practical turbulent modeling is currently based on solving the time-averaged Navier–Stokes equations, in which the Reynolds stress must be solved by introducing additional simplified and semiempirical closure forms through either algebraic or differential equations.

2.3.3   Reynolds-Averaged Navier–Stokes Equations

While the Navier–Stokes equations are applicable to the instantaneous description of a turbulent flow, it is often impractical and unnecessary to resolve all the fluctuating components of the turbulent quantities. Instead, a time-averaged approach is often used to formulate a set of governing equations for the time-averaged quantities in turbulent flow. From Reynolds decomposition, any instantaneous variable, ϕ, can be divided into a time-averaged[1] quantity and a fluctuating part as:

$$\phi(t) = \overline{\Phi} + \phi'(t), \tag{2.3-5}$$

where:

$$\overline{\Phi} = \frac{1}{\tau_t} \int_0^{\tau_t} \phi(t)dt. \tag{2.3-5a}$$

The integral time period, τ_t, should be sufficiently short in duration when compared to the characteristic time scale of the overall system.

The time-averaged continuity can be obtained by taking the time average over Eqs. (2.2-10) and (2.2-11), which gives:

$$\frac{\partial \overline{\rho}}{\partial t} + \frac{\partial}{\partial x_j}\left(\overline{\rho}\,\overline{U_j} + \overline{\rho' u_j'}\right) = 0. \tag{2.3-6}$$

The time-averaged ith-component momentum equation is obtained as

$$\frac{\partial}{\partial t}\left(\overline{\rho}\,\overline{U_i} + \overline{\rho' u_i'}\right) + \frac{\partial}{\partial x_j}\left(\overline{\rho}\,\overline{U_i}\,\overline{U_j} + \overline{\rho}\,\overline{u_i' u_j'} + \overline{U_i}\,\overline{\rho' u_j'} + \overline{U_j}\,\overline{\rho' u_i'} + \overline{\rho' u_i' u_j'}\right)$$

$$= -\frac{\partial \overline{\sigma}_{ij}}{\partial x_j} + \sum_j \overline{F}_{ji}, \tag{2.3-7}$$

where $\overline{\rho}\,\overline{u_i' u_j'}$ is defined as the turbulent Reynolds stress and the time-averaged stress tensor, σ_{ij}, is given by:

$$\overline{\sigma}_{ij} = \overline{p}\delta_{ij} - \overline{\mu}\left(\frac{\partial \overline{U_i}}{\partial x_j} + \frac{\partial \overline{U_j}}{\partial x_i}\right) + \frac{2}{3}\left(\overline{\mu}\frac{\partial \overline{U_m}}{\partial x_m} + \overline{\mu'\frac{\partial u_m'}{\partial x_m}}\right)\delta_{ij} - \mu'\left(\overline{\frac{\partial u_i'}{\partial x_j} + \frac{\partial u_j'}{\partial x_i}}\right). \tag{2.3-7a}$$

In Eq. (2.3-7a), μ' denotes the viscosity fluctuation, which may result from the temperature fluctuation.

For simplicity, consider the incompressible and isothermal turbulent flows where $\rho' = 0$ and $\mu' = 0$. The continuity equation is thus simplified from Eq. (2.3-6) to:

$$\frac{\partial \overline{U_j}}{\partial x_j} = 0 \tag{2.3-8}$$

The ith-component momentum equation is reduced from Eq. (2.3-7) to:

$$\frac{\partial}{\partial t}\left(\overline{\rho}\,\overline{U_i}\right) + \frac{\partial}{\partial x_j}\left(\overline{\rho}\,\overline{U_i}\,\overline{U_j}\right) = -\frac{\partial \overline{p}}{\partial x_i} + \frac{\partial}{\partial x_j}\left(\overline{\mu}\left(\frac{\partial \overline{U_i}}{\partial x_j} + \frac{\partial \overline{U_j}}{\partial x_i}\right) - \overline{\rho u_i' u_j'}\right) + \sum_j \overline{F}_{ji}. \tag{2.3-9}$$

Equation (2.3-9) is the Reynolds-averaged Navier–Stokes (RANS) equation. It has a similar form to the original N-S equation, except that the averaged flow quantities replace their instantaneous counterparts in the original N–S equation and an additional term, the Reynolds stress, appears in the RANS equation. To render Eq. (2.3-9)

---

[1] More precisely, the RANS is based on the statistical approach with ensemble average, which can be approximated by time average, especially for steady turbulent flows.

solvable, it is necessary to provide a closure model for the turbulent Reynolds stress.

For isotropic turbulent flows, a scalar turbulent eddy viscosity, μ_e, is defined using the Boussinesq formulation as:

$$- \overline{\rho u_i' u_j'} = \mu_e \left(\frac{\partial \overline{U}_i}{\partial x_j} + \frac{\partial \overline{U}_j}{\partial x_i} \right). \tag{2.3-9a}$$

It should be noted that the turbulent viscosity is a function of the turbulence structure in addition to the physical properties of the fluid. An effective turbulent viscosity, μ_{eff}, may be further introduced as:

$$\mu_{eff} = \mu + \mu_e. \tag{2.3-9b}$$

Thus, Eq. (2.3-9) can be expressed as:

$$\frac{\partial}{\partial t} \left(\overline{\rho} \overline{U}_i \right) + \frac{\partial}{\partial x_j} \left(\overline{\rho} \overline{U}_i \overline{U}_j \right) = -\frac{\partial \overline{p}}{\partial x_i} + \frac{\partial}{\partial x_j} \left(\overline{\mu}_{eff} \left(\frac{\partial \overline{U}_i}{\partial x_j} + \frac{\partial \overline{U}_j}{\partial x_i} \right) \right) + \sum_j \overline{F}_{ji}. \tag{2.3-10}$$

For anisotropic turbulent flows, such as strong swirling flows or buoyant flows, the turbulent viscosity is a tensor instead of a scalar; therefore, all components of the Reynolds stress need to be modeled individually.

2.3.4   Turbulence Modeling

In order to solve the RANS equations, such as those in Eq. (2.3-9) or (2.3-10), closure relations must be provided for the Reynolds stress or μ_e in the case of isotropic turbulence. This is typically achieved by using a turbulence model, which can be a set of transport equations or constitutive relations that simulate higher-order correlations using lower-order correlations or time-averaged quantities. Since the physics of turbulence is not fully understood at present, there is no universal turbulence model available today. However, a number of existing closure turbulence models with semiempirical correlations have been developed; they are demonstrably successful for particular types of turbulent flows within acceptable tolerances for most engineering applications. In this section, several of the most popular models are introduced, including the mixing length model (one-equation model) and the k–ε model (two-equation model) for isotropic turbulent flows, and the Reynolds stress model (seven-equation model) for anisotropic turbulent flows.

2.3.4.1   Mixing Length Model

The mixing length model was originally proposed by Prandtl (1925). The model is based on an analogy between the fluctuating motion of turbulent eddies and the molecular motion of a gas. The turbulent viscosity is analogous to the laminar (molecular) viscosity of the gas. According to the molecular kinetic theory of gases, the viscosity is the product of gas density, the length of the mean-free path, and mean fluctuating velocity. Hence, the averaged turbulent viscosity is expressed as a product of the averaged density, the mixing length in eddy transport, and the mean fluctuating velocity of turbulent eddies. This gives:

$$\mu_e = \overline{\rho} \, l_m < u >. \tag{2.3-11}$$

In a simple thin shear turbulent flow, the ratio of mean fluctuating velocity to the mixing length of turbulent eddies is approximated as being equal to the local gradient of velocity. Thus it gives:

$$\frac{< u >}{l_m} = \left| \frac{\partial \overline{U}}{\partial y} \right|. \tag{2.3-12}$$

Combining Eq. (2.3-11) with Eq. (2.3-12) yields:

$$\mu_e = \overline{\rho} \, l_m^2 \left| \frac{\partial \overline{U}}{\partial y} \right|, \tag{2.3-13}$$

where l_m is the turbulent mixing length, which is determined empirically and depends on the type of flow.

The mixing length model is simple, with μ_e defined by an algebraic expression. For some simple scenarios, such as jet flows, flows of the boundary layer type, and pipe flows, l_m can be reasonably linked to the associated flow characteristic length. For flows with recirculation regions or with complex geometries, determination of l_m may be impossible. The selection of l_m for various simple flows can be found in the literature (Launder and Spalding, 1972). Notably, the mixing length model indicates that the effect of turbulence will vanish where the strain, based on the averaged velocity, is zero (*i.e.*, near the end of a recirculation region or near a symmetric axis). However, this is not always true.

2.3.4.2   *k-ε* Model

The number of independent scaling quantities that are required to characterize the local state of the turbulent flow can be determined from a dimensional analysis. The dimensions of the Reynolds stress, the rate of strain, and the turbulent viscosity are as follows:

$$\left[\overline{\rho u_i' u_j'} \right] = \frac{[M]}{[L][T]^2}; \qquad \left[\frac{\partial \overline{U}_i}{\partial x_j} \right] = \frac{1}{[T]}; \qquad [\mu_e] = \frac{[M]}{[L][T]}, \tag{2.3-14}$$

where [M] denotes the mass unit; [L] represents the length unit; and [T] is the time unit. In order to correlate the Reynolds stress with the rate of strain, a length scale and a time scale, or any two independent combinations of the length and time scales (with $\overline{\rho}$ as the mass scale) are needed to characterize the local turbulent momentum transfer. Therefore, two more governing equations are required to describe the transport of these two scaling quantities.

A common choice for the two scaling parameters is the kinetic energy of turbulence, k, and its dissipation rate, ε; defined respectively as:

$$k = \frac{1}{2} \overline{u_i' u_i'} \tag{2.3-15}$$

$$\varepsilon = \frac{\overline{\mu}}{\overline{\rho}} \frac{\partial u_i'}{\partial x_j} \frac{\partial u_i'}{\partial x_j}. \tag{2.3-16}$$

From the dimensional analysis, it can be shown that k and ε are the two independent scaling parameters of time and length scales:

$$[T] = \frac{[k]}{[\varepsilon]}; \quad [L] = \frac{[k]^{\frac{3}{2}}}{[\varepsilon]}. \tag{2.3-17}$$

Thus, the turbulent eddy viscosity μ_e may be related to k and ε in the form:

$$\mu_e = C_\mu \overline{\rho} \frac{k^2}{\varepsilon}, \tag{2.3-18}$$

where C_μ is an empirical constant.

For incompressible and isothermal flows, the transport equation of the turbulence kinetic energy can be derived as:

$$\frac{\partial}{\partial t} (\overline{\rho} k) + \frac{\partial}{\partial x_j} (\overline{\rho} \overline{U}_j k) = -\overline{\rho u_i' u_j'} \frac{\partial \overline{U}_i}{\partial x_j} + \frac{\partial}{\partial x_j} \left(\overline{\mu} \frac{\partial k}{\partial x_j} - \frac{1}{2} \overline{\rho u_i' u_i' u_j'} - \overline{p' u_j'} \right). \tag{2.3-19}$$

The first two terms on the right-hand side represent the production and dissipation of turbulence kinetic energy, respectively. The third term on the right-hand side represents the transport of k due to molecular and turbulent diffusion. Hence, with an analogy to the laminar transport, this term may be expressed as:

$$\overline{\mu} \frac{\partial k}{\partial x_j} - \frac{1}{2} \overline{\rho u_i' u_i' u_j'} - \overline{p' u_j'} = \frac{\mu_{eff}}{\sigma_k} \frac{\partial k}{\partial x_j}, \tag{2.3-19a}$$

where σ_k is an empirical constant. Thus, Eq. (2.3-19) takes the form:

$$\frac{\partial}{\partial t} (\overline{\rho} k) + \frac{\partial}{\partial x_j} (\overline{\rho} \overline{U}_j k) = \frac{\partial}{\partial x_j} \left(\frac{\mu_{eff}}{\sigma_k} \frac{\partial k}{\partial x_j} \right) + \mu_e \frac{\partial \overline{U}_i}{\partial x_j} \left(\frac{\partial \overline{U}_i}{\partial x_j} + \frac{\partial \overline{U}_j}{\partial x_i} \right) - \overline{\rho} \varepsilon. \tag{2.3-20}$$

Equation (2.3-20) is called the k-equation. The k-equation can also be written in the vector form:

$$\frac{\partial (\overline{\rho} k)}{\partial t} + \nabla \cdot \left(\overline{\rho} \overline{U} k \right) = \nabla \cdot \left(\frac{\mu_{eff}}{\sigma_k} \nabla k \right) + 2\mu_e \overline{\mathbf{S}} : \overline{\mathbf{S}} - \overline{\rho} \varepsilon, \tag{2.3-21}$$

where the rate of deformation tensor $\mathbf{S}$ is given by Eq. (2.2-1).

The transport equation for the dissipation rate of turbulence kinetic energy, known as the ε-equation, can be derived in a similar manner as:

$$\frac{\partial (\overline{\rho} \varepsilon)}{\partial t} + \nabla \cdot \left(\overline{\rho} \overline{U} \varepsilon \right) = \nabla \cdot \left(\frac{\mu_{eff}}{\sigma_\varepsilon} \nabla \varepsilon \right) + 2\mu_e C_1 \frac{\varepsilon}{k} \overline{\mathbf{S}} : \overline{\mathbf{S}} - C_2 \overline{\rho} \frac{\varepsilon^2}{k}. \tag{2.3-22}$$

For the five empirical constants in the k-ε model, it is suggested (Launder and Spalding, 1972) that $C_\mu = 0.09$; $C_1 = 1.44$; $C_2 = 1.92$; $\sigma_k = 1.0$; and $\sigma_\varepsilon = 1.22$. With k and ε values calculated from Eqs. (2.3-21) and (2.3-22), and μ_e found from Eq. (2.3-18), the RANS equations with the k-ε model is fully closed.

2.3.4.3 Reynolds Stress Model (RSM)

For an anisotropic turbulent flow, a scalar-based turbulence viscosity is conceptually insufficient to cross-link the flow stress tensor to the strain tensor. Hence each component in the Reynolds stress tensor may need to be modeled individually. The transport equation of the Reynolds stress component $\overline{u_i'u_j'}$ can be expressed by (Bradshaw, 1978):

$$\frac{\partial \overline{u_i'u_j'}}{\partial t} + \overline{U}_k \frac{\partial \overline{u_i'u_j'}}{\partial x_k} = D_{ij} + P_{ij} + G_{ij} + \Phi_{ij} - E_{ij}. \tag{2.3-23}$$

D_{ij} is the diffusion of the Reynolds stress by turbulent, pressure, and molecular transport, and it is defined by:

$$D_{ij} = -\frac{\partial}{\partial x_k} \left(\overline{u_i'u_j'u_k'} + \frac{\overline{p'}}{\rho}(u_i'\delta_{jk} + u_j'\delta_{ik}) - \frac{\mu}{\rho}\frac{\partial \overline{u_i'u_j'}}{\partial x_k} \right). \tag{2.3-23a}$$

Since the turbulent diffusion is much stronger than the molecular diffusion for most turbulent flows where the Peclet number is large, the molecular diffusion terms are normally neglected in RSM. The pressure diffusion term in most RSMs is simply neglected, although this simplification is hardly justified (Launder $et\ al.$, 1975). Assuming isotropic transport, the turbulent diffusion of the Reynolds stress can be simplified as (Daly and Harlow, 1970):

$$\overline{u_i'u_j'u_k'} = -C_s \frac{k}{\varepsilon} \overline{u_k'u_l'} \frac{\partial \overline{u_i'u_j'}}{\partial x_l}, \tag{2.3-23b}$$

where C_s is an empirical constant (e.g., $C_s = 0.25$).

P_{ij} is the shear production of the Reynolds stress, and it is defined by:

$$P_{ij} = -\left(\overline{u_i'u_k'} \frac{\partial \overline{U}_j}{\partial x_k} + \overline{u_j'u_k'} \frac{\partial \overline{U}_i}{\partial x_k} \right). \tag{2.3-23c}$$

G_{ij} is the buoyancy production, and it is represented as:

$$G_{ij} = -\frac{g}{\rho} \left\{ \overline{\rho'u_i'}\delta_{jz} + \overline{\rho'u_j'}\delta_{iz} \right\}, \tag{2.3-23d}$$

where subscript z denotes the vertical coordinate. For incompressible turbulent flows, the buoyancy production term vanishes. Using the Boussinesq formulation, the buoyancy fluxes are related to the mean density gradient by:

$$\overline{\rho'u_i'} = -\left(\frac{\overline{u_i'^2}}{2k/3} \right) D_e \frac{\partial \overline{\rho}}{\partial x_i}, \tag{2.3-23e}$$

where $\overline{u_i'^2}$ denotes the i-component of turbulence kinetic energy, not to be confused with Einstein summation notation; D_e is the turbulent diffusivity.

Φ_{ij} is the pressure-strain contribution to Reynolds stress, and it is defined by:

$$\Phi_{ij} = -\frac{\overline{p'}}{\rho} \left(\frac{\partial u_i'}{\partial x_j} + \frac{\partial u_j'}{\partial x_i} \right). \tag{2.3-23f}$$

The pressure-strain term can be significantly influenced by the presence of a wall boundary. For turbulent flows far from any walls, the pressure-strain term is due to three effects: the turbulence interactions, the averaged-strain effects, and the buoyancy effects. For wall-bounded turbulent flows, the rigid boundary reduces the length scale of the fluctuations, reflects the pressure fluctuations, and imposes no-slip in velocity condition on the walls. Hence, the pressure-strain term may be expressed by (Gibson and Launder, 1978):

$$\Phi_{ij} = \Phi_{ij,1} + \Phi_{ij,2} + \Phi_{ij,3} + \Phi_{ij,w}, \tag{2.3-23g}$$

where:

$$\Phi_{ij,1} = -C_1 \left(\frac{\varepsilon}{k} \overline{u_i' u_j'} - \frac{2}{3} k \delta_{ij} \right) \tag{2.3-23h}$$

$$\Phi_{ij,2} = -C_2 \left(P_{ij} - \frac{1}{3} P_{kk} \delta_{ij} \right) \tag{2.3-23i}$$

$$\Phi_{ij,3} = -C_3 \left(G_{ij} - \frac{1}{3} G_{kk} \delta_{ij} \right) \tag{2.3-23j}$$

$$\Phi_{ij,w} = \frac{k^{3/2}}{C_w \, \varepsilon \, x_n} \left\{ C_4 \frac{\varepsilon}{k} \left(\overline{u_n'^2} \delta_{ij} - \frac{3}{2} \overline{u_n' u_i'} \delta_{nj} - \frac{3}{2} \overline{u_n' u_i'} \delta_{nj} \right) + C_5 \left(\Phi_{nn,2} \delta_{ij} \right. \right.$$
$$\left. \left. - \frac{3}{2} \Phi_{in,2} \delta_{nj} - \frac{3}{2} \Phi_{jn,2} \delta_{ni} \right) + C_6 \left(\Phi_{nn,3} \delta_{ij} - \frac{3}{2} \Phi_{in,3} \delta_{nj} - \frac{3}{2} \Phi_{jn,3} \delta_{ni} \right) \right\}. \tag{2.3-23k}$$

In Eq. (2.3-23k), x_n is the normal distance from a wall, and the subscript n denotes the normal direction to the wall. The empirical constants in Eq. (2.3-23g) are suggested to take the following values: $C_1 = 1.8$; $C_2 = 0.6$; $C_3 = 0.6$; $C_4 = 0.5$; $C_5 = 0.3$; $C_6 = 0$; and $C_w = 2.5$.

E_{ij} is the viscous dissipation of the Reynolds stress, and it is defined by:

$$E_{ij} = -2 \frac{\mu}{\rho} \overline{\frac{\partial u_i'}{\partial x_k} \frac{\partial u_j'}{\partial x_k}}. \tag{2.3-23l}$$

The viscous dissipation can be modeled by assuming local isotropy, such that:

$$E_{ij} = \frac{2}{3} \varepsilon \, \delta_{ij}. \tag{2.3-23m}$$

The Reynolds stress tensor is symmetric; thus, six components are to be determined from Eq. (2.3-23). In order to solve those six equations, ε needs to be solved from another transport equation (2.3-22); k can be directly calculated from its definition Eq. (2.3-15). The turbulent eddy viscosity, μ_e, required to solve the ε equation, can be calculated using Eq. (2.3-18). With the Reynolds stress tensor solved from the RSM model, the RANS equations for anisotropic turbulent flows can be readily closed.

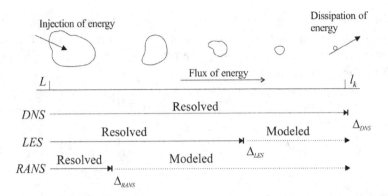

Figure 2.3-3 Resolved length scales and requirements on grid size Δ.

2.3.5   Large Eddy Simulation

The kinetic energy of turbulence is generated and carried in large eddies, and this energy is transferred to successively smaller eddies. If this transfer process creates eddies equal to the smallest eddies on the Kolmogorov scale, then the prior kinetic energy of turbulence dissipates as heat through molecular viscosity. Momentum, mass, and energy are chiefly transported by the large-scale flow structures, which are more dependent on the particular flow problem features such as geometry and boundary conditions. Conversely, the small-scale flow structures are comparatively more isotropic and less affected by the macroscopic flow features. Large eddy simulation (LES) computes the transport in the large-scale flow structures (large eddies) explicitly, and uses models only for the small eddies that are still somewhat energetic. The demarcation between the resolved large eddies and unresolved small eddies depends on a critical length scale, which lies in the inertial subrange shown in Figure 2.3-2. In contrast, a direct numerical simulation (DNS) resolves all scales down to the Kolmogorov scale without any approximation, and RANS applies statistical modeling for all length scales.

The comparison between DNS, LES, and RANS is schematically illustrated in Figure 2.3-3. The rationale behind LES is that by directly resolving the large-scale structures, less error will be introduced as compared to other types of turbulence modeling; the small eddies are more isotropic and universal, thus they require less effort to model. In general, LES can be considered as an intermediate simulation approach, in terms of both accuracy and computational cost, which falls between RANS and DNS strategies.

The governing equations of LES are obtained by applying a filtering operation to the Navier–Stokes equation. A filtered variable $\bar{f}$ is defined by:

$$\bar{f}(x) = \int_D f(x')G(x - x')dx', \tag{2.3-24}$$

where D is the fluid domain, and G is the filter function. Commonly used filter functions in LES include the sharp Fourier cutoff filter, the top-hat filter, and the Gaussian filter. The effect of these filter functions is to preserve the low-frequency (large-scale) information of the flow field while simultaneously filtering out the high-frequency (small-scale) eddies. After filtering, the governing equations for an incompressible Newtonian flow become:

$$\frac{\partial \overline{\rho}}{\partial t} + \frac{\partial}{\partial x_j}\left(\overline{\rho}\,\overline{U}_j\right) = 0 \tag{2.3-25}$$

$$\frac{\partial}{\partial t}\left(\overline{\rho}\,\overline{U}_i\right) + \frac{\partial}{\partial x_j}\left(\overline{\rho}\,\overline{U}_i\overline{U}_j\right) = -\frac{\partial \overline{p}}{\partial x_i} + \frac{\partial}{\partial x_j}\left(\mu\left(\frac{\partial \overline{U}_i}{\partial x_j} + \frac{\partial \overline{U}_j}{\partial x_i}\right)\right) - \frac{\partial \hat{\tau}_{ij}}{\partial x_j} + \sum_j \overline{F}_{ji}.$$

$$\tag{2.3-26}$$

The variables with overbars are called the filtered, or resolved, large-scale variables, and they describe the evolution of the large-scale structures of the fluid motion. The effect of the unresolved small-scale eddies are represented in the filtered momentum equation by a sub-grid scale (SGS) stress:

$$\hat{\tau}_{ij} = \rho\overline{U_iU_j} - \rho\overline{U}_i\overline{U}_j. \tag{2.3-26a}$$

The sub-grid scale stresses account for the unresolved sub-grid scale dissipation that removes energy from the resolved scales. It involves an undetermined correlated term; thus, it must be modeled. In such models, the eddy-viscosity (Boussinesq) assumption is often used, and the SGS stress term is expressed as:

$$\hat{\tau}_{ij} = -2\mu_e\overline{S}_{ij} + \frac{1}{3}\hat{\tau}_{kk}\delta_{ij}, \tag{2.3-26b}$$

where $\overline{S}_{ij}$ is the filtered strain rate tensor, computed from large-scale velocities by:

$$\overline{S}_{ij} = \frac{1}{2}\left(\frac{\partial \overline{U}_i}{\partial x_j} + \frac{\partial \overline{U}_j}{\partial x_i}\right), \tag{2.3-26c}$$

and μ_e is the eddy viscosity. A simple and widely used eddy viscosity model is the Smagorinsky model, which was also the first SGS model introduced in LES. Similar to Prandtl mixing length theory in Eq. (2.3-13), the eddy viscosity can be modeled by:

$$\mu_e = \rho(C_s\Delta x)^2(2\overline{S}_{ij}\overline{S}_{ij})^{1/2}. \tag{2.3-26d}$$

The constant C_s can be adjusted to match the sub-grid energy dissipation rate to ε. It has been found that a value of 0.1 often yields satisfactory results for many flow problems, including channel flows and free-shear flows. The length scale Δx can be represented by the local grid size.

The standard Smagorinsky SGS model has proven to be successful in many flow conditions, and it appears at present to be the most popular SGS model for engineering applications. However, this model encounters difficulties in some situations; for example, its predictions are too dissipative with regard to the presence of a boundary. Close to the wall, the eddy viscosity has to be reduced due to the anisotropic nature of the turbulence. For these cases, ad hoc corrections are often employed, such as damping functions (van Driest, 1956; Piomelli *et al.*, 1988). In addition, the Smagorinsky model cannot be applied to transitional flows because it gives a nonzero eddy viscosity, even when laminar flows are considered. The limitations with the Smagorinsky model have motivated the development of other more advanced SGS models.

2.4 Flows in Porous Media

A permeable porous medium can be approximated as an interlinked matrix of individual solids, where the topology and geometry of the solid matrix determine the degree of permeability. A porous medium is isotropic if all parameters are independent of orientation; otherwise it is anisotropic. A porous medium is defined as homogeneous if the averaged parameters are independent of location; otherwise it is heterogeneous.

For viscous fluids passing through homogeneous porous media at low velocities, Darcy (1956) proposed a law that correlates the pressure drop per unit bed length to the bed porosity, fluid viscosity, and flow velocity. Darcy's law can be interpreted from various theoretical models (Scheidegger, 1960; Bear, 1972). However, Darcy's law does not consider the effects of inertia, which become significant at high fluid velocities. Darcy's law also fails when the porosity becomes very large. To take the inertial effects into consideration, Ergun (1952) presented a semiempirical equation for flows in packed beds of solid spheres, which has been validated over a wide range of flow conditions (Ergun and Orning, 1949; Ergun, 1952). When there is a boundary between a porous medium and a free fluid, two boundary conditions must be satisfied at the free-porous interface: these are the continuity of the fluid velocity and the shear stress. However, Darcy's law alone is insufficient to capture both conditions at the interface. To facilitate the boundary condition modeling, Brinkman (1947) modified and generalized Darcy's equation, known as the Brinkman equation, by introducing a semiempirical parameter (effective viscosity) for the shear stress matching at the interface.

2.4.1 Darcy's Law

Based on the measurements of pressure drop in viscous flows through columns filled with fibrous media, Darcy (1956) suggested the following equation, known as Darcy's law:

$$U = -\frac{\kappa}{\mu}\frac{dp}{dz}, \tag{2.4-1}$$

where U is the superficial velocity of the fluid; μ is the viscosity of fluid; and κ is the specific permeability of the porous medium. Equation (2.4-1) can be interpreted based on the capillaric model, or the Kozeny theory, which treats the porous medium as a solid that is embedded with various tortuous tubes. Assuming a porous medium made of a bundle of straight capillaries of parallel distance to the flow, with a tube diameter of δ and a porosity of α, the specific permeability is given by:

$$\kappa = \frac{\alpha \delta^2}{32}.$$

(2.4-1a)

It is noted that straight capillary tubes may not portray the complex structure of the porous medium. Thus, in practice, the factor 32 in Eq. (2.4-1a) is commonly replaced by an empirical parameter known as tortuosity, which accounts for the tortuous paths of the porous medium. The tortuosity can be further linked to the voidage of the porous medium by the Kozeny theory. The Kozeny theory assumes that the porous medium can be represented by an ensemble of channels of various cross sections of definite length (Kozeny, 1927). In general, the specific permeability can be expressed by:

$$\kappa = \frac{c}{S_0^2} \frac{\alpha^3}{(1 - \alpha)^2},$$

(2.4-1b)

where S_0 is the surface area exposed to the fluid per unit volume of solid materials; c is known as the Kozeny constant, which depends only on the shape of the cross section.

The pressure drop in Darcy's law is entirely due to the viscous dissipation between the moving fluid and the wetted solid surface, which implies that Darcy's law becomes invalid when the inertial effects of the flow become important or when the porosity becomes very large (Carman, 1956).

2.4.2   Ergun's Equation

When modeling flow in porous media, the porous structure of the solid matrix can be either simplified as an assembly of torturous tubes (Kozeny, 1927) or as an assembly of interlinked solid spheres (Burke and Plummer, 1928). The former treats the flow through a bundle of capillary tubes, whereas the later regards the flow around a cluster of solid spheres. As suggested by Reynolds (1900), for a general flow through a packed bed, the pressure drop can be expressed by the sum of two terms: one that is proportional to the first power of the fluid velocity due to viscous energy loss, and another that is proportional to the product of the density of the fluid and the second power of the fluid velocity, which is due to the kinetic energy loss. This leads to the expression:

$$-\frac{dp}{dz} = aU + bU^2,$$

(2.4-2)

where a and b are coefficients to be determined.

Burke and Plummer (1928) proposed that the total resistance of the packed bed can be treated as the sum of the resistances of the individual particles. At high flow velocities, the drag force acting on an isolated spherical particle is:

$$F_D = f\rho d_p^2 U^2,$$

(2.4-3)

where f is a constant friction factor. For flow passing through a packed column filled with spheres, the rate of work done by the drag force, or the power, is equal to the

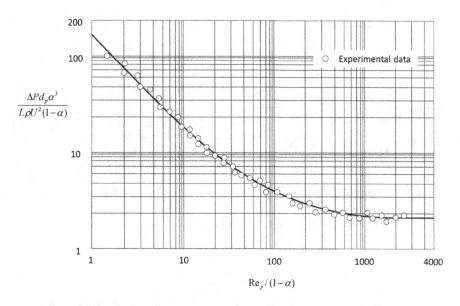

Figure 2.4-1 Pressure drop in fixed beds.

product of the interstitial velocity and the force acting on the particles. Hence, it can be shown that (Ergun and Orning, 1949):

$$-\frac{dp}{dz} = \frac{f}{\pi} S_0 \rho U^2 \frac{(1-\alpha)}{\alpha^3}. \tag{2.4-4}$$

For porous media made of mono-sized solid spheres, the specific surface area can be expressed as a function of the sphere diameter:

$$S_0 = \frac{6}{d_p}. \tag{2.4-4a}$$

Ergun (1952) suggested the Kozeny equation, Eq. (2.4-1) with Eq. (2.4-1b), be used for viscous energy loss and Burke and Plummer's relation, Eq. (2.4-4), be used for kinetic energy loss in the general Reynolds equation of pressure drop, shown in Eq. (2.4-2). Combining those equations, the Ergun equation is obtained as:

$$-\frac{dp}{dz} = 150\frac{(1-\alpha)^2}{\alpha^3}\frac{\mu U}{d_p^2} + 1.75\frac{(1-\alpha)}{\alpha^3}\frac{\rho U^2}{d_p}, \tag{2.4-5}$$

where the coefficients of 150 and 1.75 are based on the curve fitting of experimental data with Eq. (2.4-5), as shown in Figure 2.4-1 (Ergun, 1952). For Re < 10, the Kozeny equation is a good approximation of Ergun's equation, while for Re < 1,000, the Burke–Plummer equation can fit the data reasonably well.

It is noted that Ergun's equation is obtained from a packed bed of spherical particles. It is less useful for other types of porous media that have starkly nonuniform structures, such as those made of fibrous materials. In that case, a more general correlation, known as the Forchheimer equation, can be used (Papathanasiou *et al.*, 2001):

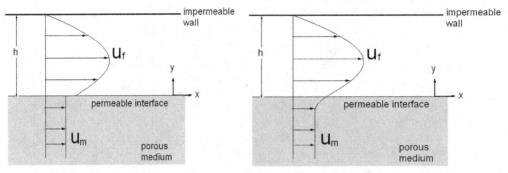

(a) Virtual velocity slip at interface   (b) Actual velocity and shear stress continuity at interface

Figure 2.4-2 Velocity near a permeable interface of a porous medium.

$$-\frac{dp}{dz} = \frac{\mu}{\kappa}U + \beta\rho U^2.$$ (2.4-6)

Gas flow in a porous medium at very low pressure deviates from Darcy's law due to the gas slippage, which needs to be accounted for by the Klinkenberg correction (Klinkenberg, 1941).

2.4.3   Brinkman Equation

Difficulties with Darcy's equation often arise at the boundary of the porous medium. For example, it is impossible to apply the no-slip condition for the tangential velocity at an impermeable boundary because Darcy's equation predicts a uniformly nonzero velocity profile under a pressure drop. Consider another illustrative case where a parallel flow exists both inside a porous medium and in the adjacent free fluid, as shown in Figure 2.4-2. The steady incompressible flow in the free stream is bounded by a two-dimensional channel, with a permeable wall at $y = 0$ and an impermeable wall at $y = h$. In the creeping flow regime, the free flow is governed by the Stokes equation. Meanwhile, the flow inside the porous medium is governed by Darcy's equation, which leads to a constant velocity:

$$U = U_m = -\frac{\kappa}{\mu}\frac{dp}{dx}.$$ (2.4-7)

Coupling between flows in the free stream and in the porous medium requires continuity of both the fluid velocity and the shear stress. However, Darcy's law fails to meet both conditions at the interface due to insufficient consideration of viscous shearing of the fluid flow through the porous media.

Beavers and Joseph (1967) proposed an empirical relationship to specify the boundary conditions at the permeable interface:

$$\frac{du}{dy}\bigg|_{y=0^+} = \frac{\alpha}{\sqrt{\kappa}}(U - U_m).$$ (2.4-8)

In this approach, the velocity gradient calculated from the flow within the channel is related to a slip velocity at the interface, as shown in Figure 2.4-2(a). The constant, α, is a dimensionless quantity that depends on the material property of the porous medium but not on the viscosity of the fluid. It has been demonstrated that the flow rates calculated from such a treatment can be satisfactorily correlated with experimental data. However, the discontinuity of the velocity across the permeable interface may not be physically sound. In reality, the effect of viscous stress penetrates a short distance into the porous medium and creates a thin boundary layer region below the permeable interface, as shown in Figure 2.4-2(b).

To avoid the confusion regarding the boundary treatment in Darcy's equation, Brinkman (1947) proposed to add a viscous resistance term in the Darcy equation. The Brinkman equation is expressed by:

$$\nabla p = -\frac{\mu}{\kappa}\overline{U} + \tilde{\mu}\nabla^2\overline{U}, \qquad (2.4\text{-}9)$$

where the effective viscosity, $\tilde{\mu}$, is not the actual fluid viscosity but an artificial parameter used for shear stress matching at the interface, and it is defined by:

$$\mu\left.(\nabla\overline{U}\cdot\mathbf{n})\right|_{A^+} = \tilde{\mu}\left.(\nabla\overline{U}\cdot\mathbf{n})\right|_{A^-}. \qquad (2.4\text{-}9a)$$

In the above equation, A^+ and A^- refer to boundary regions of the interface in the free-fluid and the porous medium, respectively. It should be noted that the Brinkman equation reduces to Darcy's equation for a small permeability and to the form of Stokes' equation for a large permeability. With the viscous stress term, the Brinkman equation yields a consistent velocity profile on both sides of the permeable interface. The new set of boundary conditions, shown in the example in Figure 2.4-2, is expressed by:

$$U = 0 \qquad\qquad\qquad \text{at } y = h \qquad (2.4\text{-}9b)$$

$$U = \overline{U} \qquad\qquad\qquad \text{at } y = 0 \qquad (2.4\text{-}9c)$$

$$\mu\left.\frac{dU}{dy}\right|_{y=0^+} = \tilde{\mu}\left.\frac{d\overline{U}}{dy}\right|_{y=0^-} \qquad \text{at } y = 0 \qquad (2.4\text{-}9d)$$

$$\overline{U} = U_m = -\frac{\kappa}{\mu}\frac{dp}{dx}. \qquad \text{at } y \to -\infty. \qquad (2.4\text{-}9e)$$

Among the above boundary conditions, Eq. (2.4-9c) and Eq. (2.4-9d) ensure continuity of the tangential velocity and shear stress at the permeable interface. Although the Brinkman equation is semiempirical in nature, it has been validated numerically by solving Stokes' equations in regions near the free-porous interface (Martys *et al.*, 1994). Necessity for the Brinkman correction can also be shown by volume averaging Stokes' equations over the boundary region of the free-porous interface (Ochoa-Tapia and Whitaker, 1995), although rigid formulae and closure of the Brinkman correction term are yet to be established.

The Brinkman equation can also be applied for flow through wall-bounded porous media, such as flows through columns packed with solid spheres in applications of granular filters, heat exchangers, absorbers, or chemical reactors. Due to the wall effect, the flow velocity and temperature profiles become strongly nonuniform near the wall region in a cross-section. For instance, consider an isothermal incompressible fluid flow passing through a column filled with spheres. Assuming the flow is axially symmetric and fully developed, the hydrodynamic behavior is then the continuity equation and the extended Brinkman equation, respectively, as (Winterberg *et al.*, 2000; Maußner *et al.*, 2017)

$$Q = \int_0^R 2\pi r U dr = \pi R^2 U_m \tag{2.4-10}$$

$$-\frac{dp}{dz} = 150\frac{(1-\alpha)^2}{\alpha^3}\frac{\mu U}{d_p^2} + 1.75\frac{(1-\alpha)}{\alpha^3}\frac{\rho\,U^2}{d_p} + \frac{\mu_e}{r}\frac{\partial}{\partial r}\left(r\frac{\partial U}{\partial r}\right), \tag{2.4-11}$$

where Q is the fluid volumetric flow rate; the radial distributions of volume fraction and effective viscosity are expressed by empirical correlations, such as suggested by Giese *et al.* (1998) as:

$$\alpha(r) = 0.4\left\{1 + 1.36\exp\left(-5\frac{R-r}{d_p}\right)\right\} \tag{2.4-11a}$$

$$\frac{\mu_e}{\mu} = 2\exp\left(0.002\frac{\rho U_m d_p}{\mu}\right). \tag{2.4-11b}$$

The boundary conditions for Eq. (2.4-11) are given as:

$$\frac{\partial U}{\partial r} = 0 \qquad \text{at } r = 0 \tag{2.4-11c}$$

$$U = 0 \qquad \text{at } r = R \tag{2.4-11d}$$

$$p = p_{in} \qquad \text{at } z = 0. \tag{2.4-11e}$$

The abovementioned model can be further extended to provide the flow field information for the heat and mass transfer in packed beds. The wall heat transfer in packed beds with fluid flow is typically analogous to the convective heat transfer of single-phase pipe flows, in which the heat transfer rate is expressed by Newton's cooling law with empirical heat transfer coefficient and a temperature jump between wall and averaged fluid flow. Such an analog does not reflect the gradual changes in temperature near the wall region that are responsible for the wall heat transfer, nor does the heat transfer coefficient reflect the true heat conduction nature within the wall region. Hence, an alternative modeling approach, using the extended Brinkman equation to describe near-wall transport behaviors, is established. Such an approach is represented by a simple model of heat and mass transfer of cylindrical packed beds of spheres, with correlations for heat transfer coefficients of near-wall transport properties (Winterberg *et al.*, 2000).

2.5   Kinetic Theory of Collision-Dominated Granular Flows

Granular material belongs to a special "continuum" medium that flows under inertia or external field forces, in which the effect of fluids (such as gas) is negligibly small in comparison to the interparticle collisions. The dynamic motions of granular materials can be approximately governed by the extended kinetic theory of gases. In the following, the regimes of granular flow are discussed first. Then, for granular flows in the inertia regime, the transport theorem is derived, which leads to the governing equations and the constitutive relations for the granular transport. This kinetic theory modeling of granular flow is constantly implemented into the modeling of solids-laden multiphase flows in which the interparticle collisions cannot be ignored.

2.5.1 Regimes of Granular Flows

A granular material is an assembly of numerous discrete solid particles. Common examples of granular materials include rice grains, coal powders, sand, and medicine tablets. Under certain conditions, the collective motion of individual particles, when viewed macroscopically, behaves as a flow of a fluid. In such circumstances, the bulk granular material can be considered as a pseudo-continuous medium.

The void between the individual particles is often filled with either a gas or a liquid, which also contributes to the rheology of the solid–fluid mixture. The extent to which the interstitial fluid affects the mixture can be characterized by the dimensionless Bagnold Number (Ba) given by (Bagnold, 1954):

$$\text{Ba} = \sqrt{\lambda} \frac{d_p^2 \rho_m}{\mu} \left(\frac{dU}{dy} \right) = \frac{\text{particle inertial force}}{\text{fluid viscous force}}, \tag{2.5-1}$$

where ρ_m is the mixture density; λ is a parameter related to voidage of suspension, given by:

$$\lambda = \frac{1}{1 - (\alpha_m/\alpha)^{1/3}}, \tag{2.5-2}$$

in which α_m is the voidage in the maximum possible concentration. When Ba < 40, the fluid viscosity dominates the motion of solids so that the overall suspension behaves like a viscous fluid. This phenomenon is characteristic of the macro-viscous regime, and the shear stress is linearly related to the shear rate with a modified viscosity by:

$$\tau = \lambda^{3/2} \mu \frac{dU}{dy}. \tag{2.5-3}$$

When Ba > 450, the suspension behaves like a dry granular flow. The particle motion is largely governed by particle–particle interactions, and the shear stress depends on the square of the shear rate via:

$$\tau = \rho_m d_p^2 \lambda^2 \left(\frac{dU}{dy} \right)^2. \tag{2.5-4}$$

This regime is defined as the grain inertia regime. In this regime, the interstitial fluid effect is insignificant, and the granular flow can be studied as a special form of single-phase flow, which behaves distinctly different from the flow of a conventional fluid.

Equation (2.5-4), also known as the Bagnold friction law (Jaeger and Nagal, 1992), provides the basis of modern granular flow theory.

In a granular flow, where the effect of interstitial fluid is negligible, the flow regime can be further categorized by the nature of the interactions between particles. Detailed theory and models for these interactions are given in Chapters 4 and 5. For low-velocity flows with high particle concentrations, particles collide and the force is transmitted through force chains, which are structures that support the internal stress in the bulk granular material. This behavior is characteristic of the quasi-static regime, and it is sometimes referred to as the frictional regime; although in fact, the friction-like response comes from the rotation and collapse of the force chains rather than the frictional sliding within the bulk material (Campbell, 2006). A typical example of the quasi-static regime is the hopper flow.

With comparatively lower particle concentrations and a higher shear rate, the flow is free of force chains, and particle interaction is dominated by collisions. This behavior defines the inertial regime. An extreme case where particle interactions are dominated by instantaneous binary collisions, known as rapid granular flow, will be the focus of the remainder of this section. Theoretical modeling of rapid granular flows can be developed using an analogy between particle collisions in granular flows and molecular collisions in the kinetic theory of gases (Culick, 1964). However, this approach rigorously followed the kinetic theories of gases for solid particles and was halted due to the complexity involved with direct application of the Boltzmann equation to account for interparticle collisions. An alternative approach, using simplified kinetic theories of gases based on mechanistically derived or intuitive relationships in lieu of the Boltzmann equation, has been deemed viable (Savage and Jeffery, 1981; Jenkins and Savage, 1983; Lun *et al.*, 1984). Although most types of granular flows have a high particle concentration and are rarely encountered outside the laboratory, the kinetic theory of granular flow has been applied to many gas–solid flow systems, including fluidization and pneumatic transport (Gidaspow, 1993).

2.5.2   Transport Theorem of Collision-Dominated Granular Particles

For a system of hard, smooth, spherical particles, a transport theorem based on an analogy of the kinetic theory of dense gases (Reif, 1965) may be derived. Let $f d\mathbf{v}$ be the differential number of particles per unit volume with velocities between $\mathbf{v}$ and $\mathbf{v} + d\mathbf{v}$, and let $\mathbf{F}$ be the total external force acting on a particle per particle mass. The evolution of the single-particle velocity distribution function, $f^{(1)}(\mathbf{r}, \mathbf{v}, t)$, is governed by the Boltzmann integrodifferential equation and can be expressed as:

$$\frac{\partial f}{\partial t} + \mathbf{v} \cdot \frac{\partial f}{\partial \mathbf{r}} + \mathbf{F} \cdot \frac{\partial f}{\partial \mathbf{v}} = \left(\frac{\partial f}{\partial t}\right)_{coll}. \tag{2.5-5}$$

The above equation shows that the evolution of the particle distribution function, f, is affected by three factors: (1) change of particle location by convection, as described by the second term on the left-hand side (LHS); (2) change of particle velocity due to external force, as described by the third term on the LHS; and (3) interparticle collision, as described by the right-hand side (RHS). As a result, any property that is carried by the particle is also influenced by those three mechanisms.

If $\psi(\mathbf{v})$ is a property of a particle and is only dependent upon the particle's velocity, $\mathbf{v}$, the ensemble average of ψ is defined as:

$$\langle \psi \rangle = \frac{1}{n} \int \psi(\mathbf{v}) f^{(1)}(\mathbf{r}, \mathbf{v}, t) d\mathbf{v}, \tag{2.5-6}$$

where $n(\mathbf{r},t)$ is the particle number density. According to the definition of f, n is given by:

$$n = \int f^{(1)}(\mathbf{r}, \mathbf{v}, t) d\mathbf{v}. \tag{2.5-7}$$

Multiplying Eq. (2.5-5) by ψ and applying ensemble averaging with respect to particle velocity, $\mathbf{v}$, lead to the transport theorem accounting for the evolution of $<\psi>$. The result is known as Enskog's equation of change and can be expressed as (Jenkins and Savage, 1983):

$$\frac{\partial}{\partial t} \langle n\psi \rangle = n \left(\mathbf{F} \cdot \frac{\partial \psi}{\partial \mathbf{v}} \right) - \nabla \cdot \langle n\mathbf{v} \, \psi \rangle + C(\psi). \tag{2.5-8}$$

The last term of Eq. (2.5-8) $C(\psi)$ is the rate of change of $<\psi>$ due to particle collision. For a binary collision, the above equation can be expressed as:

$$C(\psi) = \frac{1}{2} \int_{\mathbf{v}_{12} \cdot \mathbf{k} > 0} \left[(\psi_1' - \psi_1) f^{(2)}(\mathbf{r}, \mathbf{v}_1, \mathbf{r} + d_p\mathbf{k}, \mathbf{v}_2; t) \right.$$
$$\left. + (\psi_2' - \psi_2) f^{(2)}(\mathbf{r} - d_p\mathbf{k}, \mathbf{v}_1, \mathbf{r}, \mathbf{v}_2; t) \right] (\mathbf{v}_{12} \cdot \mathbf{k}) \, d_p^2 d\mathbf{k} d\mathbf{v}_1 d\mathbf{v}_2, \tag{2.5-9}$$

in which primed and unprimed quantities refer to the values generated after and prior to the collision, respectively. In this case, $\mathbf{k}$ is the unit vector along the line from the center of particle 1 to that of particle 2, such that $\mathbf{r}_1 - \mathbf{r}_2 = d_p\mathbf{k}$, as shown in Figure 2.5-1. The relative velocity is given by $\mathbf{v}_{12} = \mathbf{v}_1 - \mathbf{v}_2$. For a collision to occur, particle 2 must lie in a cylindrical volume $d_p^2(\mathbf{v}_{12} \cdot \mathbf{k}d\mathbf{k})$ per unit time. The expression $f^{(2)}(\mathbf{v}_1, \mathbf{r}_1, \mathbf{v}_2, \mathbf{r}_2; t)$ is the pair distribution function; defined such that $f^{(2)}(d\mathbf{r}_1, d\mathbf{r}_2 d\mathbf{v}_1 d\mathbf{v}_2)$, represents the probability of finding a pair of particles in the volume elements $d\mathbf{r}_1$ and $d\mathbf{r}_2$ centered on the points $\mathbf{r}_1$, $\mathbf{r}_2$ and having velocities within the ranges $\mathbf{v}_1$ and $\mathbf{v}_1 + d\mathbf{v}_1$, and $\mathbf{v}_2$ and $\mathbf{v}_2 + d\mathbf{v}_2$, respectively.

Applying Taylor Series expansion to the pair distribution function yields:

$$f^{(2)}(\mathbf{r}, \mathbf{v}_1, \mathbf{r} + d_p\mathbf{k}, \mathbf{v}_2; t) = f^{(2)}(\mathbf{r} - d_p\mathbf{k}, \mathbf{v}_1, \mathbf{r}, \mathbf{v}_2; t)$$
$$+ \left[d_p\mathbf{k} \cdot \nabla - \frac{1}{2!} (d_p\mathbf{k} \cdot \nabla)^2 + ... \right] f^{(2)}(\mathbf{r}, \mathbf{v}_1, \mathbf{r} + d_p\mathbf{k}, \mathbf{v}_2; t). \tag{2.5-10}$$

Discarding gradient terms higher than the first order in the above equation permits the collision term, $C(\psi)$, to be explicitly expressed in terms of collision transfer contribution and a source term as (Lun et al., 1984):

$$C(\psi) = -\nabla \cdot \theta + \chi, \tag{2.5-11}$$

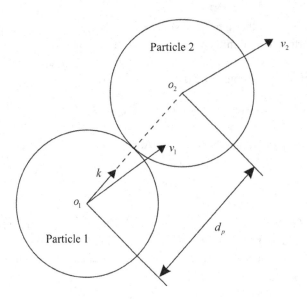

Figure 2.5-1 Binary collision of same-size particles.

where θ is the collisional transfer contribution, represented by:

$$\theta = -\frac{d_p^3}{2} \int_{\mathbf{v}_{12} \cdot \mathbf{k} > 0} (\psi_1' - \psi_1)(\mathbf{v}_{12} \cdot \mathbf{k}) \, \mathbf{k} f^{(2)} \left(\mathbf{r} - \frac{1}{2} d_p \mathbf{k}, \mathbf{v}_1, \mathbf{r} + \frac{1}{2} d_p \mathbf{k}, \mathbf{v}_2; t \right) d\mathbf{k} d\mathbf{v}_1 d\mathbf{v}_2,$$

(2.5-11a)

and χ is the source term, given by:

$$\chi = \frac{d_p^2}{2} \int_{\mathbf{v}_{12} \cdot \mathbf{k} > 0} (\psi_1' + \psi_2' - \psi_1 - \psi_2)(\mathbf{v}_{12} \cdot \mathbf{k}) \, f^{(2)} \left(\mathbf{r} - d_p \mathbf{k}, \mathbf{v}_1, \mathbf{r}, \mathbf{v}_2; t \right) d\mathbf{k} d\mathbf{v}_1 d\mathbf{v}_2.$$

(2.5-11b)

Finally, for a system of hard, smooth, spherical particles controlled by binary collisions, the transport theorem is expressed as:

$$\frac{\partial}{\partial t} < n\psi > = n \left\langle \mathbf{F} \cdot \frac{\partial \psi}{\partial \mathbf{v}} \right\rangle - \nabla \cdot < n\mathbf{v} \, \psi > - \nabla \cdot \theta + \chi.$$

(2.5-12)

2.5.3 Governing Equations

Fundamental quantities in a collision include particle mass, momentum, and kinetic energy. Governing equations for the macroscopic flow of a granular material are obtained by applying the transport theorem, Eq. (2.5-12), to those basic quantities. For conserved quantities such as mass and total momentum, there are no source terms in their transport equations. However, the total kinetic energy of the colliding particles is not conserved due to inelastic collision and heat dissipation.

(1) *Conservation of Mass*
    Setting $\psi = m$ in Eq. (2.5-12) leads to:

$$\frac{\partial}{\partial t} (\alpha_p \rho_p) + \nabla \cdot (\alpha_p \rho_p \, \mathbf{U}_p) = 0,$$

(2.5-13)

where $\alpha_p \rho_p = nm$ and $\mathbf{U}_p = <\mathbf{v}>$ is the macroscopic (hydrodynamic) velocity of the particle phase.

(2) *Conservation of Momentum*

By taking $\psi = m\mathbf{v}$, the local form of the balance equation of the linear momentum can be obtained as:

$$\frac{\partial}{\partial t}(\alpha_p \rho_p \mathbf{U}_p) + \nabla \cdot (\alpha_p \rho_p \mathbf{U}_p \mathbf{U}_p) = -\nabla \cdot \mathbf{P}_p + \alpha_p \rho_p \mathbf{F}, \tag{2.5-14}$$

where the total stress tensor, $\mathbf{P}_p$, is the sum of the kinetic part, $\mathbf{P}_k$, and the collisional part, $\mathbf{P}_c$. $\mathbf{P}_k$ is given by:

$$\mathbf{P}_k = <\alpha_p \rho_p \mathbf{u}\mathbf{u}>, \tag{2.5-14a}$$

where $\mathbf{u} = \mathbf{v} - <\mathbf{v}> = \mathbf{v} - \mathbf{U}_p$ is the fluctuating component of the particle velocity. Using the velocity relation for binary particle collision introduced in Section 5.3, $\mathbf{P}_c$ is given by:

$$\mathbf{P}_c = \frac{md_p^3}{4}\int_{\mathbf{v}_{12}\cdot\mathbf{k}>0}(1+e)(\mathbf{v}_{12}\cdot\mathbf{k})^2\,\mathbf{k}\mathbf{k}\,f^{(2)}\left(\mathbf{r}-\frac{1}{2}d_p\mathbf{k},\mathbf{v}_1,\mathbf{r}+\frac{1}{2}d_p\mathbf{k},\mathbf{v}_2;t\right)d\mathbf{k}d\mathbf{v}_1d\mathbf{v}_2. \tag{2.5-14b}$$

(3) *Equation of Fluctuating Kinetic Energy*

Taking ψ to be $\frac{1}{2}mv^2$ in Eq. (2.5-12) yields the equation of the total kinetic energy. Subtracting from the total kinetic energy balance, which is based on the mean particle velocity $\mathbf{U}_p$, the fluctuating kinetic energy equation can be expressed by:

$$\frac{3}{2}\left[\frac{\partial}{\partial t}(\alpha_p \rho_p T_c) + \nabla \cdot (\alpha_p \rho_p T_c \mathbf{U}_p)\right] = -\mathbf{P}_p : \nabla \mathbf{U}_p - \nabla \cdot \mathbf{q}_p$$
$$+ \gamma + \alpha_p \rho_p \langle \mathbf{F} \cdot \mathbf{v} \rangle - \alpha_p \rho_p \mathbf{F} \cdot \mathbf{U}_p, \tag{2.5-15}$$

where T_c is defined as the granular temperature, expressed as:

$$\frac{3}{2}T_c = \frac{1}{2}<\mathbf{u}\cdot\mathbf{u}>. \tag{2.5-15a}$$

$\mathbf{q}_p$ is the heat flux of fluctuation energy, which consists of the kinetic contribution, $\mathbf{q}_k$, and the collisional contribution, $\mathbf{q}_c$. The kinetic part, $\mathbf{q}_k$, is given by:

$$\mathbf{q}_k = \frac{1}{2}<\alpha_p \rho_p u^2 \mathbf{u}>. \tag{2.5-15b}$$

The collisional contribution, $\mathbf{q}_c$, is given by the following:

$$\mathbf{q}_c = \frac{md_p^3}{4}\int_{\mathbf{v}_{12}\cdot\mathbf{k}>0}\left[(1+e)(\mathbf{v}_1\cdot\mathbf{k}) + \frac{1}{4}(1+e)^2(\mathbf{v}_{12}\cdot\mathbf{k})^2\right](\mathbf{v}_{12}\cdot\mathbf{k})\,\mathbf{k}$$
$$\cdot f^{(2)}\left(\mathbf{r}-\frac{1}{2}d_p\mathbf{k},\mathbf{v}_1,\mathbf{r}+\frac{1}{2}d_p\mathbf{k},\mathbf{v}_2;t\right)d\mathbf{k}d\mathbf{v}_1d\mathbf{v}_2. \tag{2.5-15c}$$

γ is the collisional rate of dissipation per unit volume, and it is given by:

$$\gamma = \frac{1}{8}m(e^2-1)d_p^2\int_{\mathbf{v}_{12}\cdot\mathbf{k}>0}(\mathbf{v}_{12}\cdot\mathbf{k})^3\,f^{(2)}\left(\mathbf{r}-d_p\mathbf{k},\mathbf{v}_1,\mathbf{r},\mathbf{v}_2;t\right)d\mathbf{k}d\mathbf{v}_1d\mathbf{v}_2. \tag{2.5-15d}$$

The last two terms of Eq. (2.5-15) denote the change of fluctuating kinetic energy due to the work of the external force.

2.5.4   Constitutive Relations

In order to evaluate the kinetic parts of the transport, P_k and q_k, as well as the collisional integrals, P_c, q_c, and γ, it is necessary to explicitly define the pair distribution function, $f^{(2)}(r_1, v_1; r_2, v_2; t)$. A relationship between the pair distribution function, $f^{(2)}$, and the single-particle velocity distribution function, $f^{(1)}$, can be established by introducing a configurational pair-correlation function, $g_0(r_1, r_2)$, such that:

$$f^{(2)}(r_1, v_1; r_2, v_2; t) = g_0(r_1, r_2) f^{(1)}(r - \frac{1}{2}d_p k, v_1, t) f^{(1)}(r + \frac{1}{2}d_p k, v_2, t)$$

$$= g_0 \cdot \left[f^{(1)}(r, v_1; t) f^{(1)}(r, v_2; t) + \frac{1}{2}d_p f^{(1)}(r, v_1; t) f^{(1)}(r, v_2; t) k \cdot \nabla \left(\ln \frac{f_2}{f_1} \right) \right].$$

(2.5-16)

The pair-correlation function, g_0, is suggested to take the form (Lun and Savage, 1986):

$$g_0 = \left(1 - \frac{\alpha_p}{\alpha_{p, \max}} \right)^{-5\alpha_{p, \max}/2}.$$

(2.5-16a)

Assuming the gradients in the mean flow variables of velocity, granular temperature, and bulk density are small, which implies that the particles are nearly elastic, the particle distribution function can be approximated by a perturbation, ϕ with $\phi \ll 1$, to the Maxwellian equilibrium velocity distribution:

$$f^{(1)}(r, v, t) = f^{(0)}(1 + \phi).$$

(2.5-17)

The local Maxwellian equilibrium distribution is determined by the local number density, $n(r, t)$, and hydrodynamic velocity, U_p:

$$f^{(0)}(r, v, t) = \frac{n}{(2\pi T_c)^{3/2}} \exp \left[-\frac{(v - U_p)^2}{2T_c} \right].$$

(2.5-17a)

The perturbation can be constructed as:

$$\phi = a_1 \left(uu - \frac{1}{3}u^2 I \right) : \nabla U_p + a_2 \left(\frac{5}{2} - \frac{u^2}{2T_c} \right) u \cdot \nabla \ln T_c + a_3 \left(\frac{5}{2} - \frac{u^2}{2T_c} \right) u \cdot \nabla \ln n,$$

(2.5-17b)

where the coefficients a_1, a_2, and a_3 are found by satisfying the moments in Eq. (2.5-12) for mass, momentum, and energy (Lun et al., 1984).

Based on Eq. (2.5-17), the first-order approximation of the total stress tensor, P, can be obtained as:

$$P = P_k + P_c = \left(P_s - \lambda_s \nabla \cdot U_p \right) I + \tau_s,$$

(2.5-18)

where P_s is the solid pressure, defined by:

$$P_s = \alpha_p \rho_p \left[1 + 2\alpha_p (1 + e) g_0 \right] T_c,$$

(2.5-18a)

and λ_s is the solid bulk viscosity, given by:

$$\lambda_s = \frac{4}{3}\rho_p \alpha_p^2 g_0 d_p (1+e)\sqrt{\frac{T_c}{\pi}}. \tag{2.5-18b}$$

$\boldsymbol{\tau}_s$ is the solid shear stress tensor, defined as:

$$\boldsymbol{\tau}_s = 2\mu_s \left[\frac{1}{2}\left(\nabla \mathbf{U}_p + (\nabla \mathbf{U}_p)^T \right) - \frac{1}{3}(\nabla \cdot \mathbf{U}_p)\mathbf{I} \right], \tag{2.5-18c}$$

in which the solid shear viscosity is given by:

$$\mu_s = \frac{5\pi}{24}\frac{\rho_p d_p}{(1+e)(3-e)g_0}\sqrt{\frac{T_c}{\pi}}\left[1 + \frac{4}{5}(1+e)\alpha_p g_0\right]\left[1 + \frac{2}{5}(1+e)(3e-1)\alpha_p g_0\right] + \frac{3}{5}\lambda_s. \tag{2.5-18d}$$

Similarly, the first-order approximation of the total flux of fluctuation energy, **q**, can be expressed as:

$$\mathbf{q} = \mathbf{q}_k + \mathbf{q}_c = -k_s \nabla T_c - k'\nabla \alpha_p, \tag{2.5-19}$$

where the granular thermal conductivity is given as:

$$k_s = \frac{25\rho_p d_p \sqrt{\pi T_c}}{16 g_0 \eta (41 - 33\eta)}\left\{ \left[1 + \frac{12}{5}\eta \alpha_p g_0\right]\left[1 + \frac{12}{5}\eta^2 (4\eta - 3)\alpha_p g_0\right]\right. $$
$$\left. + \frac{64}{25\pi}(41 - 33\eta)(\eta \alpha_p g_0)^2 \right\}, \tag{2.5-19a}$$

and

$$k' = \frac{15\sqrt{\pi}(2\eta - 1)(\eta - 1)}{4(41 - 33\eta)}\frac{\rho_p d_p}{\alpha_p g_0}T_c^{3/2}\left(1 + \frac{12}{5}\alpha_p g_0 \eta\right)\frac{d}{d\alpha_p}\left(\alpha_p^2 g_0\right), \tag{2.5-19b}$$

in which:

$$\eta = \frac{1}{2}(1+e). \tag{2.5-19c}$$

The rate of energy dissipation γ is obtained from Eq. (2.5-15d) as:

$$\gamma = 12(1 - e^2)\frac{\rho_p \alpha_p^2 g_0}{d_p \sqrt{\pi}}T_c^{3/2}. \tag{2.5-20}$$

Substituting the constitutive relations (Eqs. (2.5-18), (2.5-19), and (2.5-20)) into the governing Eqs. (2.5-13), (2.5-14), and (2.5-15) produces five equations for the five unknowns α_p, $\mathbf{U}_p$, and T_c. Hence, the closure problem is resolved for the kinetic theory of granular flows.

2.5.5   Advancement in Kinetic Theory for Granular Flow

The derivation of the constitutive equations of the kinetic theory for granular flow in the previous section is based on an analog relating rapid granular flow to molecular gas flow. However, there are several fundamental differences, which result in dissimilar dynamic behavior. The particle interactions in the granular flow are essentially dissipative, meaning kinetic energy is no longer conserved. Additionally, the inherent inelastic collision produces a dissipation term in the energy equation. Thus, the particle distribution function in granular flows does not have a Maxwellian equilibrium, which is a fundamental characteristic of molecular gas flows. Depending on how the Maxwellian in Eq. (2.5-17) is perturbed, it can yield slightly different forms of the constitutive relations for **P** and **q**. For instance, the finite size of granular particles requires different pair-distribution functions from those of molecular gas, and various choices for the pair-distribution function have been attempted (van Wachem *et al.*, 2001). The rough surface of the granular particles causes rotation, and the rotational motion can be characterized by a rotational granular temperature that may be solved from an additional rotational kinetic energy equation (Lun, 1991). In addition, the cohesive forces resulting from the weak van der Waals force, electrostatic force, or liquid bridge force need to be accounted for in the case of cohesive powders (Kim and Arastoopour, 2002). Particle shape effect is often a factor that is beyond the scope of the above derivation; nonspherical particles are expected to behave differently from spherical ones. Also, the particle size distribution needs to be accounted for in realistic solid flow modeling; it has a significant effect on phenomena such as segregation, and efforts have been made to extend the kinetic theory of monodispersed spheres to a form of polydispersed particles (Benyahia, 2008). The kinetic theory for granular flows has enjoyed prolific application in multiphase gas–solid flows, and coupling between the solid particles and the carrier fluid is discussed further in Chapter 6.

2.6   Case Studies

The following case studies serve for further discussion with applications of the theories of Chapter 2. Examples include the modeling closure, the smallest scale in continuum validation, and simple applications of flows in porous media and electroosmotic flows.

2.6.1   Model Closure of a Multicomponent Single-Phase Reacting Flow

Problem Statement: Consider the modeling of a multicomponent single-phase reacting flow. List major assumptions and discuss the model closure.

Analysis: Assume that the flow is laminar, and the reaction is one-step and stoichiometric.

For an n-component (species) single-phase flow, there are $(6 + n)$ independent variables: p, ρ, U, T, Y_i ($i = 1, n$). The velocity is a vector, with three velocity components.

The independent equations include continuity equation (*e.g.*, Eq. (2.2-10)), momentum equation (*e.g.*, Eq. (2.2-12)), with three component equations along each coordinate), the energy equation (*e.g.*, Eq. (2.2-16)), the equation of state

(Eq. (2.2-18)), and the species-conservation equation (*e.g.*, Eq. (2.2-11) for the *i*th-species, $i = 1, n$). The total number of independent equations matches the total number of independent variables, and hence the model is closed.

Comments: (1) Whether one considers the reaction rate as an independent variable or a parameter depends on if the equation of reaction is considered as an independent equation or a constitutive correlation for the reaction term. In the former case, the total number of independent variables becomes $(6+2n)$; namely, p, ρ, U, T, Y_i ($i = 1$, n), γ_i ($i = 1, n$). Then the equation of reaction of each species should be added to the total number of independent equations. (2) If the equation of conservation of mass fraction of all species (Eq. (2.2-11a)) is used as an independent equation, then the number of independent species-conservation equations will be $(n - 1)$. The same treatment is also applied to the equations of species reaction rate.

2.6.2 Smallest Characteristic Length of a Continuum-Based CFD

Problem Statement: For air at atmospheric conditions, estimate the smallest characteristic length of fluid element with a tolerance of 1%. For this case, what is the flow system size that a current CFD code can simulate while ensuring continuum fluid mechanics is still valid?

Analysis: Based on the tolerance of 1% and Eq. (2.2-31), the minimum number of molecules per fluid element should be $N = 10^4$. For an ideal gas under standard temperature and pressure, 1 mole of the gas has a volume of 22.4 liters, containing 6.023×10^{23} molecules, the smallest fluid element occupies a volume of 3.72×10^{-22} m^3. Therefore, the characteristic length of the fluid element can be estimated from Eq. (2.2-32) as:

$$l_{min} = 7.2 \times 10^{-8} \text{ m} = 72 \text{ nm}$$

Let us consider the minimum characteristic length of a fluid element as the finest grid length of the CFD simulation. Based on the Kn criterion of Eq. (2.2-33), using l_{min} to approximate l_{mfp}, the smallest system size is estimated as:

$$\text{System size} = 7.2 \text{ } \mu m$$

Comments: Continuum fluid mechanics modeling can be used for gas flow in a micron-sized system (about 10 μm), but it may not be valid for nano-sized systems.

2.6.3 Flow into a Spherical Cavity in an Infinite Porous Medium

Problem Statement: A spherical cavity (drilling hole) is formed in an unbounded porous medium to extract fluid, as shown in Figure 2.6-1. If the difference between the pressure P_c inside the cavity and ambient pressure P_∞ (at a very far distance) of the porous media is fixed and given, estimate the total flow rate from the drilling tube using Kozeny theory.

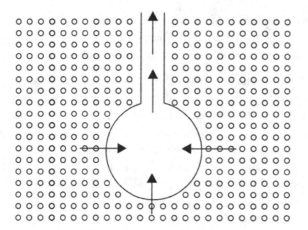

Figure 2.6-1 Flow into a spherical cavity in a porous medium.

Analysis: Assume that (1) the drilling tube cross-sectional area is much smaller than the surface area of the cavity; (2) the gravity effect on pressure in the porous medium can be neglected; and (3) the fluid and porous medium properties are constant. Thus, the problem is approximated as spherically symmetric, and the radial flow through the porous medium becomes one-dimensional. With constant fluid viscosity and constant specific permeability, the pressure gradient along the radial direction can be related to the radial flow velocity (and hence flow rate), based on the Kozeny theory, as:

$$\frac{dp}{dr} = \frac{\mu}{\kappa} U = \frac{\mu Q}{4\pi r^2 \kappa}. \tag{2.6-1}$$

Note that the total flow rate, Q, is constant along the radial direction. By integrating Eq. (2.6-1) from cavity radius to infinite, it gives:

$$Q = 4\pi r_c \frac{\kappa}{\mu} (P_\infty - P_c). \tag{2.6-2}$$

It can be assumed that the flow in the drilling tube is fully developed laminar flow and the exit pressure is atmospheric pressure P_a. The flow rate can then be estimated from the Hagen–Poiseuille equation as:

$$Q = \frac{\pi r_d^4}{8\mu} (P_c - P_a - \rho g H). \tag{2.6-3}$$

Combining Eq. (2.6-2) and Eq. (2.6-3) yields:

$$Q = \frac{P_\infty - P_a - \rho g H}{\left(\frac{\mu}{4\pi r_c \kappa} + \frac{8\mu}{\pi r_d^4} \right)}. \tag{2.6-4}$$

Comments: (1) The Kozeny equation is applied to the interfacial area between the porous medium and the porous-free medium, assuming there is no tangential flow. In regions with tangential flow, the Brinkman equation should be substituted. This

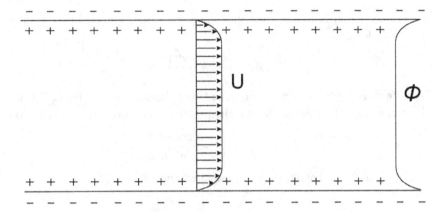

Figure 2.6-2 Principle of electroosmotic flow.

example simply illustrates a rough engineering approach for a quick estimation. (2) Due to the fluid gravity and gravitational settling of porous media, the hydraulic pressure distribution within the media and its properties (such as permeability) are unlikely to be spherically symmetric to the cavity. However, simplifications are adopted here in order to demonstrate an efficient solution that still accounts for the elementary transport mechanisms.

2.6.4   Electroosmotic Flow

Problem Statement: The principle of electroosmotic flow (EOF) is shown in Figure 2.6-2. In a microchannel with charged surfaces, a charge distribution is established in the liquid with counterions concentrated near the walls. This arrangement is known as the electrical double layer. The electrical potential associated with the double layer is described by (Bruus, 2008) as:

$$\Phi(z) = \zeta \frac{\cosh(z/\lambda_D)}{\cosh(h/2\lambda_D)}, \tag{2.6-5}$$

where ζ is the zeta potential of the channel wall, and λ_D is the Debye length. When an external electric field, E_x, is applied parallel to the channel direction, the liquid will flow due to the electric force. Derive an expression for the velocity profile in the channel.

Analysis: In a steady Stokes flow with an electric force as the body force, the momentum equation can be written as:

$$0 = -\nabla p + \mu \nabla^2 \mathbf{U} + \rho_e \mathbf{E}, \tag{2.6-6}$$

where ρ_e and $\mathbf{E}$ are the charge density and the external electric field, respectively. According to Gauss's law, the electrical charge density is obtained from the potential as:

$$\rho_e = -\varepsilon \nabla^2 \Phi. \tag{2.6-7}$$

Since the flow profile is invariant along the x direction, the above equation can be simplified to:

$$0 = -\frac{\partial p}{\partial x} + \mu \frac{\partial^2 U}{\partial z^2} - \varepsilon E_x \frac{\partial^2 \Phi}{\partial z^2}. \tag{2.6-8}$$

In absence of a pressure gradient, and using the expression in Eq. (2.6-7) with no-slip conditions at the walls, the velocity profile can be solved from Eq. (2.6-8) as:

$$U(z) = \left[1 - \frac{\cosh(z/\lambda_D)}{\cosh(h/2\lambda_D)}\right] U_{EO}. \tag{2.6-9}$$

U_{EO} is the Helmholtz–Smoluchowski velocity, defined by:

$$U_{EO} = \frac{\varepsilon \zeta}{\mu} E_x. \tag{2.6-10}$$

Comments: For microchannels with $h \gg \lambda_D$, Eq. (2.6-9) is approximated as:

$$U(z) = U_{EO} = \frac{\varepsilon \zeta}{\mu} E_x. \tag{2.6-11}$$

Therefore, electroosmosis generates a plug flow within a microchannel. However, in a nanochannel whose height is sufficiently small such that the thickness of the double layer is no longer negligible, the velocity distribution will deviate from that of a plug flow.

2.7   Summary

This chapter provides a detailed account of the transport theories (with derived formulation of governing equations) on continuum single-phase flows that include the laminar or turbulent flows of viscous fluids, the viscous fluid flows through porous media, and flows of granular particles. The most essential one is the continuum modeling of laminar flows of Newtonian fluids, which is the foundation of the Eulerian modeling approach of multiphase flows.

The distinctly different transport mechanisms between a laminar flow and a turbulent flow as well as the flow regime transition conditions are presented. The essential aspects of the modeling of a turbulent flow of Newtonian fluid are also given, which include:

- Reynolds decomposition in the time-averaging approach;
- time-averaged transport equations (*e.g.*, RANS equations) containing various turbulent transport properties with formulation closed using additional turbulence models;
- simplifications used in various turbulent models, especially on the length-scale consideration for eddy transport in the turbulent flow;
- commonly used turbulence models (such as the mixing length model, the k–ε model, the Reynolds stress model, and the large eddy simulation model) and their applicability in CFD simulations of turbulent flows.

The chapter describes the basic theories and their applicability of flows through porous media, represented by Darcy's law, Ergun's equation, and Brinkman's equation. It also illustrates the continuum theory of granular flow, represented by the kinetic theory of granular flows whose transport mechanisms are dominated by inertia and interparticle collisions. It is noted that the kinetic theory of granular flows can be particularly useful in the modeling of dense-phase multiphase flows where the effect of interparticle collisions on phase transport is prominent.

Nomenclature

A	Area
a	Speed of sound
Ba	Bagnold number
C	Molecular concentration
	Rate of change by collision
c	Kozeny constant
c_p	Specific heat at constant pressure
c_v	Specific heat at constant volume
D	Diffusivity
D_{ij}	Diffusion of Reynolds stress
d_p	Particle diameter
E	Electric field
E	Total energy per unit mass
	Elastic modulus
E_{ij}	Viscous dissipation of Reynolds stress
e	Internal energy per unit mass
	Restitution coefficient
F	Force vector
F	General variable
	Force
F_D	Drag force
f	Field force per unit mass
f	General quantity
	Friction factor
$f^{(1)}$	Single-particle velocity distribution function
$f^{(2)}$	Pair distribution function
G	Filter function
G_{ij}	Buoyance production to Reynolds stress
g	Gravity acceleration
g_0	Pair-correlation function
h	Enthalpy per unit mass
I	Identity tensor
J_q	Heat flux vector

J	Jacobian determinant
J_e	Rate of heat generation per unit volume
K	Thermal conductivity
k	Turbulence kinetic energy
L	Characteristic length
$[L]$	Length unit
l	Length
M	Molar mass
Ma	Mach number
$[M]$	Mass unit
m	Mass
m	Mass flux
$\boldsymbol{n}$	Unit normal vector
n	Particle number density
$\boldsymbol{P}$	Total stress tensor of granular particles
P	pressure
P	Shear production of Reynolds stress
p	Pressure
$\boldsymbol{q}$	Heat flux vector of fluctuation energy
Q	Volumetric flow rate
R	Gas constant
	Radius of curvature
Re	Reynolds number
$\boldsymbol{r}$	Position vector
r	Coordinate in r-space
$\boldsymbol{S}$	Strain rate
S	Surface area
$\boldsymbol{T}$	Total stress tensor of fluid
T	Temperature
$[T]$	Time unit
t	Time
$\boldsymbol{U}$	Velocity vector
U	Superficial velocity
u	Instant velocity
	Interstitial velocity
v	Volume
$\boldsymbol{v}$	Particle velocity vector
Y	Mass fraction

Greek symbols

α	Porosity
γ	Specific heat ratio

Rate of dissipation

Rate of reaction

δ Tolerance

Capillary tube diameter

δ_{ij} Kronecker delta

ε Dissipation rate of turbulence kinetic energy

Permittivity

θ Contact angle

Collision transfer to $C(\psi)$

κ Curvature

Specific permeability

μ Viscosity

μ' Second viscosity

Viscosity fluctuation

ξ Coordinate in ξ-space

ρ Density

σ Surface tension

τ Shear stress tensor

τ_t Integral time period

Φ Rate of generation of f per unit volume

Electrical potential

Φ_{ij} Pressure-strain contribution to Reynolds stress

ϕ Dissipation function

χ Source term to $C(\psi)$

ψ Flux of f

Subscripts

0 Initial condition

∞ Ambient

a Atmospheric

c Collision

e Eddy

Electron

eff Effective

f Fluid

i Species i

Vector component in i-direction

j Vector component in j-direction

k Phase k

Turbulence kinetic energy

m Mixing or mixture

n Normal

p	Particle	
s	Solid	
	Interfacial surface	
t	Tangential	

Superscripts

*	Dimensionless	
'	Fluctuation	
T	Transpose	

Problems

P2.1 Discuss the model closure of a single-component single-phase laminar flow.

P2.2 Based on the flow compressibility discussed in Section 2.2.5.2, judge if the following flow should be considered as compressible or incompressible: (a) an air jet of 50 m/s into stagnant air; and (b) a water jet of 100 m/s into stagnant water.

P2.3 How much air temperature elevation can be caused by an impingement of an air jet of 50 m/s on a solid surface? How much air density change may the impingement lead to?

P2.4 For incompressible and isothermal single-component turbulent flows, list the independent variables and equations to closure the problem. Use the k–ε turbulence model in the discussion.

P2.5 For an incompressible and isothermal single-component turbulent flow with strong rotation, list the independent variables and equations to closure the problem. Use the Reynolds stress turbulence model in the discussion.

P2.6 Find the typical values (or correlations) of critical Reynolds number (Re_{cr}) of transition of laminar flows to turbulent flows for the following flows:
(a) Jet flows;
(b) Rotating flow (such as tornado; flow in cyclone or centrifugal fan; flow in a rotating blender);
(c) Flow over a reacting or evaporating object.

P2.7 Summarize the applicability, advantages, and disadvantages of the following turbulent models:
(a) Mixing length model;
(b) k-ε model;
(c) Reynolds stress model
Give three examples for each model.

P2.8 Select suitable turbulent models for the following single-component turbulent flows. Explain your selection. For each case, list the independent variables and equations required for problem closure.

(a) jet flow into an unbounded fluid medium;

(b) pipe flow in a 90° bend;

(c) flow with a sudden-expansion inlet;

P2.9 Consider gravity-driven water drainage through a porous plate. Given an initial water level, assuming the same ambient pressure above the water and below the porous plate, estimate the drainage characteristics (water level as a function of time).

P2.10 Using the Hagen–Poiseuille equation of flow through a straight tube, show that Eq. (2.4-1a) of simple the straight capillary model is a special case of the Kozeny model, Eq. (2.4-1b), in which the Kozeny constant $c = 1/2$.

P2.11 Consider a flow pumped into a nonuniform porous medium. Assume that both porosity and grain size are inversely proportional to the depth along the gravity direction. The flow can be regarded as one-dimensional (along the gravity direction) with a given diverse angle. Estimate the pressure distribution along the flow path. To achieve a desired pressure at a given depth, estimate the pumping power required. Discuss the effect of gravity to your model.

P2.12 Consider a power-law fluid motion in a tubular annulus with a moving inner tube. The inner tube moves at a velocity, V, and drives the fluid flow. Calculate the velocity profile and the volumetric flow rate.

P2.13 Consider gravity-driven slurry (liquid–solids suspension) drainage through a porous plate. Given an initial water level, assuming the same ambient pressure above the water and below the porous plate, estimate the drainage characteristics (slurry level as a function of time).

P2.14 Show that the dissipation function defined by Eq. (2.2-14a) is always positive. Then simplify the expression for a simple shear flow.

P2.15 The simplest capillaric model is the one representing a porous medium by a bundle of straight capillaries parallel to the flow.

(a) For viscous flow passing through a circular tube of diameter δ, show that the flow rate through the tube can be expressed by the Hagen–Poiseuille equation as

$$Q = \frac{\pi \, \delta^4}{128 \, \mu} \frac{dp}{dz}. \qquad \text{(P2.15-1)}$$

(b) Assume that the number of capillaries per unit area of cross section, n, can be related to the voidage α by

$$\alpha = \frac{\pi}{4} n \delta^2. \qquad \text{(P2.15-2)}$$

Show that the specific permeability κ in the porous medium is given by Eq. (2.4-1a).

P2.16 In the configuration in Figure 2.4-2(b), the Brinkman's solution inside the porous medium can be found analytically to be

$$U = -\frac{\kappa}{\mu}\frac{dp}{dx}\left[1 + \frac{h^2/2\kappa - 1}{1 + h/\sqrt{\kappa}}\exp(y/\sqrt{\kappa})\right] (y < 0). \tag{P2.16-1}$$

If the boundary layer thickness δ is defined to be the depth at which the flow velocity is reduced to 1.01 times the Darcy velocity, find out the expression for δ. Calculate the value of δ when permeability $\kappa = 10^{-4}$ cm^2, and $h = 20$ cm.

P2.17 A narrow channel filled with a porous medium is bounded by two impermeable walls. The permeability of the porous medium is κ. Derive expressions of the velocity profile and the flow rate when the flow is driven by a constant pressure drop.

P2.18 The charge density near a charge surface in a microchannel is described by the Poisson–Boltzmann distribution

$$\rho_e(r) = -2Zec_0 \sinh\left[\frac{Ze}{k_BT}\Phi(r)\right], \tag{P2.18-1}$$

in which r is the distance from the surface, Z is the valence, e is the elementary charge, k_B is the Boltzmann constant, and T is temperature.
1) Develop an equation for the electrical potential near the surface and the boundary conditions
2) Under the assumption of thin Debye layer $Ze\zeta \ll k_BT$, solve the electric potential. (This is called the Debye–Huckel approximation.)

P2.19 Consider a flow in a microchannel with height $h = 10$ μm, width $w = 100$ μm, and length $L = 5$ mm. The zeta potential on the top and bottom walls is $\zeta = 100$ mV. The liquid viscosity is $\mu = 0.001$ Pa · s. Permittivity of the liquid is 6.9×10^{-12} F/m. The required flow rate is $Q = 10^{-6}$ ml/s.
1) If the flow is driven by an electroosmotic pump, calculate the electric voltage drop along the channel.
2) If the flow is driven by pressure drop, calculate the pressure drop along the channel.

References

Bagnold, R. A. (1954). Experiments on a gravity-free dispersion of large solid spheres in a Newtonian fluid under shear. *Proc. R. Soc. London*, **A225**, 49.

Bear, J. (1972). *Dynamics of Fluids in Porous Media*. New York: American Elsevier.

Beavers, G. and Joseph, D. (1967). Boundary conditions at a naturally permeable wall. *J. Fluid Mech.*, **30**, 197–207.

Benyahia, S. (2008). Verification and validation study of some polydisperse kinetic theories. *Chem. Eng. Sci.*, **63**, 5672–5680.

Bradshaw, P. (1978). Turbulence. *Top. Appl. Phys.*, **12** (ed. P. Bradshaw), Berlin Heidelberg: Springer-Verlag.

Brinkman, H. C. (1947). A calculation of the viscous force exerted by a flowing fluid on a dense swarm of particles. *Appl. Sci. Res.*, **A1**, 27–34.

Bruus, H. (2008). *Theoretical Microfluidics*. Oxford: Oxford University Press.

Burke, S. P., and Plummer, W. B. (1928). Suspension of macroscopic particles in a turbulent gas stream. *I & EC.*, **20**, 1200.

Campbell, C. S. (2006). Granular material flows – An overview. *Powder Technol.*, **162**, 208–229.

Carman, P. C. (1956). *Flow of Gases through Porous Media*. London: Butterworths.

Culick, F. E. C. (1964). Boltzmann equation applied to a problem of two-phase flow. *Phys. Fluids*, **7**, 1898.

Daly, B. J., and Harlow, F. H. (1970). Transport equation in turbulence. *Phys. Fluids*, **13**, 2634.

Darcy, H. (1856). *Les Fontaines Publiques de la Ville de Dijon*. Paris: Victor Dalman.

Ergun, S. (1952). Fluid flow through packed columns. *Chem. Eng. Prog.*, **48**, 89.

Ergun, S., and Orning, A. A. (1949). Fluid flow through randomly packed columns and fluidized beds. *I & EC.*, **41**, 1179.

Gibson, M. M., and Launder, B. E. (1978). Ground effects on pressure fluctuations in the atmospheric boundary layer. *J. Fluid Mech.*, **86**, 491–511.

Gidaspow, D. (1993). Hydrodynamic modeling of circulating and bubbling fluidized beds. *Particulate Two-Phase Flow* (ed. M. C. Roco). Boston: Butterworth-Heinemann.

Giese, M., Rottschäfer, K., and Vortmeyer, D. (1998). Measured and modeled superficial flow profiles in packed beds with liquid flow. *AIChE J.*, **44**(2), 484–490.

Jaeger, H. M., and Nagel, S. R. (1992). Physics of granular states. *Science*, **255**, 1524.

Jenkins, J. T., and Savage, S. B. (1983). A theory for the rapid flow of identical, smooth, nearly elastic, spherical particles. *J. Fluid Mech.*, **130**, 187.

Kim, H., and Arastoopour, H. (2002). Extension of kinetic theory to cohesive particle flow, *Powder Technol.*, **122**, 83–94.

Klinkenberg, L. J. (1941). The permeability of porous media to liquids and gases. *Drilling and Production Practice*, American Petroleum Inst., 200–213.

Kozeny, J. (1927). Über kapillare Leitung des Wassers im Boden (Aufstieg, Versickerung und Anwendung auf die Bewässerung). *Ber. Wien. Akad.*, **136a**, 271.

Launder, B., and Spalding, D. (1972). *Mathematical Models of Turbulence*. Waltham: Academic Press.

Launder, B. E., Reece, G. J., and Rodi, W. (1975). Progress in the development of a Reynolds-stress turbulence closure. *J. Fluid Mech.*, **68**, 537–566.

Lumley, J. L. (1970). *Introduction to Homogeneous Turbulence*, Cambridge: Massachusetts Institute of Technology Press.

Lun, C. K. K. (1991). Kinetic theory for granular flow of dense, slightly inelastic, slightly rough spheres, *J. Fluid Mech.*, **233**, 539–559.

Lun, C. K. K., and Savage, S. B. (1986). The effects of an impact velocity dependent coefficient of restitution on stresses developed by sheared granular materials. *Acta Mechanica*, **63**, 15–44.

Lun, C. K. K., Savage, S. B., and Jeffery, D. J. (1984). Kinetic theories for granular flow: inelastic particles in Couette flow and slightly inelastic particles in a general flow field. *J. Fluid Mech.*, **140**, 223.

Martys, N., Bentz, D. P., and Garboczi, E. J. (1994). Computer simulation study of the effective viscosity in Brinkman's equation. *Phys. Fluids*, **6**, 1434.

Maußner, J., Pietschak, A., and Freund, H. (2017). A new analytical approximation to the extended Brinkman equation. *Chem. Eng. Sci.*, **171**, 495–499.

Ochoa-Tapia, J. A., and Whitaker, S. (1995). Momentum transfer at the boundary between a porous medium and a homogeneous fluid – I. Theoretical development. *Int. J. Heat Mass Transfer*, **38**, 2635–2646.

Papathanasiou, T. D., Markicevic, B., and Dendy, E. D. (2001). Computational evaluation of the Ergun and Forchheimer equations for fibrous porous media. *Phys. Fluids*, **13**, 2795–2804.

Piomelli, U., Moin, P., and Ferziger, J. H. (1988). Model consistency in large eddy simulation of turbulent channel flows. *Phys. Fluids*, **31**, 1884–1891.

Prandtl, L. (1925). Bericht uber Untersuchung zur ausgebildeten Turbulenze. *Z. A. Math. Mech.* **5**(1): 136–139.

Reif, F. (1965). *Fundamentals of Statistical and Thermal Physics*. New York: McGraw-Hill.

Reynolds, O. (1900). *Papers on Mechanical and Physical Subjects*. Cambridge: Cambridge University Press.

Savage, S. B., and Jeffery, D. J. (1981). The stress tensor in a granular flow at high shear rates. *J. Fluid Mech.*, **110**, 255.

Scheidegger, A. E. (1960). *The Physics of Flow through Porous Media*. Toronto: University of Toronto Press.

van Driest, E. R. (1956). On turbulent flow near a wall. *J. Aeronautical Sci.*, **23**(11), 1007–1011.

van Wachem, B. G., Schouten, M. J. C., van den Bleek, C. M., Krishna, R., and Sinclair, J. L. (2001). Comparative analysis of CFD models of dense gas–solid systems. *AIChE J.*, **47**, 1035–1051.

Winterberg, M., Tsotsas, E., Krischke, A., and Vortmeyer, D. (2000). A simple and coherent set of coefficients for modeling of heat and mass transport with and without chemical reaction in tubes filled with spheres. *Chem. Eng. Sci.*, **55**, 967–979.

3 Transport of Isolated Objects: Solid Particles, Droplets, and Bubbles

3.1 Introduction

Transport properties of an isolated object in a fluid continuum, such as a solid particle, a droplet, or a bubble, can be characterized by the phase interactions. The fundamental behavior of their interactions can be quantified by the Lagrangian trajectory modeling of multiphase flows. The momentum, heat, and mass transfer between the particle and the surrounding Newtonian fluid medium, occurring as a result of relative motion or reaction, are discussed in this chapter.

The momentum transfer between a particle and a surrounding fluid medium, in terms of various interfacial forces on the particle, is introduced in Section 3.2. To begin the discussion, the drag force is first introduced by considering the hydrodynamic force on a rigid spherical particle moving at a constant linear relative velocity in an unbound uniform fluid medium. Modifications of drag coefficients are then discussed for the following situations: a nonspherical particle, a permeable sphere, a deformable particle with internal motion, blowing from an evaporating or burning particle, wall boundary, Cunningham slip due to noncontinuity in the fluid medium, and compressibility at a high Mach number. When a particle is accelerating or decelerating rectilinearly, two additional hydrodynamic forces, the carried mass force (also known as added mass or virtual mass force) and the Basset force, act on the moving particle. For a particle moving in a nonuniform flow field with velocity, pressure, and/or temperature gradients, there exists additional gradient-related forces, such as the Saffman force. The rotation of a particle will effectuate a net force perpendicular to the particle moving direction, known as the Magnus force. Besides the interfacial forces, some additional field forces on the particle are also included in this section.

The heat and mass transfer of a spherical particle are introduced, in Sections 3.3 and 3.4, respectively. The treatment begins with heat conduction of a sphere in a stagnant fluid. The resulting discussion also provides the limiting conditions of lumped heat capacity modeling as well as the limiting Nusselt number of heat convection. Basic models of heat convection and thermal radiation are briefly reviewed. Most of the mass transfer in multiphase flows can be divided into two categories: the mass transfer by physical phase change, such as the evaporation of droplets, and the mass transfer involving chemical reactions, such as gaseous phase reactions and solid fuel combustion. The basic concepts of molecular diffusion of multicomponent gases and

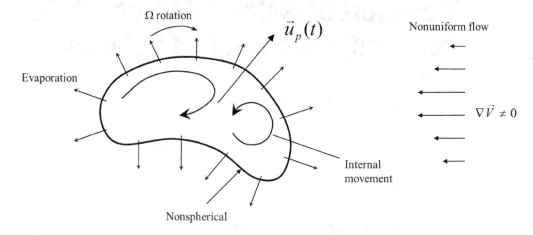

Figure 3.2-1 A particle moving in a fluid flow.

Stefan flow for mass transfer are first introduced. The simple models of droplet evaporation without gas phase reactions are also discussed, along with their solutions that lead to the d^2-law of diffusive evaporation of droplets in Section 3.4. Mass transfer with reactions involves much more complicated and coupled mechanisms than hydrodynamics alone, and will be introduced later in Chapter 12.

In Section 3.5, the equation of motion of a single particle in the Stokes regime is described. It is described by the arithmetic sum of all hydrodynamic and field forces, as exemplified by the Basset–Boussinesque–Oseen (BBO) equation. Modifications and generalizations of the equation of motion beyond the Stokes regime are briefly discussed. Some advanced topics are further discussed in Section 3.6, including the nature of deformation of a fluid particle in motion and the instability of the particle motion (particularly due to an asymmetric wake and wake shedding). Finally the case studies are given in Section 3.7.

3.2   Momentum Transfer

The motion of a single particle in a surrounding fluid may be generally described by a nonspherical particle accelerating and rotating in an unbound nonuniform flow field. The flowing particle can be rigid or deformable, solid or porous, and with or without mass transfer on the surface, as schematically illustrated in Figure 3.2-1. The wall boundary may also have a significant influence on the particle movement.

The total hydrodynamic force on a particle consists of two types of interfacial stresses that are exerted on the particle surface by the surrounding flow field: One is due to pressure, and the other is from viscous stress. The combined particle–fluid interaction force can be expressed as:

$$\mathbf{F}_{pf} = \oint_S p\,\mathbf{n}\,dS + \oint_S \tau \cdot \mathbf{n}\,dS \qquad (3.2\text{-}1)$$

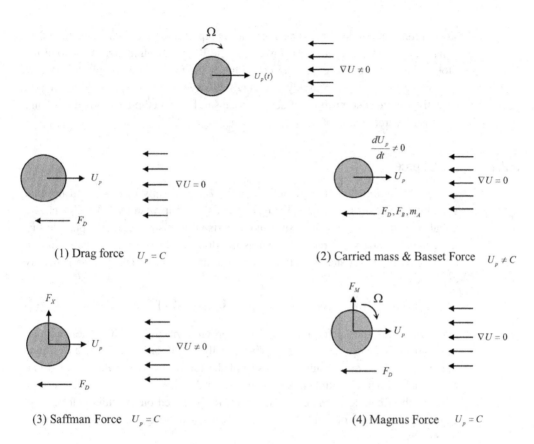

Figure 3.2-2 Conceptual decomposition of a general motion at low Re_p.

Fundamental models of fluid–particle interactions are typically predicated on the transport of a solid sphere in an unbound flow field. Complex movement of a solid sphere can be regarded as a superposition of four types of simple motions: (1) a sphere moving with a constant velocity in a uniform flow field; (2) a sphere accelerating in a uniform flow field; (3) a sphere moving with a constant velocity in a nonuniform flow field; and (4) a sphere rotating with a constant angular velocity in a uniform flow field, as shown in Figure 3.2-2. Various hydrodynamic forces, such as the drag force, the Basset force, the carried mass force, the Saffman force, and the Magnus force, correspond to each of these simple motions. The effects of nonsphericity, porosity, deformations, evaporation, and wall boundary interaction, as well as other effects due to turbulence, rarefied gas, and flow compressibility are generally treated as additional modifications to the corresponding particle–fluid interactions on a solid sphere in an unbound flow.

It should be noted that such decomposition of a particle motion is true only when the governing equation for the flow field around the particle is linear, which requires the particle Reynolds numbers to be very low. Under such conditions, a simple summation of the individual force components represents the actual resultant hydro-dynamic force on a particle. At higher particle Reynolds numbers, far beyond the Stokes flow regime, the governing equation for the flow around a sphere becomes highly nonlinear and even highly unsteady with asymmetric wake shedding. In that case, the simple superposition of decomposed sub-force method embodied in Figure 3.2-2 is rendered unreliable.

3.2.1   Drag Force

In a particle–fluid flow, the particle velocity, U_p, generally differs from the fluid velocity, U. The difference in velocity, defined as the slip velocity $(U - U_p)$, causes unbalanced distributions of pressure and imparts viscous stresses on the particle surface; the resulting net force is known as the drag force. The drag force acting on a single particle at a steady-state motion in a uniform flow field can be generally expressed by:

$$\mathbf{F}_D = C_D A \frac{\rho}{2} \left| \mathbf{U} - \mathbf{U}_p \right| (\mathbf{U} - \mathbf{U}_p),\qquad (3.2\text{-}2)$$

where A is the exposed frontal area of the particle to the direction of the incoming flow and C_D is the drag coefficient. C_D is typically a function of the particle Reynolds number and the local turbulent intensity of the fluid flow. It should be noted that the particle Reynolds number, Re_p, is based on the slip velocity and particle size, whereas the Reynolds number of fluid flow, Re, is based on the fluid velocity and characteristic length of fluid flow system. Each of these quantities is respectively defined as:

$$Re_p = \frac{\rho \left| \mathbf{U} - \mathbf{U}_p \right| d_p}{\mu}\qquad (3.2\text{-}2a)$$

$$Re = \frac{\rho\, U_L}{\mu}\qquad (3.2\text{-}2b)$$

Re_p dictates the flow characteristics around a particle; Re dominates the local characteristics of ambient fluid flow such as turbulent intensity.

Based on a wide range of measurements on terminal velocities of settling spheres, drag coefficient for a single sphere in a uniform and undisturbed flow at various Re_p can be empirically plotted into a unified curve (Lapple and Shepherd, 1940; Schlichting, 1979). Realizing the existence of many unreliable points in the original data set, due to various effects of measurements such as the wall effects, fluid particle deformation and vibration at higher Re_p, and insufficient settling, Brown and Lawler (2003) recollected a set of high-quality and modified data to replot the curve. This modified curve, as shown in Fig. 3.2-3, can be expressed by a combination of the Stokes equation and an empirical correlation (Cheng, 2009):

$$C_D = \frac{24}{Re_p} \qquad Re_p < 0.1 \qquad (3.2\text{-}3a)$$

$$C_D = \frac{24}{Re_p}\left(1 + 0.27\, Re_p\right)^{0.43} + 0.47\left[1 - \exp\left(-0.04\, Re_p^{0.38}\right)\right] \quad 0.1 < Re_p < 2 \times 10^5.$$

$$(3.2\text{-}3b)$$

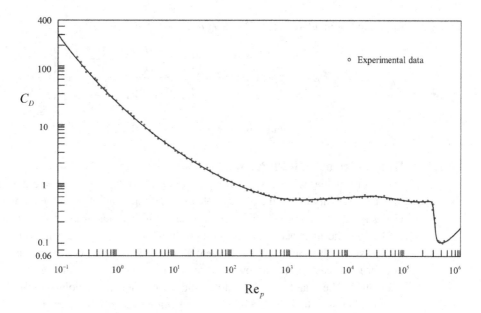

Figure 3.2-3 Drag coefficient versus Re_p for a sphere. (from Schlichting, 1979; Brown and Lawler, 2003; Cheng, 2009)

As Figure 3.2-3 illustrates, the sharp reduction in the drag coefficient at high Re_p (e.g., 3×10^5) is due to the transition from a laminar to a turbulent boundary layer over the particle. This transition is due to the change in the pressure distribution on the particle surface caused by the change of the wake structure trailing the particle in the turbulent regime. However, for most multiphase flows, it is very rare for Re_p to reach the transition or turbulent regimes, whereas ambient fluid flows are routinely found in the turbulent flow regime.

For nonspherical particles, computing the drag coefficient requires the particle's sphericity and its orientation to the flow to be considered, in addition to the particle Reynolds number. The sphericity of a nonspherical particle is defined by the ratio of surface area of a sphere having the same volume as the nonspherical particle to the actual surface area of this particle, expressed by:

$$\varphi = \frac{\pi}{A_s} \left(\frac{6V}{\pi} \right)^{2/3} \tag{3.2-4}$$

While the orientation affects the frontal area in the calculation of drag, it also influences the stability of the particle transport. Drag coefficient correlations can be plotted with respect to sphericity and the particle Reynolds number.

3.2.1.1 Effect of Permeability

For a permeable and porous solid sphere, the fluid not only flows around the sphere but also penetrates it. Consequently, the drag force of a permeable sphere is lower than that of a solid sphere in identical flow conditions. For a sphere with a homogeneous porous and permeable structure moving in the creeping flow regime, the drag force can be calculated from (Neale *et al.*, 1973; Vanni, 2000):

$$\mathbf{F}_D = 3\,\pi\,\mu\,d_P\,(\mathbf{U} - \mathbf{U}_p)\left\{\frac{2\,\beta^2\left(1 - \frac{\tan h\,\beta}{\beta}\right)}{2\,\beta^2 + 3\left(1 - \frac{\tan h\,\beta}{\beta}\right)}\right\}, \tag{3.2-5}$$

where $\beta = d_p/\left(2\sqrt{\kappa}\right)$ and κ is the permeability, defined by Eq. (2.4-1).

3.2.1.2 Drag Force of a Fluid Particle

For a deformable particle, such as a droplet or a bubble, the drag force depends on the fluid viscosity inside the deformable particle, μ_d, in addition to the ambient flow viscosity, μ. A viscous fluid sphere immersed in another viscous fluid is considered here. The interface between the two fluids is given to be clean, so that it has a fixed surface tension, and the tangential stress is assumed continuous across the interface. A steady uniform flow is passed over the fluid sphere, inducing an internal circulation. The drag force of a deformable particle at low Reynolds numbers can be described by the Hadamard–Rybcyznski equation (Soo, 1990) as:

$$\mathbf{F}_D = 3\pi\,\mu\,d_p\,(\mathbf{U} - \mathbf{U}_p)\left\{\frac{1 + \frac{2\mu}{3\mu_d}}{1 + \frac{\mu}{\mu_d}}\right\}. \tag{3.2-6}$$

Equation (3.2-6) shows that, with a very high viscosity of the fluid particle, the drag force approaches that of a solid sphere. Conversely, with a very low viscosity, as in the case of a gas bubble in a liquid, Eq. (3.2-6) leads to:

$$\mathbf{F}_D = 2\pi\,\mu\,d_p\,(\mathbf{U} - \mathbf{U}_p). \tag{3.2-7}$$

Hence, comparing to the drag of a solid sphere, the drag of a low-viscous fluid sphere is reduced by 1/3.

For Reynolds numbers exceeding unity, the flow over a fluid sphere may exhibit different behavior depending on the viscosity ratio, μ_d/μ, and interface contamination. With a large viscosity ratio or a contaminated surface, the fluid sphere surface is effectively subjected to the no-slip boundary condition, and internal recirculation is suppressed. Moreover, flow separation occurs when Re_p is sufficiently high. As a result, the flow exhibits a similar pattern to that over a solid sphere, and the Schiller–Naumann correlation is popularly adopted for the estimation of drag force (Schiller and Naumann, 1933):

$$C_D = \begin{cases} \dfrac{24}{Re_p}(1 + 0.15\,Re_p^{0.687}) & Re_p < 1{,}000 \\ 0.44 & Re_p > 1{,}000 \end{cases}. \tag{3.2-8}$$

Alternatively, the drag force may be calculated by the revised formula Eq. (3.2-3). In contrast, in the limiting case of a low viscosity ratio and uncontaminated surface, the interface is effectively handled as a free-slip surface. There is strong internal recirculation inside the sphere, which suppresses wake separation in the external flow. In this limiting case, the drag coefficients can be estimated by (Moore, 1965; Mei et al., 1994):

$$
C_D = \begin{cases}
\dfrac{16}{\mathrm{Re}_p} \left\{ 1 + \left[\dfrac{8}{\mathrm{Re}_p} + \dfrac{1}{2} \left(1 + \dfrac{3.315}{\sqrt{\mathrm{Re}_p}} \right) \right]^{-1} \right\} & 1 < \mathrm{Re}_p < 100 \\[3ex]
\dfrac{48}{\mathrm{Re}_p} \left(1 - \dfrac{2.21}{\sqrt{\mathrm{Re}_p}} \right) & \mathrm{Re}_p > 100
\end{cases}
\tag{3.2-9}
$$

For intermediate values of the viscosity ratio, various experiments and direct numerical simulations have demonstrated that the drag on a clean fluid sphere decreases monotonically as the viscosity ratio decreases. While closed-form analytical solutions are difficult to obtain because of possible flow separation, some correlations have been developed that account for the viscosity ratio effect on drag force (Clift et al., 1978; Loth, 2008).

3.2.1.3 Blowing Effect

Blowing from the surface of an evaporating droplet or from that of a burning particle affects the viscous shear and pressure distributions in the boundary layer that contribute to the drag force. In the creeping flow regime, the drag coefficient of a droplet subjected to Stefan convection may be expressed by (Sirignano, 2010):

$$
C_D = \frac{24}{\mathrm{Re}_{pm}} \left(1 + \frac{Y_0 - Y_\infty}{1 - Y_0} \right)^{-1},
\tag{3.2-10}
$$

where Y_0 and Y_∞ stand for the vapor molar fractions at the droplet surface and the ambient, respectively; Re_{pm} is the particle Reynolds number, with the gas-film viscosity defined by:

$$
\mu_m = \frac{2}{3}\mu_0 + \frac{1}{3}\mu_\infty.
\tag{3.2-11}
$$

For evaporating droplets at high Reynolds numbers ($30 < \mathrm{Re}_{pm} < 200$), the drag coefficient can be estimated by (Sirignano, 2010):

$$
C_D = \frac{24.432}{\mathrm{Re}_{pm}^{0.721}} \left(1 + \frac{h_\infty - h_0}{L} \right)^{-0.27},
\tag{3.2-12}
$$

where h is the specific enthalpy and L is the latent heat.

3.2.1.4 Wall Boundary Effect

When a sphere moves near a wall, the drag force depends on the distance from the wall, in the moving direction of the sphere, and the particle Reynolds number. In the creeping flow regime, when particle diameter is much smaller than the distance from the wall (i.e., $d_p \ll y$), the drag coefficient of a sphere moving normal to the wall can be estimated, to the first order, as (Brenner, 1962):

$$
C_D = \frac{24}{\mathrm{Re}_p} \left(1 + \frac{1}{2}\frac{d_p}{y} \right),
\tag{3.2-13}
$$

whereas the drag coefficient of a sphere moving parallel to the wall can be estimated, to the first order, as (Faxen, 1923):

$$C_D = \frac{24}{Re_p} \left(1 - \frac{9}{32} \frac{d_p}{y} \right)^{-1}. \tag{3.2-14}$$

When the particle Reynolds number becomes very large, the drag force of a sphere is dominated by the form drag, which is determined from potential flow theory. The drag coefficient of a sphere moving normal to the wall is thus given by (Zhang et al., 1999):

$$C_D = \frac{3}{64} \left(\frac{d_p}{y} \right)^4. \tag{3.2-15}$$

3.2.1.5 Cunningham Slip Effect

When the sphere size is on the same order of, or smaller than, the molecular mean free path of the fluid (such as particles in the rarefied gas or submicron aerosols in an air stream), slip between the sphere and the fluid occurs. Taking the Cunningham slip effect into consideration, the drag coefficient can be generally expressed in the Knudsen–Weber form as (Knudsen and Weber, 1911):

$$C_D = \frac{24}{Re_p} \left(1 + Kn \left\{ \alpha + \beta \exp \left(-\frac{\gamma}{Kn} \right) \right\} \right)^{-1}, \tag{3.2-16}$$

where Kn is the Knudsen number, defined by $Kn = 2\lambda/d_p$; λ is the molecular mean free path; α, β, and γ are empirical constants that weakly depend on material properties of aerosols. For $Kn < 83$: $\alpha = 1.165$; $\beta = 0.483$; and $\gamma = 0.997$ (Kim et al., 2005).

3.2.1.6 Effect of Flow Compressibility

In compressible flow, the drag force on a sphere is affected by the flow's compressibility, which is characterized by the particle Mach number, Ma, defined by the ratio of relative velocity to speed of sound. The drag coefficient of a sphere in a subsonic flow may be estimated by (Henderson, 1976):

$$C_D = \frac{24}{Re_p + Ma\sqrt{\frac{\gamma}{2}} \left(4.33 + 1.567 \exp \left[-0.247 \frac{Re_p}{Ma} \sqrt{\frac{2}{\gamma}} \right] \right)}$$

$$+ \exp \left(-\frac{Ma}{2Re_p} \right) \left(\frac{4.5 + 0.38 \left(0.03 Re_p + 0.48\sqrt{Re_p} \right)}{1 + 0.03 Re_p + 0.48\sqrt{Re_p}} + 0.1\,Ma^2 + 0.2\,Ma^8 \right)$$

$$+ 0.6\,Ma\sqrt{\frac{\gamma}{2}} \left\{ 1 - \exp \left(-\frac{Ma}{Re_p} \right) \right\}, \tag{3.2-17}$$

where γ is the specific heat ratio. The drag coefficient of a sphere in a supersonic flow with Ma > 1.75 may be estimated by (Hughes and Gilliland, 1952):

$$C_D = \frac{0.9 + \dfrac{0.34}{\text{Ma}^2} + 1.86\sqrt{\dfrac{\text{Ma}}{\text{Re}_p}}\left(2 + \dfrac{1.058}{\text{Ma}\sqrt{\dfrac{\gamma}{2}}} + \dfrac{2}{\left(\text{Ma}\sqrt{\dfrac{\gamma}{2}}\right)^2} - \dfrac{1}{\left(\text{Ma}\sqrt{\dfrac{\gamma}{2}}\right)^4}\right)}{1 + 1.86\sqrt{\dfrac{\text{Ma}}{\text{Re}_p}}}.$$

(3.2-18)

3.2.1.7 Effect of Non-Newtonian Fluid

For spheres moving in non-Newtonian fluids, the drag force depends on the rheological characteristics in addition to the particle size and relative velocity of the non-Newtonian fluid. For example, the particle Reynolds number for a sphere moving in a shear thinning power-law fluid is computed from:

$$\text{Re}_{pn} = \frac{\rho d_p^n |\mathbf{U} - \mathbf{u}_p|^{2-n}}{\kappa},$$

(3.2-19)

where κ is the fluid consistency, and n is the power-law index. The drag coefficient can be generally expressed as:

$$C_D = \frac{24}{\text{Re}_{pn}} X_n,$$

(3.2-20)

where X_n is a drag correction factor as a function of n and Re_{pn}, as given by (Ceylan et al., 1999):

$$X_n = \begin{cases} 3^{2n-3}\left(\dfrac{n^2 - n + 3}{n^{3n}}\right) & \text{Re}_{pn} < 10^{-5} \\[3mm] 3^{2n-3}\left(\dfrac{n^2 - n + 3}{n^{3n}}\right) + \dfrac{1 - n^2}{3n + 1}\log(10^3 \text{Re}_{pn}) & 10^{-5} < \text{Re}_{pn} < 10^{-3}. \\[3mm] 3^{2n-3}\left(\dfrac{n^2 - n + 3}{n^{3n}}\right) + \dfrac{n^4}{6\,\text{Re}_{pn}^{\left(\frac{n-3}{3}\right)}} & 10^{-3} < \text{Re}_{pn} < 10^3 \end{cases}$$

(3.2-20a)

Equation (3.2-20a) is applicable for liquids with a power-law index in the range of $0.5 < n < 1$.

3.2.1.8 Effect of Turbulence in Ambient Flow

The drag coefficient curve, shown in Figure 3.2-3, was obtained for a uniform and undisturbed ambient flow. The particle Reynolds number is only associated with the relative velocity; it does not classify the flow regime of the ambient flow (i.e., laminar or turbulent). However, in most multiphase flows, the ambient flow itself is often turbulent, even when the relative velocity and the resulting particle Reynolds number are low. If the ambient flow is turbulent, the local turbulent intensity and the length scale of the turbulent eddies will affect the drag force or the relative motion of the sphere. For a free-flowing sphere in a turbulent flow, the portion of the turbulence spectrum having an amplitude greater than the diameter of the sphere will cause the sphere to oscillate while the portion of the turbulence spectrum having an amplitude smaller than the diameter of the sphere will affect the total drag force via modified surface friction in the turbulent boundary layer and modified form drag due to the change of turbulent flow separation.

Hence, the effect of turbulence on the deviations in C_D from the standard curve depends on the classification and characteristics of turbulence in the ambient flow in addition to the diameter of sphere and averaged particle Reynolds number (Soo, 1990). Due to the lack of predictability of a turbulent flow, no general conclusive relation between C_D and the local turbulence characteristics is available today; although in principle, Direct Numerical Simulation (DNS) or Large-Eddy Simulation (LES) of a sphere in various turbulent flows may be capable of producing such a correlation.

3.2.2   Basset Force and Carried Mass

Once a particle is accelerating or decelerating in a fluid, the particle acceleration, along with its corresponding history, will impart additional fluid resistance to the particle motion. These forces are known as the carried mass force and Basset force. For an accelerating sphere moving in the creeping flow regime, the total fluid resistance on the sphere motion can be expressed by (Lovalenti and Brady, 1993; Fan and Zhu, 1998):

$$\mathbf{F} = 3\pi\mu d_p \left(\mathbf{U} - \mathbf{U}_p\right) + C_A \frac{\rho d_p^3}{6} \frac{d}{dt}\left(\mathbf{U} - \mathbf{U}_p\right) + \frac{3\sqrt{\pi}}{2} d_p^2 \sqrt{\frac{\pi}{\rho}} \int_0^t \frac{\frac{d}{d\tau}\left(\mathbf{U} - \mathbf{U}_p\right)}{\sqrt{t - \tau}} d\tau.$$
(3.2-21)

On the right-hand side (RHS), the first term is the Stokes drag force, the second term is the carried mass force, and the last term is the Basset force, which is also known as the Boussinesq history integral force.

The carried mass force originates from the need to accelerate the fluid surrounding the accelerating particle. The carried mass, also known as the added mass or the virtual mass, is the equivalent mass of fluid that moves with the same acceleration as the particle. In general, the volume of the carried mass depends on the particle geometry and orientation to the flow. For a spherical particle, the virtual mass coefficient, C_A, is equal to 0.5. For nonspherical particles, C_A depends on shape and orientation. The shapes that are comparatively slender in the stream-wise direction generally have lower virtual mass coefficients than fatter ones. Thus, shapes that minimize the drag force are also often found to result in less resistance during acceleration because of their smaller virtual mass force (Batchelor, 1967; Vogel, 1994).

The Basset force, or the Basset history integral, accounts for the effect of the particle's acceleration history on the resistance. The Basset force can be substantial when

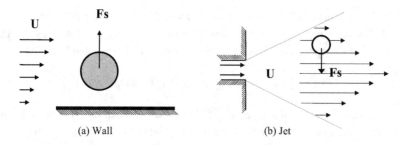

(a) Wall   (b) Jet

Figure 3.2-4 Saffman force on a particle in (a) wall boundary flow and (b) jet flow.

the particle is accelerated at a high rate. The total force on an accelerating particle can be much higher than the drag force of a particle moving at an identical velocity without acceleration (Hughes and Gilliland, 1952). In a simple model with constant acceleration, the ratio of the Basset force to the Stokes drag may be estimated by (Rudinger, 1980):

$$\frac{F_B}{F_D} = \sqrt{\frac{18}{\pi} \frac{\rho}{\rho_p} \frac{\tau_S}{t}}, \tag{3.2-22}$$

where τ_S is the Stokes relaxation time, defined as:

$$\tau_S = \frac{\rho_p d_p^2}{18\,\mu}. \tag{3.2-22a}$$

For particle Reynolds numbers beyond the Stokes regime, the original Basset term can be modified by introducing a correction factor (Odar and Hamilton, 1964; Clift *et al.*, 1978). Modifications of the Basset force are also needed for a sphere in an ambient flow with a small fluctuation (Mei and Adrian, 1992), for a viscous sphere with interface slip (Michaelides and Feng, 1995), and for a porous sphere (Looker and Carnie, 2004). For arbitrarily oriented finite-length cylinders and spheroids, approximated formulas of added mass and Basset force are available for the creeping flow regime (Loewenberg, 1993). The Basset force may be neglected when the fluid–particle density ratio is small or when the time change is much longer than the Stokes relaxation time (*i.e.*, the acceleration rate is low).

3.2.3   Saffman Force and Other Gradient-Related Forces

When a sphere moves in a flow where a velocity gradient, pressure gradient, or temperature gradient exists, additional related forces can be as significant as the drag force.

For a sphere moving in a velocity gradient, for instance near a wall or in a high-shear region, the Saffman lift force is produced. The direction of the Saffman force is always perpendicular to the relative velocity of particle motion, as shown in Figure 3.2-4.

The Saffman force was originally derived for the motion of a sphere at constant velocity in a simple shear flow at low Re_p (Saffman, 1965; Saffman, 1968). The shear flow over the sphere produces a lift force, which is represented by:

$$\mathbf{F}_S = \frac{K}{4} d_p^2 (\rho\mu)^{\frac{1}{2}} |\boldsymbol{\omega}|^{\frac{1}{2}} \left[(\mathbf{U} - \mathbf{U}_p) \times \boldsymbol{\omega} \right], \tag{3.2-23}$$

where $\boldsymbol{\omega}$ is the curl of the velocity field of the carrier fluid ($\boldsymbol{\omega} = \nabla \times \mathbf{U}$), and K is an integrated numerical factor, determined to be 6.46 for creeping flows at low shear rates. In a simple shear flow, Eq. (3.2-23) can be simplified to:

$$\mathbf{F}_S = \frac{K}{4} \mu d_p \left| \mathbf{U} - \mathbf{U}_p \right| \sqrt{Re_G} \left(\frac{\partial \mathbf{U}}{\partial y} \right), \tag{3.2-23a}$$

where the shear Reynolds number is defined by:

$$Re_G = \frac{\rho d_p^2}{\mu} \left| \frac{\partial \mathbf{U}}{\partial y} \right|. \tag{3.2-23b}$$

Thus, the Saffman force is aligned with the direction of the velocity gradient for the carrier fluid if the particle velocity is lower than the fluid velocity. Conversely, it is directed against the velocity gradient of the carrier phase if the particle velocity is higher than the fluid velocity. The Saffman lift force was deduced where $Re_p \ll \sqrt{Re_G} \ll 1$. Extensions to other conditions, for instance where $Re_p > \sqrt{Re_G}$ or $Re_p > 1$, have been made by McLaughlin (1991), Dandy and Dwyer (1990), and Mei (1992).

The ratio of the Saffman force to the Stokes drag is estimated by:

$$\frac{F_S}{F_D} = \frac{K}{12\pi} \sqrt{Re_G} \tag{3.2-24}$$

Equation (3.2-24) indicates that the Saffman force is negligible at very small shear rates or very low Re_p when compared to the Stokes drag criterion. However, since the Saffman force is orthogonal to the drag force, it may still be quite important in some cases even if the Saffman force is smaller than the Stokes drag.

Where there is a pressure gradient field in the flow, a pressure-gradient force will act on the spherical particle, which can be expressed by:

$$\mathbf{F}_P = -\frac{\pi d_p^3}{6} \nabla p. \tag{3.2-25}$$

This force becomes significant in a number of situations, such as when a shock wave propagates through a gas–solid suspension.

Forces acting on a particle due to a temperature gradient in a gas (thermophoresis) or from nonuniform radiation (photophoresis) are known as radiometric forces. For a particle diameter much larger than the mean free path of a gas, the force due to the temperature gradient is given by (Hettner, 1926; Epstein, 1929):

$$\mathbf{F}_T = -\frac{9\pi}{2} \frac{K\mu^2 d_p R_g}{(2K + K_p)p} \nabla T, \tag{3.2-26}$$

where R_g is the gas constant, and K and K_p are the thermal conductivities of the gas and the solid, respectively. Equation (3.2-26) indicates that the force is directed toward the lower-temperature region. At room conditions and in ambient air, considering the radiometric forces is meaningful only for the analysis of submicrometer-sized particles. However, under conditions of high temperatures or large temperature gradients, such as those found in plasma coating, the thermophoresis effect may be significant for larger particles.

3.2.4 Magnus Force

Particle rotation may result from collisions of the particle with a solid wall, colliding with other particles, or from a velocity gradient in a nonuniform flow region. The particle rotation leads to fluid entrainment, resulting in an increase in the velocity on one side of the particle and a decrease in the velocity on the other side. Thus, a "lift" force is established that moves the particle toward the region of higher velocity (and lower pressure). This phenomenon is known as the Magnus effect (Magnus, 1852), and the lift force is called the Magnus force.

Consider a steady and slow motion of a sphere in an incompressible fluid at relative velocity, $\mathbf{U} - \mathbf{U}_p$, and angular velocity, $\mathbf{\Omega}$. Assuming no flow separation, the Magnus force at low Reynolds number can be obtained (Rubinow and Keller, 1961) as:

$$\mathbf{F}_M = -\frac{\pi}{8} d_p^3 \rho \, \mathbf{\Omega} \times (\mathbf{U} - \mathbf{U}_p). \tag{3.2-27}$$

Hence, the direction of the Magnus force depends on the vector product of the relative velocity and the angular velocity, as shown in Figure 3.2-5. The ratio of the Magnus force to the Stokes drag is obtained by:

$$\frac{F_M}{F_D} = \frac{d_p^2}{24} \frac{\rho}{\mu} \Omega \tag{3.2-28}$$

Therefore, the lift force due to the particle spin is negligibly small when compared to the drag force if the particle size is small or the spin velocity is low. At high Reynolds numbers, the rotation of a sphere yields a flow separation with an asymmetric wake shedding, as shown in Figure 3.2-5(b). In this case, the theoretical analysis of the Magnus force and the drag force becomes rather complex because of the difficulties associated with obtaining the expressions for the pressure and velocity distributions around the surface of the sphere. Thus, determining the lift force and the drag force primarily requires the experimental approach or a numerical simulation.

3.2.5 Field Forces

Field forces, also known as the body forces, are long-range forces exerted by various fields outside a flow system. Typical field forces in a multiphase flow system include the gravitational force, the electric force, and the magnetic force.

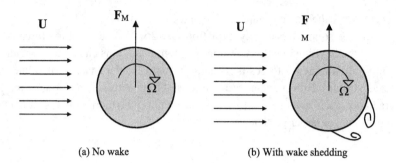

(a) No wake   (b) With wake shedding

Figure 3.2-5 Magnus force and drag force on a rotating sphere.

When an external electric field is applied to a charged particle, the electric force $\mathbf{F}_E$ is expressed by:

$$\mathbf{F}_E = q\mathbf{E} \tag{3.2-29}$$

where q is the charge carried by a particle and $\mathbf{E}$ is the electric field intensity. Driven by this force, particles carrying positive charges tend to move toward the cathodic side while negatively charged particles move to the anodic side.

In a nonuniform electric field, the electric force acting on a charged particle includes both the electrostatic effect due to the net electric charge and the polarization effect due to the permanent or induced electric dipole. This is captured in the expression:

$$\mathbf{F}_E = q\mathbf{E} + \left(\mathbf{p}_d \cdot \nabla\right)\mathbf{E}, \tag{3.2-30}$$

where $\mathbf{p}_d$ is the dipole moment for dielectric materials, given by (Pethig, 2010):

$$\mathbf{p}_d = \frac{\pi}{2}\frac{(\varepsilon_r - 1)}{(\varepsilon_r + 2)}d_p^3 \varepsilon_m \mathbf{E}. \tag{3.2-30a}$$

In Eq. (3.2-30a), ε_m is the absolute permittivity of the surrounding material and $\varepsilon_r = \varepsilon_p/\varepsilon_m$ is the relative permittivity, or dielectric coefficient, of the dielectric particle. The second term in Eq. (3.2-30) is known as the dielectrophoretic force, which results from the nonuniformity of the electric field, as shown in Figure 3.2-6. It should be noted that the existence of a dielectrophoretic force does not require a net charge on the particle.

Equation (3.2-30) also applies to a dielectric particle in a dielectric fluid and exposed to a nonuniform electric field. When a dielectric particle becomes polarized, it is compelled by the dielectrophoretic force toward the extrema (maximum or minimum) of the electric field; the tendency depends on both the fluid and particle dielectric properties. The time-average dielectrophoretic (DEP) force in a DC or stationary AC field with magnitude E can be expressed as:

$$\langle \mathbf{F}_{DEP} \rangle = \frac{\pi}{4}\varepsilon_m d_p^3 \mathrm{Re}\left[f_{CM}\right]\nabla|\mathbf{E}|^2, \tag{3.2-31}$$

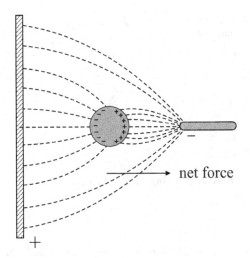

Figure 3.2-6 Dielectrophoretic force of a dielectric particle in an electric field.

where:

$$f_{CM} = \frac{\varepsilon_p^*(\omega) - \varepsilon_m^*(\omega)}{\varepsilon_p^*(\omega) + 2\varepsilon_m^*(\omega)}. \tag{3.2-31a}$$

f_{CM} is the Clausius–Mossotti factor whose value depends on the dielectric constants of the ambient fluid and the particle as well as the frequency of the applied field. $\text{Re}[f_{CM}]$ is the real component of the complex parameter f_{CM}. Equation (3.2-31) shows that the direction of the dielectrophoretic force is aligned with $\nabla|E|^2$. When $\text{Re}[f_{CM}] > 0$, particles tend toward the maximum of an electric field, due to the positive dielectrophoretic force. Conversely, when $\text{Re}[f_{CM}] < 0$, particles move to the minimum of an electric field, due to the negative dielectrophoretic force. If an AC field has a spatially dependent phase, the dielectric particle experiences an additional force, known as the traveling wave dielectrophoresis (twDEP) force. As a result, particles are subjected to a force in the direction tangential to the electrodes in addition to the conventional DEP force toward or away from the electrodes (Wang et al., 1994).

If particles are magnetic, they will experience a force once they are exposed to a heterogeneous magnetic field. This force is known as the magnetophoretic force, $\mathbf{F}_m$, and it is given by (Pamme, 2005):

$$\mathbf{F}_m = \frac{\pi}{6}d_p^3\frac{(\chi_p - \chi_m)}{\mu_0}(\mathbf{B} \cdot \nabla)\,\mathbf{B}, \tag{3.2-32}$$

where $\mathbf{B}$ is the magnetic flux density; μ_0 is the magnetic permeability in a vacuum; χ_p and χ_m are the magnetic susceptibilities of the particle and surrounding medium, respectively. Since most magnetic materials are highly conductive, the electrostatic effect in magnetic materials can usually be neglected.

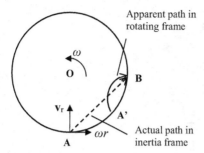

Figure 3.2-7 llustration of Coriolis force in a rotating reference frame.

3.2.6   Coriolis Force

When a particle is observed in a rotational reference frame (such as the Earth rotating on its axis), the apparent, or relative, acceleration, a_r, can be expressed as:

$$a_r = a - 2\omega \times v_r - \omega \times (\omega \times r). \tag{3.2-33}$$

a is the acceleration in the inertia reference frame, ω is the angular velocity vector of the rotational reference frame, v_r is the relative particle velocity to the rotational reference frame, and r is the particle position. The second term on the right-hand side of the equation is referred to as the Coriolis acceleration, and the last term on the right-hand side is the centrifugal acceleration. Such terms are the result of the relative motion between the rotational and inertial reference frames. In the rotational reference frame, the net effect is perceived as additional forces acting on the particle. Figure 3.2-7 illustrates the effect of the Coriolis force on a particle moving from point A on a rotating disk with a relative velocity, v_r, directed toward the center of the disk. In the inertial reference frame, the particle moves in a straight path along chord AB, which is determined by the vector sum of v_r and $\omega \times r$. However, in the rotational reference frame, the Coriolis force creates apparent particle motion along a deflected curve, A'B. Since typical discussions of multiphase flows are on transport phenomena occurring on Earth (*i.e.*, in a rotating reference frame), the Coriolis force is included and expected to influence the flow dynamics, functioning to trigger instability in the flow pattern because of asymmetric effects.

3.3   Heat Transfer

In a multiphase flow, heat can be transferred from a phase to its neighboring phases via interfacial contact or to its surroundings via thermal radiation. The most basic model is the heat conduction of a spherical particle in a quiescent fluid. A detailed analysis of such a case leads to some important limiting conditions of heat transfer in a particle-containing flow, such as the limiting conditions of uniform temperature inside a particle (for lumped heat transfer model of suspended particles) or limiting

Nusselt number as the relative velocity between the particle and fluid approaches zero.

3.3.1 Heat Conduction of a Sphere in Quiescent Fluid

The simplest mode of heat transfer in a multiphase flow is heat conduction between a suspended solid sphere and the surrounding fluid with no relative motion. For simplicity, it is assumed that the temperature distribution throughout the sphere is radially symmetric, that there is no heat generation inside the sphere, and that no thermal radiation occurs between the sphere and the surrounding environment. Thus, the transient heat balance in the particle reduces to:

$$\rho_p c \frac{\partial T_p}{\partial t} = \frac{1}{R^2} \frac{\partial}{\partial R} \left(R^2 K_p \frac{\partial T_p}{\partial R} \right). \tag{3.3-1}$$

The boundary and initial conditions are:

$$\frac{\partial T_p}{\partial R} = 0 \qquad\qquad \text{at } R = 0$$

$$K_p \frac{\partial T_p}{\partial R} = h \left(T_\infty - T_p \right) \qquad \text{at } R = \frac{d_p}{2} \tag{3.3-1a}$$

$$T_p = T_{p0} \qquad\qquad \text{at } t = 0,$$

where h is the convective heat transfer coefficient. The above equations can be reduced to dimensionless forms as:

$$\frac{\partial T_p^*}{\partial \mathrm{Fo}} = \frac{1}{R^{*2}} \frac{\partial}{\partial R^*} \left[R^{*2} \frac{\partial T_p^*}{\partial R^*} \right], \tag{3.3-2}$$

with the boundary conditions:

$$\frac{\partial T_p^*}{\partial R^*} = 0 \qquad\qquad \text{at } R^* = 0$$

$$\frac{\partial T_p^*}{\partial R^*} + \mathrm{Bi} T_p^* = 0 \qquad \text{at } R^* = 1 \tag{3.3-2a}$$

$$T_p^* = T_p^*(R^*) \qquad\qquad \text{at } \mathrm{Fo} = 0,$$

where the dimensionless variables are defined as:

$$R^* = \frac{2R}{d_p} \qquad\qquad T_P^* = \frac{T_p - T_\infty}{T_{p0} - T_\infty}$$

$$\tag{3.3-2b}$$

$$\mathrm{Bi} = \frac{h d_p}{2K_p} \qquad\qquad \mathrm{Fo} = \frac{4K_p t}{\rho_p c d_p^2}.$$

The dimensionless solution can be obtained as (Bergman *et al.*, 2011):

$$T_p^* = \frac{2}{R^*} \sum_{n=1}^{\infty} \exp\left(-\lambda_n^2 \text{Fo}\right) \frac{4\left(\sin \lambda_n - \lambda_n \cos \lambda_n\right)}{2\lambda_n - \sin\left(2\lambda_n\right)} \frac{\sin(\lambda_n R^*)}{\lambda_n R^*},$$

(3.3-2c)

where λ_n's are the eigenvalues of:

$$\lambda_n \cot \lambda_n + \text{Bi} - 1 = 0.$$

(3.3-2d)

Fo and Bi, known as the Fourier number and Biot number, respectively, are two important dimensionless parameters in heat conduction. Fo can be interpreted as:

$$\text{Fo} = \frac{\text{Rate of heat conduction}}{\text{Rate of heat storage}}.$$

(3.3-3)

Bi can be described by:

$$\text{Bi} = \frac{\text{Rate of heat convection}}{\text{Rate of heat conduction}}.$$

(3.3-4)

Based on Eq. (3.3-2c), the temperature distribution in the sphere can be approximated as uniform with an error of less than 5% when $\text{Bi} < 0.1$. In transient heat transfer processes, where the particle–fluid contact time is very short, the criterion $\text{Fo} \geq 0.1$ is also required if the internal thermal resistance within the particles to be neglected (Gel'Perin and Einstein, 1971).

For quasi-steady-state heat conduction between an isothermal sphere and an infinitely large quiescent fluid medium, the temperature distribution in the fluid phase is governed by:

$$\frac{1}{R^2} \frac{d}{dR}\left(R^2 K \frac{dT}{dR}\right) = 0,$$

(3.3-5)

with boundary conditions:

$$T = T_p \qquad at\ R = \frac{d_p}{2}$$

$$T = T_\infty \qquad at\ R \to \infty,$$

(3.3-5a)

which leads to the solution:

$$T = \frac{d_p}{2R}(T_p - T_\infty) + T_\infty.$$

(3.3-5b)

The corresponding heat flux at the surface of the sphere is given by:

$$J_q = \frac{2K(T_p - T_\infty)}{d_p}.$$

(3.3-5c)

Note that the heat flux can also be expressed as:

$$J_q = h(T_p - T_\infty).$$

(3.3-5d)

Thus, the limiting Nusselt number at $\text{Re}_p \to 0$ is expressed by:

$$\text{Nu}_p = \frac{hd_p}{K} = 2.$$

(3.3-5e)

Equation (3.3-5e) indicates that, for an isolated and isothermal solid sphere in an infinite fluid, Nu_p approaches a value of 2 in the limiting cases of negligible forced and natural convection.

3.3.2   Convective and Radiant Heat Transfer of a Sphere

For forced convective heat transfer over a sphere in a uniform flow beyond the Stokes flow regime, the Ranz–Marshall correlation is frequently used (Ranz and Marshall, 1952). This correlation is expressed by:

$$Nu_p = 2 + 0.6(Re_p)^{1/2}(Pr)^{1/3}. \tag{3.3-6}$$

The first term on the right-hand side reveals the theoretical requirement that $Nu_p = 2$ at $Re_p = 0$, and the second term may be regarded as an enhancement factor due to forced convection. Equation (3.3-6) is applicable to a range of Re_p up to 10^4 and $Pr \geq 0.7$.

Consider a situation in which a particle is suddenly exposed to a radiant flux from a thermal environment, such as a heated wall or surrounding objects. For simplicity, it is assumed that the radiant flux is isotropic and the temperature distribution inside the particle is uniform. Absorption of radiation by the gas is considered negligible. Moreover, all the physical properties are to remain constant. The transient behavior of the particle temperature is then governed by the energy balance, assuming diffusive radiation, as:

$$mc\frac{dT_p}{dt} = hS(T_\infty - T_p) + S\varepsilon\sigma_b(T_w^4 - T_p^4), \tag{3.3-7}$$

where S and ε are the surface area and emissivity of the particle, respectively.

When the temperature difference between the particle and its thermal environment is trivial, the radiation term can be linearized as:

$$T_w^4 - T_p^4 \approx 4T_w^3(T_w - T_p). \tag{3.3-7a}$$

Thus, a solution is readily obtained in the form:

$$T_p = T_{p0}e^{-Bt} + A(1 - e^{-Bt}), \tag{3.3-7b}$$

where the coefficients, A and B, are given by:

$$A = \frac{hT_\infty + 4\varepsilon\sigma_b T_w^4}{h + 4\varepsilon\sigma_b T_w^3} \qquad B = \frac{S}{mc}\left(h + 4\varepsilon\sigma_b T_w^3\right). \tag{3.3-7c}$$

For particle heating, with $T_w \gg T_p$, it is reasonable to neglect particle radiation to the surroundings. In this scenario, Eq. (3.3-7) yields a solution with the exact same form as Eq. (3.3-7b), but the coefficients, A and B, now are given by:

$$A = T_\infty + \frac{4}{h}\varepsilon\sigma_b T_w^4 \qquad B = \frac{hS}{mc}. \tag{3.3-7d}$$

3.4   Mass Transfer

Mass transfer occurs when there is a concentration gradient of a species, a physical phase change (*e.g.*, evaporation), a chemical reaction (*e.g.*, fuel combustion), or a combination of the above situations. Thus, mass transfer can occur either within the same dynamic phase (*e.g.*, by species diffusion) or across the interfaces between different dynamic phases. Notably, if there is a phase change or a chemical reaction, the mass transfer is always coupled with the associated heat transfer of such processes.

3.4.1   Mass Fluxes in a Multicomponent Fluid

Flows with inherent mass transfer or chemical reactions often involve multiple constituent components. The total density is the sum of the partial densities of each species, and is given by:

$$\rho = \sum_s \rho_s.$$ (3.4-1)

For a multicomponent gas, it is common to assume approximate behavior as a mixture of ideal gases so that, based on Dalton's law, the total pressure is the sum of the partial pressures of each species as given by:

$$p = \sum_s p_s.$$ (3.4-2)

The total mass flux of a species includes diffusive mass flux and convective mass flux components. In general, three velocities are involved in the mass transfer of each species at any location: these are the mixture velocity of all species within the same dynamic phase (*i.e.*, the phase velocity) $\mathbf{U}$, the individual velocity of each species $\mathbf{U}_s$, and the diffusion velocity of each species relative to the mixture v_s, so that:

$$\mathbf{v}_s = \mathbf{U}_s - \mathbf{U}.$$ (3.4-3)

From Fick's law, the species diffusion flux can be expressed as:

$$\mathbf{J}_{ms} = \rho_s \mathbf{v}_s = -\rho D_s \nabla Y_s = -\rho D_s \nabla \left(\frac{\rho_s}{\rho} \right),$$ (3.4-4)

where D_s is the molecular diffusivity of s-species, and Y_s is the mass fraction of s-species. The species mass flux and the total mass flux of gases are thus given by:

$$\dot{m}_s = \rho_s \mathbf{U}_s = J_{ms} + Y_s \rho U$$

$$\dot{m} = \rho U = \sum_s (J_{ms} + Y_s \rho U) = \sum_s J_{ms} + \rho U,$$ (3.4-5)

where $\dot{m}$ denotes mass flux. Equation (3.4-5) yields:

$$\sum_s \rho_s v_s = \sum_s J_{ms} = 0.$$ (3.4-6)

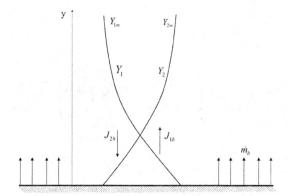

Figure 3.4-1 Schematic mass transfer with Stefan flux on a nonpermeable boundary.

Equations (3.4-5) and (3.4-6) indicate that the species mass flux is the sum of diffusion mass flux and mass-fraction-weighted total mass flux. Moreover, the sum of all diffusion mass fluxes is zero, whereas the linear sum of diffusion velocities of species is not zero.

3.4.2 Stefan Flux

For multicomponent flows with a species produced or removed at a nonpermeable boundary surface, such as in the case of water evaporation into the air or the combustion of solid carbon in oxygen, there exists a net mass flux normal to the surface, which is called Stefan flux.

The origin and physical mechanisms of the Stefan flux can be derived from the mass transfer of a two-species system with a nonpermeable boundary, as shown in Figure 3.4-1. According to Eq. (3.4-5), the mass fluxes of both species on the boundary surface are given by:

$$\dot{m}_{1b} = -D\rho(\nabla Y_1 \cdot \mathbf{n})_b + Y_{1b}\rho(\mathbf{U} \cdot \mathbf{n})_b$$

$$\dot{m}_{2b} = -D\rho(\nabla Y_2 \cdot \mathbf{n})_b + Y_{2b}\rho(\mathbf{U} \cdot \mathbf{n})_b,$$

(3.4-7)

where the subscript b indicates the boundary surface, and the total mass flow normal to the boundary surface, $(\rho U \cdot \mathbf{n})_b$, is the Stefan flux. Since the boundary is nonpermeable, one species mass flux (e.g., of species "2") must be zero; that is:

$$\dot{m}_{2b} = -D\rho(\nabla Y_2 \cdot \mathbf{n})_b + Y_{2b}\rho(\mathbf{U} \cdot \mathbf{n})_b = 0.$$
(3.4-8)

Noting that $Y_1 + Y_2 = 1$, substituting Eq. (3.4-8) in Eq. (3.4-7) yields:

$$\dot{m}_{1b} = \rho(\mathbf{U} \cdot \mathbf{n})_b = -\frac{D\rho}{1 - Y_{1b}}(\nabla Y_1 \cdot \mathbf{n})_b = D\rho(\nabla \ln(1 - Y_1) \cdot \mathbf{n})_b.$$ (3.4-9)

Equation (3.4-9) indicates that the total mass flux at the nonpermeable boundary surface is the Stefan flux, rather than the species diffusion flux alone. In addition,

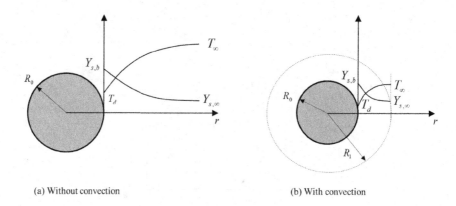

(a) Without convection   (b) With convection

Figure 3.4-2 Evaporation of a droplet without reaction.

the direction of Stefan flow depends on the profile of species mass concentration on the surface: If the flow is from a surface to a gas medium, such as with water evaporation and carbon combustion, a positive direction is assigned; conversely, a negative direction is assigned if the flow is from a gas medium to a surface, such as with the solid-state combustion of magnesium in air to produce magnesium oxide (Zhou, 1993).

3.4.3   Evaporation of a Droplet

In a gaseous mixture, liquid droplets evaporate in an unsaturated vapor environment or condense in a supersaturated vapor environment. If the initial temperature of a droplet is below the vapor saturation temperature, the droplet undergoes a subcooled evaporation process, in which the heat transfer to the droplet elevates the droplet temperature in addition to promoting evaporation. If the initial temperature of a droplet is near the vapor saturation temperature, the heat transfer to the droplet merely compensates for the latent heat required for the evaporation. The droplet evaporation rate, in general, depends on mass diffusion and convection between the ambient environment and the droplet surface in addition to heat transfer, which requires the coupled mass conservation equations of each species and the energy equation in the ambient flow environment and inside the droplet to be solved.

3.4.3.1   Diffusive Evaporation of a Saturated Droplet

An elementary model of droplet evaporation is used to predict the evaporation rate and droplet life time in a stagnant gas medium at a given temperature and pressure. To develop the formulation, consider a spherical droplet suspended in a binary gaseous mixture consisting of the liquid droplet vapor and another gaseous species, as shown in Figure 3.4-2(a).

The evaporation is considered to be quasi-steady, due to the slow regression rate of the droplet. The droplet temperature is assumed to be uniform and to be at the vapor saturation temperature. The relative motion between the droplet and the gas mixture is assumed to be negligibly small such that the evaporation can be reasonably approximated as spherically symmetric. The continuity equation of the gaseous mixture is then given by:

$$\frac{d}{dR}\left(\rho U R^2\right) = 0.$$

(3.4-10)

The equation of vapor-species conservation is captured by the expression:

$$\rho U \frac{dY_s}{dR} = \frac{1}{R^2}\frac{d}{dR}\left(R^2 D\rho \frac{dY_s}{dR}\right).$$

(3.4-11)

The boundary conditions of Y_s are:

$$D\rho \frac{dY_s}{dR} = (\delta_{sd} - Y_{sb})(\rho U)_b \qquad \text{at } R = \frac{d_d}{2},$$

(3.4-11a)

where δ_{sd} is the Kronecker delta, and:

$$Y_s|_{R\to\infty} = Y_{s\infty}.$$

(3.4-11b)

The energy equation can be written as:

$$\rho U c_p \frac{dT}{dR} = \frac{1}{R^2}\frac{d}{dR}\left(R^2 K \frac{dT}{dR}\right),$$

(3.4-12)

where K is the thermal conductivity of the binary gaseous mixture. The boundary conditions here are:

$$K\frac{dT}{dR} = (\rho U)_b L \qquad \text{at } R = \frac{d_d}{2},$$

(3.4-12a)

where L is the latent heat of evaporation, and:

$$T|_{R\to\infty} = T_\infty.$$

(3.4-12b)

The evaporation rate is determined, by solving Eqs. (3.4-10) to (3.4-12), as:

$$\dot{m}_v = \pi d_d^2(\rho U)_b = 2\pi d_d D\rho \ln\left(1 + \frac{Y_{db} - Y_{d\infty}}{1 - Y_{db}}\right) = 2\pi d_d \frac{K}{c_p}\ln\left(1 + \frac{c_p(T_\infty - T_{db})}{L}\right).$$

(3.4-13)

In the analogy between mass transfer and heat transfer, a dimensionless group, known as the Lewis number, is defined for the comparison of transport properties as:

$$\text{Le} = \frac{D}{\left(\frac{K}{\rho c_p}\right)}.$$

(3.4-14)

Equation (3.4-12) indicates that, for cases where Le=1, the general transfer number B, defined below, can be expressed as:

$$\text{B} = \frac{c_p(T_\infty - T_{db})}{L} = \frac{Y_{db} - Y_{d\infty}}{1 - Y_{db}}.$$

(3.4-15)

Hence, based on Eq. (3.4-12), the equation of mass balance for a droplet can be expressed as:

$$\frac{d}{dt}\left(\frac{\pi}{6}\rho_d d_d^3\right) = -2\pi d_d D\rho \ln(1 + \text{B}).$$

(3.4-16)

Integrating Eq. (3.4-16) gives:

$$d_0^2 - d_d^2 = 8 \frac{\rho}{\rho_d} D \ln (1 + B) \, t. \tag{3.4-16a}$$

Equation (3.4-16a) is known as the d^2-law of diffusive evaporation for a droplet without reaction. The evaporation time is proportional to the square of the droplet diameter, thus:

$$t_v = \frac{\rho_d d_0^2}{8 D \rho \ln (1 + B)}. \tag{3.4-16b}$$

Once the surface temperature, T_{db}, and mass fraction, Y_{db}, are known, the transfer number, B, evaporation rate, and evaporation time can be calculated from Eqs. (3.4-14), (3.4-13), and (3.4-16b), respectively. The values of T_{db} and Y_{db} are found by solving Eq. (3.4-15) with the Clausius–Clapeyron equation.

3.4.3.2   Effects of Convection

When an evaporating droplet is moving in a fluid, flow convection leads to an asymmetric boundary for the evaporation. In case of a large Reynolds number for the particle, there is asymmetric and unsteady wake shedding; this complicates the analysis of droplet evaporation, if not rendering it impossible.

In situations where the mass fraction of evaporated species is very low (*i.e.*, $Y_{d\infty} < Y_{db} \ll 1$), mass transfer is dominated by diffusion. Hence the evaporation rate of a droplet is proportional to the difference of the mass fraction, which can be expressed by (Crowe, 2011):

$$\frac{d}{dt} \left(\frac{\pi}{6} \rho_d d_d^3 \right) = -\text{Sh} \pi d_d D \rho \, (Y_{db} - Y_{d\infty}), \tag{3.4-17}$$

where the proportional constant, Sh, is the Sherwood number, defined as the ratio of mass transfer rate to diffusion rate. As the particle Reynolds number approaches zero, the limiting value of Sh is 2, which can be proven directly from Eq. (3.4-16) with the imposed condition of the transfer number approaching zero. The Sherwood number can be empirically correlated to the particle Reynolds number and the Schmidt number, such as embodied by the Ranz–Marshall correlation (Ranz and Marshall, 1952):

$$\text{Sh} = 2 + 0.6 \, \text{Re}_d^{1/2} \, \text{Sc}^{1/3}, \tag{3.4-17a}$$

where the Schmidt number is defined as the ratio of kinematic viscosity to diffusivity.

For cases where the transfer number is not very low, an alternative approximation approach can be introduced that is based on the equivalence in total heat transfer between convection-affected evaporation and diffusive evaporation through a stagnant spherical shell, as shown in Figure 3.4-2(b). Based on the conservation of total heat flow through the shell, it yields:

$$- 4\pi R^2 K \frac{dT}{dR} = \text{const} = 2\pi K \left(\frac{T_\infty - T_d}{d_1^{-1} - d_d^{-1}} \right). \tag{3.4-18}$$

The boundary condition of Eq. (3.4-12a) thus becomes:

$$(\rho U)_b L = h_0 (T_\infty - T_d) = \frac{2K}{d_d^2} \left(\frac{T_\infty - T_d}{d_1^{-1} - d_d^{-1}} \right), \tag{3.4-18a}$$

where h_0 denotes the convective heat transfer coefficient, not including evaporation. Thus, the shell diameter is calculated from:

$$d_1 = d_d \frac{\text{Nu}_0}{(\text{Nu}_0 - 2)}, \tag{3.4-18b}$$

where Nu_0 is the Nusselt number, with no evaporation as determined by Eq. (3.3-5e).

Using Eq. (3.4-18a) to replace Eq. (3.4-12) and following the same modeling approach in Section 3.4.3.1, the evaporation rate with convective contributions becomes:

$$\dot{m}_v = \frac{2\pi}{(d_d^{-1} - d_1^{-1}) c_p} K \ln(1 + B). \tag{3.4-19}$$

3.5 Equation of Motion

The most convenient mathematical description for particle motion in a flow field is obtained using Lagrangian coordinates, with the origin at the center of the moving particle. For the linear motion of a particle at very low particle Reynolds numbers, particle motion is governed by the BBO equation, originated by Basset (1888), Boussinesq (1903), and Oseen (1927). To describe the general motion of particles subjected to various forces, the governing equation may be approximately formulated by the superposition of all applicable forces. This general formulation for particle motion forms the basis of the Lagrangian trajectory modeling of multiphase flows.

3.5.1 Basset–Boussinesq–Oseen Equation

The Basset–Boussinesq–Oseen (BBO) equation delineates a sphere in a linear motion at low Reynolds numbers and can be expressed as (Maxey and Riley, 1983; Gatignol, 1983):

$$\begin{aligned}
\frac{\rho_p \pi d_p^3}{6} \frac{d U_p}{dt} &= 3\pi \mu d_p (\mathbf{U} - \mathbf{U}_p) - \frac{\pi d_p^3}{6} \nabla p + \frac{\rho \pi d_p^3}{12} \frac{d}{dt} (\mathbf{U} - \mathbf{U}_p) \\
&+ \frac{3\sqrt{\pi}}{2} d_p^2 \sqrt{\frac{\pi}{\rho}} \int_0^t \frac{\frac{d}{d\tau}(\mathbf{U} - \mathbf{U}_p)}{\sqrt{t - \tau}} d\tau + \sum_i \mathbf{f}_i,
\end{aligned} \tag{3.5-1}$$

where d/dt is the substantial derivative following the particle flow. The five terms on the right-hand side of Eq. (3.5-1) are, in the order from left to right, the forces due to the Stokes drag, pressure gradient, added mass, Basset historic integral, and other external forces (*e.g.*, gravitational force and electrostatic force). The pressure gradient may be estimated from the Navier–Stokes equation of a single-phase fluid by (Corrsin and Lumley, 1956):

$$-\nabla p = \rho \frac{D\mathbf{U}}{Dt} - \mu \nabla^2 \mathbf{U} = \rho \left(\frac{d\mathbf{U}}{dt} + (\mathbf{U} - \mathbf{U}_p) \cdot \nabla \mathbf{U} \right) - \mu \nabla^2 \mathbf{U}, \quad (3.5\text{-}1a)$$

where D/Dt is the substantial derivative following the gas flow. With the pressure gradient estimated by Eq. (3.5-1a), Eq. (3.5-1) becomes:

$$\left(2\frac{\rho_p}{\rho} + 1 \right) \frac{d\mathbf{U}_p}{dt} = 3\frac{d\mathbf{U}}{dt} - \frac{2\mu}{\rho} \nabla^2 \mathbf{U} + \frac{36\mu}{\rho d_p^2} (\mathbf{U} - \mathbf{U}_p) + 2 (\mathbf{U} - \mathbf{U}_p) \cdot \nabla \mathbf{U}$$

$$+ \frac{18}{d_p} \sqrt{\frac{\mu}{\rho \pi}} \int_0^t \frac{\frac{d}{d\tau}(\mathbf{U} - \mathbf{U}_p)}{\sqrt{t - \tau}} d\tau + \frac{12}{\rho \pi d_p^3} \sum_i \mathbf{f}_i.$$

$$(3.5\text{-}1b)$$

In multiphase flows well beyond the Stokes regime, the effect of convective acceleration by the flow surrounding the particle may be required. To incorporate this effect into the preceding formulation, modifications of the expressions for the Stokes drag, carried mass, and Basset force in the BBO equation are necessary. The modified BBO equation takes the form (Hansell et al., 1992):

$$\frac{d\mathbf{U}_p}{dt} = \frac{3C_D}{4d_p} \frac{\rho}{\rho_p} |\mathbf{U} - \mathbf{U}_p| (\mathbf{U} - \mathbf{U}_p) + \frac{1}{2} \frac{\rho}{\rho_p} C_A \frac{d}{dt} (\mathbf{U} - \mathbf{U}_p) + \frac{\rho}{\rho_p} \frac{D\mathbf{U}}{Dt}$$

$$(3.5\text{-}2)$$

$$+ \frac{9}{d_p} \frac{\rho}{\rho_p} C_B \sqrt{\frac{\mu}{\rho \pi}} \int_0^t \frac{\frac{d}{d\tau}(\mathbf{U} - \mathbf{U}_p)}{\sqrt{t - \tau}} d\tau + \left(\frac{\rho}{\rho_p} - 1 \right) \mathbf{g},$$

where C_D, C_A, and C_B are the drag, carried mass, and Basset coefficients, respectively. For simple harmonic motion at $Re_p < 62$, correlations of C_A and C_B are suggested as (Odar and Hamilton, 1964):

$$C_A = 2.1 - \frac{0.132 \text{An}^2}{1 + 0.12 \text{An}^2} \qquad C_B = 0.48 + \frac{0.52 \text{An}^3}{(1 + \text{An})^3}, \qquad (3.5\text{-}2a)$$

where An is the acceleration number, defined by (Faeth, 1983):

$$\text{An} = \frac{\left| \frac{d(\mathbf{U} - \mathbf{U}_p)}{dt} \right|}{|\mathbf{U} - \mathbf{U}_p|^2} d_p. \qquad (3.5\text{-}2b)$$

In general, C_A and C_B are functions of the particle Reynolds number, Re_p, and An.

3.5.2   General Equation of Motion

A general formulation for particle motion should include additional forces such as the Saffman force, Magnus force, and electrostatic force. Assuming that all forces applied on the moving particle are additive, the equation of motion for a particle in an arbitrary flow can be expressed by:

$$m\frac{d\mathbf{U}_p}{dt} = \mathbf{F}_D + \mathbf{F}_A + \mathbf{F}_B + \mathbf{F}_S + \mathbf{F}_M + \mathbf{F}_C + \cdots, \qquad (3.5\text{-}3)$$

where $\mathbf{F}_D$ is the drag force; $\mathbf{F}_A$ is the added (or carried) mass force; $\mathbf{F}_B$ is the Basset force; $\mathbf{F}_S$ is the Saffman force; $\mathbf{F}_M$ is the Magnus force; $\mathbf{F}_C$ is the collisional force; and "…" denotes other forces such as the electrostatic force, the van der Waals force, the gravitational and buoyancy force, the magnetic force, and forces associated with various field gradients due to shock wave propagation, electrophoresis, thermophoresis, and photophoresis phenomena.

It is typically not feasible to intrinsically add the forces in Eq. (3.5-3) in a linear fashion. The drag force, Basset force, Saffman force, and Magnus force all depend on the same flow field, which is described by the Navier–Stokes equation. The inertia term in the Navier–Stokes equation, $\nabla \cdot (\mathbf{UU})$, is not generally negligible, making the equation highly nonlinear. The governing equations can only be characterized as linear for fluid flows corresponding to very low Reynolds numbers, representing the only reasonable condition where forces on the solids may be added directly. In all other cases, linear superposition of each flow mode in the flow field should only be considered a rough approximation, as the equation will actually be nonlinear.

3.6   Advanced Topics

Two topics are further discussed in this section. One is focused on the characteristics and regime classifications of fluid particles that deform in shape during their dynamic transport, which are linked to the complex interactions between the inertia force, viscous stresses, and the surface tension. The other is orientation and path of motions of nonspherical particles.

3.6.1   Characteristics and Shape Regime of Fluid Particles

A fluid particle can be either a gas bubble or a liquid droplet. It can be conceptualized as an internal fluid that is separated from an external fluid by a deformable interface. Intrinsic coupling between deformation and transport defines the distinctive dynamic characteristics of a given fluid particle. To illustrate this maxim, a comparison in buoyancy-driven motions between a rigid hollow sphere and a deformable bubble in liquid is given in Figure 3.6-1. As the hollow solid sphere rises in an essentially linear motion, the bubble rises in a wobbly motion with shedding wake vortices and changes its shape through an internal recirculation flow.

The shape of fluid particles depends on their size, speed, and physical properties of both the dispersed and continuous phases. In an unbound domain, the basic shape regimes of fluid particles include spherical, oblate ellipsoidal, and spherical cap shapes, as schematically depicted in Figure 3.6-2. For highly viscous fluids, other shapes, such as skirted silhouettes, may also appear. It should be noted that fluid particles in non-Newtonian fluids may embody shapes that are drastically different from those common for Newtonian fluids. Surface contamination also has a considerable impact on the shape of fluid particles. By partially or completely immobilizing the surface, the contaminant makes the fluid particle more rigid and difficult to deform.

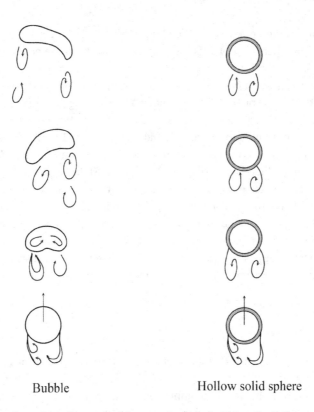

Bubble   Hollow solid sphere

Figure 3.6-1 Buoyancy-driven motions of a bubble and a hollow solid sphere.

This rationale explains why 2–3 mm bubbles in contaminated water are found to be comparatively more spherical than those in purified water (Loth, 2008). When the particle size is commensurate with the dimensions of the domain, the presence of the wall also has a significant effect on the shape of the fluid particles, as in the case of Taylor bubbles in tubes or channels.

The shape variation and the wobbly motion of a fluid particle are the result of the complex interactions between the inertia force, viscous stresses, and the surface tension. To characterize the effects of such forces on the hydrodynamics of a fluid particle, a number of dimensionless groups are often employed. Table 3.6-1 lists some of the typical dimensionless groups frequently used as criteria to estimate the shape, wake structure, pattern of movement, and various hydrodynamic forces acting on the fluid particles. Taking the shape regime in Figure 3.6-2 as an example, the fluid particles with a small Weber number (We $\ll$ 1) are usually spherical. With moderate We, they evolve into oblate ellipsoidal shapes. In cases of large We ($\gg$ 1), the shape becomes a spherical cap or oblate ellipsoidal cap, but this also depends on the particle Reynolds number Re_p. With large Re_p, the asymmetric wake shedding leads to unsteady and unbalanced hydrodynamic forces on the fluid particle and results in oscillatory motion.

Table 3.6-1 Nondimensional numbers for fluid particles

Dimensionless group	Definition	Comments
Particle Reynolds number (Re_p)	$Re_p = \dfrac{\rho_f d_p |\mathbf{U}_f - \mathbf{U}_p|}{\mu_f}$	Ratio of inertia to viscous force. Strong influence on drag and wake structure.
Weber number (We)	$We = \dfrac{\rho_f d_p |\mathbf{U}_f - \mathbf{U}_p|}{\sigma}$	Ratio of inertia to surface tension force. Strong influence on shape of fluid particle.
Eötvös number (Eo)	$Eo = \dfrac{g d_p^2 |\rho_f - \rho_p|}{\sigma}$	Ratio of buoyancy to surface tension force. Strong influence on deformation of fluid particle.
Strouhal number (St)	$St = \dfrac{f d_p}{|\mathbf{U}_f - \mathbf{U}_p|}$	Characterization of oscillating frequency of flows such as the vortex shedding in particle wake.

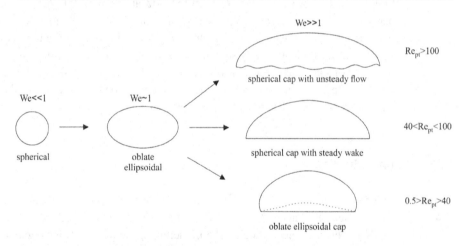

Figure 3.6-2 Typical shape modes of drops and bubbles (from Loth, 2008).

In Table 3.6-1, the subscripts f and p refer to the carrying fluid and the dispersed particle, respectively. For nonspherical particles, the volumetrically equivalent diameter is used. It should be noted that some common dimensionless groups can be dependent on the other dimensionless groups. For instance, Capillary number, Ca, is equivalent to the ratio of We to Re_p, and Morton number, Mo, is equivalent to a combination of We, Eo, and Re_p.

3.6.2   Orientation and Path Instability of Nonspherical Particles

While spherical particles typically follow steady rectilinear paths at low Reynolds numbers, a nonspherical particle may exhibit dynamic variations in its orientation

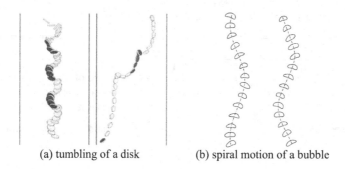

(a) tumbling of a disk   (b) spiral motion of a bubble

Figure 3.6-3 Periodic oscillation and gliding and spiral motion.

and moving direction, even under a steady driving force such as gravity or buoyancy. For example, a rigid nonspherical particle in free fall with a large Re_p (*i.e.,* > 100) exhibits significant unsteady motions, such as periodic oscillation and glide-tumbling, as shown in Figure 3.6-3(a) (Mandø and Rosendahl, 2010). For a fluid particle, such as an oblate ellipsoidal bubble in liquid, oscillations in shape may occur in addition to rising in a spiral or zigzag path with varied orientation (Veldhuis *et al.*, 2008), as schematically shown in Figure 3.6-3(b). Orientation and path instability of nonspherical particles is often observed for particles with a high aspect ratio and high Re_p. For fluid particles, surface contamination can immobilize the surface and suppress the oscillation in their path and shape.

3.7   Case Studies

The following case studies serve for further discussion with simple applications of the theories of Chapter 3. Examples include the particle trajectories in a rotating fluid flow or in an electric field. The modeling of the motion of a parachuted object illustrates not only the large-scale particle–fluid interaction but also the coupling nature of drag force formulation in terms of multicomponent velocities of the particle and fluid flow. Lastly, a simple model on the motion of an evaporating droplet is presented to illustrate the coupled governing mechanisms among momentum, heat, and mass transfer during the droplet transport.

3.7.1   Particle Trajectory in a Rotating Fluid

Problem Statement: Consider a rotating flow in the void formed by two coaxial cylinders of different diameters with small particles injected into the flow radially from a slot on the inner cylinder. The outer cylinder is rotating at a constant angular velocity, ω, while the inner cylinder is stationary. Assume that the flow can be approximated as rigid body rotating with the same constant angular velocity, ω, as the outer cylinder, and the only driving force is the Stokes drag (*i.e.*, ignore other forces,

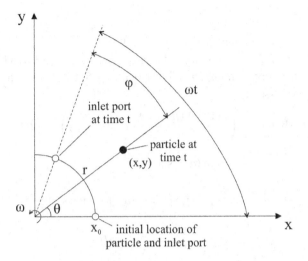

Figure 3.7-1 Coordinate system for particle motion in a rotating flow.

including gravitational force). Develop the equations for the particle trajectory in this rotating flow and discuss the effect of particle sizes on the trajectory.

Analysis: Assume that the flow can be treated as two-dimensional. Select a two-dimensional Cartesian coordinate system with the origin at the axis of rotation and the x-axis coinciding with a line connecting the origin and the initial location of the particle, as shown in Fig. 3.7-1.
The flow velocity is given by:

$$\mathbf{U} = \boldsymbol{\omega} \times \mathbf{r}. \tag{3.7-1}$$

The equation of particle motion is given by:

$$\frac{d\mathbf{U}_p}{dt} = \frac{\mathbf{U} - \mathbf{U}_p}{\tau_S}, \tag{3.7-2}$$

where τ_S is the Stokes relaxation time. Note that:

$$\frac{d}{dt} = \omega\left(\frac{d}{d\theta}\right); \qquad \mathbf{U}_p = \frac{d\mathbf{r}}{dt}. \tag{3.7-2a}$$

Equation (3.7-2) can be expressed as:

$$\frac{d^2x}{d\theta^2} + \frac{1}{\omega\tau_S}\frac{dx}{d\theta} + \frac{y}{\omega\tau_S} = 0 \tag{3.7-2b}$$

$$\frac{d^2y}{d\theta^2} + \frac{1}{\omega\tau_S}\frac{dy}{d\theta} - \frac{x}{\omega\tau_S} = 0 \tag{3.7-2c}$$

The initial conditions for particles being injected radially into the flow are defined by:

$$x|_{\theta=0} = x_0; \qquad y|_{\theta=0} = 0 \tag{3.7-3a}$$

$$\left.\frac{dx}{d\theta}\right|_{\theta=0} = \frac{\dot{x}_0}{\omega}; \qquad \left.\frac{dy}{d\theta}\right|_{\theta=0} = x_0. \tag{3.7-3b}$$

Equations (3.7-2b), (3.7-2c), and initial conditions given by Eq. (3.7-3) can be solved analytically using the Laplace transformation (Kriebel, 1961). The resulting particle trajectory can be expressed by:

$$x = x_0 e^{-\psi} \left\{ \left(\cosh(a\psi) + \left(\frac{a(1+2AB) + 2Ab}{c^2} \right) \sinh(a\psi) \right) \cos(b\psi) \right.$$
$$\left. + \left(\frac{b(1+2AB) - 2Aa}{c^2} \right) \cosh(a\psi) \sin(b\psi) \right\} \tag{3.7-4a}$$

$$y = x_0 e^{-\psi} \left\{ \left(\frac{a}{2} + \frac{a(1+4AB)}{2c^2} \right) \sin(b\psi) \cosh(a\psi) + \sinh(a\psi) \right.$$
$$\left. \times \sin(b\psi) + \left(\frac{b}{2} - \frac{b(1+4AB)}{2c^2} \right) \sinh(a\psi) \cos(b\psi) \right\}, \tag{3.7-4b}$$

where:

$$A = \omega\tau_S; \quad B = \frac{\dot{x}_0}{\omega x_0}; \quad \psi = \frac{\theta}{2A}$$

$$a^2 = \sqrt{\frac{1}{4} + 4A^2} + \frac{1}{2}; \quad b^2 = \sqrt{\frac{1}{4} + 4A^2} - \frac{1}{2}; \quad c^2 = \sqrt{1 + 16A^2}. \tag{3.7-4c}$$

Based on the solution (3.7-4), the trajectories for different values of A and B can be shown as spirals, which start from the injection point and end at the cylinder wall (Kriebel, 1961). The angular position of the particle lags behind the rotating fluid by an angle, ϕ, given by:

$$\varphi = \omega t - \tan^{-1}\left(\frac{y}{x}\right). \tag{3.7-4d}$$

For very small particles, $A \ll 1$. Eqs. (3.7-4a) and (3.7-4b) reduce to:

$$x \approx x_0 (1 + AB) e^{A\theta} \cos\left(\left(1 - 2A^2\right)\theta\right) \tag{3.7-4e}$$

$$y \approx x_0 (1 + AB) e^{A\theta} \sin\left(\left(1 - 2A^2\right)\theta\right). \tag{3.7-4f}$$

For large particles, $A \gg 1$. The particle trajectory is given by:

$$x \approx x_0 + Bx_0\theta \tag{3.7-4g}$$

$$y \approx x_0\theta. \tag{3.7-4h}$$

Comments: (1) The above solution ignores viscous boundary effects for coaxial cylindrical flow. For low rotational speeds, the viscous boundary layer effects should not be ignored. For the viscous laminar flow, the fluid velocity can be approximated by

$$U(r) = \frac{R_2^2\omega}{R_2^2 - R_1^2}\left(r - \frac{R_1^2}{r}\right), \tag{3.7-4i}$$

where R_1 and R_2 are the radii of inner and outer cylinders, respectively. At high rotating speeds, the flow may be turbulent. (2) Beyond the Stokes flow regime, the drag

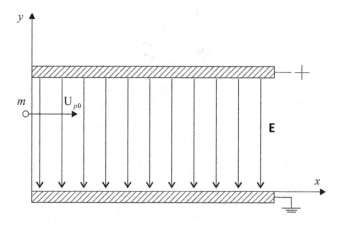

Figure 3.7-2 A charged particle in parallel-plate electrostatic precipitator.

force components cannot be decoupled from other velocity components because the drag force depends on the norm of relative velocity. Thus, the analytical methodology described may not be applicable to cases beyond the Stokes flow regime. (3) Due to the unsteady motion of particles and nonuniform flow field, established theory suggests there will be additional hydrodynamic forces (such as carried mass force, Basset force, and Saffman force) acting on the particle. The above discussion is valid only when the Stokes drag dominates other forces.

3.7.2 Motion of a Charged Particle between Parallel Electric Plates

Problem Statement: Consider a charged particle injected into a uniform electric field created by two parallel electric plates, as shown in Figure 3.7-2. The direction of the applied electric field is perpendicular to the direction of particle injection. The fluid in the electric field between the plates is stagnant. Develop the modeling equations for particle motion and particle trajectory. Develop a criterion for the collection of particles by the electric plates.

Analysis: Assume that particle motion is governed by the Coulomb force and Stokes drag only. The equation for the particle motion is then given by:

$$m\frac{d\mathbf{U}_p}{dt} = q\mathbf{E} - 3\pi\mu d_p\mathbf{U}_p, \tag{3.7-5}$$

where m is the particle mass and q is the electric charge carried by the particle. Equation (3.7-5) can be divided into component equations and expressed by:

$$m\frac{dU_p}{dt} = -3\pi\mu d_p U_p \tag{3.7-5a}$$

$$m\frac{dV_p}{dt} = qE - 3\pi\mu d_p V_p. \tag{3.7-5b}$$

The initial conditions are given by:

$$x|_{t=0} = 0; \qquad\qquad y|_{t=0} = y_0 \tag{3.7-6a}$$

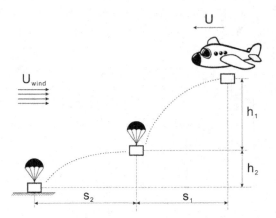

Figure 3.7-3 A parachuted object dropped from an airplane.

$$U_p\big|_{t=0} = \frac{dx}{dt}\bigg|_{t=0} = U_{p0}; \qquad V_p\big|_{t=0} = \frac{dy}{dt}\bigg|_{t=0} = 0. \qquad (3.7\text{-}6b)$$

The particle velocity can be obtained by integrating Eq. (3.7-5) with Eq. (3.7-6) to yield:

$$U_p = U_{p0}\,\exp\left(-\frac{t}{\tau_S}\right) \qquad (3.7\text{-}7a)$$

$$V_p = \frac{qE}{3\pi\mu d_p}\left(1 - \exp\left(-\frac{t}{\tau_S}\right)\right) = V_m\left(1 - \exp\left(-\frac{t}{\tau_S}\right)\right), \qquad (3.7\text{-}7b)$$

where V_m is the migration velocity, representing the terminal velocity of particle toward the electric plates, and is defined mathematically as:

$$V_m = \frac{qE}{3\pi\mu d_p}. \qquad (3.7\text{-}7c)$$

Comments: (1) In reality, the electric field likely becomes nonuniform near both ends of the parallel plates; (2) This model ignores the impact of charged particle(s) on the electrostatic field. The deposition of charged particles on the collecting plate may alter the electric field's intensity and distribution.

3.7.3   Motion of a Parachuted Object from an Airplane

Problem Statement: A packed parachute is dropped by an airplane moving horizontally only, as shown in Figure 3.7-3. After an initial unfettered decent, h_1, the parachute is opened at h_2. Assuming a constant opposing horizontal wind, develop a model to determine (i) the landing velocity of the box; (ii) the horizontal distance between initial drop and landing locations.

Analysis: Assume that motion of the box is governed only by the drag force and the gravitational force. Due to the small density ratio of air to that of the box, the carried mass force and Basset force may be neglected. Thus, the equation for the motion of the box is given by:

$$m\frac{d\mathbf{U}_p}{dt} = \frac{1}{2}C_D\rho A \left|\mathbf{U} - \mathbf{U}_p\right|(\mathbf{U} - \mathbf{U}_p) + m\mathbf{g},\qquad(3.7\text{-}8)$$

where A is the frontal area to the relative velocity $(\mathbf{U} - \mathbf{U}_p)$, and C_D is the drag coefficient that depends on the characteristic length, the shape of the box, and the particle Reynolds number.

Define Stage-1 as the interval between initial drop and just prior to opening the parachute. For simplicity, the box is approximated as a sphere during initial drop but before the parachute opens, so that C_D is given by Eq. (3.2-3). The horizontal and vertical components of Eq. (3.7-8) then provide, respectively:

$$m\frac{dU_b}{dt} = \frac{\pi}{8}C_{Db}\rho d_b^2 \left|\mathbf{U} - \mathbf{U}_b\right|(U - U_b)\qquad(3.7\text{-}8a)$$

$$m\frac{dV_b}{dt} = \frac{\pi}{8}C_{Db}\rho d_b^2 \left|\mathbf{U} - \mathbf{U}_b\right|(V - V_b) - mg,\qquad(3.7\text{-}8b)$$

where the subscript "b" stands for "box," and:

$$\left|\mathbf{U} - \mathbf{U}_b\right| = \sqrt{(U - U_b)^2 + (V - V_b)^2}.\qquad(3.7\text{-}8c)$$

Note that Eq. (3.7-8c) is also used to evaluate Re_b in the calculation for C_{Db}.

Four independent equations (Eq. (3.7-8a), Eq. (3.7-8b), Eq. (3.7-8c), and Eq. (3.2-3)) are now given for U_b, V_b, $|\mathbf{U}-\mathbf{U}_b|$ and C_{Db}. Thus, with the initial condition and wind information, the Stage-1 model is established and can be solved. The solution from the Stage-1 model provides the initial location and velocity conditions for the Stage-2 model.

Define Stage-2 as the interval between the parachute opening and landing. It is assumed that (1) after the opening of the parachute, the horizontal drag component, C_D, is given by Eq. (3.2-3) and substitutes the parachute diameter as the characteristic length; (2) the vertical drag component, C_D, is a constant C (typically a value between 0.75 and 1.40, a characteristic of the specific parachute used). Similar to the approach for the Stage-1 modeling, based on Eq. (3.7-8), the component equations of motion for Stage-2 in the horizontal and vertical directions are given, respectively, by:

$$m\frac{dU_p}{dt} = \frac{\pi}{8}C_{Dp}\rho d_p^2 \left|\mathbf{U} - \mathbf{U}_p\right|(U - U_p)\qquad(3.7\text{-}9a)$$

$$m\frac{dV_p}{dt} = \frac{\pi}{8}C\rho d_p^2 \left|\mathbf{U} - \mathbf{U}_p\right|(V - V_p) - mg,\qquad(3.7\text{-}9b)$$

where the subscript "p" stands for "parachute," and:

$$\left|\mathbf{U} - \mathbf{U}_p\right| = \sqrt{\left(U - U_p\right)^2 + \left(V - V_p\right)^2}.\qquad(3.7\text{-}9c)$$

As in Stage-1, Eq. (3.7-9c) is again used to evaluate Re_p in the calculation for C_{Dp}. Hence, the Stage-2 model is established that can be used to solve for U_p, V_p, $|\mathbf{U}-\mathbf{U}_p|$, and C_{Dp} using four coupled equations (Eq. (3.7-9a), Eq. (3.7-9b), Eq. (3.7-9c), and Eq. (3.2-3)).

Comments: (1) Here, a relatively crude model for A and C_D is adopted to describe the opened parachute, which may oversimplify the coupling effect between the horizontal and vertical motions; (2) Ignoring the added mass and Basset forces when the parachute opens (where a dramatic change in acceleration occurs) may decrease the mathematical model's accuracy.

3.7.4   Motion of an Evaporating Droplet

Problem Statement: A water droplet is suspended and evaporates in air. (1) Calculate the evaporation time of the droplet if there is no relative motion between the droplet and the air. (2) Develop a model to estimate the evaporation time if the droplet is settling at its terminal velocity in the air.

Analysis:

1) Assume that Lewis number $= 1$ and that the droplet undergoes a saturated evaporation. Since there is no relative motion between the droplet and the air, the droplet evaporates by diffusion only and obeys the d^2-law presented in Eq. (3.4-16a). Thus, the evaporation time is computed from Eq. (3.4-16b) as:

$$t_v = \frac{\rho_L c_p d_0^2}{8k \ln \left(1 + \frac{c_p(T_\infty - T_s)}{L} \right)}. \tag{3.7-10}$$

2) When a droplet is settling by gravity at its terminal velocity, the drag force balances the gravitational force, leading to:

$$\frac{\pi}{8} C_D \rho d_d^2 U_{dt}^2 = \frac{\pi}{6} g \rho_d d_d^3, \tag{3.7-11}$$

where C_D, determined by Eq. (3.2-3), is a function of the droplet Reynolds number at terminal velocity Re_{dt}. Re_{dt} is defined by:

$$Re_{dt} = \frac{\rho d_d U_{dt}}{\mu}. \tag{3.7-12}$$

Based on the saturated evaporation condition, dominated by heat transfer rate, it gives

$$\frac{d}{dt} \left(\frac{\pi}{6} \rho_d d_d^3 \right) = \frac{h \pi d_d^2 (T_\infty - T_s)}{L}, \tag{3.7-13}$$

where the convective heat transfer coefficient is estimated by the Ranz–Marshall correlation as:

$$\frac{h d_d}{k} = 2.0 + 0.6 \, Re_{dt}^{0.5} Pr^{1/3}. \tag{3.7-14}$$

Thus, the five unknowns (d_d, U_{dt}, h, C_D, and Re_{dt}) can be solved simultaneously using the five coupled equations (Eqs. 3.7-11 to 3.7-14; Eq. 3.2-3), yielding the time dependency parameter, d_d. Setting $d_d = 0$, corresponding to the time at the end of evaporation, and solving determines the process evaporation time.

Comments: (1) The equation for C_D, Eq. (3.2-3), was developed for a solid sphere. For an evaporating droplet, a formulation that accounts for internal fluid motion and blowing effects may be more suitable. However, both effects diminish when the transfer number, B, is very small. (2) The most significant velocity parameter in this case study was the relative velocity between the droplet and surrounding air. However, many real-world applications exist, such as a jet spray of water into air or a jet injection of liquid fuel in a combustor, where gravitational droplet settling velocity does not dominate the model.

3.8   Summary

This chapter provides basic formulation of various fluid–particle interactions of an isolated object (*e.g.*, a solid particle, a droplet, or a bubble) that has a relative motion in a fluid flow and in the absence of any interactions with other transported objects in the same fluid flow. Most essential formulation of these fluid–particle interactions is derived with the Newtonian fluid flowing over a rigid sphere and under the creeping flow conditions (*i.e.*, $Re_p \ll 1$), where the governing equations describing the flow around an object can be reasonably approximated to be linear so that it becomes possible to use a superposition of individually imposed fluid–particle interactions of simple motions to represent the fluid–particle interaction of a complex motion between the fluid and particle. This approximated method leads to the basic formulation of the Lagrangian modeling approach for the discrete phase transport in a multiphase flow.

The chapter describes the distinctly different transport mechanisms governing the fluid–particle interactions, their basic mathematical formulas, and the corresponding ranges of validation. The most essential interactions are represented by:

- drag force;
- carried mass (also known as added mass or virtual mass);
- Basset or history integral force;
- Saffman force, pressure gradient force, and radiometric forces;
- Magnus force;
- convective heat transfer;
- Stefan flux;
- d^2-law of diffusive evaporation

The application of the fluid–particle interaction formula to account for the corresponding governing equations (such as the BBO equation) for the transport of isolated objects in a carrying fluid flow is illustrated. The usefulness of the order-of-magnitude analysis of the transport mechanisms in modeling simplification is discussed.

Nomenclature

A	Frontal area
An	Acceleration number
$\boldsymbol{a}$	Acceleration
$\boldsymbol{B}$	Magnetic flux density
B	Transfer number
Bi	Biot number
C	Coefficient
c	Specific heat
D	Diffusivity
d	Diameter
$\boldsymbol{E}$	Electric field intensity
E	Activation energy
Eo	Eötvös number
$\boldsymbol{F}$	Force
Fo	Fourier number
g	Gravity acceleration
h	Specific enthalpy
	Convective heat transfer coefficient
J_q	Heat flux
J_m	Mass diffusion flux
K	Thermal conductivity
Kn	Knudsen number
L	Characteristic length
	Latent heat
Le	Lewis number
Ma	Mach number
m	Mass
$\dot{m}$	Mass flux
$\boldsymbol{n}$	Normal unit vector
n	Power-law index
Nu	Nusselt number
Pr	Prandtl number
p	Pressure
$\boldsymbol{p}_d$	Dipole moment
Q	Reaction heat
q	Electric charge
R	Radial coordinate in spherical system
R_g	Gas constant
Re	Reynolds number
Re_G	Shear Reynolds number
$\boldsymbol{r}$	Position vector
Sc	Schmidt number

Sh	Sherwood number
St	Strouhal number
S	Surface area
T	Temperature
t	Time
$\boldsymbol{u}$	Instant velocity
V	Volume
v	Diffusion velocity
w	Reaction rate
We	Weber number
Y	Mass fraction
y	Distance from wall

Greek symbols

γ	Specific heat ratio
δ	Kronecker delta
ε	Emissivity
	Permittivity
κ	Fluid consistency
	Permeability
μ	Dynamic viscosity
ρ	Density
σ	Surface tension
σ_b	Stefan–Boltzmann constant
τ	Viscous stress
τ	Relaxation time
φ	Sphericity
Ω	Angular velocity
ω	Curl of velocity

Subscript

0	Initial
∞	Ambient
A	Added or carried mass
B	Basset
b	Boundary surface
C	Collision
D	Drag
d	Droplet
E	Electric

f	Fluid
	Fuel
M	Magnus
m	Magnetic
n	Power-law fluid
P	Pressure
pf	Particle-fluid
p	Particle
r	Relative
S	Saffman
	Stokes
s	Species
T	Thermophoresis
v	Evaporation
w	Wall

Problems

P3.1 Consider an incompressible and viscous fluid flowing over a spherical particle in the creeping flow regime ($\mathrm{Re}_p \ll 1$). Derive the Stokes drag force.

P3.2 Estimate the size and terminal velocity of settling solid spheres of density 2,500 kg/m^3 in air and water for (a) $\mathrm{Re}_p = 200$ and (b) $\mathrm{Re}_p = 1$.

P3.3 Calculate the sphericity of a six-sphere agglomerate, which is constructed in as follows: Four glass spheres are placed on a flat surface and glued such that each sphere just makes point contact with two others. Another sphere of the same size is glued on top of the four spheres. These five spheres are then inverted and a sixth sphere of identical size is glued to the underside.

P3.4 Find the equations of carried mass for three different nonspherical objects such as elliptical, cubical, and cylindrical particles, and make a table comparison.

P3.5 A small sphere initially at rest is released into a fully developed pipe flow. Neglecting gravitational effect and given an initial location away from the centerline, describe qualitatively the trajectory of the sphere for the following cases:
(a) Flow is laminar;
(b) Flow is turbulent.

P3.6 Assume that the Stokes relaxation time can be used as a characteristic time in the evaluation of the importance of Basset history force compared to the drag force in accelerating motions of particles. Based on Eq. (3.2-22), estimate the ratio of Basset force to the drag force under the following conditions: (a) solids settling in gas, (b) solids settling in liquid, and (c) bubble rising in liquid. All are assumed in Stokes flow regime.

P3.7 In a welding application of plasma jet, the radial distribution of tempera-ture follows Gaussian distribution. Estimate the minimum injection velocity of a particle at a given injection location, needed to ensure reaching to the center of the plasma jet. Assume that the gravitational force and Basset force can be neglected.

P3.8 Establish a model to describe the trajectory of a soccer ball kicked into the air. Assume that the initial rotation vector of the ball is in the direction against gravity and will remain unchanged during the flying path of the ball. The air velocity is known.

P3.9 Make a contour map of particle Reynolds number at terminal velocity as a function of particle diameter (up to 5 mm) and the density difference ($\rho_p - \rho_f$) in the following fluid: (a) air; (b) water. Assume that the particle is a solid sphere with a density of 2,000 kg/m^3.

P3.10 For an agglomerate consisting of four spherical glass beads of 1 mm and density of 2,500 kg/m^3, determine the following equivalent diameters:
(a) Surface-equivalent diameter;
(b) Volume-equivalent diameter;
(c) Hydrodynamic-equivalent diameter in water (*i.e.*, having the same terminal velocity).

P3.11 A hail particle starts growing from a spherical agglomerate of ice crystals with initial size of 2 μm at the altitude of 2,000 m by falling through a 1,000 m thick cloud and capturing ice particles of average 1 μm in diameter. Derive the differential equation for particle velocity for the following cases, assuming the density of the hail particle does not change and there is no wind.
(a) The concentration in the cloud is 10^{10} particles per cubic centimeter (cc);
(b) The concentration of the cloud is 10^4 particles per cc.

P3.12 Consider a power-law liquid slowly moving around a sphere. Show that the drag force on the sphere can be represented by (Haider and Levenspiel, 1989):

$$F_D = 2\pi a^2 k \left(\frac{81}{4}\right)^{n-1/2} \left(\frac{V_\infty}{a}\right)^n \frac{(n^2 - n + 3)}{n^{3n}}. \tag{P3.12-1}$$

P3.13 Consider a rotating flow in a coaxial cylindrical chamber, with small particles injected radially from a slot on the inner cylinder into the flow. The outer cylinder is rotating at a constant angular velocity ω while the inner cylinder is stationary. Assume that the flow is laminar and the only driving force is the Stokes drag (*i.e.*, ignore other forces including gravitational force). Develop a model for predicting the particle trajectory.

P3.14 A suspended and charged particle is released at a centerline location into a fully developed laminar flow between two parallel electric plates where a uni-form electric field also exists. Assuming the Stokes drag as the only hydrodynamic force and ignoring the gravity effect, develop the equations of particle motion and a criterion for the collection of particles by the electric plates.

P3.15 Repeat P3.14, with driving force including the Stokes drag, carried mass, and Saffman forces but excluding gravitational force. The fluid in the electric field is turbulent and fully developed. Develop the equations of particle motion and particle trajectory. Develop a criterion for the collection of particles by the electric plates.

P3.16 In an aerial spray drift study, pesticide droplets with mean initial diameter of 300 μm and density of 10^3 kg/m^3 are released from a plane flying at the speed of 177 km/h and 3 m above the canopy. If the average evaporation rate of the droplet is 3×10^{-4} mg/s, estimate the liquid reside percentage when spray reaches the ground? What if the plane is at the height of 30 m above ground?

P3.17 Estimate the volume of an air bubble formed through a vertical nozzle immersed in water at low gas flow rate. The nozzle diameter is 1.3 mm. The air and water densities are 1.2 kg/m^3 and 1,000 kg/m^3, respectively. The air–water surface tension is 7.28×10^{-2} N/m.

P3.18 Consider an air bubble formed through a vertical nozzle immersed in water at high gas flow rate.
(1) Establish the force balance equations for the bubble;
(2) Obtain the expression for the bubble diameter;
(3) Based on (2), estimate the bubble size under the following conditions: the nozzle diameter is 1.3 mm. The air and water densities are 1.2 kg/m^3 and 1,000 kg/m^3, respectively. The air and water viscosities are 1.81×10^{-5} kg/m-s and 10^{-3} kg/m-s, respectively. The air–water surface tension is 7.28×10^{-2} N/m. The air flow rate is 6.75×10^{-6} m^3/s.

P3.19 Consider the heat transfer of a small solid sphere settling in a stagnant fluid. Using the Ranz–Marshall correlation between the settling velocity and convective heat transfer coefficient, develop a model for the solid–fluid heat transfer as a function of time.

P3.20 The initial growth of a spherical bubble in an incompressible viscous liquid is driven by the pressure difference inside the bubble and in the liquid. Derive the expression for the radial expansion of the bubble (Rayleigh–Plasset equation).

P3.21 In the later stage of bubble growth in a superheated liquid, the pressure difference between the bubble and the liquid becomes small, and bubble growth due to the evaporation on the vapor–liquid surface is controlled by the heat transfer from the liquid to the bubble. Assume that the inertia force and surface tension force are negligible.
(1) Develop an equation that dictates the change of the bubble radius;
(2) Assuming the heat transfer in the liquid film at the interface can be approximated using the Ranz–Marshall correlation, find out the relation between bubble radius and time.

References

Basset, A. B. (1888). *A Treatise on Hydrodynamics*, **2**. Cambridge: Deighton, Bell and Co.

Batchelor, G. K. (1967). *An Introduction to Fluid Dynamics*. London: Cambridge University Press.

Bergman, T. L., Lavine, A. S., Incropera, F. P., and Dewitt, D. (2011). *Introduction to Heat Transfer*, 6th ed., John Wiley & Sons, Inc.

Boussinesq. J. (1903). *Theorie Analytique de la Chaleur*, **2**. Paris: Gauthier-Villars.

Brenner, H. (1962). Effect of finite boundaries on the Stokes resistance of an arbitrary particle. *J. Fluid Mech.*, **12**, 35–48.

Brown, P. P., and Lawler, D. F. (2003). Sphere drag and settling velocity revisited. *J. Environ. Eng.*, **129**, 222–231.

Ceylan, K., Herdem, S., and Abbasov, T. (1999). A theoretical model for estimation of drag force in the flow of non-Newtonian fluids around spherical solid particles. *Powder Technol.*, **103**, 286–291.

Cheng, N.-S. (2009). Comparison of formulas for drag coefficient and settling velocity of spherical particles. *Powder Technol.*, **189**, 395–398.

Clift, R., Grace, J. R., and Weber, M. E. (1978). *Bubbles, Drops, and Particles*. New York: Academic Press.

Corrsin, S., and Lumley, J. (1956). On the equation of motion for a particle in turbulence fluid. *Appl. Sci. Res.*, **6A**, 114–116.

Dandy, D. S., and Dwyer, H. A. (1990). A sphere in shear flow at finite Reynolds number: effect of particle lift, drag and heat transfer. *J. Fluid Mech.*, **216**, 381–410.

Epstein, P. S. (1929). Zur theorie des radiometers. *ZS. f. Phys.*, **54**, 537–563.

Faeth, G. M. (1983). Evaporation and combustion of sprays. *Prog. Energy Combust. Sci.*, **9**, 1–76.

Fan, L. S., and Zhu, C. (1998). *Principles of Gas-Solid Flows*. Cambridge: Cambridge University Press.

Faxen, H. (1923). Die Bewegung einer starren Kugel längs der Achse eines mit zahrer flussigkeit gefullten rohres. *Arkiv Mat. Astron. Fys*, **17**, 1–28.

Gatignol, R. (1983). The Faxen formulas for a rigid particle in an unsteady non-uniform Stokes-flow. *J. de Mecanique Theorique et Appliquee*, **2**, 143–160.

Gel'Perin, N. I., and Einstein, V. G. (1971). Heat transfer in fluidized beds. In *Fluidization*. Ed. Davison and Harrison. New York: Academic Press.

Haider, A., and Levenspiel, O. (1989). Drag coefficient and terminal velocity of spherical and non-spherical particles. *Powder Technol.*, **58**, 68–70.

Hansell, D., Kennedy, I. M., and Kollmann, W. (1992). A simulation of particle dispersion in a turbulent jet. *Int. J. Multiphase Flow*, **18**, 559–576.

Henderson, C. B. (1976). Drag coefficient of spheres in continuum and rarefied flows. *AIAA J.*, **14**, 707–708.

Hettner, G. (1926). Zur theorie der photophorese. *ZS. f. Phys.*, **37**, 179–192.

Hughes, R. R., and Gilliland, E. R. (1952). The mechanics of drops. *Chem. Eng. Prog.*, **48**, 497–504.

Kim, J. H., Mulholland, G. W., Kukuck, S. R., and Pui, D. Y. H. (2005). Slip correction measurements of certified PSL nanoparticles using a nanometer differential mobility analyzer (Nano-DMA) for Knusen number from 0.5 to 83. *J. Res. Natl. Inst. Stand. Tech.*, **110**, 31–54.

Knudsen, M., and Weber, S. (1911). Air resistance for the slow motion of small spheres. *Ann Phys.* **36**. 981–994.

Kriebel, A. R. (1961). Particle trajectories in a gas centrifuge. *Trans. ASME J. Basic Eng.*, **83D**, 333–339.

Lapple, C. E., and Shepherd, C. B. (1940). Calculation of particle trajectories. *Ind. Eng. Chem.*, **32**, 605–617.

Loewenberg, M. (1993). Stokes resistance, added mass, and Basset force for arbitrarily oriented finite-length cylinders. *Phys. of Fluids A*, **5**, 765–767.

Looker, J. R., and Carnie, S. L. (2004). The hydrodynamics of an oscillating porous sphere. *Phys. of Fluids*, **16**, 62–72.

Loth, E. (2008). Quasi-steady shape and drag of deformable bubbles and drops. *Int. J. Multiphase Flow*, **34**, 523–546.

Lovalenti, P. M., and Brady, J. F. (1993). The hydrodynamic force on a rigid particle undergoing arbitrary time-dependent motion at small Reynolds number. *J. Fluid Mech.*, **256**, 561–605.

Magnus, G. (1852). Über die Abweichung der Geschosse. *Abhandlungen der Königlichen.* Akademie der Wissenschaften zu Berlin, 1–23.

Mandø, M., and Rosendahl, L. (2010). On the motion of non-spherical particles at high Reynolds number. *Powder Technol.*, **202**, 1–13.

Maxey, M. R., and Riley, J. J. (1983). Equation of motion for a small rigid sphere in a nonuniform flow. *Phys. Fluids*, **26**, 883–890.

McLaugnlin, J. B. (1991). Inertial migration of small sphere in linear shear flows. *J. Fluid Mech.*, **224**, 261–274.

Mei, R. (1992). An approximate expression for the shear lift on a spherical paricle at finite Reynolds number. *Int. J. Multiphase Flow*, **18**, 145–147.

Mei, R., and Adrian, R. J. (1992). Flow past a sphere with an oscillation in the free-stream and unsteady drag at finite Reynolds number. *J. Fluid Mech.,* **237**, 323–341.

Mei, R., Klausner, J., and Lawrence, C. (1994). A note on the history force on a spherical bubble at finite Reynolds number. *Phys. Fluids*, **6**, 418–420.

Michaelides, E. E., and Feng, Z.-G. (1995). The equation of motion of a small viscous sphere in an unsteady flow with interface slip. *Int. J. Multiphase Flow*, **21**, 315–321.

Moore, D. W. (1965). The velocity rise of distorted gas bubbles in a liquid of small viscosity. *J. Fluid Mech.*, **23**, 749–766.

Neale, G., Epstein, N., and Nader, W. (1973). Creeping flow relative to permeable spheres. *Chem. Eng. Sci.*, **28**, 1865–1874.

Odar, F., and Hamilton, W. S. (1964). Forces on a sphere accelerating in a viscous fluid. *J. Fluid Mech.*, **18**, 302–314.

Oseen, C. W. (1927). *Hydrodynamik.* Leipzig: Akademische Verlagsgescellschafe.

Pamme, N. (2006). Magnetism and microfluidics. *Lab Chip*, **6**, 24–38.

Pethig, R. (2010). Review Article – Dielectrophoresis: Status of the theory, technology, and applications. *Biomicrofluics*, **4**, 022811.

Ranz, W. E., and Marshall, W. R. Jr. (1952). Evaporation from drops, Part I and Part II. *Chem. Eng. Prog.*, **48**, 141–146 (Part I); 173–180 (Part II).

Rubinow, S. I., and Keller, J. B. (1961). The transverse force on a spinning sphere moving in a viscous fluid. *J. Fluid Mech.*, **11**, 447–459.

Rudinger, G. (1980). *Fundamentals of Gas-Particle Flow.* Amsterdam: Elsevier Scientific.

Saffman, P. G. (1965). The lift on a small sphere in a slow shear flow. *J. Fluid Mech.*, **22**, 385–400.

Saffman, P. G. (1968). Corrigendum. *J. Fluid Mech.*, **31**, 624.

Schiller, L., and Naumann, A. Z. (1933). Über die grundlegenden Berechungen bei der Schwerkraftaufbereitung. *Ver. Deut. Ing.*, **77**, 318–320.

Schlichting, H. (1979). *Boundary Layer Theory*, 7th ed. New York: McGraw-Hill.

Shukman, Z. P. (1975). *Convection Heat and Mass Transfer in Rheological Liquids* (in Russian). Moscow: Energia.

Sirignano, W. A. (2010). *Fluid Dynamics and Transport of Droplets and Sprays*. 2nd ed. Cambridge, UK: Cambridge University Press.

Soo, S. L. (1990). *Multiphase Fluid Dynamics*. Beijing: Science Press.

Vanni, M. (2000). Creeping flow over spherical permeable aggregates. *Chem. Eng. Science*, **55**, 685–698.

Veldhuis, C. H. J., Biesheuvel, A., and van Wijngaarden, L. (2008). Shape oscillations on bubbles rising in clean and in tap water. *Phys. Fluids*, **20**, 040705.

Vogel, S. (1994). *Life in Moving Fluids: The Physical Biology of Flow*. Princeton: Princeton University Press.

Wang, X.-B., Huang, Y., Becker, F. F., and Gascoyne, P. R. C. (1994). A unified theory of dielectrophoresis and travelling wave dielectrophoresis, *J. Phys. D Appl. Phys.*, **27**, 1571–1574.

Zhang, J. P., Fan, L. S., Zhu, C., Pfeffer, R., and Qi, D. (1999). Dynamic behavior of collision of elastic spheres in viscous fluids. *Powder Technol.*, **106**, 98–109.

Zhou, L. (1993). *Theory and Numerical Modeling of Turbulent Gas-Particle Flows and Combustion*. Beijing: Science Press and CRC Press.

4 Interactions of Particles, Droplets, and Bubbles

4.1 Introduction

The discrete phase of a multiphase flow typically involves a large number of solid or fluid particles. In such a case, a particle's dynamic behavior may be influenced by particle–particle interactions and will be modified from that of an isolated particle, as described in Chapter 3. Particles can interact with one another through many contact or noncontact mechanisms in a multiphase flow. Noncontact interactions can occur in two ways: the hydrodynamic interactions through the changes in continuous phase flow, or interactions via other non-hydrodynamic means. Hydrodynamic interactions include four subcategories: (1) through local flow deviation caused by an adjacent particle, (2) wake from a leading particle to the trailing particles, (3) a change of flow passage created by volume fraction variations of neighboring particles in a confined flowing system, and (4) by a pressure wave propagating through, and often colliding with, suspended particles, such as sound wave alternation in a particle–fluid flow (Soo, 1990). Non-hydrodynamic interactions are exemplified by the electrostatic forces among many charged particles in a confined flow and the thermal irradiation of suspended particles in a furnace. The direct contact interaction, typically by way of collision, may lead to the elastic or plastic deformation of colliding particles in addition to the expected interfacial transfers of mass, momentum, and energy. Collisions may also cause the change of particle size distribution via particle regrouping. The collision-induced particle breakup can be found in the case with solid particle attrition or bubble/droplet splitting. The combination or merger of particles occurs in the examples of agglomeration of solid particles, adhesion of fine powders, or coalescence of bubbles or droplets.

As long as the particle interactions play a significant role in dictating the dynamic transport of a multiphase flow, the flow is categorized as dense phase transport. When the particle interactions is the dominant mechanism of particle transport, the flow is classified as collision-dominated dense phase transport. In this chapter, various types of noncontact and collision-induced particle–particle interactions are introduced. Sub-models, chiefly based on the controlling physical mechanisms of each interaction mode, are established. These sub-models are the basis for the integrated theories and modeling concepts that will be introduced in Chapters 5 and 6.

4.2   Transport Properties of a Cloud of Particles

This section discusses noncontact particle–particle interactions in the absence of direct interparticle contact or collisions, including those via the carrying fluid and via external or self-induced fields (*e.g.*, thermal radiation and electrostatic charges). Such noncontact particle–particle interactions on the particle transport lead to the modification or reformulation of drag forces, heat transfer coefficients, thermal radiation intensity, and electrostatic forces.

4.2.1   Hydrodynamic Forces of a Pair of Spheres

Describing the flow interactions between a particle of interest and its adjacent surrounding particles necessitates modification of the drag and other hydrodynamic forces. Correction factors are introduced to account for the deviation of the flow field from the isolated particle scenario. In general, this correction factor is a function of the particle Reynolds number, the ratio of particle to fluid density, as well as the separation distance and orientation between the pair of particles.

The most significant interaction between a pair of particles is represented in cases where a particle approaches another head-on or is located in its wake. Two identical spheres approaching head-on can be modeled as a symmetric problem where the particles have trajectories normal to the plane of symmetry, as shown in Figure 4.2-1. For particles in the creeping flow regime and when the particle diameter is much smaller than the distance from the symmetric plan (*i.e.*, $d_p << y$), the drag coefficient of a sphere moving normal to the plane of symmetry can be estimated to the first order as (Brenner, 1962):

$$C_D = \frac{24}{Re_p}\left(1 + \frac{1}{2}\frac{d_p}{y}\right). \tag{4.2-1}$$

As two identical spheres approach head-on in a colinear fashion, the particles decelerate as the fluid pressure increases. Additional hydrodynamic forces, such as the added mass force and fluid pressure force, will influence the dynamic head-on approach and resulting collision. An approximated analysis of the added mass and the fluid pressure force can be obtained from the related case of a sphere moving normally toward a wall in an inviscid fluid. It is noted that without viscous energy dissipation, the total kinetic energy of the approaching spheres and the inviscid liquid is conserved, which can be expressed as (Milne-Thomson, 1968):

$$\frac{1}{2}mU^2 + \frac{1}{4}m'\left(1 + \frac{3}{64}\frac{d_p^3}{y^3}\right)U^2 = \text{constant}, \tag{4.2-2}$$

where m is the mass of sphere, m' is the mass of liquid displaced by the sphere, and y is the distance between the center of a sphere and the symmetrical plane of the two spheres. The second term on the left-hand side (LHS) of the equation is obtained by

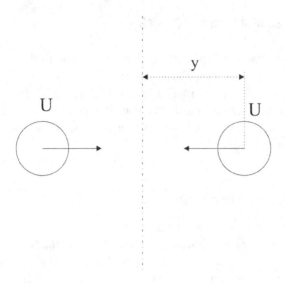

Figure 4.2-1 Head-on approach of two identical spheres.

integrating the kinetic energy of the fluid over the entire fluid phase. Note that:

$$U = -\frac{dy}{dt}; \qquad \frac{dU}{dt} = -U\frac{dU}{dy}. \tag{4.2-2a}$$

The equation of motion of the sphere moving toward the symmetric plane can be obtained by differentiating Eq. (4.2-2) with respect to t, as:

$$\left(m + \frac{1}{2}m'\left(1 + \frac{3}{64}\frac{d_p^3}{y^3}\right)\right)\frac{dU}{dt} = -\frac{9}{256}\frac{m'd_p^3}{y^4}U^2. \tag{4.2-3}$$

Hence, the added mass is obtained by:

$$m_A = \frac{1}{2}m'\left(1 + \frac{3}{64}\frac{d_p^3}{y^3}\right). \tag{4.2-3a}$$

The pressure force from the fluid due to the sphere approaching the symmetric plane is obtained by:

$$F_P = \frac{9}{256}\frac{m'd_p^3}{y^4}U^2. \tag{4.2-3b}$$

Equations (4.2-1), (4.2-3a), and (4.2-3b) are regarded as the first-order estimations of the hydrodynamic force corrections for the head-on approach of a pair of particles in the creeping flow regime. In a rebounding process (occurring after a collision or in the case of a colinear separating process of a pair of identical particles), the fluid pressure force will be negative, and it will tend to resist the separation. The above model for a spherical particle approaching the symmetric plane may be extended to capture the hydrodynamic behavior of a spherical particle approaching a wall.

However, such an extension can only be an approximation because the tangential velocity along the symmetric plane is not restricted by viscous stagnation, as it should be for the wall.

The wake-induced particle interactions have prominent influence on the drag forces of particles. For a pair of near-contact spheres of the same size moving in the same direction, the drag force of the trailing sphere can be less than 20% of the drag force of an isolated particle, but the drag force of the leading sphere decreases moderately, typically by only a few percentage points (Chiang and Sirignano, 1993; Liang et al., 1996). The drag coefficient of a trailing particle, under the wake influence of a leading particle, may be expressed by an empirical correlation as (Zhu et al., 1994):

$$\frac{C_D}{C_{D0}} = 1 - (1 - A)\exp\left(-B\frac{l}{d_p}\right), \tag{4.2-4}$$

where C_{D0} is the drag coefficient from the standard curve shown in Figure 3.2-3; l is the distance between the two interacting particles; A and B are empirical coefficients that can be a function of Re_p. For example, when $20 < Re_p < 150$, it yields:

$$
\begin{aligned}
A &= 1 - \exp(-0.483 + 3.45 \times 10^{-3}Re_p - 1.07 \times 10^{-5}Re_p^2) \\
B &= 0.115 + 8.75 \times 10^{-4}Re_p - 10^{-7}Re_p^2.
\end{aligned}
\tag{4.2-4a}
$$

Equation (4.2-4) shows that the wake-induced interaction becomes significant when the interrelating distance is within about two particle diameters. Based on the cubic-structure model for a multiparticle suspension, the critical volume fraction of spheres can be roughly estimated to be 2%; beyond this value, a correction on the drag coefficient should be utilized.

4.2.2   Hydrodynamic Forces on a Sphere in a Swamp of Spheres

For a creeping flow through an unbound and uniformly distributed suspension of spheres, the drag force on each individual sphere can be estimated via the Happel's free-surface cell model (Happel, 1958), which assumes the surrounding fluid over each individual sphere is equivalent to a spherical shell engulfing the sphere; a uniform flow velocity is defined at the shell's outer boundary, as shown in Figure 4.2-2.

The drag force is obtained as:

$$F_D = 3\pi \mu d_p U\Omega, \tag{4.2-5}$$

where U is the superficial fluid velocity flowing through the swarm; Ω is the drag force correction factor due to the effect of neighboring spheres, which is given by:

$$\Omega = \frac{2 + \frac{4}{3}\eta^5}{2 - 3\eta + 3\eta^5 - 2\eta^6}, \tag{4.2-5a}$$

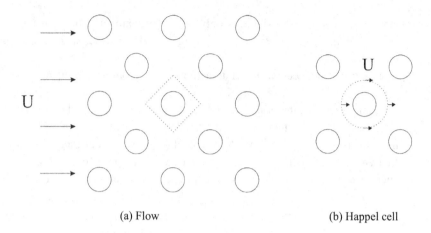

(a) Flow   (b) Happel cell

Figure 4.2-2 Happel's free-surface cell model.

where η is the ratio of diameter of the sphere to the outer diameter of the spherical shell, which can be linked to the suspension voidage or fluid volume fraction, α, through:

$$\eta = \sqrt[3]{1-\alpha}. \tag{4.2-5b}$$

For solid spheres that are porous and permeable, Eq. (4.2-5a) must be modified to account for the permeability of spheres (Neale *et al.*, 1973). It then yields:

$$\Omega = \frac{2\beta^2 + \frac{4}{3}\beta^2\eta^5 + 20\eta^5 - \frac{\tanh \beta}{\beta}(2\beta^2 + 8\beta^2\eta^5 + 20\eta^5)}{\left(\begin{array}{c} 2\beta^2 - 3\beta^2\eta + 3\beta^2\eta^5 - 2\beta^2\eta^6 + 90\beta^{-2}\eta^5 + 42\eta^5 - 30\eta^6 + 3 \\ -\frac{\tanh \beta}{\beta}(-3\beta^2\eta + 15\beta^2\eta^5 - 12\beta^2\eta^6 + 90\beta^{-2}\eta^5 + 72\eta^5 - 30\eta^6 + 3) \end{array}\right)}. \tag{4.2-5c}$$

Here, β is a parameter defined by:

$$\beta = \frac{d_p}{2\sqrt{k}}, \tag{4.2-5d}$$

where κ is the permeability of single permeable sphere. For very large β, or very low κ, the hydrodynamic behavior of a permeable sphere existing in a swarm of permeable spheres is similar to that of a solid sphere in a swarm of solid spheres; therefore, as $\beta \to \infty$, Eq. (4.2-5c) reduces to Eq. (4.2-5a), which is Happel's formula for a solid sphere in a swarm of solid spheres.

Movement of suspended particles in a confined domain will induce flow of the surrounding fluid due to the particle occupancy and confinement by the walls. The hydrodynamic behavior of suspended particles in a confined chamber can be gleaned from the particle settling experiments (Richardson and Zaki, 1954), resulting in the following expression for the terminal settling velocity:

$$\frac{U_{pt}}{U_{pt0}} = \alpha^n. \tag{4.2-6}$$

The Richardson and Zaki index, n, is an empirical parameter that generally depends on the Archimedes number and ratio of particle to column diameter, as evidenced by:

$$\frac{4.64 - n}{n - 2.4} = 0.043 \, \text{Ar}^{0.57} \left(1 - 2.4 \left(\frac{d_p}{D} \right)^{0.27} \right),$$

(4.2-6a)

where the Archimedes number is defined by:

$$\text{Ar} = \frac{d_p^3 \rho (\rho_p - \rho) g}{\mu^2}.$$

(4.2-6b)

For motions analogous to a swarm of bubbles, a correction to the added mass force of an isolated bubble is necessary. The correction factor, Ω_A, to the added mass force can be estimated by (Zuber, 1964):

$$\Omega_A = \frac{3 - 2\alpha}{\alpha}.$$

(4.2-7)

4.2.3   Heat Transfer of Suspended Particles

For heat conduction between a stagnant fluid and an unbound suspension of uniformly distributed solid spheres, the heat conducted from the surrounding fluid to each solid sphere can be classified as heat conduction through a spherical shell of fluid with an inner diameter, d_p, and an outer diameter, d_f. It can be shown that:

$$\text{Nu}_p = \frac{2}{1 - \frac{d_p}{d_f}} = \frac{2}{1 - \sqrt[3]{1 - \alpha}}.$$

(4.2-8)

Equation (4.2-8) can be reduced to Eq. (3.3-5e) as d_f becomes infinitely large. For a packed bed of monodispersed spheres arranged in a cubic particle array, the bed voidage is about 0.476, leading to $\text{Nu}_p \approx 10$; this result indicates that the Nu_p of a single particle, which is due to thermal conduction in a particle suspension, varies roughly from 2 to 10.

For heat convection of a particle suspension in pipe flow, one simple model is the pseudo-continuum one-phase flow model, which assumes (1) local thermal equilibrium exists between the two phases; (2) particles are evenly distributed; (3) flow is uniform; and (4) heat conduction is dominant in the cross-stream direction. Therefore, the heat balance leads to a single-phase energy equation that is based on effective fluid–particle properties, averaged temperatures, and averaged velocities. For an axisymmetric flow heated by a cylindrical heating surface at T_w, the heat balance equation can be written as:

$$\rho_m c_{pe} U \frac{\partial T}{\partial z} = \frac{K_e}{r} \frac{\partial}{\partial r} \left(r \frac{\partial T}{\partial r} \right),$$

(4.2-9)

with boundary conditions of:

$$T|_{r=R} = T_w; \quad \left. \frac{\partial T}{\partial r} \right|_{r=0} = 0; \quad T|_{z=0} = T_i,$$

(4.2-9a)

where K_e is the effective thermal conductivity; U is the superficial relative velocity of fluid. The temperature distribution is solved from:

$$T = T_w + 2(T_i - T_w) \sum_{n=1}^{\infty} \frac{J_0(\lambda_n r)}{\lambda_n R J_1(\lambda_n R)} \exp\left(-\frac{K_e \lambda_n^2 z}{U \rho_m c_{pe}}\right), \tag{4.2-9b}$$

where J_0 and J_1 are the zeroth and first-order Bessel functions of the first kind, respectively; λ_n's are the eigenvalues, obtained from the eigen equation:

$$J_0(\lambda_n R) = 0. \tag{4.2-9c}$$

The heat transfer coefficient is then obtained as:

$$h = \frac{2K_e}{R} \sum_{n=1}^{\infty} \exp\left(-\frac{K_e \lambda_n^2 z}{U \rho_m c_{pe}}\right). \tag{4.2-9d}$$

The length-averaged Nusselt number is computed from:

$$\mathrm{Nu} = \frac{\overline{h}L}{K_e} = \frac{2U \rho_m c_{pe}}{R K_e} \sum_{n=1}^{\infty} \lambda_n^{-2} \left\{ 1 - \exp\left(-\frac{K_e \lambda_n^2 L}{U \rho_m c_{pe}}\right)\right\}. \tag{4.2-9e}$$

The applicability of the preceding pseudo-continuum approach to convective heat transfer of fluid–particle systems without heat sources depends on the reasonableness of a local thermal equilibrium assumption and the validity of a phase continuum approximation. Local thermal equilibrium may be assumed only if the particle-heating surface residence time, τ_{ps}, is much greater than the particle thermal diffusion time, τ_{pd}. Here, τ_{ps} and τ_{pd} may be estimated by:

$$\tau_{pd} = \frac{d_p^2}{D_T} \qquad \tau_{ps} = \frac{L}{U} \tag{4.2-10}$$

Furthermore, a modified Fourier number can be defined as:

$$\mathrm{Fo}_m = \frac{\tau_{ps}}{\tau_{pd}} = \frac{L D_T}{U d_p^2}. \tag{4.2-10a}$$

A study of heat transfer in gas–solid flows suggests that the pseudo-continuum model is appropriate when $\mathrm{Fo}_m > 1$ and $\alpha < 0.9$ (Hunt, 1989). For cases where the heat transfer between particles and local fluid must be considered, a more sophisticated model can be developed based on a combined approach of Eq. (4.2-8) and Eq. (4.2-9).

Thermal radiation with a particle cloud may be conveniently studied using a simple model in which the energy is radiated from one heating surface to another through a uniform particulate medium. The particulate medium is treated as a pseudo-continuum where particles absorb, emit, and scatter the radiative heat fluxes. Here, scattering represents the combined effects of reflection, refraction, diffraction, and transmission of thermal radiation by the particles. Thermal radiation in the particle-carrying fluid is either transparent to the radiation (such as in gas–solid flows) or combined with the particulate medium (such as in solid–liquid flows).

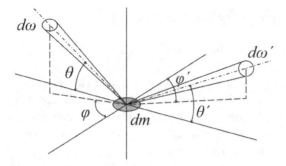

Figure 4.2-3 Coordinate system for scattering function.

Radiation in a dilute suspension is dominated by single scattering, such as when a single ray of radiative beam traversing a medium is scattered only once before leaving the medium. In this case, the scattered intensity of the suspension is equal to the scattered intensity from a single particle multiplied by the total number of particles. However, with radiation in a dense suspension, multiple scattering becomes significant. The classification between single and multiple scattering in a particulate medium may be based on the value of the characteristic optical path length, L_s. An accepted criterion for single scattering was suggested by Bayvel and Jones (1981) as:

$$L_s = \int_0^L \sigma_s \, dl < 0.1 \tag{4.2-11}$$

where L is the characteristic length of the system, and σ_s is the scattering coefficient. Radiation dependency on neighboring particles is another concern that complicates the radiation transport equations. Independent scattering refers to scattering from a single particle in a cloud that is not affected by the neighboring particles, whereas dependent scattering refers to scattering that is affected by the neighboring particles (Tien and Drolen, 1987). In this text, all scattering is assumed to be independent for the sake of simplicity.

Radiation through a particle medium, in general, is governed by the *Radiation Transport Equation*. Consider the energy balance of radiant flux through a scattering medium of monodispersed spherical particles where absorption, emission, and scattering by spherical particles are included. As shown in Figure 4.2-3, the equation for the change in monochromatic radiant intensity, with respect to distance, along a pencil of rays at s, θ, ϕ can be expressed by (Love, 1968):

$$\frac{dI(s,\theta,\varphi)}{ds} = -\sigma_{e\lambda} I(s,\theta,\varphi) + \sigma_{a\lambda} I_{b\lambda}(s)$$

$$+ \frac{\sigma_{s\lambda}}{4\pi} \int_0^{2\pi} \int_0^{\pi} I(s,\theta,\varphi) S\theta,\varphi,\theta',\varphi' \sin\theta' \, d\theta' \, d\varphi', \tag{4.2-12}$$

where I represents the local radiant intensity; S is the scattering phase function; $\sigma_{e\lambda}$, $\sigma_{s\lambda}$, and $\sigma_{a\lambda}$ are the monochromatic coefficients for extinction, scattering, and absorption, respectively; $I_{b\lambda}$ is the Planck intensity function, expressed by:

$$I_{b\lambda} = \frac{C_1 \lambda^{-5}}{e^{C_2/\lambda T} - 1},\qquad(4.2\text{-}12a)$$

where λ is wavelength (μm); T is absolute temperature (K); $C_1 = 3.743 \times 10^8$ (W$\cdot\mu$m^4/m^2); and $C_2 = 1.4387 \times 10^4$ μm$\cdot$K. On the right-hand side of Eq. (4.2-12), the first term represents the extinction of the ray in accordance with the Bourger–Beer relation; the second term is the increase in intensity due to thermal-energy emission from the particles; and the last term is the energy scattered into (θ, φ) from rays traversing in all directions. On the basis of Kirchhoff's law, the monochromatic emission coefficient is assumed to be equal to the absorption coefficient. The scattering function S is defined in such a way that the expression:

$$\sigma_{s\lambda} I(s, \theta', \varphi') S(\theta, \varphi, \theta', \varphi') \frac{d\omega'}{4\pi} df\, dm\, d\omega\qquad(4.2\text{-}12b)$$

represents the rate at which radiant energy is scattered by the differential element of mass, dm, from the pencil of ray enclosed in the solid angle, $d\omega'$, having an intensity, $I(s, \theta', \varphi')$, within the frequency range f and $f + df$ into the solid angle, $d\omega$, characterized by θ and φ. Note that the scattering function, S, satisfies the condition:

$$\frac{1}{4\pi} \int_0^{4\pi} S(\theta, \varphi, \theta', \varphi') d\omega = 1.\qquad(4.2\text{-}12c)$$

In general, solving for transmitted and reflected energy from a multiple scattering medium requires a numerical solution of the integral-differential transport equation, Eq. (4.2-12). In the case of single scattering, the equation is greatly simplified because the last term in Eq. (4.2-12) vanishes.

4.2.4   Mass Transfer of a Cluster

Mass transfer in a cluster of particles exhibits distinct behavior from that of an isolated particle. For example, the evaporation and combustion rates of fuel droplets in the dense region of a spray are affected by their collective interactions. In particular, many studies show that evaporation (or burning rate) decreases monotonically with decreasing droplet spacing, and current theories for dilute spray evaporation considerably overestimate the evaporation rate in a concentrated cluster.

The effect of droplet interaction on the mass transfer can be investigated using a simplified model, as shown in Figure 4.2-4, that employs uniformly distributed monodisperse droplets within a spherical cluster. Each individual droplet is assumed to be surrounded by a fictitious sphere of influence. Thus, the cluster is an ensemble of closely packed spheres. For a droplet, energy and mass conservation are applied to calculate its evaporation rate at the droplet surface. Two sets of equations are solved for the gas phase: the first set describes evaporation within the sphere of influence; the other set applies to the space outside the sphere of influence, but still within the overall cluster. The liquid phase and different gas phase regions are coupled at their interfaces. When the cluster is exposed to convective flow, a portion of the

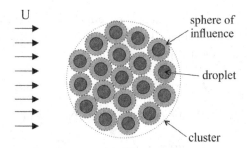

U

sphere of influence

droplet

cluster

Figure 4.2-4 Physical picture of a cluster of droplets evaporating in a convective flow.

flow penetrates the cluster as if percolating through a porous medium. Since it is impractical to ascertain exterior and internal flow pattern specifics, only the most relevant aspects of the interactions between the cluster and the surrounding flow can be considered; this includes the extent of flow penetration into the cluster, the relative velocity between the cluster and the flow, and the relationship between the evaporation rate with and without convection (Bellan and Harstad, 1987a).

This model demonstrates that the mass transfer mechanism in very dense clusters is controlled by diffusion; in very dilute clusters, convection dominates. Between these extremes, there exists a convection–diffusion controlled regime where both phenomena are simultaneously necessary to determine the rate of evaporation. Generally, the control parameters for fuel droplet evaporation in dense clusters are in the following order of importance (1) the initial droplet temperature, (2) the initial gas temperature, (3) the initial fuel concentration in the gas, and (3) the relative velocity between the cluster and the gas. In contrast, for dilute clusters the order of importance for different control parameters is (1) the initial temperature of surrounding gas, (2) the relative velocity between the cluster and the gas, (3) the initial droplet temperature, and (4) the initial fuel concentration in gas phase (Bellan and Harstad, 1987b).

4.2.5   Charge Effect due to Interparticle and Particle–Wall Interactions

For two charged particles in an infinite vacuum, the interactive electrostatic force can be described by Coulomb's law as:

$$F_e = -C_e \frac{q_{e1} q_{e2}}{r^2}, \qquad (4.2\text{-}13)$$

where C_e is the proportionality constant, also known as the Coulomb's law constant; q_e is the quantity of charge on particle; and r is the distance of separation between the two charged particles. Equation (4.2-13) can be regarded as a special case of Eq. (3.2-29) where a charged particle is situated in the electric field induced by another charged particle. When there are many charged particles suspended in a fluid medium, assuming that there is no electrification or polarization for the fluid by the presence of particle charges, the electric field is governed by the space distribution of electrostatic charges. The electrostatic field intensity, E, can be conveniently defined in terms of the gradient of electric potential, Φ, as:

$$E = -\nabla\Phi. \tag{4.2-14}$$

Thus, the governing equation of the electric potential can be expressed by:

$$\nabla \cdot (\varepsilon\nabla\Phi) = -\rho_e, \tag{4.2-15}$$

where ρ_e is electrostatic charge density; ε is electromagnetic permeability. The typical boundary conditions for the electric potential distribution are:

(i) Wall at a given voltage (or electric potential distribution)

$$\Phi = \text{constant} \tag{4.2-15a}$$

(ii) Wall with a given surface charge density distribution

$$\nabla\Phi \cdot \mathbf{n} = -\frac{\sigma_e}{\varepsilon}, \tag{4.2-15b}$$

where σ_e is the surface charge density. It is noted that ρ_e depends on the space distribution of particles, whereas σ_e depends on the surface deposition of charged particles and surface electrification or discharge. Hence, both ρ_e and σ_e must be obtained from the particle trajectories in the multiphase flow transport.

The above discussion clearly shows that the transport and surface depositions of charged powders in flow chambers are complicated by the coupling effects of flow hydrodynamics, particulate dynamics, and external and/or charge-induced internal electrostatics. Although the effects of electrostatics and hydrodynamics on particle movement can be superposed, coupling effect between particulate flow dynamics and charge-induced electrostatics must be considered, often requiring an iterative computation between the two fields until achieving solution convergence of residual quantities.

4.3   Collision of a Pair of Solid Spheres

During a collision, contacting solids undergo both elastic and inelastic (or plastic) deformations. These deformations are accompanied by changes of stresses and strains; the magnitudes and gradients of which are functions of the material properties of the solids and the applied external forces. Theories for predicting the elastic deformations of two elastic bodies in contact are introduced in this section, including Hertzian theory for frictionless contact and Mindlin's approach for frictional contact. A few contested theories currently exist for inelastic deformation, but these are primarily extensions of elastic contact theory.

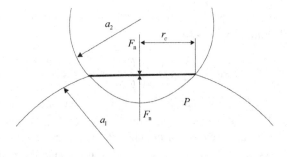

Figure 4.3-1 Hertzian contact.

4.3.1 Hertzian Contact of Frictionless Spheres

Hertzian contact refers to the contact between two elastic and frictionless spherical bodies under compression, a case first investigated by Hertz in 1881. Consider two frictionless spheres in contact, as shown in Fig. 4.3-1. When r/a_1 and r/a_2 are sufficiently small, based on the theory of elastic contact of solids, it gives (Fan and Zhu, 1998):

$$\frac{1}{\pi E^*} \iint P dA = s - \frac{r^2}{2a^*}, \tag{4.3-1}$$

where s is an approaching distance between the two centers of the spheres, E^* is the contact modulus, and a^* is the relative radius. E^* and a^* are defined as:

$$\frac{1}{E^*} = \frac{1-v_1^2}{E_1} + \frac{1-v_2^2}{E_2} \qquad \frac{1}{a^*} = \frac{1}{a_1} + \frac{1}{a_2}. \tag{4.3-1a}$$

The pressure distribution that satisfies Eq. (4.3-1) for any r can be expressed as:

$$P = \frac{3}{2} \frac{F_n}{\pi r_c^3} \sqrt{r_c^2 - r^2} \tag{4.3-1b}$$

where F_n is the normal loading force, and r_c is the contact radius. Substituting Eq. (4.3-1b) into Eq. (4.3-1) yields:

$$\frac{3}{8} \frac{F_n}{E^* r_c^3} \left(2r_c^2 - r^2\right) = s - \frac{r^2}{2a^*} \tag{4.3-1c}$$

Since Eq. (4.3-1c) is valid for any small r, the corresponding terms on both sides of the equation should be equal. This rationale advances two important expressions for the radius of the surface of contact, r_c, and the centers approaching distance, s. These are, respectively, given by:

$$r_c = \sqrt[3]{\frac{3}{4} F_n \frac{a^*}{E^*}} \tag{4.3-1d}$$

$$s = \sqrt[3]{\left(\frac{3}{4} \frac{F_n}{E^*}\right)^2 \frac{1}{a^*}}. \tag{4.3-1e}$$

The maximum pressure, P_{max}, at the center of the contact is thus obtained by substituting Eq. (4.3-1d) into Eq. (4.3-1b) as:

$$P_{\max} = \frac{3}{2\pi} \sqrt[3]{\frac{16}{9} F_n \frac{E^{*2}}{a^{*2}}}. \qquad (4.3\text{-}1f)$$

Equations (4.3-1d) to (4.3-1f) show that, for a compression contact of two elastic spheres (also known as Hertzian contact), the radius of contact (r_c), the approaching distance (s), and the maximum pressure of contact (P_{max}) can be calculated using the contact force, the elastic material properties of the spheres, and the radii of the spheres.

4.3.2   Frictional Contact of Spheres

For oblique (or rotating) contact between two spheres, contact friction leads to tangential force (or torsion forces in the case of rotation), and resulting tangential (or angular for rotating case) displacements. Oblique contact mechanics for the contact of two frictional elastic bodies was explored by Mindlin (1949); the basic theory on contact with torsion can be found in the references (Johnson, 1985).

4.3.2.1   Oblique Contact of Spheres

In oblique contact, two contact objects can either remain relatively stationary or in tangential relative motion. In the former case, the tangential force is called the static frictional force, and its magnitude is less than the limiting frictional force. In cases with relative tangential motion, the frictional force on the contact interface is proportional to the loading force normal to the contact area; this maxim is the essence of Amontons' law of sliding friction. The proportional constant, or coefficient of kinetic friction, depends on the material composition and the physical conditions of the contact surface (*e.g.*, surface roughness).

When an applied tangential force is less than the limiting tangential force of "static friction," no sliding motion is produced. Nevertheless, tangential traction will still occur. An analysis of the force and displacement in the contact area reveals that it consists of a core region ($r < c$) for no-slip motion and an annulus region ($c < r < r_c$) for micro-slip. The distribution of traction is utilized to ensure that all surface points within the no-slip region undergo the same tangential displacement; the distribution of traction is obtained from (Johnson, 1985; Fan and Zhu, 1998):

$$q(r) = \begin{cases} \dfrac{f_s P_{\max}}{r_c} \left(\sqrt{r_c^2 - r^2} - \sqrt{c^2 - r^2} \right) & r < c \\[3mm] \dfrac{f_s P_{\max}}{r_c} \sqrt{r_c^2 - r^2} & c \leq r \leq r_c \end{cases}. \qquad (4.3\text{-}2)$$

Here, f_s is the coefficient of static friction; P_{max} is the maximum pressure, defined in Eq. (4.3-1f). The tangential force, F_t, is obtained by integrating Eq. (4.3-2) over the contact area, to yield:

$$F_t = f_s F_n \left(1 - \frac{c^3}{r_c^3} \right). \qquad (4.3\text{-}2a)$$

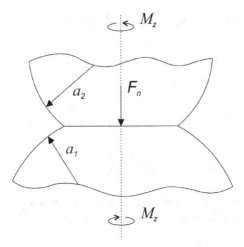

Figure 4.3-2 Contact with compression and twist.

Equation (4.3-2a) can be used to estimate the radius of the core region, c, from the given tangential force. The equation also shows that the area of the core region decreases with increasing tangential force, while the normal force remains constant. Once the tangential force reaches the limiting static friction force ($F_t = f_s F_n$), the core region shrinks to a single point, so that the sliding motion is imminent.

For sliding contact of spheres, assuming that Amontons' law of sliding friction can be applied at each elementary area of the interface, the tangential traction can be expressed as:

$$q(r) = \frac{3 f_k F_n}{2 \pi r_c^2} \sqrt{1 - \frac{r^2}{r_c^2}}, \tag{4.3-3}$$

where f_k is the coefficient of kinetic friction. The total tangential force F_t in this case becomes:

$$F_t = f_k F_n. \tag{4.3-3a}$$

It should be noted that the sliding friction force is usually much smaller than the corresponding limiting force of static friction.

4.3.2.2 Contact with Torsion

In addition to their role in oblique contact, tangential displacements are also be produced by a compressional twist, as shown in Fig. 4.3-2. Similar to the case of no-slip oblique contact, the contact area in nonspin contact with torsion consists of micro-slip in an annular region ($c \leq r \leq r_c$) and a no-slip core region ($r \leq c$).

The surface distribution of traction can be given as (Lubkin, 1951):

$$q(r) = \begin{cases} \dfrac{3f_s F_n}{\pi^2 r_c^2}\sqrt{1 - \dfrac{r^2}{r_c^2}}\left(\dfrac{\pi}{2} + [K(\chi) - E(\chi)]\,K(\lambda) - K(\chi)E(\lambda)\right) & r < c \\[4mm] \dfrac{3f_s F_n}{\pi^2 r_c^2}\sqrt{1 - \dfrac{r^2}{r_c^2}} & c \leq r \leq r_c, \end{cases}$$

(4.3-4)

where λ and χ are defined by:

$$\lambda = \frac{c}{r_c}; \quad \chi = \sqrt{1 - \lambda^2},$$

(4.3-4a)

and $K(\chi)$ and $E(\chi)$ are the complete elliptic integrals of the first and second kinds of modulus of χ, respectively; the elliptic integrals are defined as:

$$K(\chi) = \int_0^{\frac{\pi}{2}} \left(1 - \chi^2 \sin^2\theta\right)^{-\frac{1}{2}} d\theta$$

$$E(\chi) = \int_0^{\frac{\pi}{2}} \left(1 - \chi^2 \sin^2\theta\right)^{\frac{1}{2}} d\theta.$$

(4.3-4b)

Thus, the twisting moment is obtained by integrating Eq. (4.3-4) over the contact area, to yield:

$$\frac{M_z}{f_s F_n r_c} = \frac{3\pi}{16} + \frac{3\chi^2 K(\chi)}{4\pi}\left(2\lambda - \frac{\sin^{-1}\lambda}{\chi} + \int_0^{\frac{\pi}{2}} \frac{\sin^{-1}(\lambda\sin\theta)}{\left(1 - \lambda^2\sin^2\theta\right)^{\frac{3}{2}}} d\theta\right)$$
$$+ \frac{K(\chi) - E(\chi)}{4\pi}\left(\lambda\left(4\lambda^2 - 3\right) - 3\int_0^{\frac{\pi}{2}} \frac{\sin^{-1}(\lambda\sin\theta)}{\left(1 - \lambda^2\sin^2\theta\right)^{\frac{1}{2}}} d\theta\right).$$

(4.3-4c)

4.3.3   Normal Collision of Elastic Spheres

When two particles with smooth surfaces collide at low velocity, any resulting deformation is typically elastic. In this case, the collision can be classified as quasi-static with a frictionless impact of elastic spheres. The elastic deformation process during a collision can be described using Hertzian contact theory.

Consider a concentric collision between two frictionless elastic spheres such that only normal forces and normal velocities are involved (*i.e.*, in the absence of any tangential forces or tangential velocities). This process is shown in Fig. 4.3-3.

From Newton's law, it gives:

$$m_1 \frac{dU_1}{dt} = -F_C \qquad m_2 \frac{dU_2}{dt} = F_C,$$

(4.3-5)

where U_1 and U_2 correspond to particle velocities, and F_C is the collisional force. The approaching distance between the centers of the particles O_1 and O_2 during the collision can be linked to the relative collision velocity by:

$$\frac{ds}{dt} = U_1 - U_2.$$

(4.3-6)

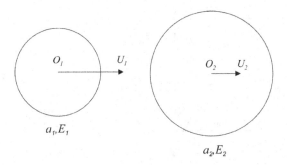

Figure 4.3-3 Normal collision of elastic spheres.

Combining Eqs. (4.3-1e), (4.3-5), and (4.3-6) yields:

$$\frac{d^2 s}{dt^2} = -\frac{4E^*}{3m^*}\sqrt{a^*}\,s^{\frac{3}{2}}, \tag{4.3-7}$$

where m^* is the relative mass, defined by:

$$\frac{1}{m^*} = \frac{1}{m_1} + \frac{1}{m_2}, \tag{4.3-7a}$$

The initial conditions are given by:

$$\left(\frac{ds}{dt}\right)_{t=0} = U_{10} - U_{20} = U_C \qquad s_{t=0} = 0. \tag{4.3-7b}$$

Integration of Eq. (4.3-7) yields:

$$\frac{ds}{dt} = \left(U_C^2 - \frac{16E^*}{15m^*}\sqrt{a^*}\,s^{\frac{5}{2}}\right)^{\frac{1}{2}}. \tag{4.3-7c}$$

The maximum deformation in an elastic collision occurs where $ds/dt = 0$. Using Eqs. (4.3-1d) to (4.3-1e) with (4.3-7c), the maximum approaching distance and the maximum radius of collisional contact area during the collision can be obtained, respectively, as:

$$s_m = \left(\frac{15}{16}\frac{m^* U_C^2}{E^*\sqrt{a^*}}\right)^{\frac{2}{5}} \tag{4.3-7d}$$

$$r_{cm} = \sqrt{a^*}\left(\frac{15}{16}\frac{m^* U_C^2}{E^*\sqrt{a^*}}\right)^{\frac{1}{5}}. \tag{4.3-7e}$$

The corresponding maximum collisional force and the maximum pressure can be computed from Eqs. (4.3-1b) and (4.3-1f) as:

$$F_{Cm} = \frac{4}{3}E^*\sqrt{a^*}\left(\frac{15}{16}\frac{m^* U_C^2}{E^*\sqrt{a^*}}\right)^{\frac{3}{5}} \tag{4.3-7f}$$

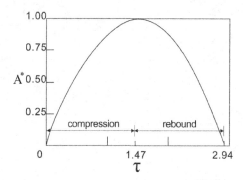

Figure 4.3-4 Change of contact area with time.

$$P_{Cm} = \frac{2}{\pi} \frac{E^*}{\sqrt{a^*}} \left(\frac{15}{16} \frac{m^* U_C^2}{E^* \sqrt{a^*}} \right)^{\frac{1}{5}}. \tag{4.3-7g}$$

Considering the symmetric nature of the compression and rebounding effects in a collision process, the total collision duration can be obtained from Eq. (4.3-7c) as:

$$t_C = \frac{2}{U_C} \int_0^{s_m} \left\{ 1 - \left(\frac{s}{s_m} \right)^{\frac{5}{2}} \right\}^{-\frac{1}{2}} ds = \frac{2.94}{U_C} \left(\frac{15}{16} \frac{m^* U_C^2}{E^* \sqrt{a^*}} \right)^{\frac{2}{5}}. \tag{4.3-7h}$$

The dependence of the contact area, A_C, on compression process time can be obtained from Eq. (4.3-7c) and Eq. (4.3-1d) as:

$$\frac{dA_C}{dt} = \left((\pi U_C a^*)^2 - \frac{16}{15\pi} \frac{E^*}{m^*} A_C^{\frac{5}{2}} \right)^{\frac{1}{2}}. \tag{4.3-7i}$$

The solution of Eq. (4.3-7i) is illustrated in Figure 4.3-4, where the dimensionless area of contact and the dimensionless time of contact are defined, respectively, by:

$$A^* = \frac{A_C}{A_{cm}} = \frac{r_c^2}{r_{cm}^2} \qquad \tau = \left(\frac{16E^*}{15m^*} \right)^{\frac{2}{5}} (a^* U_C)^{\frac{1}{5}} t. \tag{4.3-7j}$$

4.3.4   Oblique and Rotational Collisions

While the collision of frictionless spheres is modeled using Hertzian contact theory, collisions including frictional forces require Mindlin's approach. It is commonly assumed that normal compression is unaffected by tangential momentum transfer. Thus, the contact area, compression load, and collision time can be independently determined using the normal collision velocities. Tangential momentum transfer is typically governed by the linear momentum balance in the tangential direction, the angular momentum balance, and the intrinsic correlation between the tangential friction force and the normal compression force that depends on the tangentially static, sliding, or rolling modes of contact. To begin with the simplified analysis, consider the impact of two spheres in coplanar motion without torsion, as shown in Fig. 4.3-5.

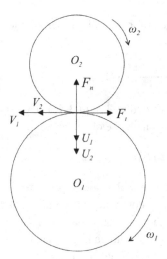

Figure 4.3-5 Coplanar oblique impact of two spheres.

The momentum equation in the tangential direction is given by:

$$m_1 \frac{d}{dt}(V_1 + \omega_1 a_1) = F_t; \qquad m_2 \frac{d}{dt}(V_2 - \omega_2 a_2) = -F_t, \tag{4.3-8}$$

where F_t is the frictional force; V_1 and V_2 are the tangential velocities at the contact point; ω_1 and ω_2 represent the angular velocities. Because the contact area is small and no external torque exists, angular momentum about the center of contact is conserved for each sphere. Hence, it gives:

$$\frac{d}{dt}\left[m_1 V_1 a_1 + m_1 \omega_1 \left(a_1^2 + k_1^2\right)\right] = 0 \qquad \frac{d}{dt}\left[m_2 V_2 a_2 - m_2 \omega_2 \left(a_2^2 + k_2^2\right)\right] = 0, \tag{4.3-9}$$

where k_1 and k_2 are the radii of gyration of the spheres about their centers of mass. Based on the preceding equations, it yields:

$$\frac{dV_{12}}{dt} = \frac{F_t}{\hat{m}^*}, \tag{4.3-10}$$

where V_{12} is the relative tangential velocity ($V_2 - V_1$), and:

$$\frac{1}{\hat{m}^*} = \frac{1 + a_1^2/k_1^2}{m_1} + \frac{1 + a_2^2/k_2^2}{m_2}. \tag{4.3-10a}$$

The change in tangential velocity, as governed by Eq. (4.3-10), requires further information about the tangential deformation, which is affected by the presence of micro-slip, the magnitude of normal compressional force F_C, the coefficient of friction, and the loading history (Johnson, 1985).

For the case depicted in Fig. 4.3-5, the tangential velocities of the colliding spheres are not equal to each other before the impact, an alternate scenario that would likely lead to sliding on the contact interface. If it is further assumed that there is no relative rotation and sliding is maintained in the collision, Eq. (4.3-10) can be simplified using the Hertzian relations for the normal deformation. Thus, it gives:

$$\frac{dV_{12}}{dt} = -\frac{4f_k E^* \sqrt{a^*}}{3\hat{m}^*} s^{\frac{3}{2}}.$$ (4.3-10b)

Note that the relation between s and t is governed by Eq. (4.3-7c). Thus, the change of relative tangential velocity after the collision can be obtained as:

$$\Delta V_{12} = 2f_k U_{12},$$ (4.3-10c)

which shows that the change of relative tangential velocity is related only to the surface friction coefficient and normal velocity component; it is independent of other material properties. Equation (4.3-10c) further indicates that, to ensure the sliding assumption is valid, the relative incident angle should be larger than $\tan^{-1}(2f_k)$.

It is clear that the collision between two elastic, but frictional, spheres is inelastic because of the kinetic energy loss through frictional work; sliding at contact is inevitable. Furthermore, the preceding analyses of both Hertzian collision and frictional collision can be applied to particle–wall collisions where the radius of the wall is simply regarded as infinitely large.

4.3.5   Collision of Inelastic Spheres

Any solid material has an upper limit of elastic deformation under either normal or tangential stresses. Once the stresses exceed this limit, known as the yield stress, plastic deformation will occur. Three criteria are commonly used to estimate the onset yield stresses of solid materials. The Tresca criterion is based on maximum shear stress and is given as:

$$\max\{|\sigma_1 - \sigma_2|, |\sigma_2 - \sigma_3|, |\sigma_3 - \sigma_1|\} = Y,$$ (4.3-11a)

where σ_1, σ_2, and σ_3 are the principal stresses; Y is the yield stress in simple compression or tension. The von Mises criterion considers the shear strain–energy, which is calculated by:

$$(\sigma_1 - \sigma_2)^2 + (\sigma_2 - \sigma_3)^2 + (\sigma_3 - \sigma_1)^2 = 2Y^2.$$ (4.3-11b)

The third criterion is obtained from the maximum reduced stress and is expressed as:

$$\max\{|\sigma_1 - \sigma|, |\sigma_2 - \sigma|, |\sigma_3 - \sigma|\} = \frac{2}{3}Y,$$ (4.3-11c)

where $\sigma = (\sigma_1 + \sigma_2 + \sigma_3)/3$.

No significant difference is found in the prediction of the yield stress by the three methods. However, Tresca's criterion is more widely used than the other two due to its inherent simplicity. When two solid spherical particles are in contact, the principal stresses along the normal axis through the contact point can be obtained from the

Hertzian elastic contact theory as (Johnson, 1985):

$$\sigma_r = \sigma_\theta = -P_{max}\left\{(1+v)\left[1 - \frac{z}{r_c}\tan^{-1}\left(\frac{r_c}{z}\right)\right] - \frac{r_c^2}{2(r_c^2 + z^2)}\right\}$$

$$\sigma_z = -\frac{P_{max}r_c^2}{r_c^2 + z^2},$$

(4.3-12)

where z denotes the depth inside the sphere along the axis of symmetry. It can be proven from Eq. (4.3-12) that, for materials with Poisson's ratio of 0.3 (true for most solids), the maximum shear stress, $|\sigma_z - \sigma_r|$, occurs at $z/r_c = 0.48$. According to Tresca's criterion, the yield stress, Y, in a simple compression is $0.62\,P_{max}$. Therefore, when the hardness or the yield stress, Y, of the particle material is less than 0.62 of the maximum contact pressure, the sphere will likely undergo plastic deformation. From the elastic collision of two solid spheres, the maximum contact pressure is given by Eq. (4.3-7g). Thus, the relation between the critical normal collision velocity, U_{CY}, and the yield stress is given by:

$$U_{CY} = 10.3\sqrt{\frac{a^{*3}Y^5}{m^*E^{*4}}}.$$

(4.3-13)

For solids with Poisson's ratio in the range of 0.27 to 0.36, the critical normal collision velocity, estimated using Eq. (4.3-13), is with a deviation margin less than 5%.

Once the relative impact velocity between two colliding spheres is higher than the critical yield velocity, plastic deformation must occur. Heat loss is another phenomenon often coupled with such collisions. Collisions with plastic deformation are referred to as inelastic collisions. All the energy transfer in the form of plastic deformation and heat loss during an inelastic collision is considered to be a loss of kinetic energy.

The recoverability, or restitution, of kinetic energy during a normal collision between two solid objects is quantified by the coefficient of restitution, defined by the ratio of relative rebounding velocity to the relative collision velocity. Note that the coefficient of restitution cannot be used as a criterion to judge whether a collision is elastic or not unless the collision is strictly a normal one. For example, sliding at contact for the collision between two elastic spheres renders the collision inelastic, but the value of the coefficient of restitution is equal to 1. The prediction of the restitution coefficient has been a challenging research topic for decades. No reliable and accurate prediction method has been expounded to date. Fortunately, some simplified utilitarian models, with implied limits, have been developed. One example is the elastic-plastic impact model where the compression process is assumed to be plastic with part of the kinetic energy being stored for later elastic rebounding; the rebound process is considered to be completely elastic. Based on this model, the coefficient of restitution is obtained as:

$$e = \left|\frac{U_C'}{U_C}\right| = 4.08\left(\frac{a^{*3}Y^5}{E^{*4}m^*U_C^2}\right)^{\frac{1}{8}}.$$

(4.3-14)

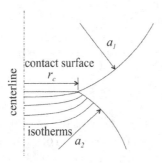

Figure 4.3-6 Conductive heat transfer in a Hertzian collision.

The elastic-plastic model reveals that the coefficient of restitution also depends on the relative impact velocity in addition to the material properties. Equation (4.3-8) indicates that the restitution coefficient decreases with increasing impact velocity by an exponent of 1/4, which is supported by experimental findings (Goldsmith, 1960).

4.3.6   Heat Transfer by Collision of Solids

In a Hertzian collision of solid spheres at different temperatures, heat transfer occurs via conduction at the contact area. Since the duration of the impact is short-lived, the temperature change of the colliding particles is confined to a small region around the contact area. Therefore, heat conduction between the two particles can be modeled as heat conduction between two semi-infinite media. For simplicity, it is further assumed that there is no thermal resistance between the contact surfaces, so the temperature and heat flux distributions are continuous across the contact area.

Heat conduction between the colliding solids can be expressed in cylindrical coordinates, as shown in Figure 4.3-6, by:

$$\frac{1}{D_{Ti}} \frac{\partial T_i}{\partial t} = \frac{1}{r} \frac{\partial}{\partial r} \left(r \frac{\partial T_i}{\partial r} \right) + \frac{\partial^2 T_i}{\partial z^2} \quad i = 1, 2, \tag{4.3-15}$$

where D_T's are the thermal diffusivities of solids. The associated boundary and initial conditions are given as:

$$T_1 = T_2; \quad K_{p1} \frac{\partial T_1}{\partial z} = K_{p2} \frac{\partial T_2}{\partial z} \quad at \; z = 0; \; r \leq r_c(t) \tag{4.3-15a}$$

$$\frac{\partial T_1}{\partial z} = \frac{\partial T_2}{\partial z} = 0 \quad at \; z = 0; \; r > r_c(t) \tag{4.3-15b}$$

$$\frac{\partial T_1}{\partial z} = \frac{\partial T_2}{\partial z} = 0 \quad at \; r = 0 \tag{4.3-15c}$$

$$T_1 = T_{10}; \quad T_2 = T_{20} \quad at \; t = 0. \tag{4.3-15d}$$

If contact time is very short, heat transferred by conduction from one medium to the other does not extensively penetrate the contact surface. Therefore, a one-dimensional approximation reasonably accounts for the heat transfer phenomenon, which is characterized by a small Fourier number and defined in terms of the total impact duration and maximum contact radius. Using a one-dimensional approximation, the interface heat flux can be represented by (Sun and Chen, 1988):

$$J_{qw} = \frac{T_{20} - T_{10}}{\sqrt{\pi t}} \left(\frac{\sqrt{D_{T1}}}{K_1} + \frac{\sqrt{D_{T2}}}{K_2} \right)^{-1}. \tag{4.3-15e}$$

Here, t is measured from the moment of initial contact and is expressed as a function of the radius of contact area, r_c. The total heat exchange per impact is:

$$q_c = 2\pi \int_0^{r_{cm}} \int_0^{t_c - 2t} J_{qw} dt \, r dr. \tag{4.3-16}$$

In the above equation, r_{cm} is the maximum radius of collisional contact area during the collision, as defined by Eq. (4.3-7e). With t related to r_c by Eq. (4.3-7i), the numerical integration of above equation gives:

$$q_{c0} = 2.73 \, (T_{20} - T_{10}) \, r_{cm}^2 \sqrt{t_c} \left(\frac{\sqrt{D_{T1}}}{K_1} + \frac{\sqrt{D_{T2}}}{K_2} \right)^{-1}, \tag{4.3-16a}$$

where the subscript "0" for q_c indicates applicability for the case where the Fourier number approaches zero. Equation (4.3-16a) only captures heat conduction in the z–direction when Fo is negligibly small. If heat conduction in the r–direction is not negligible, more heat would be transferred between the two colliding media. Since there is no general analytical solution for Eq. (4.3-15) for appropriate boundary and initial conditions, heat conduction with a finite Fourier number can currently only be evaluated using a numerical method to approximate the closed-form solution.

4.3.7 Charge Transfer by Collision of Solids

Similar to heat transfer in a particle collision, charge transfer can also occur when particles collide with each other, or with a solid wall. A simple analytical model for the charge transfer can be established by assuming a direct analogy between charge transfer by collisions and heat transfer by convection (Cheng and Soo, 1970). The current density through the contact area of these two particles, J_{eC}, is expressed as:

$$J_{eC} = h_{eC} \, (\Phi_2 - \Phi_1) = \frac{\sigma_1}{d_{p1}} \, (\Phi_C - \Phi_1) = \frac{\sigma_2}{d_{p2}} \, (\Phi_2 - \Phi_C), \tag{4.3-17}$$

where h_{eC} is the charge transfer coefficient; Φ_1, Φ_2, and Φ_C are the electric potentials of particles, 1, 2, and the contact surface, respectively; σ_1 and σ_2 are the electrical conductivities of the materials. The charge transfer coefficient can be obtained by eliminating the contact potential, Φ_C, as:

$$h_{eC} = \frac{\sigma_1 \sigma_2}{\sigma_1 d_{p2} + \sigma_2 d_{p1}}. \tag{4.3-17a}$$

For each isolated particle, the charging current can be related to the particle capacity, C, as well as the contact area, A_C. Thus, it gives:

$$C_1 \frac{d\Phi_1}{dt} = -C_2 \frac{d\Phi_2}{dt} = A_C h_{eC} \, (\Phi_2 - \Phi_1). \tag{4.3-18}$$

For a homogeneous spherical particle, the capacitance is expressed by:

$$C = 2\pi \varepsilon d_p, \tag{4.3-18a}$$

where ε is the permittivity of the material. The electric potential of the particle is:

$$\Phi = \phi + \frac{q}{2\pi \varepsilon d_p}, \tag{4.3-18b}$$

where ϕ is the work function, and q is the charge carried by the particle. Therefore, the equation for the charge transfer can be expressed as:

$$-\frac{d}{dt}(\Phi_2 - \Phi_1) = \frac{A_C h_{eC}}{C^*}(\Phi_2 - \Phi_1), \tag{4.3-19}$$

where C^* is the relative capacitance, defined by:

$$\frac{1}{C^*} = \frac{1}{C_1} + \frac{1}{C_2}. \tag{4.3-19a}$$

For a Hertzian collision, the relation of contact area and contact time, t, can be obtained from Eq. (4.3-7i). Thus, the transferred charge, q_{eC}, can be calculated by:

$$q_{eC} = C^* \left(\phi_2 - \phi_1 + \frac{q_{20}}{C_2} - \frac{q_{10}}{C_1} \right)(1 - e^{-B}), \tag{4.3-19b}$$

where subscript "0" denotes the initial state, and B is a constant given by:

$$B = 2\frac{s_m^2 \pi a^* h_{eC}}{U_C C^*} \int_0^1 x \left(1 - x^{\frac{5}{2}}\right)^{-\frac{1}{2}} dx. \tag{4.3-19c}$$

4.4   Other Interaction Forces between Solid Particles

Typical noncollisional forces in a particulate system include the van der Waals force, capillary force, and the electrostatic force. These forces are responsible for the cohesive behavior of fine particles. For micron-size particles, the van der Waals force is generally the dominant cohesive force. The capillary force is important for the liquid-induced particle agglomeration.

4.4.1   Van der Waals Force

The van der Waals force originates from interactions of permanent or transient electric dipole moments that are generated by fluctuations in the electron cloud of molecules. For two macroscopic bodies, the van der Waals force is usually calculated from the interaction potential, which depends on the geometry, material, and relative distance of the bodies. For two spherical particles, the van der Waals force is expressed by:

$$F_v(z) = \frac{A_H}{12z^2} \left(\frac{d_{p1} d_{p2}}{d_{p1} + d_{p2}} \right), \tag{4.4-1}$$

where z is the distance between the surface of two particles; and A_H is the Hamaker constant, which is a function of material properties. Similarly, the van der Waals force acting between a spherical particle and an infinite plate is given by:

$$F_v(z) = \frac{A_H d_p}{12z^2}.$$ (4.4-2)

The Hamaker constant's value for common liquid and solid materials is typically in the range of 10^{-20} to 10^{-19} joules (Crowe et al., 1998). For two different materials, the Hamaker constant can be estimated by:

$$A_{H12} = \sqrt{A_{H11}A_{H22}},$$ (4.4-3)

where A_{H11} and A_{H22} are defined as the Hamaker constant of each distinct material. When the two bodies are separated by a third material, the Hamaker constant can be approximated using:

$$A_{H123} = \left(\sqrt{A_{H11}} - \sqrt{A_{H33}}\right)\left(\sqrt{A_{H22}} - \sqrt{A_{H33}}\right),$$ (4.4-4)

where A_{H33} is the Hamaker constant of the third material, or medium. The van der Waals force between two macroscopic bodies is usually attractive, leading to a positive value for the Hamaker constant. However, as Eq. (4.4-4) shows, the van der Waals force can repel in some cases.

4.4.2   Liquid–Bridge Force

The capillary force, also known as the liquid bridge force, is illustrated in Figure 4.4-1. This force is the combined result of the adhesion force on the liquid–particle interface and low pressure inside the liquid volume from the curvature of the gas–liquid interface (Zhu et al., 2007). This is represented by:

$$F_c = \pi \sigma d_p \sin \phi \sin(\phi + \theta) + \frac{\pi d_p^2}{4} \Delta p \sin^2 \phi.$$ (4.4-5)

Here, σ is the surface tension; θ and ϕ are the contact angle and half filling angle, respectively. The pressure difference between the inside and outside of the liquid bridge is determined from the radius of curvature of the bridge, R_1, and medium R_2, as shown in Figure 4.4-1.

For two wettable spheres ($\theta = 0$) of identical size, the above expression simplifies to:

$$F_c = \pi \sigma d_p \sin^2 \phi + \frac{\pi d_p^2 \sigma}{4} \left(\frac{1}{R_1} - \frac{1}{R_2}\right) \sin^2 \phi.$$ (4.4-6)

4.5   Interactions between Fluid Particles

Collision processes involving droplets or bubbles, also referred to as fluid particles, manifest a host of noteworthy phenomena including deformation, disintegration, and

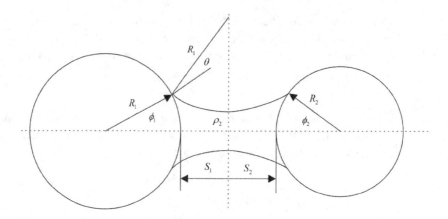

Figure 4.4-1 Illustration of the liquid bridge force.

coalescence. These occurrences influence particle size distribution and interfacial area, impacting the overarching interphase transport. In this section, a microscopic approach is presented to illustrate the collisions of individual droplets or bubbles. This knowledge forms the basis for the mechanistic collision models required for a macroscopic description of a multiphase flow system, namely the two-fluid model and the population balance model.

4.5.1   Droplet Impact on a Flat Solid Surface

A collision between a liquid droplet and a solid surface is a typical process found in many applications, such as inkjet printing, spray cooling, and spray painting. Depending on the impact conditions and fluid properties, the possible outcomes of a collision include deposition, partial rebound, or total rebound from the surface (Rioboo *et al.*, 2001), as schematically shown in Figure 4.5-1. Disintegration of droplets is usually observed, particularly at high impact velocities.

As discussed in Chapter 3, droplet dynamics are primarily governed by the interplay of inertial forces, viscous forces, and surface tension forces. The Weber number and the Reynolds number therefore characterize the dynamics. However, solid surface properties, such as wettability and surface roughness, also are prominent considerations that influence the outcome of a collision. When impact velocity is negligibly small, a droplet can be deposited on the target surface, and the capillary force dominates the process. In contrast, inertia governs dynamics for droplets at relatively high impact velocities; a high Reynolds number and Weber number are also likely in this scenario. The discussion of droplet–surface collisions in this section is focused on scenarios involving high impact velocities. Two paradigmatic occurrences are examined in this category: it begins with droplets colliding into dry surfaces and then introduces models for wet surface droplet collision.

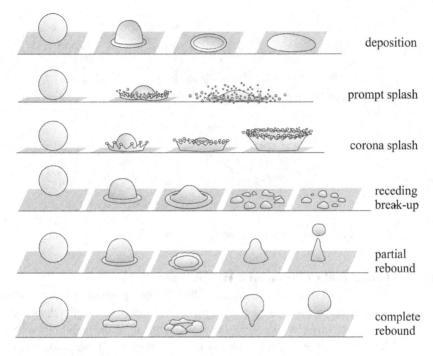

Figure 4.5-1 Morphology of drop impact on a dry surface.

4.5.1.1 Isothermal Collision on a Dry Surface

The collision of a droplet with a dry surface can be divided into a series of stages. Figure 4.5-2 illustrates the typical collision process of a partially rebounded droplet; the stages shown include initial deformation after first contact, droplet spreading, droplet recession after maximum spreading, and partial rebound from the surface (Rioboo *et al.*, 2001). When the Reynolds and Webber numbers are high, inertia dominates the initial deformation and spreading stages. However, the capillary and wettability forces control the receding stage.

Droplet disintegration can occur at different stages of the collision, and the droplet can thus present a variety of morphologies during the collision process. Sample droplet morphologies are depicted in Figure 4.5-1. At low impact velocities, a droplet is typically deposited on the surface when the receding stage concludes. At a high impact velocity, a droplet colliding with a rough surface is often subjected to prompt splash, in which it disintegrates in the initial deformation stage. With reduced surface tension, the lamella detaches from the surface, resulting in a corona splash. In the receding stage, if the droplet lacks sufficient momentum to overcome viscous dissipation and contact angle hysteresis, the receding lamella will break up into small droplets on the target surface. If the droplet maintains sufficient kinetic energy throughout the receding stage, it will be pushed upward in a partial or full rebound.

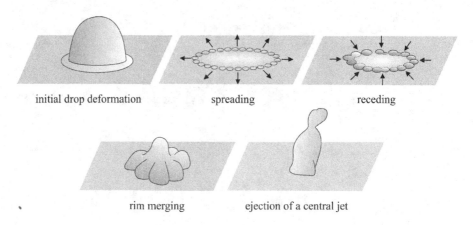

initial drop deformation   spreading   receding

rim merging   ejection of a central jet

Figure 4.5-2 Collision stages of a droplet on a dry surface.

4.5.1.2   Isothermal Collision on a Wet Surface

As shown in Figure 4.5-3, a droplet collision with a preexisting thin liquid film results in many of the same characteristic patterns as a collision with a dry surface. At low impact velocity (*i.e.*, We < 5), the droplet adheres to the liquid film with negligible deformation. For moderate velocity (*i.e.*, 5 < We < 10), the droplet rebounds from the wet surface after collision. At high velocity, when We > 10, the droplet spreads to a thin layer on the liquid film and deposits onto the film at the end of the collision, provided the splash threshold is not reached. The splash regime can be further categorized into prompt splash and corona splash. Prompt splash occurs during the initial deformation stage with secondary droplets forming at the edge of the expanding lamella. Corona splash is evidenced by the formation of a crown-shaped lamella. The splash threshold depends on the liquid properties, impact speed, and liquid film thickness (Cossali, 1997). In a collision between a droplet and a thick liquid film, a semispherical crater may form in the film, which is often associated with gas bubble entrapment or Worthington jet formation at the crater center, as exemplified in Figure 4.5-3.

4.5.2   Binary Droplet Collision

The effect of a collision between two droplets is governed by the droplet size and fluid properties, in addition to the relative velocity vector. The outcome is typically dependent on the impact parameter, B, and the Weber number, We, defined respectively by:

$$B = \frac{2\chi}{d_{d1} + d_{d2}} \qquad We = \frac{\rho_d U^2 (d_{d1} + d_{d2})}{2\sigma}. \tag{4.5-1}$$

Here, χ is the projection of the separation distance between the droplet centers in the direction normal to the relative velocity U. For two droplets formed from the same liquid and with equal diameters, the outcome of the collision is classified into

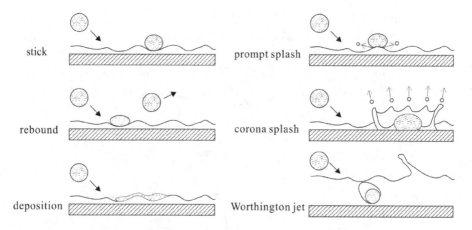

Figure 4.5-3 Regimes of a droplet collision with a wet surface (from Crowe, 2006).

the five regimes shown in Figure 4.5-4. The collision at a low Weber number is mostly influenced by the film of the remaining interstitial gas in the void between the approaching droplets. If the gas film is released to the characteristic distance of the van der Waals force, the droplets will coalesce; otherwise, they will rebound. At higher Weber numbers, near head-on collisions are dominated by the inertia force. The droplets devolve to a thin liquid film, which first expands and then retracts. If the kinetic energy of the droplet is sufficiently high, it will overcome the retraction of the liquid film and cause the droplet to split into two or more secondary droplets. If the kinetic energy is below this threshold, the deformed liquid film may experience moderate oscillations but ultimately will remain coalesced. For collisions with both a high Weber number and large obliquity, the liquid film will be severely stretched and eventually dissociated into small droplets.

4.5.3 Breakup of Fluid Particles

The fluid particle interactions with the surrounding continuous phase hydrodynamics, colliding with solid surfaces, and collisions with other particles can all cause a fluid particle to break up. Since the latter two mechanisms were addressed in Sections 4.5.1 and 4.5.2 respectively, the subsequent focus is then placed on flow-induced breakup. Among the various acting hydrodynamic forces, external stresses on the continuous phase act to destabilize the particle, while the surface tension and the internal viscous forces tend to be restorative. The flow-induced breakup mechanism can be roughly classified into four categories (Liao and Lucas, 2010):

1) Breakup due to turbulent fluctuation and eddy collision

This category represents the dominant mechanism of breakup in turbulent flows. The particle surface becomes unstable when exposed to a fluctuating turbulent flow field or when it collides with a turbulent eddy. The parent particle deforms until it

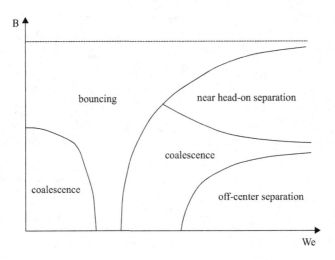

Figure 4.5-4 Regimes of collisions of two identical droplets (from Qian and Law, 1997).

divides into two or more daughter particles. The driving force of the breakup has its genesis in the fluctuating dynamic pressure of the surrounding flow. The destabilization mechanism is illuminated through a balance of the dynamic pressure and the surface tension (described by the Weber number).

2) Breakup due to viscous shear

Viscous shear stress can break up a fluid particle by inducing a velocity gradient near the particle surface. To illustrate, when only part of a bubble is in the wake of another bubble, the shear stress across the wake boundary splits the particle into dominant halves, with less pronounced satellite bubbles also resulting. In this case, the external viscous force and the surface tension force, whose ratio is characterized by the Ca number, are relevant to the analysis.

3) Breakup due to a shearing-off process

For large fluid particles in high-viscosity fluids, a velocity potential across the interface may cause small particles from the cap, or skirt region, to shear off of the parent particle. In less viscous flows with high relative velocity, the internal fluid penetrates into the external fluid film and generates small particles, as observed in the cap-shape gas bubbles in water phenomenon.

4) Breakup due to interfacial instability

Finally, breakup can be the result of the interfacial instability. When a low-density fluid is accelerated into a high-density fluid, the interface is subjected to Rayleigh–Taylor instability. When a sufficient velocity difference across the interface between two fluids is developed, Kelvin–Helmholtz instability may occur. These types of interfacial instability will cause attrition at the interface and eventually cause a breakup of the fluid particles.

Modeling the breakup of a fluid particle requires both the breakup frequency and the size distribution of the daughter particles after breakup. Models are developed for each of the four breakup mechanisms. The daughter particle size distribution is often studied based on either statistical descriptions treating the daughter particle size as a random variable, or phenomenological models with a specific profile formulated through experimental observations.

4.5.4 Coalescence of Fluid Particles

For two colliding fluid particles of diameters d_1 and d_2, the coalescence is governed by two factors: the collision frequency, $h(d_1,d_2)$, and the coalescence efficiency, $\lambda(d_1,d_2)$. The collision frequency considers the various mechanisms causing the approach and collision of two particles, while coalescence efficiency quantifies the fraction of fluid particles that coalesce after the collision.

For turbulent induced collisions, the model for the collision frequency takes into account relative motion caused by the turbulent flow. Just as in kinetic theory, the collision frequency between two fluid particles is determined by the volume swept by the collision cross-section area, with the relative velocity. Relative velocity between the particles can only result from turbulent eddies of a similar size to the particles; according to classical turbulence modeling, this can be correlated as:

$$U_r \propto \left(\sqrt[3]{d_1^2} + \sqrt[3]{d_2^2} \right)^{\frac{1}{2}} \varepsilon^{\frac{1}{3}} \tag{4.5-2}$$

where ε is the turbulent eddy dissipation rate. Thus, the collision frequency is given by:

$$h(d_1, d_2) = C \frac{\pi}{4}(d_1 + d_2)^2 \left(\sqrt[3]{d_1^2} + \sqrt[3]{d_2^2} \right)^{\frac{1}{2}} \varepsilon^{\frac{1}{3}}, \tag{4.5-3}$$

where C is a correction factor of collision cross-section area. Other collision mechanisms, such as the effects of a velocity gradient, capture in a turbulent eddy, difference in buoyant rise velocity, and wake entrainment, can be modeled to account for the specific flow conditions.

The most widely accepted model for coalescence efficiency is the film drainage model, which assumes that the contact time is sufficiently long so as to allow the drainage of the film between the particles. The coalescence efficiency thus can be estimated by:

$$\lambda(d_1, d_2) = \exp\left(-\frac{t_{\text{drainage}}}{t_{\text{contact}}}\right). \tag{4.5-4}$$

The drainage time depends on the deformation and mobility of interfaces of the colliding particles. For two rigid spheres immersed in a fluid, as shown in Figure 4.5-5(a), the drainage time is modeled by:

$$t_{\text{drainage}} = \frac{3\pi\mu}{2F} \left(\frac{d_1 d_2}{d_1 + d_2} \right)^2 \ln\left(\frac{\delta_i}{\delta_f} \right), \tag{4.5-5}$$

Table 4.5-1 Drainage time of deformable particles (Chesters, 1975; Chesters, 1991)

Interfaces	Drainage time, $t_{drainage}$
Immobile	$\dfrac{3\mu F d^2}{64\pi\sigma^2}\left(\dfrac{1}{\delta_f^2}-\dfrac{1}{\delta_i^2}\right)$
Partially mobile	$\dfrac{\pi\mu F^{1/2}}{2(4\pi\sigma/d)^{3/2}}\left(\dfrac{1}{\delta_f}-\dfrac{1}{\delta_i}\right)$
Fully mobile (Viscosity controlled)	$\dfrac{3\mu d}{4\sigma}\ln\left(\dfrac{\delta_i}{\delta_f}\right)$
Fully mobile (Inertia controlled)	$\dfrac{\rho U_t d^2}{8\sigma}$

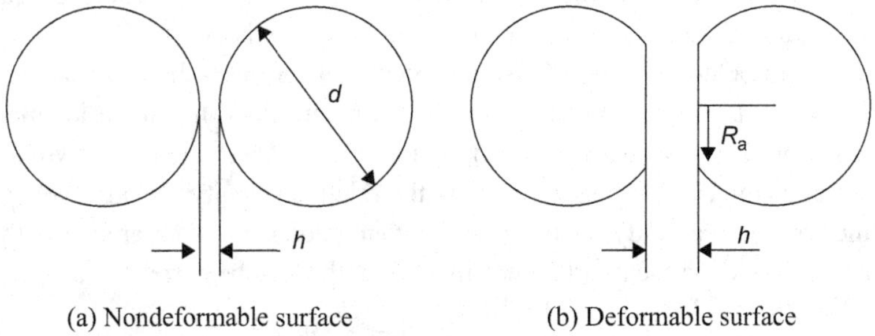

(a) Nondeformable surface (b) Deformable surface

Figure 4.5-5 Approaching surfaces in a coalescence.

where δ_i and δ_f are the initial and critical film thickness, respectively; F is the compressing force. The compression force is generally not constant because it is affected by changes in the drag force and contact area. The compression force is assumed to be proportional to the mean square velocity difference at either end of an eddy with a size equal to the particle equivalent diameter. This is expressed as:

$$F \approx \rho\varepsilon^{\frac{2}{3}}(d_1+d_2)^{\frac{2}{3}}\left(\frac{d_1 d_2}{d_1+d_2}\right)^2. \tag{4.5-6}$$

For deformable fluid particles, it is often assumed that the approaching surfaces are deformed into parallel discs, as shown in Figure 4.5-5(b), although, in reality, the pressure distribution in the film is more likely to form dimples. Expressions for film drainage time between two deformable particles of equal size have been derived for the different mobility characteristics of the colliding surfaces, as summarized in Table 4.5-1.

The most widely used expression of contact time is derived from dimensional analysis as:

$$t_{contact} = \sqrt[3]{\frac{d^2}{\varepsilon}}. \tag{4.5-7}$$

More sophisticated expressions for the compression force and contact time have been established, such as those accounting for external flow, added mass of the particles, and different particle sizes (Liao and Lucas, 2010b).

4.6 Case Studies

The following case studies serve for further discussion with simple applications of the theories of Chapter 4. The case study of settling of suspended particles in a column leads to the well-known Richardson–Zaki equation that is extensively applied for modeling of particle settling and fluidization. The case studies of wake-induced collision of a pair of interacting particles illustrate not only the applications of BBO equation and Hertzian collision theory but also the importance of various particle–particle and particle–fluid interactions in the modeling of particle motions. Lastly, a modeling example of droplet–surface collision in the Leidenfrost regime is presented to illustrate the coupled governing mechanisms among momentum, heat, and mass transfer involved in the process.

4.6.1 Settling of Suspended Particles in Column

Problem Statement: Consider a group of suspended particles settling in a very long column with a closed end. The column is prefilled with liquid. Develop a hydrodynamic model to correlate the terminal settling velocity with the particle volume fraction and column diameter. This problem was originally addressed by Richardson and Zaki in 1954.

Analysis: Assume the settling is very slow, such that (1) there are no interparticle collisions; (2) the particle-laden suspension around a particle can be treated as a mixture with apparent density and effective viscosity; and (3) settling is in the Stokes flow regime. When a particle is settling at its terminal velocity, all forces acting on the particle must be balanced. Hence, the settling force balance on the particle is:

$$C_{Da}\frac{\pi}{8}d_p^2\rho_a(U_{pt} - U)^2 = \frac{\pi}{6}d_p^3(\rho_p - \rho_a)g, \qquad (4.6\text{-}1)$$

where the apparent density is defined as:

$$\rho_a = \alpha\rho + (1 - \alpha)\rho_p. \qquad (4.6\text{-}1\text{a})$$

The apparent drag coefficient in the Stokes regime is given by:

$$C_{DS} = \frac{24}{Re_{pa}} = \frac{24\mu_e}{\rho_a d_p(U_{pt} - U)} = \frac{24\mu}{\rho_a d_p(U_{pt} - U)f(\alpha)}. \qquad (4.6\text{-}1\text{b})$$

Here, $f(\alpha)$ is a viscosity factor that depends on the fluid volume fraction in the particle-laden suspension. Equation (4.6-1) can be simplified, by substitution of Eqs. (4.6-1a) and (4.6-1b), to:

$$\frac{U_{pt} - U}{U_{pt0}} = \alpha f(\alpha), \qquad (4.6\text{-}2)$$

where U_{pt0} is the terminal velocity of an isolated particle in an unbounded stagnant fluid.

The actual fluid velocity and particle settling velocity can be defined, respectively, by their volumetric flow rates as:

$$U = \frac{Q}{A\alpha}; \quad U_{pt} = \frac{Q_p}{A(1-\alpha)}. \tag{4.6-3}$$

Because there is no net flow rate of the suspension in the closed-end column, it gives:

$$Q + Q_p = 0, \tag{4.6-4}$$

which leads to:

$$U = -U_{pt}\left(\frac{1-\alpha}{\alpha}\right). \tag{4.6-4a}$$

Thus, substitution of Eq. (4.6-4a) into Eq. (4.6-2) yields:

$$\frac{U_{pt}}{U_{pt0}} = \alpha^2 f(\alpha). \tag{4.6-5}$$

Based on experimental data and observation, Richardson and Zaki (1954) proposed that:

$$\frac{U_{pt}}{U_{pt0}} = \alpha^n, \tag{4.6-5a}$$

or:

$$f(\alpha) = \alpha^{n-2}, \tag{4.6-5b}$$

where n is the empirical Richardson and Zaki index, determined as:

$$n = \begin{cases} 4.64 & Re_p < 0.3 \\ 2.4 & Re_p > 500. \end{cases} \tag{4.6-5c}$$

Equation (4.6-5a) is the well-known Richardson–Zaki equation, which is extensively used in applications involving particle settling and fluidization.

Comments: (1) In the Richardson–Zaki model, the averaged drag force on a particle at a terminal settling velocity can be expressed, from Eq. (4.6-1), as:

$$F_D = \frac{\pi}{6}d_p^3 \alpha \left(\rho_p - \rho\right) g, \tag{4.6-1c}$$

which shows that, for a given particle, the particle–particle interactions reduce the averaged drag force at terminal setting velocity by a factor of the fluid volume fraction. (2) Based on Eq. (4.6-5a), the settling mass flux can be expressed by:

$$\dot{m}_{ps} = U_{pt}(1-\alpha)\rho_p = U_{pt0}(1-\alpha)\rho_p\alpha^n, \tag{4.6-6}$$

implying a maximum settling mass flux must be reached as the volumetric loading of particles continues to increase.

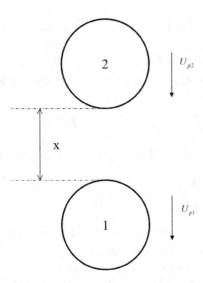

Figure 4.6-1 Wake attraction of a pair of spheres.

4.6.2 Wake-Induced Motion of a Pair of Spheres

Problem Statement: Consider a pair of spheres in an arrangement so that a trailing sphere is in the wake region of a leading sphere in a multiphase flow. Due to the relatively low pressure in the wake region, the trailing sphere approaches the leading sphere to a wake-induced collision. Develop a model of this wake-induced motion and the collision velocity.

Analysis: When the trailing sphere moves toward the leading sphere, the interaction between the two will reduce the drag forces of each spheres, with most drag reduction on the trailing one and slight reduction on the leading sphere (Zhu et al., 1994). Hence, viewed from the frame of observation on the leading sphere, the trailing sphere approaches the leading sphere in an acceleration motion, driven by the net drag force between the two. Figure 4.6-1 illustrates the wake attraction of a pair of spheres.

Assume that (1) the motion of the leading sphere is not affected by the approach of the trailing sphere; (2) spheres are equal-sized, rigid, and nonspinning; and (3) initially, the pair of spheres move at their terminal velocity and are separated by a characteristic distance (typically about two diameters of the leading sphere).

The general motion of trailing sphere is governed by the B.B.O. equation, Eq. (3.5-3). In this case of acceleration process, forces on the sphere include the drag force, the added mass force, the Basset force, and buoyancy force. Thus, the equation of motion takes the following form:

$$\left(\rho_p + \frac{1}{2}\rho\right)\frac{\pi}{6}d_p^3\frac{dU_{p2}}{dt} = (\rho_p - \rho)\frac{\pi}{6}d_p^3 g - F_{D2} - F_{B2}, \qquad (4.6\text{-}7)$$

where the drag force, F_D, is expressed by Eq. (4.2-4); and the Basset force, F_B, takes the form given in Eq. (3.2-21).

It is noted that the drag force on the leading particle is balanced against its buoyancy force, which equals the drag force of a single noninteracting sphere, so that

$$\left(\rho_p + \frac{1}{2}\rho\right)\frac{\pi}{6}d_p^3\frac{dU_{p1}}{dt} = (\rho_p - \rho)\frac{\pi}{6}d_p^3 g - F_{D1} = 0. \qquad (4.6\text{-}8)$$

Subtracting Eq. (4.6-8) from Eq. (4.6-7) yields the equation of relative motion as:

$$\frac{dU_r}{dt} = \frac{2(\rho_p - \rho)g(1-A)}{2\rho_p + \rho}\exp\left(\frac{B}{d_p}\int_t^{t_c}U_r dt\right) - \frac{18\sqrt{\pi\rho\mu}}{(2\rho_p + \rho)\pi d_p}\int_0^t \frac{\frac{dU_r}{d\tau}}{\sqrt{t-\tau}}d\tau, \qquad (4.6\text{-}9)$$

where t_C is the time when the spheres come in contact, and A and B are empirical coefficients in Eq. (4.2-4). Thus, with the given initial relative velocity and location, Eq. (4.6-9) can be solved numerically to yield the collision velocity at contact.

An approximate analytical solution of Eq. (4.6-8) can be obtained with some additional assumptions. Based on experimental measurements (Zhu et al., 1994), it may be further assumed that the relative acceleration is proportional to the relative velocity; and (2) the initial relative velocity is nonzero. Thus,

$$\frac{dU_r}{dt} = CU_r, \qquad (4.6\text{-}10)$$

where C is the proportional coefficient. Eq. (4.6-10) can also be expressed in terms of approaching distance, x, by

$$\frac{dU_r}{dx} = -C. \qquad (4.6\text{-}10a)$$

Integration of Eq. (4.6-10) gives the time-dependent relative velocity as

$$U_r = U_c \exp\left(-C(t_c - t)\right), \qquad (4.6\text{-}10b)$$

where U_C is the collision velocity at contact. Using Eq. (4.6-10a) and Eq. (4.6-10b), C can be related to the initial velocity U_0, contact velocity U_C, initial separation distance l_0 and contact time t_C as

$$C = \frac{1}{t_c}\ln\left(\frac{U_c}{U_0}\right) = \frac{U_c - U_0}{l_0}. \qquad (4.6\text{-}10c)$$

Substitution of Eq. (4.6-10b) into Eq. (4.6-9) yields

$$CU_r = \frac{2(\rho_p - \rho)g(1-A)}{2\rho_p + \rho}\exp\left(\frac{B}{d_p}\int_t^{t_c}U_r dt\right) - \frac{18U_r\sqrt{C\rho\mu}}{(2\rho_p + \rho)}\mathrm{erf}\left(\sqrt{Ct}\right), \qquad (4.6\text{-}11)$$

where $\mathrm{erf}(\phi)$ is the error function of ϕ. At the moment of contact, with Eq. (4.6-10c), Eq. (4.6-11) reduces to

$$\frac{(U_c - U_0)d_p}{l_0} = \frac{2(\rho_p - \rho)g(1-A)}{2\rho_p + \rho} - \frac{18U_c}{(2\rho_p + \rho)}\left(\frac{(U_c - U_0)\rho\mu}{l_0}\right)^{\frac{1}{2}}\mathrm{erf}\left(\sqrt{\ln\left(\frac{U_c}{U_0}\right)}\right). \qquad (4.6\text{-}11a)$$

Equation (4.6-11a) expresses the collision velocity explicitly as a function of the given flow conditions and the properties of the fluid and solid spherical particles.

Comments: (1) Wake shedding occurs at higher Reynolds numbers, which not only triggers the asymmetric and unsteady fluid flow in the wake region (that may alter the drag forces) but also leads to a rotation or spin motion of the interacting spheres (so that the addition of Magnus force and alternation in added mass force may need to be considered). (2) When spheres approach each other or a solid surface, the added mass is not a constant, as shown in Eq. (4.2-3a), and there exists additional pressure force, as shown in Eq. (4.2-3b).

4.6.3 Collision of Elastic Spheres in Fluid

Problem Statement: Consider a head-on collision of two elastic spheres in a stagnant viscous fluid. Develop a model for the dynamic motion and mechanical behaviors of the colliding spheres. A complete collision process includes a compression and a rebound, as discussed by Zhang *et al.* (1999).

Analysis: Assume the two spheres are identical so that the collision process is symmetric and in the horizontal direction. Thus only analysis on one colliding sphere is needed. During the compression process, the forces acting on the sphere include the drag force, the pressure force, the compression force due to the elastic deformation of the colliding sphere, and the Basset force. It is assumed that the particle deformation is small so that the viscous drag force and the pressure force still take the same form as for a sphere, while the compression force takes the same form as if the collision were in a vacuum. Moreover, during collision the added mass is considered constant while the Basset force takes the same form as if the sphere were in an unbounded fluid environment.

Denoting s as the approaching distance from the center of the sphere to the symmetric plane of collision, the equation of the sphere can be expressed as

$$\left(\rho_p + \frac{1}{2}\rho\right)\frac{\pi}{6}d_p^3\frac{dU_p}{dt} = -F_D - F_P - F_C - F_B, \tag{4.6-12}$$

where both the drag force, F_D, and the Basset force, F_B, take the forms given in Eq. (3.2-21); the pressure force, F_P, is expressed as in Eq. (4.2-3b); and the compression force, F_C, Eq. (4.3-1e). Substituting the detailed forms of these forces, Eq. (4.6-12) can be expressed in terms of the approaching distance, s, as

$$\left(\rho_p + \frac{1}{2}\rho\right)\frac{\pi}{6}d_p^3\frac{d^2s}{dt^2} = -3\pi\mu d_p\frac{ds}{dt} - \frac{3\pi\rho}{32}d_p^2\left(\frac{ds}{dt}\right)^2$$
$$- \frac{4E}{3(1-v^2)}\sqrt{\frac{d_p}{2}}s^{\frac{3}{2}} - \frac{3}{2}\sqrt{\frac{\pi\mu}{\rho}}d_p^2\int_0^t\frac{1}{\sqrt{t-\tau}}\left(\frac{d^2s}{d\tau^2}\right)d\tau \tag{4.6-12a}$$

where E is the Young's modulus and v is the Poisson's ratio. Equation (4.6-12a) can be solved numerically, with the initial conditions of:

$$\left(\frac{ds}{dt}\right)_{t=0} = U_C \qquad s_{t=0} = 0. \qquad (4.6\text{-}12b)$$

At maximum compression, the collision velocity is zero, which gives the condition to determine the compression time, t_C, and the maximum approaching distance, s_m, as:

$$\left(\frac{ds}{dt}\right)_{t=t_c} = U_C \qquad s_{t=t_c} = s_m. \qquad (4.6\text{-}12c)$$

The maximum collision force is thus obtained from Eq. (4.3-1e) as

$$F_{C\,\text{max}} = -\frac{4E}{3(1-v^2)}\sqrt{\frac{d_p}{2}}s_m^{\frac{3}{2}}. \qquad (4.6\text{-}12d)$$

A similar modeling approach can be extended to the rebound process. However, during the rebound, the pressure force and the compression force aid the rebound while the drag and Basset force resist the rebound. Hence the equation of motion of the sphere in rebound is expressed by

$$\left(\rho_p + \frac{1}{2}\rho\right)\frac{\pi}{6}d_p^3\frac{d^2s}{dt^2} = -3\pi\mu d_p\frac{ds}{dt} + \frac{3\pi\rho}{32}d_p^2\left(\frac{ds}{dt}\right)^2$$
$$+ \frac{4E}{3(1-v^2)}\sqrt{\frac{d_p}{2}}s^{\frac{3}{2}} - \frac{3}{2}\sqrt{\frac{\pi\mu}{\rho}}d_p^2\int_0^t\frac{1}{\sqrt{t-\tau}}\left(\frac{d^2s}{d\tau^2}\right)d\tau, \qquad (4.6\text{-}13)$$

with the initial conditions of

$$\left(\frac{ds}{dt}\right)_{t=0} = 0 \qquad s_{t=0} = s_m. \qquad (4.6\text{-}13a)$$

After solving Eq. (4.6-13), the rebound time, t_r, and the rebound velocity, U_r, can be determined, respectively, by

$$\left(\frac{ds}{dt}\right)_{t=t_r} = U_r \qquad s_{t=t_r} = 0. \qquad (4.6\text{-}13b)$$

The total collision time is simply the sum of t_C and t_r.

Comments: (1) For particle collisions with the Reynolds number far beyond the Stokes regime, both drag force and Basset force should be modified. (2) In reality, most collisions are not perfectly center-to-center. The off-center collision will lead to tangential forces and may trigger the spinning motion of colliding particles.

4.6.4 Leidenfrost Collision of a Drop with a Flat Surface

When an evaporative droplet collides with a solid surface maintained at a sufficiently high temperature, vapor generated during the collision forms a veritable cushion between the droplet and the surface that prevents direct contact. The vapor layer, with thickness typically orders of magnitude smaller than the droplet size, forestalls

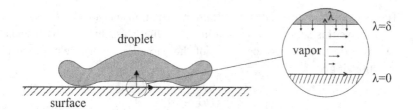

Figure 4.6-2 Droplet–surface collision in Leidenfrost regime.

the evaporation rate and protracts evaporation time. The collision is classified as within the film boiling regime, and the observed effect is termed the Leidenfrost phenomenon.

Problem Statement: Consider a typical collision between a droplet and a high-temperature surface in the Leidenfrost regime, as shown in Figure 4.6-2. Develop an analytical model for the pressure profile of the vapor layer.

Analysis: Assume the vapor layer consists solely of saturated vapor, and the vapor velocity is much larger than the rate of deformation of the vapor layer thickness and breadth. When contact between the droplet and heated surface is imminent, the vapor layer thickness is much smaller than the radius of the droplet, or:

$$\frac{\delta}{R} << 1. \tag{4.6-14}$$

In the axisymmetric cylindrical coordinate system shown in Figure 4.6-2, if the gravitational and temporal terms are neglected and based on the order of magnitude analysis, the momentum equations for the incompressible vapor flow can be written as (Ge and Fan, 2005):

$$U\frac{\partial U}{\partial \xi} + V\frac{\partial U}{\partial \lambda} = -\frac{1}{\rho}\frac{\partial p}{\partial \xi} + \frac{\mu}{\rho}\frac{\partial^2 U}{\partial \lambda^2}, \tag{4.6-15a}$$

and

$$\frac{\partial p}{\partial \lambda} = 0. \tag{4.6-15b}$$

where U and V are the vapor velocity components in the ξ and λ directions, respectively. The boundary conditions are given by:

$$\lambda = 0, \ U(\xi, 0) = V(\xi, 0) = 0$$

$$\lambda = \delta, \ U(\xi, \delta) = U_l(\xi), \ V(\xi, \delta) = V_\delta(\xi)$$

$$\xi = 0, \ \frac{\partial}{\partial \xi} = 0; \ \xi = \xi_b, \ p = p_b, \tag{4.6-15c}$$

where $U_l(\xi)$ is the tangential velocity at the droplet surface, and $V_\delta(\xi)$ is the local vapor velocity at the vapor–droplet interface due to both the droplet motion and the vapor generation, as given by

$$V_\delta(\xi) = \frac{\dot{m}}{\rho} - \frac{\partial \delta(\xi)}{\partial t}. \tag{4.6-15d}$$

The evaporation rate is calculated from the heat transfer model.

To solve this set of equations, a linear profile is assumed for the velocity in the vertical direction. Using separation of the variables for the velocity in the radial direction, it gives:

$$V(\xi, \eta) = -\eta V_\delta(\xi)$$
$$U(\xi, \eta) = \Omega(\xi)\Phi(\eta), \tag{4.6-16}$$

where $\eta = \lambda/\delta$. Thus, Eq. (4.6-15a) can be converted to an ordinary differential equation, with the solution expressed as an infinite power series. The first three terms in the series provide an acceptable accuracy for the averaged vapor flow velocity, which is obtained as:

$$\overline{U}(\xi) = \frac{1}{2}U_l(\xi) - \frac{\delta^2 \rho}{12\mu}\left(1 - \frac{3}{20}\text{Re}_\delta\right)\frac{\partial}{\partial \xi}\left(\frac{p}{\rho}\right), \tag{4.6-17}$$

where Re_δ is the local evaporation Reynolds number, which can be estimated by

$$\text{Re}_\delta = \frac{\delta(\xi)\rho V_\delta}{\mu} \cong \frac{\delta}{\mu}\dot{m} \cong \frac{\delta}{\mu}\left(\frac{K}{L}\frac{\Delta T}{\delta}\right) = \frac{K}{\mu}\frac{\Delta T}{L}, \tag{4.6-17a}$$

where L is the latent heat. Conversely, the continuity equation relates the horizontal and vertical velocities by:

$$\overline{U}(\xi) = \frac{1}{\delta(\xi)\xi}\int_0^\xi \xi' V_\delta(\xi')d\xi'. \tag{4.6-18}$$

The pressure distribution in the vapor layer can be obtained by solving Eq. (4.6-17) and Eq. (4.6-18). The final two remaining unknown variables are the thickness of the vapor layer, $\delta(\xi)$, and the boundary velocity for $V_\delta(\xi)$. Interface capturing algorithms, such as the level set method and the heat transfer model, provide efficient techniques to solve for these respective expressions. Once the vapor pressure is obtained, the vapor force is incorporated into the momentum equation as an extra interface force on the droplet surface within the contact area. The details of the CFD model and the coupling between the hydrodynamics, heat transfer, and vapor layer models can be found in the reference (Ge and Fan, 2005).

Comments: (1) The vapor layer model is predicated on continuum theory. However, the validity of this assumption must be carefully considered when the thickness of the vapor layer is reduced such that it becomes comparable to the mean free path of the gas molecules. For a subcooled droplet, however, the vapor layer thickness is much smaller. A special vapor layer model for such cases has been developed that combines the kinetic analysis and the effective slip on the boundary (Ge and Fan, 2007); (2) The pressure force from the integration of vapor pressure distribution in the vapor layer plays a major role in the droplet rebound, along with the surface-tension-induced droplet deformation.

4.7 Summary

This chapter discusses the mechanisms and formulation of various basic particle–particle interactions. The essential modes of these interactions include:

- a pair of spheres interacting by head-on approaching or by wake attraction;
- flow through uniformly suspended spheres;
- electrostatic field induced by the suspended charged particles;
- normal collision dynamics involving forces, deformation, contact area, and duration for a pair of elastic spheres;
- van der Waals force;
- capillary force due to liquid bridge between two particles.

The chapter discusses the nonidealized particle–particle interactions and associated formulation, including

- radiation transport equation for thermal radiation within a particle cloud;
- collision dynamics with tangential friction and torsional traction of elastic spheres;
- inelastic collisions (such as plastic deformation and fracture), and the concept of restitution coefficient;
- heat and charge transfer by particle collisions;
- deformation, breakup, and coalescence of fluid particles.

Knowledge of the above particle–particle interactions is critical to the model formation of dense-phase multiphase flows. These basic models on collision dynamics of a pair of solid spheres are also useful to the development of the discrete element method (DEM) described in Chapter 5.

Nomenclature

A	Area
A_H	Hamaker constant
Ar	Archimedes number
a	Radius
C	Capacity
c	Specific heat
D	Diffusivity
	Diameter of column or pipe
d	Particle diameter
E	Electric field intensity
E	Young's modulus
e	Restitution coefficient
F	Force

Fo	Fourier number
f	Friction coefficient
g	Gravity acceleration
h	Convective heat transfer coefficient
	Charge transfer coefficient
I	Radiant intensity
J_q	Heat flux
K	Thermal conductivity
L	Characteristic length
l	Distance between two interacting particles
m	Mass
$\dot{m}$	Mass flux
$\boldsymbol{n}$	Normal unit vector
n	Richardson–Zaki index
Nu	Nusselt number
Oh	Ohnesorge number
P	Pressure
Q	Volumetric flowrate
q	Electric charge
	Heat transfer per collision
R	Radius of curvature
Re	Reynolds number
s	Approaching distance
T	Temperature
t	Time
U	Velocity
V	Tangential velocity
We	Weber number
Y	Yield stress
y	Distance from symmetric plane

Greek symbols

α	Volumetric fraction
δ	Film thickness
ε	Permittivity
	Eddy dissipation rate
κ	Permeability
λ	Wavelength
μ	Dynamic viscosity
ν	Poisson's ratio
ρ	Density
σ	Electric conductivity
	Surface tension

τ	Relaxation time
Φ	Electric potential
Ω	Drag correction factor
ω	Angular velocity

Subscript

0	Isolated
A	Added or carried mass
B	Basset
C	Collision
D	Drag
e	Effective
	Electrostatic
f	Final
i	Initial
m	Mixture
n	Normal
P	Pressure
pt	Terminal
p	Particle
r	Relative
S	Stokes
s	Static
T	Thermal
t	Tangential

Problems

P4.1 Consider a head-on collision of two identical rigid spheres in a viscous fluid. Develop a model to estimate the collision velocity, assuming that the initial velocity is sufficiently high to ensure a contact of spheres.

P4.2 Develop a model to estimate the drag force of a porous and permeable agglomerate in a swamp of similar agglomerates. The agglomerate is made of solid spheres. Estimate the range of drag correction factor Ω for $0.5 < \alpha < 0.9$ (bulk fluid voidage of swamp) and $0.5 < \alpha_a < 0.99$ (fluid voidage inside an agglomerate).

P4.3 For a particle settling in a column filled with particles suspended in fluid, show that there exists a maximum settling mass flux as the volumetric loading of particles increases. Using Eq. (4.6-5a) and Eq. (4.6-1a), derive a correlation for the maximum settling mass flux to the particle properties, fluid properties, and column diameter.

P4.4 Consider the heat conduction between a solid sphere and a spherical shell filled with stagnant fluid. Show that the equivalent Nusselt number can be expressed by Eq. (4.2-8).

P4.5 Consider the heat transfer of a particle suspension flow in a pipe. The suspension flow can be approximated by uniform distributions in phase velocities and particle concentration. The suspension is heated up by the pipe wall maintained at a given temperature. Assume that the local heat transfer between the particles and the fluid can be approximated by Eq. (4.2-8). Ignore thermal radiation effects. Develop a model for temperature distributions of particles and fluid along the pipeline.

P4.6 Repeat Problem P4.5 by including the radiative heat transfer effects. The thermal radiation contributes to particle heating and that radiation is dominated by single scattering. The fluid is transparent to the thermal irradiation (*i.e.*, ignore irradiation between the fluid and pipe wall surface).

P4.7 One simple model for the coefficient of restitution is an elastic-plastic impact collision where compression process is assumed to be partially plastic (part of the kinetic energy is stored for later elastic rebounding) and the rebound process is considered to be completely elastic (Johnson, 1985). This model postulates: (1) during the plastic compression process, $s = r_c^2 / 2a^*$; (2) during the compression process, the averaged contact pressure, P_m, is a constant given by $3Y$; and (3) the elastic rebound process starts when maximum deformation is reached. Show that the coefficient of restitution can be expressed by Eq. (4.3-14).

P4.8 Consider an impact between a polyethylene particle ($d_p = 1$ cm) and a copper wall. The incident velocity is 2 m/s, and the incident angle is 30°. The friction coefficient at the interface is 0.2. The densities of polyethylene and copper are 950 and 8,900 kg/m^3, respectively. What is the contact duration for the collision? Estimate the rebound velocity of the particle. Repeat the problem for a copper particle colliding with a polyethylene wall.

P4.9 For an impact between a polyethylene particle ($d_p = 1$ cm) and a copper wall, estimate the critical normal collision velocity of the particle above which plastic deformation will occur. The yield strength for polyethylene is 2×10^7 N/m^2, and the yield strength for copper is 2.5×10^8 N/m^2. What is the critical normal collision velocity for a copper sphere colliding with the copper wall?

P4.10 Heat transfer during a collisional contact can be roughly described by a transient and one-dimensional heat transfer model with constant properties, which is defined by:

$$\frac{\partial T}{\partial t} = D_T \frac{\partial^2 T}{\partial z^2} \tag{P4.10-1}$$

$$T|_{t=0} = 0 \quad T|_{z=0} = \phi(t).$$

(1) Show that the temperature distribution can be related to its boundary condition by:

$$T = \frac{2}{\sqrt{\pi}} \int_{\frac{z}{2\sqrt{D_T t}}}^{\infty} \phi\left(t - \frac{z^2}{4 D_T \xi^2}\right) e^{-\xi^2} d\xi. \tag{P4.10-2}$$

(2) Based on the above solution, show that the temperature gradient is given by:

$$\frac{\partial T}{\partial z} = -\frac{\phi(0)}{\sqrt{D_T \pi t}} e^{-\frac{z^2}{4D_T t}} - \int_{\frac{z}{2\sqrt{D_T t}}}^{\infty} \frac{z}{D_T \xi^2 \sqrt{\pi}} \frac{\partial \phi}{\partial z} e^{-\xi^2} d\xi. \qquad (P4.10\text{-}3)$$

P4.11 Show that, for a Hertzian collision, the charge transfer equation can be expressed as:

$$-\frac{d(\Phi_2 - \Phi_1)}{\Phi_2 - \Phi_1} = \frac{\pi a^* h_{e12}}{U_{12} C^*} \left[1 - \left(\frac{s}{s_m} \right)^{\frac{5}{2}} \right]^{-\frac{1}{2}} s \, ds. \qquad (P4.11)$$

Also show that, for a complete elastic impact process, the transferred charge can be evaluated from Eq. (4.3-19b).

P4.12 Similar to the particle–wall collision in a viscous liquid, a collision between a gas bubble and a solid wall also involves a thin liquid film resulting in a repelling force. Assuming Re < 1, develop an expression for the pressure profile in the liquid film.

P4.13 In bubble suspensions, spherical bubbles tend to align themselves in horizontal pairs, while deformed bubbles tend to align vertically. Explain this phenomenon and consider the bubble interactions in the explanation.

P4.14 Consider the particle–bubble interaction between a stationary spherical particle and a colliding spherical cap bubble in liquid, assuming that the bubble moves at its terminal velocity, and the center of the bubble moves along an axis passing through the center of the particle. The particle surface is always wetted by a liquid film during the contact. Neglect the liquid motion induced by the bubble. Develop a model for the particle–bubble interaction force acting on the stationary spherical particle. Plot the variation of the force during the collision.

P4.15 The wake-induced interaction becomes significant for the interacting distance within about two particle diameters, estimate the critical volume fraction of sphere-containing suspension beyond which the interparticle effect must be considered.

P4.16 Derive an expression for the liquid bridge force between two spherical particles.

P4.17 Explain why the dry ultrafine particles (such as submicron or nanoparticles) are cohesive and easy to form agglomerates but much larger particles of the same materials are less cohesive and hard to form agglomerates.

P4.18 In a jet milling application, two nozzles are shooting particles right toward each other. What is the critical initial particle velocity for particle crack ($p_c > \sigma_{crack}$)? Assume the gas velocity is always smaller than particle velocity, which may lead to the assumption that the whole fluid field is stagnant.

References

Bayvel, L. P., and Jones, A. R. (1981). *Electromagnetic Scattering and Its Applications.* London: Applied Science.

Bellan, J., and Harstad, K. (1987a). Analysis of the convective evaporation of non-dilute clusters of drops. *Int. J. Heat Mass Transfer,* **30**, 125–136.

Bellan, J., and Harstad, K. (1987b). The details of the convective evaporation of dense and dilute clusters of drops. *Int. J. Heat Mass Transfer,* **30**, 1083–1093.

Brenner, H. (1962). Effect of finite boundaries on the Stokes resistance of an arbitrary particle. *J. Fluid Mech.,* **12**, 35–48.

Carslaw, H. S., and Jaeger, J. C. (1959). *Conduction of Heat in Solids.* 2nd ed. Oxford: Oxford University Press.

Cheng, L., and Soo, S. L. (1970). Charging of dust particles by impact. *J. Appl. Phys.,* **41**, 585–591.

Chesters, A. K., (1975). The applicability of dynamic-similarity criteria to isothermal, liquid–gas, two-phase flows without mass transfer. *Int. J. Multiphase Flow,* **2**, 191–212.

Chesters, A. K., (1991). The modeling of coalescence processes in fluid-liquid dispersions: A review of current understanding. *Chemical Engineering Research and Design: transactions of the Institution of Chemical Engineers: Part A,* **69**, 259–270.

Chiang, C. H., and Sirignano, W. A. (1993). Interacting, convecting, vaporizing fuel droplets with variable properties. *Int. J. Heat Mass Transfer,* **36**, 875–886.

Cossali, G. E. (1997). The impact of a single drop on a wetted solid surface. *Exp. Fluid,* **22**, 463–472.

Crowe, C. T. (2006). *Multiphase Flow Handbook.* Boca Raton: CRC Press.

Crowe, C. T., Sommerfeld, M., and Tsuji, Y. (1998). *Multiphase Flow with Droplets and Particles.* Boca Raton: CRC Press.

Fan, L.-S., and Zhu, C. (1998). *Principles of Gas-Solid Flows.* Cambridge: Cambridge University Press.

Ge, Y., and Fan, L.-S. (2005). Three-dimensional simulation of impingement of a liquid droplet on a flat surface in the Leidenfrost regime. *Phys. Fluids,* **17**, 027104.

Ge, Y., and Fan, L.-S. (2007). Droplet-particle collision mechanics with film-boiling evaporation. *J. Fluid Mech.,* **573**, 331–337.

Goldsmith, W. (1960). *Impact: The Theory and Physical Behavior of Colliding Solids.* London: Edward Arnold.

Happel, J. (1958). Viscous flow in multiparticle systems: slow motion of fluids relative to beds of spherical particles. *AIChE J.,* **4**, 197–201.

Hertz, H. (1881). Ueber die Berührung fester elastischer Körper. *J. für die Reine und Angewandte Mathematik,* **92**, 156–171.

Hong, T., Fan, L.-S., and Lee, D. J. (1999). Force variations on particle induced by bubble-particle collision. *Int. J. Multiphase Flow,* **25**, 477–500.

Hunt, M. L. (1989). Comparison of convective heat transfer in packed beds and granular flows. *Annual Review of Heat Transfer.* Ed. C. L. Tien. Washington: Hemisphere.

Johnson, K. L. (1985). *Contact Mechanics.* Cambridge: Cambridge University Press.

Liang, S. C., Hong, T., and Fan, L.-S. (1996). Effects of particle arrangements on the drag force of a particle in the intermediate flow regime. *Int. J. Multiphase Flow,* **22**, 285–306.

Liao, Y., and Lucas, D. (2010). A literature review on mechanisms and models for the coalescence process of fluid particles. *Chem. Eng. Sci.*, **65**, 2851–2864.

Love, T. J. (1968). *Radiative Heat Transfer*. Columbus, OH: Merrill.

Lubkin, J. L. (1951). The torsion of elastic sphere in contact. *Trans. ASME, J. Applied Mech.*, **18**, 183–187.

Milne-Thomson, L. M. (1968). *Theoretical Hydrodynamics*. New York: Dover Publications.

Mindlin, R. D. (1949). Compliance of elastic bodies in contact. *Trans. ASME., J. of Appl. Mech.*, **16**, 259–268.

Neale, G., Epstein, N., and Nadar, W. (1973). Creeping flow relative to permeable spheres. *Chem. Eng. Sci.*, **28**, 1865–1874.

Qian, J., and Law, C. K. (1997). Regimes of coalescence and separation in droplet collision. *J. Fluid Mech.*, **331**, 59–80.

Richardson, J. F., and Zaki, W. N. (1954). Sedimentation and Fluidization, Part I. *Trans. Instn. Chem. Engrs.*, **32**, 35–53.

Rioboo, R., Tropea, C., and Marengo, M. (2001). Outcomes from a drop impact on solid surfaces. *Atomization Spray*, **11**, 155–165.

Soo, S. L. (1990). *Multiphase Fluid Dynamics*. Beijing: Science Press.

Sun, J., and Chen, M. M. (1988). A theoretical analysis of heat transfer due to particle impact. *Int. J. Heat Mass Transfer*, **31**, 969–975.

Sutherland, K. L. (1948). Physical chemistry of flotation. XI. Kinetics of the flotation process, *J. Phys. Chem.*, **52**, 394–425.

Tien, C. L., and Drolen, B. L. (1987). Thermal radiation in particulate media with dependent and independent scattering. *Annual Review of Numerical Fluid Mechanics and Heat Transfer*. Ed. T. C. Chawla. Washington: Hemisphere.

Zhang, J., Fan, L.-S., Zhu, C., Pfeffer, R., and Qi, D. (1999). Dynamic behavior of collision of elastic spheres in viscous fluids. *Powder Technol.*, **106**, 98–109.

Zhu, C., Liang, S.-C., and Fan, L.-S. (1994). Particle wake effects on the drag force of an interactive particle. *Int. J. Multiphase Flow*, **20**(1), 117–125.

Zhu, H. P., Zhou, Z. Y., Yang, R. Y., and Yu, A. B. (2007). Discrete particle simulation of particulate systems: Theoretical developments. *Chem. Eng. Sci.*, **62**, 3378–3396.

Zuber, N. (1964). On the dispersed two-phase flow in the laminar flow regime. *Chem. Eng. Sci.*, **19**, 897–917.

5   Continuum-Discrete Tracking Modeling of Multiphase Flows

5.1   Introduction

This chapter introduces a continuum-discrete approach for the dispersed multiphase flow known as the Eulerian–Lagrangian model. This method solves the governing equations for the continuous phase in Eulerian coordinates, which remain fixed in space as the flow passes through. Lagrangian coordinates, which are affixed to the flowing particles and track their trajectories, are employed in concert to describe changes in the discrete phase. Tracking the discrete phase in Lagrangian coordinates has two noteworthy advantages: first, historical particle changes are inherently captured; second, it functions as a viable alternative to continuum techniques to model sufficiently diluted discrete phases, which in some cases could not otherwise be possible. In the problems addressed in this chapter, the discrete phase particle size is several orders of magnitude smaller than the domain of interest. To illustrate, for a particle-laden flow with characteristic length of 1 m and particle size of 100 μm, the particle diameter is four orders of magnitude smaller than the length scale of the system of interest. Under such conditions, surfaces of individual particles generally are not explicitly resolved. In the continuum-discrete model, the discrete phase is modeled as points representing individual particles, whereas the integral effects of various forces acting on the particle surface are modeled in the form of forces and moments acting through the center of the particles.

For a particle in a multiphase flow environment, the general expression for the momentum transfer of the particle, i, can be written in the form of Newton's second law of motion as:

$$\frac{d\left(m_i \mathbf{u}_{pi}\right)}{dt} = \int_{A_i} \mathbf{T}_f \cdot \mathbf{n} dA_i + \sum_j \mathbf{F}_{Cji} + \sum_k \int_{A_i} \sigma_k \cdot \mathbf{n} dA_i + \sum_l \int_{V_i} \mathbf{f}_{l} \rho_{pi} dV_i. \quad (5.1\text{-}1)$$

The four terms of the right-hand side of Eq. (5.1-1) correspond to the four categories of forces affecting the particle's change in momentum. The first term represents fluid–particle interaction as embodied in the occurrences between the continuous phase fluid and the discrete phase particle via the total fluid stress over the particle surface. This term can be further decomposed into drag, lift, added mass forces, and others. The second term represents the interparticle contact force, including the collision forces, van der Waals force, and liquid-bridge force. The third term represents

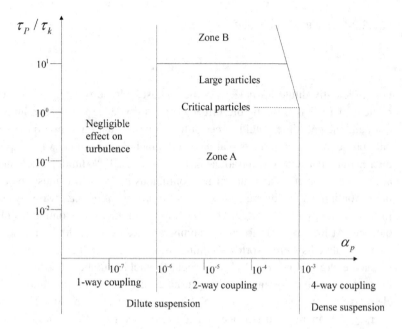

Figure 5.1-1 Classification map of particle–fluid and particle–particle phase couplings.

noncontact surface forces, including the electrostatic force, dielectric polarization force, and most gradient-based forces such as those due to electrophoresis, thermophoresis, and photophoresis. The fourth term represents noncontact mass-based forces such as gravitational force. If the initial conditions are defined and all force terms are properly modeled, the velocity and trajectory of the particle can be obtained by integrating Eq. (5.1-1). Equation (5.1-1) may take an alternative form as the equation of motion as:

$$m_i \frac{d\mathbf{u}_{pi}}{dt} = \mathbf{F}_{fi} + \mathbf{F}_{Ci} + \mathbf{F}_{NCi} + \mathbf{F}_{Gi} - \frac{dm_i}{dt}\mathbf{u}_{pi}, \qquad (5.1\text{-}2)$$

where the first four terms on the right-hand side stand for, respectively, the fluid–particle force, collision force, noncontact surface force, and field body force, corresponding to the four forces in Eq. (5.1-1), and the last term represents the effects due to changing particle mass, which may result from a reaction or evaporation during the particle's transport.

How to incorporate phase interactions into the continuum-discrete model is explored next. Selecting an approach depends on the significance of said phase interactions in the context of given flow conditions, such as the dynamic properties of the two phases and the volume fraction of the discrete phase. Figure 5.1-1 delineates where various phase coupling approaches are applicable for particle-laden turbulent flow, which chiefly depends on the volume fraction of the particle phase and time scale ratio of the particle relaxation to turbulence transport (Elghobashi, 2006). The particle relaxation time scale, τ_p, and the Kolmogorov time scale, τ_k, are adopted in the figure, in which τ_p is approximated by the Stokes relaxation time defined by

Eq. (3.2-22a) and τ_k is defined by:

$$\tau_k = \sqrt{\frac{\mu}{\rho\varepsilon}} \tag{5.1-3}$$

For sufficiently dilute flows (*e.g.*, particle phase volume fraction, $\alpha_p < 10^{-6}$), the particle phase has a negligible effect on the turbulence of the continuous phase flow, and the constituent fluid forces of the particle equation of motion represent the only phase interaction effects that need to be included (*i.e.*, one-way coupling). For intermediate particle concentrations ($10^{-6} < \alpha_p < 10^{-3}$), fluid–particle interaction affects the turbulence structure of the continuous phase, and it must be considered in the continuous and discrete phase momentum equations (*i.e.*, two-way coupling). In this regime, particles can either augment or diminish the continuous phase turbulence. At high particle Reynolds numbers (*e.g.*, $Re_p > 400$ for the steady flow over an isolated sphere), vortex shedding from the particle increases production turbulence energy (Zone B). At low particle Reynolds numbers (Zone A), the effect of particles on the continuous phase turbulence depends on the ratio of the particle relaxation time scale, τ_p, to the Kolmogorov time scale, τ_k. Small particles (*e.g.*, $\tau_p/\tau_k \leq 0.1$) increase turbulence kinetic energy and its dissipation rate, while large particles (*e.g.*, $\tau_p/\tau_k > 1$) reduce the aforementioned parameters. In dense suspensions (*e.g.*, $\alpha_p > 10^{-3}$), particle collision is significant, and particle–particle interactions must be incorporated into the model, resulting in four-way coupling.

If the local flow field of the continuous phase is known and not coupled with the motion of particles, the fluid–particle force models of Chapter 3 readily yield the particles' motion. However, the local fluid field may not always be physically predictable. For instance, for a turbulent flow, even if the time-averaged flow field can be predicted, the actual instantaneous flow velocity at a fixed location randomly fluctuates over the statistically averaged velocity. In this case, the local instantaneous fluid–particle interactions are also random in nature. Various particle trajectory models are introduced in Section 5.2, including deterministic models based on non-random fluid–particle interactions, and stochastic models with cloud tracking that account for the randomness of fluid–particle interactions. Particle–particle collision represents a major particle interaction mechanism and often employs discrete element modeling (DEM). There are two common DEM approaches: the first is the hard-sphere model, also known as the stereomechanical impact model, where collisions are assumed to be instantaneous and binary; the second is the soft-sphere model where the collision process is resolved by considering the relationship between particle deformation and various contact forces. Both approaches are described in Section 5.3. Coupling of continuous and discrete phases in the Eulerian–Lagrangian model (including the mass, momentum, heat, and charge transfers) is treated in Section 5.4.

Notably, tracking every individual particle is only feasible for very dilute multiphase flows. Most dispersed phases in real-world applications simply contain too many constituent particles to be severally tracked. For instance, with a particle volume concentration of 0.1% and an average particle size of 100 μm in a flow system

of 1 m³, there are approximately 2 billion particles that would need to be separately modeled and tracked; this is simply outside the bounds of available computing
capacity. A common practice is to assign particles to subgroups, perhaps a thou-
sand or so, and track the group instead; doing so greatly reduces the computational
power required for the tracking exercise. Group assignments are often based on a
logical relationship, such as location, particle size, and other transport properties;
groups can also be defined by various probability density functions. When grouping
is applied, it is understood that the disperse phase equations describe the dynamics
for a representative particle of the group and are not directly applied to individual
particles.

5.2   Lagrangian Trajectory Modeling

One popular (and possibly the simplest) categorization of Lagrangian trajectory
models is based on the randomness in velocity fluctuations of flows interacting with
particles of interest. Deterministic flows are perfectly predictable and without random fluctuations in fluid velocity. Conversely, stochastic flows include the instant
and random fluctuations. Two corresponding trajectory models for the discrete phase
have been developed for the deterministic and stochastic flows and are so named. The
deterministic approach considers particle motion in a laminar or time-averaged turbulent flow. Thus, the method neglects the effect of transient random flow fluctuation
(such as turbulence) on particles. For a perfectly steady flow field, the model predicts
two identical particles with the same initial conditions will follow the exact same
trajectory. On the other hand, the stochastic approaches can account for particle transport in a random turbulent flow field where the genesis of turbulence is aggregate
fluid flow or fluid–particle interactions, such as wake shedding. Turbulent transport
of particles can be simulated using two techniques: direct stochastic trajectory modeling or the particle cloud tracking. The direct stochastic model attributes random
fluctuations to particle velocity through particle–fluid interactions with random fluctuations in fluid velocity. The Monte Carlo method is employed for applications
with multifarious trajectories. The computed trajectories are interrogated to obtain
statistics for the particle motion. In the particle cloud tracking, a group of particles are represented as a cloud of particles whose radius increases over time due to
turbulent dispersion. A shape distribution of the particle positions within the cloud
is assumed; the cloud center and radius are tracked throughout its trajectory. The
three Lagrangian tracking models are schematically illustrated in Figure 5.2-1. The
deterministic model is simple, but it may overlook turbulent transport behavior. The
direct stochastic trajectory model intuitively accounts for turbulent transport of particles, but it requires tracking multitudinous trajectories to obtain meaningful statistics;
a computationally demanding task. The cloud tracking model significantly reduces
the number of particle trajectories to be calculated, but it struggles to capture cloud
behavior near its boundaries. Even in dilute flows, the actual number of particles
may still exceed reasonable capacity of available computing power. To overcome

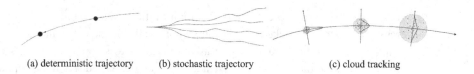

(a) deterministic trajectory   (b) stochastic trajectory   (c) cloud tracking

Figure 5.2-1 Illustration of three Lagrangian trajectories models.

this, simulations often account for far fewer particles than actually present; each particle in such an analysis represents a sample, or a parcel, of multiple actual particles with similar properties and trajectories.

5.2.1   Deterministic Trajectory Models

The basic form of a trajectory model is an expression for the motion of a single particle, such as the Basset–Boussinesq–Oseen (BBO) equation introduced in Section 3.5. Two criteria in deterministic trajectory modeling (DTM) are paramount: (1) the flow field of the carrying fluid must be without instantaneous stochastic disturbances (*i.e.*, deterministic); and (2) no random-natured particle transport (such as turbulence) can be present that would cause particle diffusion; particles with the same properties and initial conditions must follow the same trajectories. The least complex form of DTM is used for one-way coupling where only fluid effects on particles are considered, neglecting any effects of particles on the fluid flow. However, DTM can be expanded through iterative solution techniques to account for two-way coupled interactions between the particle and fluid phases and for interparticle and particle–wall collisions of four-way coupled systems.

To solve the general equation of motion for a particle, given by Eq. (5.1-1) or Eq. (5.1-2), the fluid–particle interaction force must be explicitly defined. As mentioned in Chapter 3, various hydrodynamic forces on the particle (*e.g.*, the drag force, Basset force, Saffman force, and Magnus force) are coupled with the fluid flow field, whose governing equation is often highly nonlinear. If exceptional circumstances produce a linear equation for the flow field (*e.g.*, when the particle Reynolds number is much less than unity with no change in particle phase), the individual component forces from various intrinsic hydrodynamic mechanisms can simply be added to compute the total hydrodynamic force. In this special case, the equation of motion is known as the BBO Equation, whose general expression is obtained from simplification of Eq. (5.1-1) as:

$$\frac{d\mathbf{u}_p}{dt} = \frac{1}{m}\left(\mathbf{F_D} + \mathbf{F_A} + \mathbf{F_B} + \mathbf{F_S} + \mathbf{F_M} + \mathbf{F_G} + \mathbf{F_C} + \ldots\right), \qquad (5.2\text{-}1)$$

where $\mathbf{F_D}$ is the drag force; $\mathbf{F_A}$ is the added (or carried) mass force; $\mathbf{F_B}$ is the Basset force; $\mathbf{F_S}$ is the Saffman force; $\mathbf{F_M}$ is the Magnus force; $\mathbf{F_G}$ is the gravitational force; $\mathbf{F_C}$ is the collision force; "..." stands for other forces, such as electrostatic force, van der Waals force, magnetic force, and forces due to electrophoresis or thermophoresis. The first five terms in Eq. (5.2-1) (*i.e.*, $\mathbf{F_D}, \mathbf{F_A}, \mathbf{F_B}, \mathbf{F_S}, \mathbf{F_M}$) correspond to the hydrodynamic forces acting on the particle by the carrying fluid, and they are coupling with local fluid velocity or its derivatives.

In the presence of either phase changes or chemical reactions, particle mass follows the law of mass conservation, which is expressed by the mass conservation equation as:

$$\frac{dm}{dt} = \gamma, \tag{5.2-2}$$

where γ is the rate of reaction or phase change, which is highly dependent on heat transfer rate, temperature distribution, along with the thermodynamic properties of the particle, carrying fluid, and surrounding environment. Coupling of heat transfer rate with temperature distribution requires the law of energy conservation, which is expressed by:

$$\frac{d}{dt}\left(mh_p\right) = \sum_i q_i, \tag{5.2-3}$$

where h_p is the total enthalpy per unit mass of the particle. The right-hand side of Eq. (5.2-3) represents the total rate of heat transfer to the particle, including the heat transferred via particle–fluid contact, interparticle collisions, and the thermal radiation from the surrounding environment (*e.g.*, nearby particles and surfaces of the fluid–particle transport vessel). When particles carry electrostatic charge, even in the absence of an external electric field, interactions between the charged particles and their container will induce a local electric field. Consequently, the motion of each individual particle is influenced by an electrostatic force. The details of the charge-coupled particle–particle interactions are presented in Section 5.3. The presence of an external electric field can even influence neutral particles (*i.e.*, no electrostatic charge). For example, neutral dielectric particles can be polarized in a nonuniform electric field, leading to a dipolar-induced electric force that affects the motion of even dielectric particles.

Once the particle velocity is determined, either from Eq. (5.2-1) directly or from coupling with other associated expressions, the particle trajectory is obtained by integration of the particle velocity as:

$$\mathbf{r}_p = \mathbf{r}_{p0} + \int_0^t \mathbf{u}_p \, dt, \tag{5.2-4}$$

where $\mathbf{r}_p$ is the position vector of the traced particle or the center of the subgroup of particles.

5.2.2 Stochastic Trajectory Models

The stochastic instantaneous disturbance of the continuous phase flow field can have a considerable impact on particle transport. For example, particles can be dispersed due to the random-natured flow turbulence in a phenomenon known as particle turbulent diffusion, mimicking fluid diffusion by molecular fluctuation or Brownian motion. In such cases, even if the flow field of the continuous phase is modeled with a deterministic approach on a time-averaged basis, such as the Reynolds averaged Navier–Stokes (RANS) equations with the k-ε turbulence model, the effect on particle transport due to instantaneous flow turbulence must be accounted for in the particle phase trajectory model.

Stochastic characteristics of instantaneous turbulent flow velocity can be simulated by introducing random numbers into the flow velocity calculation. To illustrate this methodology, consider a particle–fluid hydrodynamic interaction represented by the Stokes drag force with no change in phase; interparticle collisions and other forces on the particles may be ignored. A simplified expression for the instantaneous motion of a particle may be derived from Eq. (5.1-2) as:

$$\frac{d\mathbf{u}_p}{dt} = \frac{1}{\tau_S}\left(\mathbf{u} - \mathbf{u}_p\right) + \mathbf{f}, \qquad (5.2\text{-}5)$$

where τ_S is the Stokes relaxation time, defined by Eq. (3.2-22a); $\mathbf{u}$ and $\mathbf{u}_p$ are the instantaneous velocities of the fluid and the particle, respectively; $\mathbf{f}$ is the body force acting on the unit mass of the particle. The instantaneous fluid velocity $\mathbf{u}$ can be decomposed into a time-averaged component, $\mathbf{U}$, and a fluctuating component, $\mathbf{u}'$. It then yields:

$$\mathbf{u} = \mathbf{U} + \mathbf{u}'. \qquad (5.2\text{-}6)$$

The fluctuating fluid velocity vector can be related to a dimensionless random vector, ζ, and the kinetic energy from fluctuating velocity, given by:

$$\mathbf{u}' = \zeta\sqrt{\frac{2}{3}k} \qquad (5.2\text{-}6\text{a})$$

The random vector ζ is typically modeled to follow a Gaussian distribution. For the isotropic turbulent flows, the kinetic energy from fluctuating velocity is the same as the turbulence kinetic energy, k, defined in Eq. (2.3-15), which can be determined from a preferred turbulence model, such as the k–ε model. Substituting Eq. (5.2-6) into Eq. (5.2-5) yields an integrable equation for the instantaneous particle velocity. It then gives:

$$\frac{d\mathbf{u}_p}{dt} = \frac{1}{\tau_S}(\mathbf{U} + \mathbf{u}' - \mathbf{u}_p) + \mathbf{f}. \qquad (5.2\text{-}7)$$

The final stochastic trajectories of the particles are obtained by substituting the solved $\mathbf{u}_p$ into Eq. (5.2-4).

For numerical integration of a particle trajectory, selecting the integration time step is important. A typical method for choosing the integration time step relies on the interaction duration between a turbulent eddy and the particle of interest. The interaction duration, τ_i, is determined by the lesser value between eddy existence time, τ_e, and particle residence time through the eddy, τ_R. Thus:

$$\tau_i = \min\left(\tau_e, \tau_R\right). \qquad (5.2\text{-}8)$$

τ_e can be estimated from the k–ε turbulence model as:

$$\tau_e = \frac{l_e}{\sqrt{2k}} = \sqrt{\frac{3}{2}} C_\mu^{\frac{3}{4}} \frac{k}{\varepsilon}, \tag{5.2-9}$$

where l_e is the characteristic length of the turbulent eddy, defined by Eq. (2.3-2). τ_R is determined by:

$$\tau_R = -\tau_S \ln\left(1 - \frac{l_e}{\tau_S |U - u_p|}\right), \tag{5.2-10}$$

with τ_i is taken as τ_e when $l_e > \tau_S |U - u_p|$.

To use the stochastic trajectory model, a random number generator must be employed to generate the statistically distributed number, ζ, in Eq. (5.2-6a) for individual trajectory in every time step. For particles with distinct initial conditions, the trajectory computation is repeated until reliable statistics are obtained from generated sample trajectories. This schema, which runs multitudinous repeated simulations with random numbers of predefined distributions, is known as the Monte Carlo method. Due to the numerous sample trajectories that must be generated, the stochastic trajectory model requires extensive computational resources. In addition to the random nature of flow turbulence, other inherently random factors such as initial velocity, initial position, and particle size may be significant and may also be incorporated in the stochastic trajectory model.

5.2.3   Particle Cloud Tracking Models

A particle cloud tracking model follows the evolution of particle clouds in a dilute turbulent flow. Each cloud is composed of a group of particles with the same size, physical properties, and similar relative initial positions. Close proximity of the particles ensures a homologous flow environment; these conditions make it reasonable to assume the particles comprise the same cloud as it moves along a mean trajectory.

Figure 5.2-2 illustrates the effect of turbulent dispersion on the cloud radius morphology, characterized by a consistent increase, as it traverses the domain. The series of circles with increasing radii represent the spatial distributions for particles in the cloud at successive time steps. At any time, the total particle concentration at a fixed point in space includes the contribution from other clouds with different residence times. Because particle–particle interactions are ignored, the particle cloud tracking method is limited to very dilute particle-laden flows.

The spatial distribution for particles in a cloud is represented by a probability density function (PDF) of the particle location and its residence time. Assuming the particle PDF at any location and time can be represented by a multivariate Gaussian distribution, $P(\mathbf{r}_p, t)$, as:

$$P(\mathbf{r}_p, t) = \frac{1}{(8\pi)^{3/2} \prod\limits_{i=1}^{3} \sigma_i(t)} \exp\left(-\frac{1}{2} \sum_{i=1}^{3} \left(\frac{r_{p,i}(t) - <r_{p,i}(t)>}{\sigma_i(t)}\right)^2\right). \tag{5.2-11}$$

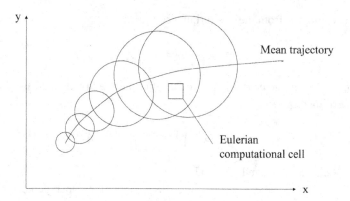

Figure 5.2-2 Two-dimensional representation of the evolution of a particle cloud.

The assembly-averaged particle position, $<\mathbf{r}_p>$, represents the most probable location for the center of the particle cloud, which is indicative of the constituent particles themselves. The statistical variance, σ_i, represents the extent of dispersion of the particles. The objective of the particle cloud tracking model is to determine $<\mathbf{r}_p>$ and σ_i, which completely describe the PDF.

The location of the cloud center can be calculated by integrating the assembly-averaged particle velocity, a task typically employed in the Lagrangian tracking models, as:

$$\langle \mathbf{r}_p(t) \rangle = \langle \mathbf{r}_p(0) \rangle + \int_0^t \langle \mathbf{u}_p(\tau) \rangle d\tau, \tag{5.2-12}$$

where the assembly-averaged particle velocity is obtained from the assembly-averaged equation of motion of the particles, expressed as:

$$\frac{d\langle \mathbf{u}_p \rangle}{dt} = \frac{1}{\tau_S} (\langle \mathbf{U} \rangle - \langle \mathbf{u}_p \rangle) + \mathbf{f}. \tag{5.2-13}$$

In the above equation, $\langle \mathbf{U} \rangle$ and $\langle \mathbf{u}_p \rangle$ are the assembly-averaged fluid and particle velocity, respectively. The assembly-averaged fluid velocity is calculated by averaging the fluid velocity in the entire cloud, weighted by the local particle population density:

$$\langle \mathbf{U} \rangle = \frac{\iiint\limits_{\text{cloud}} \mathbf{U}(\mathbf{r}) \dot{n} P(\mathbf{r}, t) d\mathbf{r}}{\iiint\limits_{\text{cloud}} \dot{n} P(\mathbf{r}, t) d\mathbf{r}}, \tag{5.2-13a}$$

where $\dot{n}$ is the number flow rate of particles in the cloud. It is noted that $<\mathbf{U}>$ is distinct from the local fluid velocity at the center of the particle cloud. The Stokes relaxation time, τ_S, is used to approximate the particle's momentum relaxation time, which is true for small particles in a dilute flow. Other forces, such as the added mass force and lift force, can be added to Eq. (5.2-13), although their contributions are usually much less significant when compared to the drag force.

The statistical variance, σ_i, in the PDF or particle positions in Eq. (5.2-11) is expressed in terms of the particle fluctuating velocity correlation, R_{pij}, as:

$$\sigma_i^2(t) = 2 \int_0^t (t - \tau) R_{pii}(\tau) d\tau. \tag{5.2-14}$$

In general, the particle fluctuating velocity correlation matrix, $R_{pij}(t_1, t_2)$, is defined as:

$$R_{pij}(t_1, t_2) \equiv \left\langle u'_{pi}(t_1) u'_{pj}(t_2) \right\rangle \tag{5.2-14a}$$

The Markovian approximation is employed so the particle correlation is dependent only on the time difference, $\tau = |t_1 - t_2|$. A simple model for the particle fluctuating velocity correlation is then given by (Smith, 1993):

$$R_{pij}(\tau) = \left\langle \left| \mathbf{u}'_p \right|^2 \right\rangle \exp\left(-\frac{\tau}{\tau_L} \right). \tag{5.2-14b}$$

The particle fluctuating velocity correlation is connected to the fluid fluctuating velocity correlation, which is further linked to the turbulence kinetic energy, by:

$$\left\langle \left| \mathbf{u}'_p \right|^2 \right\rangle = \frac{2}{3} k \left(1 - \exp\left(-\frac{\tau_e}{\tau_S} \right) \right). \tag{5.2-14c}$$

The Lagrangian integral time scale, τ_L, is approximated by:

$$\tau_L = \max(\tau_e, \tau_S). \tag{5.2-14d}$$

More sophisticated models for particle velocity correlations have been incorporated into the particle cloud tracking approach (Wang and Stock, 1993).

Once the PDF of particle position distribution is obtained, the motion of the discrete phase can be fully described. If a two-way coupling approach is employed, the contributions by the discrete phase of particles to the mass, momentum, and energy of continuous phase of the fluid are accounted for by dynamically adding source terms to the fluid transport equations in Eulerian coordinates. If the rate of change for particle property is given by $\gamma_{p\phi}$, then the total source term in an Eulerian cell is the sum of the particle clouds covering the Eulerian cell volume, $V(\mathbf{r})$, with different residence times, τ. This is expressed as:

$$s_{p\phi}(\mathbf{r}) = \int_0^\infty \int_{V(\mathbf{r})} \dot{n} \gamma_{p\phi} W(\mathbf{r}, \tau) d\mathbf{r} d\tau. \tag{5.2-15}$$

The weight function, $W(\mathbf{r}, t)$, which describes the fraction of particles inside the Eulerian cell volume, $V(\mathbf{r})$, within a cloud, is defined based on the PDF as:

$$W(\mathbf{r}, t) = \frac{P(\mathbf{r}, t)}{\int_{cloud} P(\mathbf{r}, t) d\mathbf{r}}. \tag{5.2-15a}$$

5.3 Discrete Element Method

As with other Lagrangian models of the particle phase, the discrete element method (DEM) tracks particle motion by integrating their individual equations of motion. The unique advantage of the DEM is its ability to provide microscopic dynamic information about the instantaneous particle velocity and contact forces, provided particle collisions are involved. The original formulation for the DEM was based on particle motions without considering particle–fluid interactions, essentially modeling in a vacuum. Such DEM approaches have been widely applied in granular material processes, such as separation, mixing, agglomeration, and solid transport where particle inertia dominates transport, collisions, and other nonfluid interactions (*e.g.*, external field forces and electrostatic charges). When coupled with fluid phase simulation techniques such as computational fluid dynamics (CFD), the DEM has since been extended to applications involving fluid–particle systems, such as fluidized beds and spout beds where neither the particle–fluid interactions nor particle collisions can be ignored. This section focuses solely on the most fundamental element of the DEM; namely, the explicit consideration of particle–particle interactions by collision of a pair of particles. Depending on how particle–particle contact is treated during a particle collision, the DEM approaches fall into one of two classes: these are the hard-sphere model or the soft-sphere model.

5.3.1 Hard-Sphere Model

The hard-sphere model, also known as the stereomechanical impact model, assumes that a particle collision completes devoid of particle deformation, and is instantaneous such that only momentum changes of colliding particles are considered. According to classical mechanics, momentum is simply redistributed between the particles after the collision, with some loss of kinetic energy due to inelasticity and friction. The momentum equation for the particle is given in the integral form, which does not explicitly calculate contact force. The particle collision occurs one at a time, and only binary collision is considered. Therefore, the hard-sphere model is suitable for cases where multiple-particle collisions or enduring contacts are absent, and the collision time is comparatively much shorter than the particle free flight time.

Consider a collision between two rigid spheres, as shown in Figure 5.3-1. The changes of linear and angular momenta for each of the colliding spheres can be expressed, respectively, by:

$$m_1 \left(\mathbf{u}_1' - \mathbf{u}_1 \right) = \mathbf{J} \tag{5.3-1}$$

$$m_2 \left(\mathbf{u}_2' - \mathbf{u}_2 \right) = -\mathbf{J} \tag{5.3-2}$$

$$I_1 \left(\boldsymbol{\omega}_1' - \boldsymbol{\omega}_1 \right) = a_1 \mathbf{n} \times \mathbf{J} \tag{5.3-3}$$

$$I_2 \left(\boldsymbol{\omega}_2' - \boldsymbol{\omega}_2 \right) = a_2 \mathbf{n} \times \mathbf{J}, \tag{5.3-4}$$

where $\mathbf{J}$ is the impulse force between two particles; I is the moment of inertia, for a sphere of uniform density, $I = (2/5)ma^2$; $\mathbf{n}$ is the normal vector from Particle 1 to

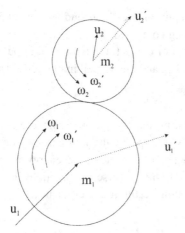

Figure 5.3-1 Collision between two rigid spheres.

Particle 2, expressed by:

$$\mathbf{n} = \frac{\mathbf{r}_2 - \mathbf{r}_1}{|\mathbf{r}_1 - \mathbf{r}_2|}. \tag{5.3-5}$$

The pre-collision relative velocity at the contact point, $\mathbf{u}_C$, is given by:

$$\mathbf{u}_C = \mathbf{u}_1 - \mathbf{u}_2 + \omega_1 \times n a_1 + \omega_2 \times n a_2. \tag{5.3-6}$$

The tangential component of relative velocity is expressed by:

$$\mathbf{u}_{Ct} = \mathbf{u}_1 - \mathbf{u}_2 - (\mathbf{u}_1 - \mathbf{u}_2 \cdot \mathbf{n})\mathbf{n} + \omega_1 \times n a_1 + \omega_2 \times n a_2. \tag{5.3-6a}$$

The tangential unit vector is then given by:

$$\mathbf{t} = \frac{\mathbf{u}_{Ct}}{|\mathbf{u}_{Ct}|}. \tag{5.3-6b}$$

The post-collision relative velocity can be defined similar to Eq. (5.3-6) by using the post-collision linear and angular velocities. With Eqs. (5.3-1) to (5.3-4), it can be shown that:

$$\mathbf{u}'_C - \mathbf{u}_C = \frac{7\mathbf{J} - 5\mathbf{n}\,(\mathbf{J} \cdot \mathbf{n})}{2m^*}, \tag{5.3-7}$$

where m^* is the reduced mass, defined by Eq. (4.3-7a).

The change of relative velocity in the normal direction is typically characterized by the restitution coefficient, which is defined by:

$$e = -\frac{\mathbf{u}'_C \cdot \mathbf{n}}{\mathbf{u}_C \cdot \mathbf{n}}. \tag{5.3-8}$$

For a perfectly elastic and normal collision, $e = 1$. Thus, the coefficient of restitution represents the degree of inelastic kinetic energy loss. The normal component of the impulse force, $\mathbf{J}$, is obtained from:

$$J_n = -m^* (1 + e)\,(\mathbf{u}_C \cdot \mathbf{n}). \tag{5.3-9}$$

The tangential components of the impulse force, $\mathbf{J}$, depend on contact process; of significance is whether tangential sliding continued or halted during contact. If the sliding occurs during the entire duration of contact, the tangential impulse is given by the product of the friction coefficient, f, and the normal impulse, J_n. Thus:

$$J_t = f J_n. \tag{5.3-10}$$

Conversely, a dissipation of energy during the sliding process can cause the relative tangential velocity, $\mathbf{u}_{Ct}$, to halt. At this point, the two particles do not slide, but they instead roll against each other. Further, the stored energy in the contact region causes the relative tangential motion to reverse its direction. In this case, a tangential restitution coefficient (or rolling restitution coefficient), β, is defined as (Walton, 1993):

$$\beta = -\frac{\mathbf{u}'_C \cdot \mathbf{t}}{\mathbf{u}_C \cdot \mathbf{t}}. \tag{5.3-11}$$

The value of β is in the range $0 \leq \beta \leq 1$; with $\beta = 0$ indicating the stick condition, and a positive β reveals a reversal of relative tangential velocity. Using Eq. (5.3-7) and Eq. (5.3-11), the tangential impulse under the stick/rolling condition is given by:

$$J_t = -\frac{2}{7} m^* (1 + \beta) (\mathbf{u}_c \cdot \mathbf{t}). \tag{5.3-12}$$

Since the tangential friction force must be less than or equal to the sliding frictional force, a criterion to determine the type of tangential contact can be established by comparing Eq. (5.3-11) and Eq. (5.3-12). The criterion for "slip-through" collision is thus given as:

$$-\frac{2}{7} m^* (1 + \beta) (\mathbf{u}_c \cdot \mathbf{t}) < f J_n. \tag{5.3-13}$$

In summary, the post-collision velocities can be obtained from the pre-collision velocities and the predetermined coefficients of restitution and friction (e, β, and f), as the following relationships summarize:

If $-\frac{2}{7} m^* (1 + \beta)(\mathbf{u}_C \cdot t) < f J_n$ (sliding-through)

$$\mathbf{u}'_1 = \mathbf{u}_1 - (\mathbf{n} - f\mathbf{t}) \{(\mathbf{u}_1 - \mathbf{u}_2) \cdot \mathbf{n}\} (1 + e) \frac{m^*}{m_1} \tag{5.3-14}$$

$$\mathbf{u}'_2 = \mathbf{u}_2 + (\mathbf{n} - f\mathbf{t}) \{(\mathbf{u}_1 - \mathbf{u}_2) \cdot \mathbf{n}\} (1 + e) \frac{m^*}{m_2} \tag{5.3-15}$$

$$\omega'_1 = \omega_1 + \frac{5}{2a_1} \{(\mathbf{u}_1 - \mathbf{u}_2) \cdot \mathbf{n}\} (\mathbf{n} \times \mathbf{t}) f (1 + e) \frac{m^*}{m_1} \tag{5.3-16}$$

$$\omega'_2 = \omega_2 + \frac{5}{2a_2} \{(\mathbf{u}_1 - \mathbf{u}_2) \cdot \mathbf{n}\} (\mathbf{n} \times \mathbf{t}) f (1 + e) \frac{m^*}{m_2}. \tag{5.3-17}$$

If $-\frac{2}{7} m^* (1 + \beta)(\mathbf{u}_C \cdot t) \geq f J_n$ (stick/rolling)

$$\mathbf{u}_1' = \mathbf{u}_1 - \left\{ (1 + \beta) (\mathbf{u}_1 - \mathbf{u}_2) \cdot \mathbf{n} + \frac{2}{7} (1 + \beta) (\mathbf{u}_c \cdot \mathbf{t}) \right\} \frac{m^*}{m_1} \tag{5.3-18}$$

$$\mathbf{u_2}' = \mathbf{u_2} + \left\{ (1+\beta)(\mathbf{u_1} - \mathbf{u_2}) \cdot \mathbf{n} + \frac{2}{7}(1+\beta)(\mathbf{u_c} \cdot \mathbf{t}) \right\} \frac{m^*}{m_2} \tag{5.3-19}$$

$$\boldsymbol{\omega_1}' = \boldsymbol{\omega_1} - \frac{5m^*}{7a_1 m_1}(1+\beta)(\mathbf{u_c} \cdot \mathbf{t})(\mathbf{n} \times \mathbf{t}) \tag{5.3-20}$$

$$\boldsymbol{\omega_2}' = \boldsymbol{\omega_2} - \frac{5m^*}{7a_2 m_2}(1+\beta)(\mathbf{u_c} \cdot \mathbf{t})(\mathbf{n} \times \mathbf{t}). \tag{5.3-21}$$

Assuming instantaneous and point contact allows the hard-sphere model to be efficiently solved for post-collision velocities without regard for details of the complex contact processes. All collisions are accounted for using only the momentum change in particle velocities, both before and after the collision; meaning the contact forces during the collisions do not appear in the momentum equation of the particle motion. In addition to these fundamental advantages, there are a few drawbacks and notable assumptions though. The assumption of instantaneous change in momentum by collision implies that the contact duration approaches zero while the collision force approaches infinity. An extremely high Young's modulus and yield stress would be required to avoid material failure; the particles are thus fictitiously assumed to be sufficiently hard so as to always withstand the collision. The algorithm that updates the particle velocity and position in the hard-sphere iterative model is event driven, and the incremental advance of time step, Δt, in the discrete phase computation is determined by the shortest possible time for the next collision event to take place. Additionally, the restitution coefficients, e and β, are not true material properties; instead, they are collision-dependent parameters to be independently formulated or empirically obtained.

5.3.2   Soft-Sphere Model

The soft-sphere model explicitly accounts for the process of deformation during a particle collision by incorporating basic formulations from classic elasticity theories, such as the Hertz's theory for normal contact and Mindlin's theory for tangential friction (Fan and Zhu, 1998). Hence, the soft-sphere model is capable of handling deformation from endured contact during a collision, contact forces, and contact involving more than two particles. During contact between a pair of particles, as shown in Figure 5.3-2, contact deformation can be decomposed into a normal component, δ_n, and a tangential component δ_t, which correspond to the normal and tangential impact forces, respectively. In addition, relative rotation of particles is affected by torque caused by eccentric contact between colliding particles. Thus, a key feature of the soft-sphere DEM model is the ability to process normal contact forces, tangential contact forces, and torque. Particle deformation during contact can be conceptually divided into different characteristic modes, including elastic deformation, viscous damping, and plastic deformation. Elastic deformation is reversible; whereas the viscous damping and plastic deformation are irreversible processes due to kinetic energy dissipation or absorption.

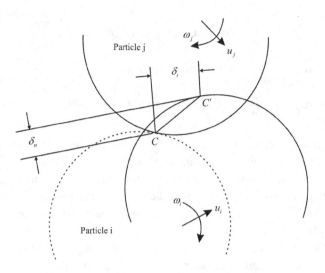

Figure 5.3-2 Deformation and velocity of two colliding particles.

5.3.2.1   Viscoelastic Contact Force Model

The most common contact force model is the spring-dashpot-slider model, which interprets the contact process as a combination of elastic deformation, kinetic energy dissipation, and tangential friction. The elastic deformation is modeled as a spring with stiffness, k; the dissipative force, which depends on the rate of deformation, is represented by a dashpot with damping coefficient, η; tangential friction is represented by a friction slider with a friction coefficient, f. A schematic representation of the spring-dashpot-slider model is shown in Figure 5.3-3. The viscoelastic models can be divided into two groups: these are the linear models that assume constant spring stiffness and the nonlinear models that are based on Hertzian theory for elastic contacts.

In the linear viscoelastic model (Cundall and Strack, 1979), the force acting on particle i caused by contact with particle j is expressed in terms of normal force, $\mathbf{F}_n$, and tangential force, $\mathbf{F}_t$, by the relationships:

$$\mathbf{F}_n = -k_n\delta_n\mathbf{n} - \eta_n(\mathbf{u}_C \cdot \mathbf{n})\mathbf{n} \tag{5.3-22a}$$

$$\mathbf{F}_t = \begin{cases} -k_t\delta_t\mathbf{t} - \eta_t\mathbf{u}_{st} & \text{if } |-k_t\delta_t\mathbf{t} - \eta_t\mathbf{u}_{Ct}| \leq f\,|\mathbf{F}_n| \quad (\text{no} - \text{sliding}) \\ -f\,|\mathbf{F}_n|\,\mathbf{t} & \text{if } |-k_t\delta_t\mathbf{t} - \eta_t\mathbf{u}_{Ct}| > f\,|\mathbf{F}_n| \qquad (\text{sliding}), \end{cases} \tag{5.3-22b}$$

where δ is the relative particle displacement; the unit vectors $\mathbf{n}$ and $\mathbf{t}$ denote the normal and tangential directions of the contact, respectively. The relative particle velocity, $\mathbf{u}_C$, is defined by:

$$\mathbf{u}_C = \mathbf{u}_i - \mathbf{u}_j. \tag{5.3-22c}$$

The slip velocity at the contact point, $\mathbf{u}_{st}$, is calculated by:

$$\mathbf{u}_{st} = \mathbf{u}_C - (\mathbf{u}_C \cdot \mathbf{n})\,\mathbf{n} + \left(a_i\boldsymbol{\omega}_i + a_j\boldsymbol{\omega}_j\right) \times \mathbf{n}. \tag{5.3-22d}$$

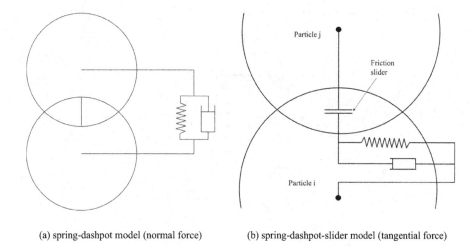

(a) spring-dashpot model (normal force)   (b) spring-dashpot-slider model (tangential force)

Figure 5.3-3 Spring-dashpot-slider model.

The nonlinear viscoelastic model describes the elastic component of deformation using the results gleaned from Hertz's theory, which propounds that the normal force is proportional to the 3/2th power of the elastic deformation. Tsuji *et al.* (1992) proposed the following model for the normal contact force:

$$\mathbf{F}_n = -\tilde{k}_n \delta_n^{3/2} \mathbf{n} - \tilde{\eta}_n (\mathbf{u}_c \cdot \mathbf{n}) \delta_n^{1/4} \mathbf{n}. \tag{5.3-23a}$$

The tangential force under the elastic collision with friction is described by Mindlin's theory (see Section 4.3.4). While it has the same form as Eq. (5.3-22b), the stiffness is no longer a constant. Instead, it is a function of the normal deformation, δ_n, as given by:

$$k_t = \tilde{k}_t \delta_n^{1/2}. \tag{5.3-23b}$$

5.3.2.2   Elastoplastic Contact Force Model

When the loading stress in a particle collision exceeds the yield stress of the colliding particles, plastic deformation takes place due to the loss in the kinetic energy and hysteresis during the unloading process; the result is permanent distortion of the particle. Hysteretic effects are represented by different degrees of spring stiffness to approximate the loading and unloading processes; this is similar to the latching spring model (Walton and Braun, 1986) shown in Figure 5.3-4. The resulting normal contact force can then be expressed as:

$$\mathbf{F}_n = \begin{cases} -k_1 \delta_n & \text{loading } (\dot{\delta}_n \geq 0) \\ -k_2 (\delta_n - \delta_{n0}) & \text{unloading } (\dot{\delta}_n < 0), \end{cases} \tag{5.3-24}$$

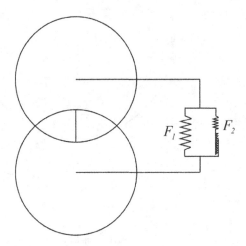

Figure 5.3-4 Latching spring model for plastic deformation.

where δ_{n0} signifies plastic deformation. The restitution coefficient and the contact time are obtained, respectively, by:

$$e = \sqrt{\frac{k_1}{k_2}} \tag{5.3-25a}$$

$$t_C = \frac{\pi \sqrt{m^*} \left(\sqrt{k_1} + \sqrt{k_2} \right)}{2\sqrt{k_1 k_2}}. \tag{5.3-25b}$$

5.3.2.3   Determination of Coefficients

Contact force models involve a number of parameters that need to be determined from the material properties or dynamic characteristics of the particles. In viscoelastic models, those parameters include the normal and tangential stiffness k_n and k_t, normal and tangential damping coefficients η_n and η_t, and friction coefficient f. Among these parameters, only the friction coefficient can be measured directly; the stiffness and damping coefficients remain undetermined.

When the simulated particles represent actual individual particles in the physical system, it is often desirable to derive the model parameters from the observed physical properties of the particles. For example, in the nonlinear viscoelastic model in Eq. (5.3-23a) and Eq. (5.3-23b), the normal and tangential stiffness can be derived from Hertz's and Mindlin's theories, respectively, as:

$$\tilde{k}_n = \frac{4}{3} E^* \sqrt{a^*} \tag{5.3-26a}$$

$$\tilde{k}_t = \frac{8\sqrt{a^*}}{\left(\frac{2-v_1}{G_1} + \frac{2-v_2}{G_2} \right)}. \tag{5.3-26b}$$

The contact modulus, E^*, and the reduced radius, a^*, are defined in Eq. (4.3-1a) as a function of Young's modulus E, Poisson ratio v, and radius, a, of the two particles.

The shear modulus, G, is defined by:

$$G = \frac{E}{2(1+\nu)}.$$ (5.3-26c)

In many situations, the Hertz's and Mindlin's theories become inaccurate or inapplicable. In these cases, the stiffness values need to be determined empirically. Such illustrative cases include the collision of nonspherical particles and nonhomogenous particles (*e.g.*, surface-coated particles). Another challenge involves a system whose particle count is so numerous that all particles must be subgrouped into various clusters while each cluster is treated as a single virtual particle in the DEM to make the simulation practical. There is currently no well-established theory to calculate the stiffness of such a virtual particle, with virtual stiffness expected to be smaller than that of actual particles. In other situations, the stiffness calculated using actual physical properties corresponds to a simulation time step that is disproportionately small when considered in the context of an appropriate time scale for the overall process of interest; when this occurs, the stiffness used in the simulation must be manipulated so that a larger time step can be used, resulting in an artificial predicted contact time (*i.e.*, protracted).

The parameters in the contact force model are adjusted so that the predicted macroscopic behaviors in a collision agree with those measured and observed. One such parameter is the coefficient of restitution, which is a measure of the irreversible loss of the kinetic energy due to viscous dissipation (spring-dashpot model) or plastic deformation (latching-spring model). If a linear spring with constant stiffness is used, the equation of motion for the particle is that of a damped harmonic oscillator; the restitution coefficient and contact time are expressed as functions of k and η. In this model, the normal damping coefficient is related to the coefficient of restitution through:

$$\eta_n = \begin{cases} \dfrac{-2\ln e\sqrt{m^*k_n}}{\sqrt{\pi^2+\ln^2 e}} & 0 < e < 1 \\ 2\sqrt{m^*k_n} & e = 0 \end{cases}.$$ (5.3-27a)

Similarly, the tangential damping coefficient is obtained by the expression:

$$\eta_t = \begin{cases} \dfrac{-2\ln e\sqrt{m'k_t}}{\sqrt{\pi^2+\ln^2 e}} & 0 < e < 1 \\ 2\sqrt{m'k_t} & e = 0 \end{cases}.$$ (5.3-27b)

Note that, for the tangential damping coefficient, the effective mass, m', is different from that in the normal damping coefficient because both translational and rotational motion must be considered. For collisions of two identical solid spheres, $m' = 2m^*/7$.

For a nonlinear spring model, the expressions of Eq. (5.3-27a) and Eq. (5.3-27b) are not valid. In the model offered by Tsuji *et al.* (1992) based on Hertzian contact, the damping coefficients are expressed by:

$$\eta_n = \eta_t = \alpha\sqrt{m^*k_n}.$$ (5.3-27c)

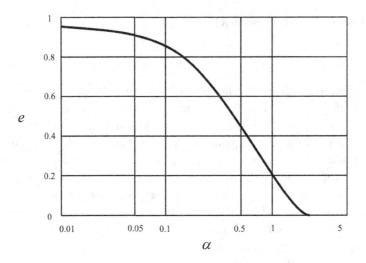

Figure 5.3-5 Relationship of α and e (from Tsuji *et al.*, 1992).

The coefficient α is a nonlinear function of the restitution coefficient e, with the relationship graphically presented in Figure 5.3-5.

5.3.2.4 Torque

A torque that contributes to particle rotation is developed because the contact forces are generally eccentric with respect to the center of mass of the particle. Tangential force and the asymmetric distribution of the normal force in the contact area contribute to the torque. Calculating the torque caused by the tangential force is straightforward and is expressed by:

$$\mathbf{T}_t = \mathbf{a} \times \mathbf{F}_t. \tag{5.3-28}$$

The torque resulting from the normal contact force, also known as the rolling friction torque, is complicated to evaluate. In some DEM formulations, the effect of rolling friction is considered to be negligible. However, the rolling friction torque is often found to be significant in many situations, such as a particle moving on a plane, or with heaping and formation of the shear band. A rolling friction torque model can be obtained by an analog to the tangential force inherent in the spring-dashpot-slider model (Iwashita and Oda, 1998). From Section 5.3.2.1, it gives:

$$\mathbf{T}_r = -\min\left\{k_r\theta_r + \eta_r\dot{\theta}_r, f_r\,|\mathbf{F}_n|\right\}\hat{\mathbf{\theta}}, \tag{5.3-29}$$

where the subscript r stands for rolling; and θ is the rolling-induced angular deformation. It is further suggested that $k_r = k_t a^2$, where k_t is the tangential sliding stiffness in Eq. (5.3-22b). The rolling damping coefficient, η_r, is primarily utilized to encourage numerical stability. Rolling friction is suggested to be equal to 1% of the sliding friction coefficient (Zhu and Yu, 2003). Currently, quantifying rolling friction torque is an active research area, and measurement of the relevant coefficients is difficult to obtain.

5.3.2.5 Equation of Motion

In the absence of mass transfer or particle breakup, the motion of particle i is determined by the total force acting on the particle. Assuming linear force addition is appropriate, the equation of linear motion is given by:

$$m_i \frac{d\mathbf{u}_{pi}}{dt} = \mathbf{F}_{Ci} + \mathbf{F}_{NCi} + \mathbf{F}_{fi} + \mathbf{F}_{Gi}. \tag{5.3-30}$$

The forces on the right-hand side include:

1) Collisional force, $\mathbf{F}_{Ci}$, given by the summation of normal force, $\mathbf{F}_{nij}$, and tangential force, $\mathbf{F}_{tij}$, for each particle j that particle i is in contact with. Thus:

$$\mathbf{F}_{Ci} = \sum_j \left(\mathbf{F}_{nij} + \mathbf{F}_{tij} \right). \tag{5.3-30a}$$

2) Noncollisional particle interaction forces, $\mathbf{F}_{NCi}$, such as the van der Waals force, electrostatic force, capillary force, and others.
3) Fluid–particle interactions, $\mathbf{F}_{fi}$, including drag, lift, pressure gradient force, added mass force, Basset force, and others.
4) Body force, $\mathbf{F}_{Gi}$, such as gravitational or electromagnetic forces.

The equation of rotational motion is given by:

$$I_i \frac{d\boldsymbol{\omega}_i}{dt} = \sum_j \left(\mathbf{T}_{tij} + \mathbf{T}_{rij} \right), \tag{5.3-31}$$

where the tangential friction torque, $\mathbf{T}_{tij}$, and rolling friction torque, $\mathbf{T}_{rij}$, are given by Eq. (5.3-28) and Eq. (5.3-29), respectively.

With the collision process in the soft-sphere model resolved, the simulation time step must be smaller than the expected particle contact duration. It is suggested that the time step be much smaller than the natural oscillation period (Tsuji *et al.*, 1993). This criterion is given by:

$$\Delta t < \frac{1}{10} \left(2\pi \sqrt{\frac{m}{k}} \right). \tag{5.3-32}$$

The above criterion typically advocates an extremely small time step when the stiffness, k, is large. As alluded to earlier in the section, this result may severely preempt computational efficiency and render DEM simulation impractical. In such cases, the stiffness is often contrived in favor of a larger simulation time step.

5.4 Coupling in Eulerian–Lagrangian Model

In a dispersed multiphase flow, mutual transport coupling of mass, momentum, and energy exists between the discrete particle phase and the continuum fluid phase. The effects of the fluid phase on the particles are reflected in the hydrodynamic forces acting on each particle and the particles' turbulent diffusion, both of which are discussed in Section 5.2. This section focuses on the effects of the particle phase on the continuous fluid phase, which distinguishes transport in the continuous phase of a multiphase flow from the single-phase flow introduced in Chapter 2. The differentiating mechanisms present themselves in three aspects: (1) because the overall flow domain is shared with the discrete phase, the continuous phase flow merely occupies a fraction of the entire domain. The effect of volume fractions is accounted for by the governing equations through averaging in the space or time domain; (2) the exchanges of mass, momentum, and energy occur through the common interface between the discrete and continuous phases. Interfacial transport is incorporated into the continuous phase model by adding source terms to the governing transport equations; (3) Turbulence of the continuous phase flow is modified by the presence of the discrete phase. This effect, referred to as turbulence modulation, should be accounted for in the turbulence model. This section centers on adding source terms in the continuous phase transport equations to account for the interfacial transport. In order to be incorporated properly, the source terms must be transformed from the Lagrangian coordinates of particles to the Eulerian coordinates of the continuous phase. The complete derivation of the averaged transport equations for the continuous phase is introduced in Chapter 6.

Because the source terms describing interfacial coupling typically depend on the condition of both phases, solutions to the coupled transport equations are often obtained through successive iterations between the Eulerian continuous phase model and the Lagrangian discrete phase model.

5.4.1   Mass Coupling

The mass of a single particle may change due to either phase change or chemical reaction as it moves along its Lagrangian trajectory. This change must be balanced by a corresponding change in the continuous phase, as required by the interfacial balance condition. For an Eulerian control volume V, the source term in the mass balance equation of the continuous phase is the surface integral of the interfacial mass flux that equates to the sum of the mass change of all individual particles within the control volume. Thus, it gives:

$$\Gamma = \frac{1}{V} \int_{A_p} \rho_p \left(\mathbf{u}_p - \mathbf{u}_s\right) \cdot \mathbf{n} dA = -\Gamma_p = -\frac{1}{V} \sum_{i=1}^{N_p} \frac{dm_{p,i}}{dt_{pi}}, \qquad (5.4\text{-}1)$$

where $\mathbf{u}_p$ and $\mathbf{u}_s$ are the velocity of the particle and the fluid velocity at the particle surface A_p, respectively. If the rate of mass change for all particles in the Eulerian control volume, V, is identical, the source term can be simplified to:

$$\Gamma = -\frac{1}{V} \sum_{i=1}^{N_p} \frac{dm_{p,i}}{dt_{pi}} = -n_p \frac{dm_p}{dt_p}, \qquad (5.4\text{-}1a)$$

where n_p is the local number density of the discrete particles. The source term due to mass coupling between the discrete phase and the continuous phase appears in the mass conservation equation of the continuous phase as:

$$\frac{\partial}{\partial t}(\alpha\rho) + \nabla \cdot (\alpha\rho\mathbf{U}) = -n_p\frac{dm_p}{dt_p},\tag{5.4-2}$$

where α is the volume fraction of fluid phase within the control volume; the quantity, d/dt_p, is the total derivative for the particles in Lagrangian coordinates, which is defined as:

$$\frac{d}{dt_p} = \frac{\partial}{\partial t} + \mathbf{u_p} \cdot \nabla.\tag{5.4-2a}$$

5.4.2 Momentum Coupling

The momentum conservation equation for the discrete particle in Eq. (5.1-2) can be written as:

$$m_p\frac{d\mathbf{u}_p}{dt_p} = -\mathbf{u_p}\frac{dm_p}{dt_p} + \mathbf{F_f} + \mathbf{F_G} + \mathbf{F_p}.\tag{5.4-3}$$

The first term on the right-hand side represents the particle momentum change due to interfacial mass transfer; $\mathbf{F_f}$ is the hydrodynamic force acting on the particle, including the steady drag, pressure gradient, added mass, lift, and Basset forces; $\mathbf{F_G}$ is the force due to external fields such as gravitational or electrical; $\mathbf{F_p}$ is the force due to particle–particle interactions, including collision and noncollision forces.

The source terms in the continuous phase momentum equation emanate from two origins. The first is the source term associated with the interfacial mass transfer, as given by:

$$\mathbf{F_\Gamma} = \frac{1}{V}\int_{S_p}\rho_p\mathbf{u_p}\left(\mathbf{u_p} - \mathbf{u_s}\right)\cdot\mathbf{n}dA = -\mathbf{F_{\Gamma p}} = -n_p\mathbf{u_p}\frac{dm_p}{dt_p},\tag{5.4-3a}$$

where S_p is the particle surface area. The second is the total hydrodynamic force acting on the continuous fluid by the particles, which is obtained by integrating the fluid stresses on the particle surface. For cases where the hydrodynamic particle–fluid interaction is represented by the drag force, added mass force, and pressure gradient force, and assuming all particles in the same control volume are subjected to the same hydrodynamic particle–fluid interactions, the source term, $\mathbf{F_s}$, due to hydrodynamic force is given by:

$$\mathbf{F_s} = \frac{1}{V}\int_{S_p}\mathbf{T_p}\cdot\mathbf{n}dA = -n_p\mathbf{F_f} = -n_p\left[\beta\left(\mathbf{U} - \mathbf{u_p}\right) - m_A\frac{d}{dt}\left(\mathbf{u_p} - \mathbf{U}\right) + V_p\nabla p\right],\tag{5.4-3b}$$

where β, m_A, and V_p are the momentum transfer coefficient, added mass, and volume of a single particle, respectively.

Thus the momentum equation of the continuous fluid with the source terms due to momentum coupling between the discrete particles and continuous fluid is given as:

$$\frac{\partial}{\partial t}(\alpha\rho\mathbf{U}) + \nabla \cdot (\alpha\rho\mathbf{U}\mathbf{U}) = \nabla \cdot (\alpha\mathbf{T}) + \alpha\rho\mathbf{f} + \mathbf{F_s} + \mathbf{F_\Gamma}.\tag{5.4-4}$$

5.4.3   Energy Coupling

Most fluids in multiphase flows can be approximated to be radiation-transparent. That is to say, thermal radiation from the surrounding environment, such as neighboring surfaces of the fluid–particle transport vessel and from the particles themselves, do not appear directly in the energy equation of fluid phase. Instead, radiation will affect the energy equation of particles and indirectly affect the fluid phase via a change in particle temperature. Hence, energy coupling between fluid and particles typically comes from two sources: heat transfer by phase change and heat transfer by convection.

The total heat generation rate via interfacial mass transfer by phase change or chemical reaction can be expressed by:

$$q_\Gamma = \frac{1}{V} \int_{S_p} \rho_p h_p \left(\mathbf{u}_p - \mathbf{u}_s \right) \cdot \mathbf{n} dA = -q_{\Gamma p} = -n_p h_p \frac{dm_p}{dt_p}, \tag{5.4-5}$$

where h_p is the specific enthalpy of the particle. The total heat generation rate from interfacial mass transfer may contribute to the fluid phase and to the particle phase itself. Convective heat transfer to the continuous phase fluid through particle surface is given by:

$$q_s = -\frac{1}{V} \int_{S_p} \mathbf{q}_p \cdot \mathbf{n} dA = -q_{sp} = -h_c S_p (T - T_p), \tag{5.4-6}$$

where h_c is the convective heat transfer coefficient, which depends on the Prandtl number and particle Reynolds number for forced convection or the Grashof number for natural convection. For a sphere in forced convection, h_c may be estimated from the Ranz–Marshall correlation discussed in Chapter 3.

The energy equation of a single particle, used to determine h_p or T_p, can be expressed from Eq. (5.2-3) as:

$$m_p c_p \frac{dT_p}{dt_p} = h_c S_p (T - T_p) + q_{rad} + (1 - f_q) h_p \frac{dm_p}{dt_p}, \tag{5.4-7}$$

where f_q is the fraction of heat gained from mass transfer distributed to the fluid; q_{rad} is the heat transferred by thermal radiation between the particle and the ambient environment, including adjacent surfaces of the fluid–particle transport vessel and surrounding particles. A detailed formulation of q_{rad} can be complex when considering various thermal radiation properties of the particle surface and the irradiation contribution of the surrounding environment. A simple expression for the radiation between a diffusive particle and surrounding ambient of radiant temperature T_∞, can be written as:

$$q_{rad} = \varepsilon_p S_p \sigma_b (T_\infty^4 - T_p^4), \tag{5.4-7a}$$

where ε_p is the emissivity of the particle surface, and σ_b is the Stefan–Boltzmann constant.

Thus, with the source terms from energy coupling between the discrete and the continuous phase, the enthalpy conservation equation of the fluid phase becomes:

$$\frac{\partial}{\partial t}(\alpha \rho h) + \nabla \cdot (\alpha \rho h \mathbf{U}) = \frac{d}{dt}(\alpha p) + \nabla \cdot (\alpha p \mathbf{U}) - \alpha \mathbf{T} : \nabla \mathbf{U}$$
$$- \nabla \cdot (\alpha \mathbf{q}) + (\alpha \rho \mathbf{U} \cdot \mathbf{f}) + q_s + f_q q_\Gamma. \tag{5.4-8}$$

Since fluid work, viscous dissipation by fluid stresses, and work done by the field forces are usually negligible when compared to the effects of heat conduction within fluid and heat transfer from particles, Eq. (5.4-8) is routinely expressed in a simplified form as:

$$\frac{\partial}{\partial t}(\alpha \rho h) + \nabla \cdot (\alpha \rho h \mathbf{U}) = -\nabla \cdot (\alpha \mathbf{q}) + q_s + f_q q_\Gamma. \tag{5.4-8a}$$

5.4.4 Coupling due to Charge-Induced Electric Field

When the particles carry electrostatic charges, an electric field is induced by the charge gradient, affecting particle phase transport. The strength of the electric field is dependent on the spatial distribution of the charges and the boundary conditions. Hence, in addition to the coupling between the discrete particle phase dynamics and the continuum fluid phase dynamics, there is also two-way coupling of the discrete particle phase dynamics and the charge-induced electric field.

The electrostatic force on a particle carrying the electrostatic charge of q_e can be expressed by:

$$\mathbf{F}_E = q_e \mathbf{E}. \tag{5.4-9}$$

The electrostatic force affects particle dynamics as described by the equation of motion, expressed by Eq. (5.1-1). The electrostatic field intensity, $\mathbf{E}$, can be related to the electric potential, Φ, as:

$$\mathbf{E} = -\nabla \Phi. \tag{5.4-10}$$

For the electric field established by numerous charged particles, electric potential is governed by:

$$\nabla \cdot (\varepsilon_e \nabla \Phi) = \rho_e = n_p q_e, \tag{5.4-11}$$

where ε_e is electromagnetic permeability of the fluid medium; ρ_e is the space charge density, which depends on the spatial distribution of particles predicted by particle trajectories.

In order to solve for the electric potential, appropriate boundary conditions must be defined. For a wall maintained at a given voltage (*e.g.*, a grounded electrically conductive surface or using a voltage power supply), the electric potential is specified by:

$$\Phi = \Phi_0. \tag{5.4-11a}$$

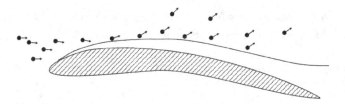

Figure 5.5-1 Schematic flow over airfoil in rain.

For a wall with a given surface charge density σ_e (*e.g.*, an electrically insulated surface), the boundary condition is defined by the normal component of the electric field at the boundary. The normal component of the electric field is defined as:

$$\nabla\Phi \cdot \mathbf{n} = -\frac{\sigma_e}{\varepsilon_e}. \tag{5.4-11b}$$

The surface charge density may be coupled with charge transfer mechanisms through charged particle deposition onto the wall surface or by collisions between the particles and the wall.

5.5   Case Studies

The following case studies serve for further discussion with simple applications of the theories of Chapter 5, in particular using the Lagrangian trajectory model to describe the dynamic transport of particles. The examples include rain over an airfoil, inhalation of ultrafine aerosols, solar harvesting via particulate flows, and charged particle dispersion. These examples illustrate various particle–fluid and particle–field interactions in the Lagrangian modeling of particle transport.

5.5.1   Flow over Airfoil in Rain

Problem Statement: Consider droplets of rain striking an airfoil, as sketched in Figure 5.5-1. Some part of the colliding droplets is splashed or bounced away from the airfoil while the remainder forms a thin moving film of water on the airfoil surface. Both splashing and moving film may significantly alter the lift and drag of air flow over the airfoil. Establish a multiphase flow model that can be used to determine the lift and drag forces in rain.

Analysis: The following is a simplified model based on the work of Wu and Cao (2015). The mechanistic model consists of three parts: (1) droplet transport onto the outer boundary of thin film; (2) flow over the airfoil with thin film; and (3) formation and flow of thin film. The droplet-air flow is modeled by a two-way coupled Lagrangian–Eulerian approach, while the film dynamics is from a droplet–surface interaction model suggested by Stanton and Rutland (1998).

The droplets are assumed to be noninteracting, nondeforming, nonevaporating, and isothermal. The droplets are subjected only to the drag and gravity forces so that the equation of motion of a droplet is given by

$$\frac{d\mathbf{r}_{di}}{dt} = \mathbf{u}_{di},$$ (5.5-1)

where the subscript i denotes the ith droplet; and $\mathbf{r}_d$ is the droplet position vector; and

$$\frac{d\mathbf{u}_{di}}{dt} = \frac{3\mu C_{Di} Re_{di}}{4\rho_d d_{di}^2}(\mathbf{U} - \mathbf{u}_{di}) + \mathbf{g},$$ (5.5-2)

where Re_d is the droplet Reynolds number using relative velocity; C_D is the drag coefficient of the droplet, expressed by Eq. (3.2-8) or Eq. (3.2-9). In Eq. (5.5-2), the instant local velocity of air is approximated by the time-averaged local velocity of air.

Neglecting the volumetric effect of droplets to air, the air flow is basically modeled as a single-phase flow, with a momentum interaction term from the discrete transport of droplets. The effect of turbulence is based on a standard k-ε model, without turbulence modulation with droplets. Equations of a steady-state air flow are given by Eq. (2.3-8) and Eq. (2.3-10) as:

$$\nabla \cdot (\rho\mathbf{U}) = 0$$ (5.5-3)

$$\nabla \cdot (\rho\mathbf{UU}) = -\nabla p + \nabla \cdot \left(\mu_{eff}\left[\nabla\mathbf{U} + (\nabla\mathbf{U})^T\right]\right) - \sum_i \frac{3\mu C_{Di} Re_{di}}{4\rho_d d_{di}^2}(\mathbf{U} - \mathbf{u}_{di})mJ_{di}\Delta t,$$ (5.5-4)

where J_d is the droplet mass flow rate per unit volume; and μ_{eff} is the effective viscosity determined from the k-ε model (*i.e.*, Eq. (2.3-18), Eq. (2.3-21) and Eq. (2.3-22)). The last term in Eq. (5.5-4) is the momentum transfer from droplet–air interaction.

The thin film of moving liquid on the airfoil surface is mainly developed from droplet–surface impingement, which not only results in a direct momentum transfer from droplet impingement but also alters the boundary layer conditions to the air flow over the airfoil. Consider the uniform incident droplet impingement and the adhered portion spread uniformly over the wall cell. The film continuity equation is given by

$$\frac{\partial\delta}{\partial t} + \frac{1}{A_w}\sum_{i=1}^{N_{side}}(\mathbf{U}_f \cdot \mathbf{n})_i \delta_i l_i = \frac{s_d}{\rho_f A_w},$$ (5.5-5)

where δ is the film thickness, A_w is the area of wall cell, N_{side} is the number of sides in each wall cell, l_i is the length of side i, $\mathbf{n}$ is the normal direction of cell side, and δ_i is the film thickness at side i. $\mathbf{U}_f$ is the film velocity, ρ_f is the film density, and s_d is the source term denoting the mass flux of droplets impinging the film or secondary droplets leaving the film due to splashing. The momentum equation is given by

$$\frac{\partial \left(\delta \mathbf{U}_f\right)}{\partial t} + \frac{1}{A_w}\sum_{i=1}^{N_{side}} \mathbf{U}_f\left(\mathbf{U}_f \cdot \mathbf{n}\right)_i \delta_i l_i \varphi_i = -\frac{\sum_{i=1}^{N_{side}}\left(P\mathbf{n}\right)\delta_i l_i}{\rho_f A_w} + \mathbf{g}\delta + \frac{\mathbf{M}_t}{\rho_f A_w} + \frac{\boldsymbol{\tau}_w + \boldsymbol{\tau}_{fg}}{\rho_f},$$

(5.5-6)

where P is a pressure defined as the sum of the free-stream air pressure and the dynamic pressure resulting from incident drop impingement and splashing, $\boldsymbol{\tau}_w$ is the wall shear and $\boldsymbol{\tau}_{fg}$ is the shear at the liquid–gas interface. ϕ is a parameter defined by

$$\phi = \frac{1}{1 - \delta_t/\delta} - \frac{\Theta}{\delta(1 - \delta_t/\delta)^2},$$

(5.5-6a)

where the displacement thickness, δ_t, and the momentum thickness, Θ, can be calculated once the velocity profile in the cross-film direction has been chosen. $\mathbf{M}_t$ is the tangential momentum contribution to the film from impingement and splashing of secondary droplets, which is given by

$$\mathbf{M}_t = \sum_{i=1}^{N_{drop}} \left(m_i \mathbf{U}_{\tau d}\right) - \sum_{j=1}^{N_{splah}} \left(m_j \mathbf{U}_{\tau j}\right),$$

(5.5-6b)

where N_{drop} is the number of incident droplets, m_i is the mass of incident droplet i, $\mathbf{U}_{\tau d}$ is the tangential velocity of incident droplet, N_{splash} is the number of secondary droplets, and $\mathbf{U}_{\tau d}$ is the tangential velocity of the jth secondary droplet.

Thus, the problem closure is successfully established with nine coupled equations (*i.e.*, Eq. (5.5-1)–Eq. (5.5-6), Eq. (2.3-18), Eq. (2.3-21), and Eq. (2.3-22)) for nine coupled unknowns (*i.e.*, $\mathbf{r}_d$, $\mathbf{u}_d$, $\mathbf{U}$, P, μ_{eff}, k, ε, δ, and $\mathbf{U}_f$). Phase coupling is represented by terms such as the drag forces in Eq. (5.5-2) and Eq. (5.5-4), the source terms of S_d in Eq. (5.5-5), and $\mathbf{M}_t$ in Eq. (5.5-6). The set of equations can be solved numerically to obtain the solutions for the air flow and the film flow. The stresses on the surface of the airfoil can then be integrated to calculate the drag and lift forces.

Comments: (1) There can be strong turbulence modulation between droplets and air, especially near the boundary layer of the airfoil or film zone. The boundary condition or wall function for the turbulence model can be affected by the surface movement of film, especially during the droplet impingement and splashing; (2) the effect of droplet size distribution and detailed simulation results can be referred to the work of Wu and Cao (2015).

5.5.2   Inhalation of Ultrafine Particulates

Problem Statement: The inhalation of ultrafine aerosols, typically of sizes from submicron to less than 0.1 μm or even a few nanometers, can cause these particulates to deposit in the lung or upper respiratory airways. The transport mechanisms are dominated by Brownian motion and turbulent dispersion. Develop a model to describe the transport characteristics and surface deposition of such aerosols.

Analysis: The following is a simplified model based on the work of Longest and Xi (2007). For ultrafine aerosols with high density ratio of particle to gas, the effect of acceleration and gravity can be ignored. Hence, the equation of motion for an aerosol particle with $\text{Re}_p \ll 1$ can be expressed by

$$\frac{d\mathbf{u}_{pi}}{dt} = \frac{1}{C_c \tau_S}(\mathbf{U} - \mathbf{u}_{pi}) + \frac{\zeta_i}{m_i}\sqrt{\frac{1}{D_B}\frac{2k_B^2 T^2}{\Delta t}}, \tag{5.5-7}$$

where the two terms on the right-hand side represent the drag force and Brownian force, respectively; C_c is the Cunningham correction factor, formulated by Eq. (3.2-16); $\mathbf{U}$ is time-averaged air velocity to approximate the local instant air velocity; τ_S is the Stokes relaxation time defined by Eq. (3.2-22a); ζ_i is the normally distributed random vector; Δt is the time-step for particle integration; and D_B is the Brownian diffusion coefficient, defined by

$$D_B = \frac{k_B T C_c}{3\pi \mu d_p}. \tag{5.5-7a}$$

For the modeling of inhaled air flow, the nature of turbulence regime is important to select a proper turbulence model. In a typical inhalation, the inlet Reynolds number for the oral airway varies approximately from 900 to 8,000, indicating the flow regime covering from laminar to transitional and weak turbulent flows. For such transitional and turbulence flow, the Low Reynolds Number (LRN) version of Wilcox k-ω model (Wilcox, 2006; 2008) provides a better flow predictability toward laminar flow transition, especially near wall boundary, as the turbulent viscosity diminishes.

For incompressible and LRN turbulent flow, the conservation equations of mass and momentum can be simplified from Eq. (2.3-8) and Eq. (2.3-10), respectively, as:

$$\nabla \cdot (\mathbf{U}) = 0 \tag{5.5-8}$$

$$\nabla \cdot (\rho \mathbf{U}\mathbf{U}) = -\nabla p + \nabla \cdot \left((\mu + \mu_T)\left[\nabla \mathbf{U} + (\nabla \mathbf{U})^T\right]\right), \tag{5.5-9}$$

where the turbulent viscosity is related to the turbulence kinetic energy, k, and the specific dissipation rate, ω, by

$$\mu_T = \alpha^* \frac{k}{\omega}, \tag{5.5-10}$$

where the coefficient α^* is given by

$$\alpha^* = \frac{0.024 + \rho k/(6\mu\omega)}{1 + \rho k/(6\mu\omega)}. \tag{5.5-10a}$$

The k-equation is expressed by

$$\frac{\partial k}{\partial t} + \nabla \cdot (\mathbf{U}k) = \nabla \cdot \left(\left(\frac{\mu}{\rho} + \frac{\mu_T}{2\rho}\right)\nabla k\right) + \frac{2\mu_T}{\rho}\mathbf{S}:\mathbf{S} - \beta^*\omega k, \tag{5.5-11}$$

where the rate of deformation tensor $\mathbf{S}$ is given by Eq. (2.2-1), and the coefficient β^* is given by

$$\beta^* = 0.09\left(\frac{0.262 + (\rho k/(8\mu\omega))^4}{1 + (\rho k/(8\mu\omega))^4}\right). \tag{5.5-11a}$$

The ω-equation is expressed by

$$\frac{\partial \omega}{\partial t} + \nabla \cdot (\mathbf{U}\omega) = \nabla \cdot \left(\left(\frac{\mu}{\rho} + \frac{\mu_T}{2\rho} \right) \nabla \omega \right) + \gamma \frac{\omega}{\rho k} (2\mu_T \mathbf{S} : \mathbf{S}) - \frac{3\omega^2}{40} + \frac{\sigma_d}{\omega} \nabla k \cdot \nabla \omega,$$

(5.5-12)

where the coefficients, γ and σ_d, are given, respectively, by

$$\gamma = \frac{13}{25\alpha^*} \left(\frac{0.11 + \rho k/(2.61\mu\omega)}{1 + \rho k/(2.61\mu\omega)} \right),$$

(5.5-12a)

and

$$\sigma_d = \begin{cases} 0, & \text{for } \nabla k \cdot \nabla \omega \le 0 \\ \frac{1}{8} & \text{for } \nabla k \cdot \nabla \omega > 0 \end{cases}.$$

(5.5-12b)

Thus, five equations of air flow (*i.e.*, Eqs. (5.5-10)–(5.5-12)) for the five coupled variables of air flow ($\mathbf{U}$, P, μ_T, k, ω) are given. Thus, with Eq. (5.5-7) for $\mathbf{u}_p$, the closure of model is established for the transport of the inhaled aerosols.

Comments: (1) The model above is solved for time-averaged air flow while the stochastic fluctuations of air flow are ignored. The trajectories of aerosols are from the stochastic Lagrangian model, with the random fluctuation resulting from the Brownian force. Thus, to yield the local particle concentration, a statistical average of trajectories with a sufficiently large number of tracking is required; (2) The transport of submicron aerosols in air flow is commonly modeled as a gas species in a single-phase flow using the Eulerian framework. Such a modeling approach is convenient to track a large quantity of aerosols, especially in terms of particle concentration. However, this method completely ignores the particle inertia, which can be important in phenomena such as particle surface deposition; (3) Detailed simulation examples with deposition rate calculations can be referred to the work of Longest and Xi (2007).

5.5.3   Solar-Absorbing Particulate-Laden Flow

As a promising alternative method of solar energy harvesting, the solid particle solar receiver (SPR) is being developed. The principle of SPR is to use solid particles, suspended and carried by a gas flow, to directly absorb concentrated solar radiation from a heliostat mirror field. The solar energy absorption by suspended particles can be more energy conversion efficient due to enhanced available surface area, more uniform heat transfer in the gas, and much reduced heat loss to the ambient (Romero *et al.*, 2002).

Problem Statement: Develop a mechanistic model for the transport of particle-laden flow with solar energy absorption.

Analysis: The following is a simplified model based on the work of Rahmani *et al.* (2018). The model uses a direct numerical simulation (DNS) approach of turbulent flow that interacts with point particles subjected to a radiative heating source.

The particles are small, monodispersed, noninteracting, and without phase change so that the particles are subjected only to the Stokes drag. The equation of motion of a particle i is given by

$$\frac{d\mathbf{r}_{\text{pi}}}{dt} = \mathbf{u}_{\text{pi}}, \tag{5.5-13}$$

where $\mathbf{r}_p$ is the particle position vector; and

$$\frac{d\mathbf{u}_{\text{pi}}}{dt} = \frac{\mathbf{u}(\mathbf{r}_{\text{pi}}) - \mathbf{u}_{\text{pi}}}{\tau_S}, \tag{5.5-14}$$

where $\mathbf{u}(\mathbf{r})$ is the instant gas velocity at the position of $\mathbf{r}$; and τ_S is the Stokes relaxation time. Further, assume that the particle surfaces are gray and diffusive, and all particles are subjected to the same irradiation. The energy equation of the particle is given by

$$\frac{\pi}{6}\rho_p d_p^3 c_p \frac{dT_{\text{pi}}}{dt} = \frac{\pi}{4}d_p^2 \varepsilon_p I_R + \pi d_p^2 h_c(T(\mathbf{r}_{\text{pi}}) - T_{\text{pi}}), \tag{5.5-15}$$

where ε_p is the particle emissivity, I_R is radiation flux, h_c is the convective heat transfer coefficient, and $T(\mathbf{r})$ is the gas temperature at the position $\mathbf{r}$.

The conservation equations of gas flow are given, in Eulerian description, by

$$\frac{\partial \rho}{\partial t} + \nabla \cdot (\rho \mathbf{u}) = 0 \tag{5.5-16}$$

$$\frac{\partial}{\partial t}(\rho \mathbf{u}) + \nabla \cdot (\rho \mathbf{u}\mathbf{u}) = -\nabla p + \nabla \cdot \left(\mu \left[\nabla \mathbf{u} + (\nabla \mathbf{u})^T\right]\right) + \mathbf{F}_s(\mathbf{r}, t), \tag{5.5-17}$$

where $\mathbf{F}_s(\mathbf{r}, t)$ is the momentum coupling per unit volume between particles and gas phase, which is expressed by

$$\mathbf{F}_s(\mathbf{r}, t) = \sum_{i=1}^{N} \frac{\pi}{6} \frac{\rho_p d_p^3}{\tau_S} \left(\mathbf{u}_{\text{pi}} - \mathbf{u}(\mathbf{r})\right) \delta(\mathbf{r} - \mathbf{r}_{pi}), \tag{5.5-17a}$$

where N is the total number of particles in the domain. The energy conservation equation is given by

$$\frac{\partial}{\partial t}(\rho c T) + \nabla \cdot (\rho c T \mathbf{u}) = \nabla \cdot (K \nabla T) + \sum_{i=1}^{N} \pi d_p^2 h_c \left(T_{\text{pi}} - T(\mathbf{r})\right) \delta(\mathbf{r} - \mathbf{r}_{pi}). \tag{5.5-18}$$

The gas density is related to the static pressure and gas temperature by the ideal gas law:

$$P = \rho R T. \tag{5.5-19}$$

Thus, the closure of model is established, with seven equations (i.e., Eqs. (5.5-13)–(5.5-19)) for the seven coupled variables of particle–air flow $(\mathbf{r}_p, \mathbf{u}_p, T_p, \mathbf{u}, P, T, \rho)$.

Comments: (1) The above model can be extended to flows with polydispersed particles; (2) Detailed simulation examples can be referred to the work of Rahmani *et al.* (2018).

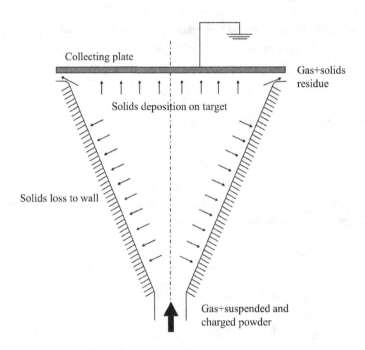

Figure 5.5-2 Schematic dispersion of charged particles by jet in a chamber.

5.5.4   Transport of Charged Particles in Chamber

Problem Statement: Consider a charged-particle-laden gas jet dispersed into a chamber, as shown in Figure 5.5-2. A grounded top cover serves as a particle-collecting plate on which most particles are discharged and deposited. The chamber wall is electrically insulated from outside while the inner wall surface may be deposited by some charged particles. The gas flow exits the chamber with the undeposited particles. Establish a model for the particle transport and surface deposition. This problem was addressed by Zhu *et al.* (2010).

Analysis: The model is based on a Lagrangian–Eulerian approach, with Lagrangian trajectory modeling of particle transport and Eulerian field description of coupled gas flow and electrostatic field intensity. For simplification, assume that there is no re-entrainment from particles already deposited on the plate or wall. The flow is isothermal and turbulent, and the injected particles are equally sized and charged, with no charge redistribution on the suspended particles during the transport process. The instant local velocity of air is approximated by the time-averaged local velocity of air.

    Due to the very large density ratio of particle to gas and assuming the particle acceleration is moderate, the added mass and Basset force are ignored compared to other forces such as the drag force. In this case, only drag force, gravitational force, and electrostatic force are considered so that the equation of motion of an individual particle is simplified from Eq. (5.1-2) and expressed by

$$\frac{\pi}{6}\rho_p d_p^3 \frac{d\mathbf{u}_{\text{pi}}}{dt} = \frac{\rho\pi}{8} C_D d_p^2 \left|\mathbf{U} - \mathbf{u}_{\text{pi}}\right| (\mathbf{U} - \mathbf{u}_{\text{pi}}) + q_e \nabla \Phi + \frac{\pi}{6}\rho_p d_p^3 \mathbf{g},\qquad (5.5\text{-}20)$$

where C_D is the drag coefficient, expressed by Eq. (3.2-3), as a function of particle Reynolds number; q_e is the electrostatic charge carried by each particle; and Φ is the electric potential. Based on the different initial conditions from the jetted particle velocity and location distribution, various particle trajectories can be solved from Eq. (5.5-20).

The electric potential is governed by the conservation equation of electric field intensity as:

$$\nabla \cdot (\varepsilon_e \nabla \Phi) = -q_e n_p,\qquad (5.5\text{-}21)$$

where ε_e is the electric permeability. The boundary conditions depend on the surface charge or ground conditions, such as

$$\Phi = \text{constant} \qquad \text{for surfaces at given electric potential} \qquad (5.5\text{-}21a)$$
$$\varepsilon_e \nabla \Phi \cdot \mathbf{n} = -\sigma_e \qquad \text{for surfaces with given charge density,} \qquad (5.5\text{-}21b)$$

where σ_e is the surface charge density. In the current case, σ_e depends on the particle trajectories terminated at the surfaces that are electrically insulated. The particle number density, n_p, is governed by the particle mass conservation and expressed in Eulerian form:

$$\frac{\partial n_p}{\partial t} + \nabla \cdot (n_p \mathbf{u}_p) = 0.\qquad (5.5\text{-}22)$$

The gas velocity in Eq. (5.5-20) must be solved from dynamic transport of gas flow. For simplicity, it is assumed that the particle dispersion is extremely dilute so that the gas flow is not affected by the particle transport (*i.e.*, one-way coupling between particle dynamics and fluid flow). Furthermore, the jetting velocity is not very high so that the gas flow is incompressible and isothermal, with isotropic turbulence that can be sufficiently described using k-ε turbulent model. Therefore, based on Eq. (2.3-8), Eq. (2.3-10), Eq. (2.3-18), Eq. (2.3-21), and Eq. (2.3-22), five equations for five unknowns of gas flow $(\mathbf{U}, P, \mu_e, k, \varepsilon)$ are given.

Combining sub-models of particle transport, gas flow, and electrostatic field, eight equations (Eq. (5.5-20), Eq. (5.5-21), Eq. (5.5-22), Eq. (2.3-8), Eq. (2.3-10), Eq. (2.3-18), Eq. (2.3-21), and Eq. (2.3-22)) for eight coupled unknowns $(\mathbf{u}_p, n_p, \Phi, \mathbf{U}, P, \mu_e, k, \varepsilon)$ are obtained. Thus, the model closure is established.

Comments: (1) Due to the time-dependent particle surface deposition on ungrounded walls, the electrostatic field will be highly time-dependent, resulting in the particle trajectories being time-dependent. This time-dependency is true even when using a deterministic trajectory model and a time-averaged gas flow field. (2) The actual local gas velocity is randomly varied due to turbulence, leading to a more dispersive particle transport; (3) Case simulation examples can be found in the literature (Zhu *et al.*, 2010).

5.6   Summary

This chapter delineates the Eulerian–Lagrangian modeling approach, which is commonly used for modeling of dilute-phase multiphase flows. The most essential part is the formulation of a Lagrangian trajectory model to describe the transport of individual particles, which requires proper sub-model selection or treatment of particle–fluid and particle–particle interactions in the applications. The basic considerations in the Eulerian–Lagrangian modeling include:

- type of Lagrangian trajectory models: such as deterministic trajectory model, stochastic trajectory model, particle-cloud tracking model, or discrete element method (*e.g.*, hard-sphere or soft-sphere model);
- degree of phase coupling between Lagrangian models of discrete phases and Eulerian model of continuum phase: such as one-way particle–fluid coupling, two-way particle–fluid coupling, or four-way particle–fluid and particle–particle coupling flow;
- field coupling with particle transport: such as the electrostatic field induced by the suspended charged particles or thermal radiation among particles;
- turbulence modulation between particles and eddy transport;

While the Lagrangian model of particle motion is defined in a general equation given in Eq. (5.1-1), this equation is not usually used directly since the accurate representation of fluid–particle interactions requires an instant and details-resolved solution of flow field around the particle. For its use, a simplification involving the approximation of additivity of individual modes of fluid–particle interactions as in the BBO equation or Eq. (5.2-1) is employed in the Lagrangian modeling of discrete phase transport.

Nomenclature

$<\ >$	Assembly-averaging
A	Area
a	Radius
C_c	Cunningham coefficient
C_D	Drag coefficient
c	Specific heat
D	Diffusivity
d	Diameter
E	Electric field intensity
E	Young's modulus
e	Restitution coefficient
F	Force
f	Body force

f	Friction coefficient
G	Shear modulus
g	Gravity acceleration
h	Specific enthalpy
h_c	Convective heat transfer coefficient
I	Moment of inertia
$\boldsymbol{J}$	Impulse force
J_d	Droplet flow per unit volume
K	Thermal conductivity
k	Turbulence kinetic energy Stiffness
k_B	Boltzmann constant
l	Characteristic length
$\mathbf{M}$	Moment
m	Mass
$\boldsymbol{n}$	Normal unit vector
n	Number density
P	Probability density function
	Static or thermodynamic pressure
p	Hydrodynamic pressure
q	Heat flow
q_e	Electric charge
R	Gas constant
R_{ij}	Correlation of i and j
Re	Reynolds number
$\boldsymbol{r}$	Position vector
$\mathbf{S}$	Rate of deformation tensor
S	Surface area
s	Source term
$\boldsymbol{T}$	Stress tensor
	Torque
T	Temperature
$\boldsymbol{t}$	Tangential unit vector
t	Time
U	Time-averaged velocity
$\boldsymbol{u}$	Instant velocity
V	Volume
W	Weigh function

Greek symbols

α	Volume fraction
β	Momentum transfer coefficient

	Tangential or rolling restitution coefficient
Γ	Interfacial mass flow
γ	Rate of change
δ	Deformational displacement
	Film thickness
	Delta function
δ_{ij}	Kronecker delta
ε	Dissipation rate of k
	Emissivity
ε_e	Electromagnetic permeability
ζ	Random number
η	Damping coefficient
μ	Dynamic viscosity
ν	Poisson ratio
ρ	Density
σ	Noncontact stress tensor
σ	Statistical variance
σ_b	Stefan–Boltzmann constant
σ_E	Surface charge density
τ	Relaxation time
ω	Angular velocity
ω	Specific dissipation rate

Subscript

0	Unloaded
∞	Ambient
A	Added mass
B	Basset
	Brownian
C	Collision
D	Drag
d	Droplet
E	Electric
e	Eddy
f	Fluid
	Film
G	Gravitational
k	Kolmogorov
L	Lagrangian
M	Magnus
NC	Noncontact

n	Normal
P	Pressure
p	Particle
R	Residence
r	Rolling
rad	Radiation
S	Saffman
	Stokes
s	Surface
st	Slip
T	Turbulent
t	Tangential
w	Wall

Superscript

′	Fluctuating
-	Time-averaging
.	Rate
*	Reduced
T	Transpose

Problems

P5.1 Consider a metal bullet fired from a gun barrel to a target. Assume the after-combustion temperature and pressure inside the gun barrel are known. The bullet is accelerated and heated up inside the barrel and then cooled down by convective heat transfer once it is outside the barrel. The friction force inside the barrel is proportional to the bullet velocity. The drag force is proportional to the square of relative velocity between bullet and air.

(1) Establish a model for the bullet motion and heat-up inside the barrel, and estimate the bullet temperature and velocity at gun barrel exit;

(2) Assuming the air is stagnant, establish a model for the bullet trajectory and cooldown outside the barrel. Estimate the bullet temperature and velocity at the target.

(3) How does the wind affect the bullet trajectory and temperature if the windy air should be considered?

P5.2 An atomized spray of liquid is injected into an air cross flow. A droplet-collecting plate is located at a short distance downstream and normal to the cross flow, which is to determine the droplet size distribution of the spray. Establish a model to correlate the droplet trajectory to the droplet size and droplet injection velocity as well as to predict the droplet deposition position on the plate. Assume that the ambient cross flow is uniform.

P5.3 For a binary impact of two spherical particles, a sliding occurs when the impact angle is larger than the critical impact angle θ_c. Otherwise the collision is of the sticking type. Show that the critical impact angle can be estimated by

$$\theta_c = \tan^{-1}\left(\frac{7}{2}\frac{f(1+e)}{1+\beta}\right). \tag{P5.3}$$

P5.4 Establish a Lagrangian–Eulerian model to describe the wind-induced dispersion, transport, and deposition of volcano ash discharged from an active volcano. The initial ash is in form of lava droplets, and the ash will undergo a solidification process to become a cold solid. Wind is cold, dry, turbulent, and horizontal.

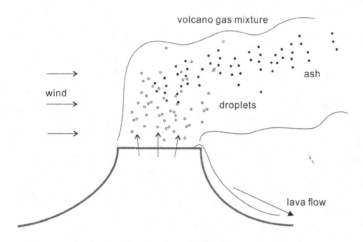

P5.5 In order to design geometric and operation conditions for slurry drying, establish a model for the transport and drying of a slurry droplet in a hot gas dryer. List the independent equations and variables to ensure the problem closure.

P5.6 Establish a model for the transport and evaporation of a well-mixed multicomponent droplet (such as a crude oil droplet) in a hot gaseous flow. Specify the boundary conditions. List the independent equations and variables to ensure the problem closure.

P5.7 For an atomized spray of firefighting, establish a Lagrangian trajectory model to estimate the coverage area as a function of shooting distance and angle. Assume that the spray injection characteristics such as velocity, size, and mass flowrate of droplets are all given. Consider the wind effect but may ignore the droplet evaporation.

P5.8 Consider a large cloud of snow flakes (highly porous flat ice agglomerates) that is falling from a certain height in the sky. Assuming the wind is strong and cold enough so that there is neither melting nor further agglomeration of the snow flakes, develop a hydrodynamic model of the snow transport and discuss the model closure.

P5.9 Consider a large cloud of icy snow (highly porous ice agglomerates) that is falling from a certain height in the sky. Assuming the wind is strong and warm enough so that snow melts to become rain during its falling, develop a model of the snow/rain transport and discuss the model closure.

P5.10 Consider a cross-flow spray drying process of slurry in a turbulent flow of a hot gas chamber. Assume that all solids in a slurry droplet remain inside the droplet and the chamber wall is hydrophobic (no wall deposition). Develop a model to estimate the drying completeness of the slurry droplet. Modeling of droplet evaporation and multicomponent mixing (between vapor and gas) must be included.

P5.11 To investigate the hydrodynamic mixing and interaction of a liquid fuel spray injected into a combustion chamber (*e.g.*, fuel-injection engine), develop a multiphase flow model by listing all independent variables and the number of independent equations to ensure model closure. Note that the gas phase is a multicomponent mixture; the flow is turbulent, and the droplet phase is dilute, with evaporation. It is assumed that the gas-phase reaction can be ignored during the hydrodynamic transport and interaction.

P5.12 Consider a gas–solid jet injected into a stagnant gas medium. Develop a model to describe the phase distributions of gas and solids. Alternatively, you may find and explain a model from literature survey.

P5.13 A series of gas jets is injected into a liquid column to generate a batch operation of bubbly columns. Develop a model to describe the dynamic motions of liquid and bubbles.

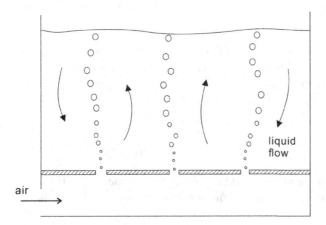

P5.14 Oil and natural gas released accidentally from ocean floor may cause severe environmental pollution. The difference in transport of oil and gas can be a strong function of water depth as well as the current drift. Due to the density stratification, high pressures, and low temperatures found in deep water, the deepwater significantly reduces the buoyancy of the plume and keeps large amounts of the oil submerged for an extended time. However, in the shallow water, the buoyancy and current drift play a major role in the location and spreading of the oil spill, as well as gas–oil separation. Establish a transport model for gas and oil in a deepwater spill accident.

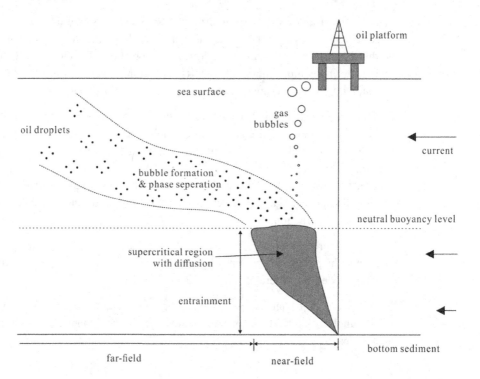

P5.15 A submicron particle with mass m suspended in a fluid medium is subjected to a random force $\mathbf{F}_B(t)$, which results from its instantaneous collision with fluid molecules. The random force has zero mean and no correlation between two different instants

$$\langle \mathbf{F}_B(t) \rangle = 0 \tag{P5.15-1}$$

$$\langle \mathbf{F}_B(t) \cdot \mathbf{F}_B(t') \rangle = c\delta(t - t'), \tag{P5.15-2}$$

where c is a constant coefficient that describes the intensity of the fluctuating force. Determine the value of the coefficient c, and the expectation of the mean squared displacement after time t.

P5.16 Develop a model for the transport of aerosols in a ventilated indoor air. The aerosol particles are typically of sizes in a range of submicrons to a few microns.

P5.17 Consider a mono-droplet generator that relies on an atomized nozzle spray of liquid with cross-flow air blowing. Develop a model of droplet size that is correlated to the characteristics of the nozzle injection and the cross flow of air.

References

Cundall, P. A., and Strack, O. D. (1979). A discrete numerical model for granular assemblies. *Geotechnique*, **29**, 47–65.

Elghobashi, S. (2006). An updated classification map of particle-laden turbulent flows. *IUTAM Symposium on Computational Approaches to Multiphase Flow*, 3–10.

Fan, L.-S., and Zhu, C. (1998). *Principles of Gas-Solid Flows*. Cambridge: Cambridge University Press.

Iwashita, K., and Oda, M. (1998). Rolling resistance at contacts in simulation of shear band development by DEM. *J. Eng. Mech.*, **124**, 285–292.

Longest, P. W., and Xi, J. (2007). Effectiveness of direct Lagrangian tracking models for simulating nanoparticle deposition in the upper airways. *Aerosol Sci. Technol.*, **41**, 380–397.

Rahmani, M., Geraci, G., Iaccarino, G., and Mani, A. (2018). Effects of particle polydispersity on radiative heat transfer in particle-laden turbulent flows.

Romero, M., Buck, R., and Pacheco, J. E. (2002). An update on solar central receiver systems, projects, and technologies. *J. Sol. Energy Eng.*, **124**(2), 98–108.

Smith, P. J. (1993). Three-dimensional turbulent particle dispersion submodel development. *Final Report to DOE*, PC/90094-T5.

Stanton, D. W., and Rutland, C. J. (1998). Multi-dimensional modeling of thin liquid films and spray-wall interactions resulting from impinging sprays. *Int. J. Heat Mass Transfer*, **40**, 3037–3054.

Tsuji, Y. Kawaguchi, T., and Tanaka, T. (1993). Discrete particle simulation of two-dimensional fluidized bed. *Powder Technol.*, **77**, 79–87.

Tsuji, Y., Tanaka, T., and Ishida, T. (1992). Lagrangian numerical simulation of plug flow of cohesionless particles in a horizontal pipe. *Powder Technol.*, **71**, 239–250.

Walton, O. R. (1993). Numerical simulation of inelastic frictional particle-particle interactions, *Particulate Two-Phase Flow*, ed. M. C. Roco, Boston: Butterworth-Heinemann.

Walton, O. R., and Braun, R. L. (1986). Stress calculation for assemblies of inelastic spheres in uniform shear. *Acta Mechanica*, **63**, 73–86.

Walton, O. R., and Braun, R. L. (1986). Viscosity, granular-temperature, and stress calculations for shearing assemblies of inelastic, frictional disks. *J. Rheol.*, **30**, 949–980.

Wang, L.-P., and Stock, D. E. (1993). Dispersion of heavy particles by turbulent motion. *J. Atmos. Sci.*, **50**, 1897–1913.

Wilcox, D. C. (2006). *Turbulence Modeling for CFD*, 3rd edition. La Canada, CA: DCW Industries, Inc.

Wilcox, D. C. (2008). Formulation of the k-omega turbulence model revisited. *AIAA J.*, **46**(11), 2823–2838.

Wu, Z., and Cao, Y. (2015). Numerical simulation of flow over an airfoil in heavy rain via a two-way coupled Eulerian-Lagrangian approach. *Int. J. Multiphase Flow*, **69**, 81–92.

Zhu, C., Wang, D. W., and Lin, C.-H. (2010). Jet dispersion and deposition of charged particles in confined chambers. *Particuology*, **8**, 28–36.

Zhu, H. P., and Yu, A. (2003). The effect of wall and rolling resistance on the couple stress of granular materials in vertical flow. *Physica A*, **325**, 347–360.

6 Continuum Modeling of Multiphase Flows

6.1 Introduction

As another major modeling approach, the Eulerian method of multiphase flows can be advantageous over the Lagrangian approach due to the employment of fixed spatial coordinates for field description and hence, computationally less demanding. To be applicable, the transport quantities of all phases must be continuous throughout the computational domain. However, such an assumption is not always true in reality, since at any instant in time, a point in space cannot be simultaneously occupied by all phases. Hence, a "continuum" for each phase that collectively shares the same space and time domain needs to be constructed to extend the traditional Eulerian description of a single-phase flow to a multiphase flow application. Averaging theorems are essential to the creation of said shared-continuum of a multiphase media.

This chapter begins with an introduction of volume averaging and progresses to averaging theorems of derivatives in Section 6.2. It then presents the development of the volume-averaged equations in Section 6.3. Volume-time-averaged equations are given in Section 6.4 to account for the inherent time-dependent fluctuations, such as those induced by fluid turbulence or stochastic collisions among particles. The resulting averaged transport equations comprise many terms, including constitutive relations and interfacial transport rates that require further treatment for model closure. Closure techniques for such terms are discussed in Section 6.5. Starting from the general form of the averaged multiphase transport equation, Section 6.6 develops particular forms of the averaged equations for fluid–fluid flows and fluid–solid flows, respectively. Section 6.7 illustrates the phenomena having a significant impact on the multiphase flow behavior, but have been considerably difficult to describe with the Eulerian description. These topics illuminate opportunities for a future research on multiphase flows. Finally, in Section 6.8, several application examples of continuum modeling are offered as case studies.

6.2 Averages and Averaging Theorems

Averaging is essential to the construction of a continuum model for a multiphase flow, where each constituent phase may have discrete spatial and temporal distributions. The continuum for each phase may be constructed using any one of the

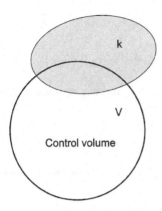

Figure 6.2-1 Concept of volume average.

following approaches: volume averaging as a function of volume fractions (Slattery, 1967a, 1967b; Whitaker, 1969; Delhaye and Archard, 1976; Jackson, 2000), time averaging as a function of fractional residence time (Ishii, 1975), ensemble averaging as a function of the probability of different realizations in the flow configuration (Drew and Lahey, 1993), and volume-time averaging with time averaging computed after volume averaging (Soo, 1989). In multiphase flows, the extensive dynamic and thermodynamic properties of a mixture increase proportionally with volume fraction but not necessarily with the fractional residence time (Soo, 1989). Moreover, a priori time averaging, based on fractional residence time, may eliminate the identity of different dynamic phases that would otherwise exist due to different inherent flow velocities. In this text, volume averaging is the preferred approach to construct a continuum for each phase, and time averaging is considered a subsequent step that may be employed for flows with high-frequency fluctuations.

6.2.1 Phase and Intrinsic Averaging

For the control volume, V, shown in Figure 6.2-1, the volume average for any quantity of phase k, $<\psi_k>$, can be written as:

$$\langle \psi_k \rangle = \frac{1}{V} \int_{V_k} \psi_k dV, \tag{6.2-1}$$

where ψ_k is a representative quantity of phase k, which can be a scalar, vector, or tensor; V_k is the volume occupied by phase k within control volume, V. $<\psi_k>$ is known as the phase average of ψ_k. The phase average apportions a pseudo-property or quantity that depends on how the control volume is selected. Hence, phase averages permit construction of an overall continuum of constituent phases, which can be described using the Eulerian approach.

However, real physical properties and quantities, such as density and velocity of phase k, require an intrinsic average, computed for the actual phase occupying the control volume. Thus, it yields:

$$^i \langle \psi_k \rangle = \frac{1}{V_k} \int_{V_k} \psi_k dV. \tag{6.2-2}$$

The phase average can readily be related to the intrinsic average via the volume fraction of phase k, α_k, as:

$$\langle \psi_k \rangle = \frac{V_k}{V}{}^i \langle \psi_k \rangle = \alpha_k{}^i \langle \psi_k \rangle. \tag{6.2-3}$$

The intrinsic average of the density of phase k is directly obtained from Eq. (6.2-3) as:

$$\langle \rho_k \rangle = \alpha_k{}^i \langle \rho_k \rangle \cong \alpha_k \rho_k. \tag{6.2-4}$$

Equation (6.2-4) assumes all particles, or elements, of phase k within V_k have the same material density, reasonable for most cases of multiphase flow. Unless otherwise noted, this text regards this approximation as valid.

The intrinsic velocity average of phase k is defined in the context of the mass flux of phase k as:

$$^i \langle \mathbf{u}_k \rangle = \frac{\int_{V_k} \rho_k \mathbf{u}_k dV}{{}^i \langle \rho_k \rangle V_k} = \frac{\langle \rho_k \mathbf{u}_k \rangle}{\alpha_k \rho_k}. \tag{6.2-5}$$

Similarly, the intrinsic average of the internal energy of phase k is given by:

$$^i \langle e_k \rangle = \frac{\int_{V_k} \rho_k e_k dV}{{}^i \langle \rho_k \rangle V_k} = \frac{\langle \rho_k e_k \rangle}{\alpha_k \rho_k}. \tag{6.2-6}$$

The temperature can further be defined from the average of internal energy of phase k, as:

$$^i \langle T_k \rangle = \frac{\int_{V_k} \rho_k e_k dV}{{}^i \langle \rho_k \rangle c_{vk} V_k} = \frac{\langle \rho_k e_k \rangle}{\alpha_k \rho_k c_{vk}}. \tag{6.2-6a}$$

The intrinsic averages defined above are mass-weighted and are not number-weighted or mass-independent. Hence, the thermodynamic significance for extensive properties of mass, momentum, and energy is preserved.

6.2.2   Volume-Averaging Theorems

Volume-averaging theorems use integrals, derivatives, or derivatives in terms of the integrals to express volume averages or derivatives of the volume averages. A change in volume average with respect to the change of the control volume center is used to interrogate the spatial derivative of a volume-averaged property. As a simplification, it is often assumed that the control volume, defined for volume averaging, is a sphere of fixed radius with its center moving along an arbitrary but continuous path s, as shown in Figure 6.2-2.

The region occupied by phase k in the control volume, V_k, is bounded by an inter-facial area, A_k, and a surface area, A_{ke}. From the general transport theorem embodied

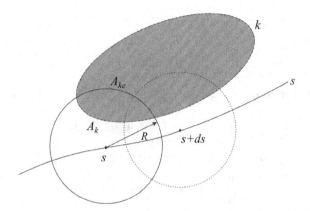

Figure 6.2-2 Volume averaging in a multiphase mixture.

in Eq. (2.2-6), the change of the integral over V_k as a function of s may be expressed as:

$$\frac{d}{ds}\int_{V_k}\psi_k dV = \int_{V_k}\frac{\partial\psi_k}{\partial s}dV + \int_{A_k\cup A_{ke}}\psi_k\left(\frac{d\mathbf{r}}{ds}\cdot\mathbf{n}_k\right)dA. \qquad (6.2\text{-}7)$$

Above, the variable t in Eq. (2.2-6) was substituted by variable s whereas $\mathbf{n}_k$ is an outwardly directed normal for the closed surface defined by A_k and A_{ke}. Note that ψ_k is a function of time and space coordinates that is independent of s; therefore, it gives:

$$\frac{\partial\psi_k}{\partial s} = 0. \qquad (6.2\text{-}7a)$$

The interfacial area, A_k, is smooth and continuous when the center of the integral sphere shifts from s to $s + ds$ such that $d\mathbf{r}/ds$ is a tangential vector to $A_k(s)$. This criterion leads to:

$$\frac{d\mathbf{r}}{ds}\cdot\mathbf{n}_k = 0. \qquad (6.2\text{-}7b)$$

However, for the surface integral over A_{ke} in Eq. (6.2-7), A_{ke} may not be continuous as the center of the sphere shifts from s to $s + ds$; it is possible for $A_{ke}(s + ds)$ to be an entirely different surface from $A_{ke}(s)$. Hence, either $d\mathbf{r}/ds$ is not a tangential vector to $A_{ke}(s)$ or $d\mathbf{r}/ds \cdot \mathbf{n}_{ke}$ is not zero. However, $\mathbf{r}(s)$ can be partitioned into $\mathbf{r}_0(s)$ and $\mathbf{R}(s)$, where $\mathbf{r}_0$ is a position vector defining the center of the sphere and $\mathbf{R}$ is a position vector relative to the center s. Considering this and the closed surface being translated, it yields:

$$\frac{d\mathbf{R}}{ds} = 0. \qquad (6.2\text{-}7c)$$

Equation (6.2-7) can then be simplified into:

$$\frac{d}{ds}\int_{V_k}\psi_k dV = \int_{A_{ke}}\psi_k\left(\frac{d\mathbf{r}_0}{ds}\cdot\mathbf{n}_{ke}\right)dA. \qquad (6.2\text{-}7d)$$

The derivative d/ds can be alternately expressed in terms of $\mathbf{r}_0(s)$ as (Whitaker, 1969):

$$\frac{d}{ds} = \frac{d\mathbf{r}}{ds} \cdot \nabla = \frac{d\mathbf{r}_0}{ds} \cdot \nabla. \tag{6.2-7e}$$

Equation (6.2-7d) then further reduces to:

$$\frac{d\mathbf{r}_0}{ds} \cdot \left(\nabla \int_{V_k} \psi_k dV - \int_{A_{ke}} \psi_k \mathbf{n}_{ke} dA \right) = 0. \tag{6.2-7f}$$

Because $\mathbf{r}_0$ is arbitrary, the preceding equation takes the final form of:

$$\nabla \int_{V_k} \psi_k dV = \int_{A_{ke}} \psi_k \mathbf{n}_{ke} dA. \tag{6.2-7g}$$

From the Gauss theorem, it gives:

$$\int_{A_{ke}} \psi_k \mathbf{n}_{ke} dA = \int_{V_k} \nabla \psi_k dV - \int_{A_k} \psi_k \mathbf{n}_k dA. \tag{6.2-7h}$$

Combining Eq. (6.2-7h) with Eq. (6.2-7g), an averaging theorem, known as the *Gauss rule*, is derived:

$$\langle \nabla \psi_k \rangle = \nabla \langle \psi_k \rangle + \frac{1}{V} \int_{A_k} \psi_k \, \mathbf{n}_k \, dA. \tag{6.2-8}$$

When $\psi_k = 1$, it can be shown that:

$$\nabla \alpha_k = -\frac{1}{V} \int_{A_k} \mathbf{n}_k \, dA. \tag{6.2-8a}$$

Similarly, it can be proven that:

$$\langle \nabla \cdot \boldsymbol{\psi}_k \rangle = \nabla \cdot \langle \boldsymbol{\psi}_k \rangle + \frac{1}{V} \int_{A_k} \boldsymbol{\psi}_k \cdot \mathbf{n}_k \, dA. \tag{6.2-9}$$

The averaging theorem for the time derivative can be obtained directly from the general transport theorem. From the definition of the phase average, it gives:

$$\left\langle \frac{\partial \psi_k}{\partial t} \right\rangle = \frac{\partial}{\partial t} \langle \psi_k \rangle - \frac{1}{V} \int_{A_k} \psi_k \mathbf{u}_s \cdot \mathbf{n}_k dA, \tag{6.2-10}$$

where $\mathbf{u}_s$ is the speed of displacement for the interface. The above expression is also referred to as the *Leibnitz rule*.

Equations (6.2-8), (6.2-9), and (6.2-10) represent specific volume-averaging theorems of derivatives for gradient, divergence, and time-derivative. Additional detailed derivations of the volume-averaging theorems can be gleaned from the references (Slattery, 1967b; Whitaker, 1969; Soo, 1989; Fan and Zhu, 1998; Jackson, 2000).

6.3   Volume-Averaged Equations

In an instantaneous configuration of a multiphase flow, each phase may be regarded as a single phase flowing within its own sub-domain. Thus, the volume-averaged transport equations for phase k can be derived by applying the averaging theorems to the governing equations for an analogous single-phase flow of phase k. During the volume-averaging process, detailed individual interface configurations are replaced by integrals over all interfaces within the control volume; the resulting volume-averaged equations are then applicable to the entire flow domain. Discrete phases remain coupled through interfacial interactions and are described by averaged interfacial jump conditions and volumetric contribution to the overall flow domain.

6.3.1 General Volume-Averaged Equations

For a physical quantity, ψ_k, associated with phase k, the instantaneous local balance can be described by the general conservation principles of Eq. (2.2-9) as:

$$\frac{\partial \psi_k}{\partial t} + \nabla \cdot (\psi_k \mathbf{u}_k) + \nabla \cdot \mathbf{J}_k - \varphi_k = 0. \tag{6.3-1}$$

The interfacial jump condition between two adjacent phases, as described in Section 2.2.4.4, is given in general form by:

$$\sum_{k=1}^{2} \{\psi_k (\mathbf{u}_k - \mathbf{u}_s) + \mathbf{J}_k\} \cdot \mathbf{n}_k = -M_{\psi 12}. \tag{6.3-2}$$

Applying volume-averaging principles with the Leibnitz and Gauss rules to the general local conservation expression of Eq. (6.3-1) results in the volume-averaged equation for ψ_k, given by:

$$\frac{\partial \langle \psi_k \rangle}{\partial t} + \nabla \cdot \langle \psi_k \mathbf{u}_k \rangle + \nabla \cdot \langle \mathbf{J}_k \rangle = -\frac{1}{V} \int_{A_k} \{\psi_k (\mathbf{u}_k - \mathbf{u}_s) + \mathbf{J}_k\} \cdot \mathbf{n}_k dA + \langle \phi_k \rangle. \tag{6.3-3}$$

Integrating the jump condition at the interface yields:

$$\frac{1}{V} \int_{A_k} \left(\sum_{k=1}^{2} \{\psi_k (\mathbf{u}_k - \mathbf{u}_s) + \mathbf{J}_k\} \cdot \mathbf{n}_k + M_{\psi 12} \right) dA = 0. \tag{6.3-4}$$

Substituting the expressions for the flux $\mathbf{J}_k$, source term ϕ_k, and interfacial jump, $M_{\psi 12}$, into the above general volume-averaged conservation equations, produces the volume-averaged equations for mass, momentum, and energy. Details of this process are illuminated in the following sections.

6.3.2 Volume-Averaged Continuity Equation

The volume-averaged continuity equation for phase k is obtained by applying the volume-averaging theorems to Eq. (2.2-10), resulting in:

$$\frac{\partial \langle \rho_k \rangle}{\partial t} + \nabla \cdot \langle \rho_k \mathbf{u}_k \rangle = \Gamma_k, \tag{6.3-5}$$

where Γ_k is the rate of mass generation of phase k per unit volume by mass transfer mechanisms through the interfacial boundary of phase k. This quantity is defined by:

$$\Gamma_k = -\frac{1}{V} \int_{A_k} \rho_k \left(\mathbf{u}_k - \mathbf{u}_s\right) \cdot \mathbf{n}_k \, dA. \tag{6.3-5a}$$

Unit "volume" in this context refers to the multiphase system. If a phase is physically defined, such as the solid phase or the gaseous phase, Γ_k may occur by chemical reactions or by physical phase changes, such as evaporation. Conversely, when a phase is dynamically defined (Soo, 1965), such as in pneumatic transport or separation of solid particles, Γ_k may also result from a change in particle size due to attrition or agglomeration as well as from chemical reactions or physical phase changes. From the expression for the averaged jump condition, a mass balance of all phases in the flow requires:

$$\sum_k \Gamma_k = 0. \tag{6.3-6}$$

Equation (6.3-5) must be further refined into terms of intrinsic averages of physically meaningful properties. Substituting Eq. (6.2-4) and Eq. (6.2-5) into Eq. (6.3-5) yields the following form of the volume-averaged continuity equation:

$$\frac{\partial}{\partial t} \left(\alpha_k \rho_k\right) + \nabla \cdot \left(\alpha_k \rho_k{}^i \langle \mathbf{u}_k \rangle\right) = \Gamma_k. \tag{6.3-7}$$

The volume fractions of all phases must follow the condition:

$$\sum_k \alpha_k = 1. \tag{6.3-7a}$$

6.3.3   Volume-Averaged Momentum Equation

Applying averaging theorems to the general momentum equation for a continuum as given by Eq. (2.2-12), produces the following volume-averaged momentum equation of phase k:

$$\frac{\partial \langle \rho_k \mathbf{u}_k \rangle}{\partial t} + \nabla \cdot \langle \rho_k \mathbf{u}_k \mathbf{u}_k \rangle = \nabla \cdot \langle \mathbf{T}_k \rangle + \langle \rho_k \mathbf{f}_k \rangle + \mathbf{F}_{Ak} + \mathbf{F}_{\Gamma k}, \tag{6.3-8}$$

where $\mathbf{F}_{Ak}$ accounts for the total stress contribution from other phases acting on the interface per unit volume, which can be written as:

$$\mathbf{F}_{Ak} = \frac{1}{V} \int_{A_k} \mathbf{T}_k \cdot \mathbf{n}_k \, dA. \tag{6.3-8a}$$

$\mathbf{F}_{\Gamma k}$ represents the momentum transfer across the interfacial boundary per unit volume due to the mass generation of the phase, which is defined by:

$$\mathbf{F}_{\Gamma k} = -\frac{1}{V} \int_{A_k} \rho_k \, \mathbf{u}_k \left(\mathbf{u}_k - \mathbf{u}_s\right) \cdot \mathbf{n}_k \, dA. \tag{6.3-8b}$$

The specific body force, **f**, includes all field forces that are proportional to phase density, such as the gravitational force. The averaged jump condition of momentum at the interface is given by:

$$\sum_{k=1}^{2} (\mathbf{F}_{Ak} + \mathbf{F}_{\Gamma k}) = \frac{1}{V} \int_{A_k} (2\sigma \kappa \mathbf{n}_s + \nabla_s \sigma) \, dA. \tag{6.3-9}$$

where ∇_s is the surface gradient, and κ is the mean curvature along $\mathbf{n}_s$ so that the right-hand side of the equation represents the net effect of surface tension.

The volume-averaged convective momentum flux of phase k can be defined in a similar fashion to the velocity expression of Eq. (6.2-5). Doing so, it yields:

$$\langle \rho_k \mathbf{u}_k \mathbf{u}_k \rangle = \langle \rho_k \rangle^i \langle \mathbf{u}_k \mathbf{u}_k \rangle. \tag{6.3-10}$$

The volume-averaged velocity tensor must be further refined into terms of intrinsic averages of physically meaningful properties. One simple approximation to facilitate this effort is to assume a direct product of intrinsic velocities is valid, so that:

$$\langle \rho_k \mathbf{u}_k \mathbf{u}_k \rangle \approx \alpha_k \rho_k^i \langle \mathbf{u}_k \rangle^i \langle \mathbf{u}_k \rangle. \tag{6.3-10a}$$

Based on Eq. (6.2-3), the volume-averaged total stress can be represented by:

$$\langle \mathbf{T}_k \rangle = \alpha_k^i \langle \mathbf{T}_k \rangle \tag{6.3-11}$$

Enhanced formulation of intrinsic total stress as a function of intrinsic averages of physically meaningful properties must be developed from constitutive mechanisms for each individual phase k, which may not follow the constitutive relations of Newtonian fluid flows.

Substituting Eq. (6.3-10a) and Eq. (6.3-11) into Eq. (6.3-8) yields a volume-averaged momentum equation of the form:

$$\frac{\partial}{\partial t} \left(\alpha_k \, \rho_k^i \langle \mathbf{u}_k \rangle \right) + \nabla \cdot \left(\alpha_k \, \rho_k^i \langle \mathbf{u}_k \rangle^i \langle \mathbf{u}_k \rangle \right) = \nabla \cdot \left(\alpha_k^i \langle \mathbf{T}_k \rangle \right) + \alpha_k \, \rho_k \, \mathbf{f}_k + \mathbf{F}_{Ak} + \mathbf{F}_{\Gamma k} \tag{6.3-12}$$

6.3.4 Volume-Averaged Energy Equation

The total energy per unit mass of phase k, E_k, is defined as the sum of internal energy and kinetic energy quantities as:

$$E_k = e_k + \frac{1}{2}(\mathbf{u}_k \cdot \mathbf{u}_k) \tag{6.3-13}$$

From the first law of thermodynamics, the energy conservation equation of phase k mimics the basic form of Eq. (2.2-13) if considered as an analog to single-phase flow. Applying the volume-averaging theorem to Eq. (2.2-13) results in the volume-averaged equation of energy conservation as a function of total energy as:

$$\frac{\partial}{\partial t} \langle \rho_k E_k \rangle + \nabla \cdot \langle \rho_k E_k \mathbf{u}_k \rangle = \nabla \cdot \langle \mathbf{T}_k \cdot \mathbf{u}_k \rangle - \nabla \cdot \langle \mathbf{J}_{qk} \rangle$$

$$+ \langle \rho_k \mathbf{u}_k \rangle \cdot \mathbf{f}_k + \langle J_{Ek} \rangle + q_{Ak} + q_{\Gamma k} + q_{Wk}, \tag{6.3-14}$$

where q_{Ak} accounts for the heat transfer across the interface into phase k, as given by:

$$q_{Ak} = -\frac{1}{V}\int_{A_k} \mathbf{J}_{qk} \cdot \mathbf{n}_k dA. \tag{6.3-14a}$$

$q_{\Gamma k}$ is the total energy transferred by mass generation of phase k resulting from phase change, which is expressed by:

$$q_{\Gamma k} = -\frac{1}{V}\int_{A_k} \rho_k \left(e_k + \frac{1}{2}(\mathbf{u}_k \cdot \mathbf{u}_k)\right)(\mathbf{u}_k - \mathbf{u}_s) \cdot \mathbf{n}_k dA. \tag{6.3-14b}$$

The work done by the total stress at the interface is written as:

$$q_{Wk} = -\frac{1}{V}\int_{A_k} (\mathbf{T}_k \cdot \mathbf{u}_k) \cdot \mathbf{n}_k dA. \tag{6.3-14c}$$

The averaged interfacial energy jump condition is given by:

$$\sum_{k=1}^{2}(q_{Ak} + q_{Wk} + q_{\Gamma k}) = \frac{1}{V}\int_{A_k} \left(\nabla_s \cdot \mathbf{q}_s - \nabla_s \cdot \sigma\mathbf{u}_s\right) dA. \tag{6.3-15}$$

The right-hand side of Eq. (6.3-15) represents the effects of interfacial heat flux and surface tension.

Alternatively, the volume-averaged energy equation can also be written as a function of internal energy as:

$$\frac{\partial}{\partial t}\langle\rho_k e_k\rangle + \nabla \cdot \langle\rho_k e_k \mathbf{u}_k\rangle = \langle\mathbf{T}_k : \nabla\mathbf{u}_k\rangle - \nabla \cdot \langle\mathbf{J}_{qk}\rangle + \langle J_{Ek}\rangle + q_{Ak} + e_{\Gamma k}, \tag{6.3-16}$$

where $e_{\Gamma k}$ is the internal energy transfer due to phase change over the interface, as defined by:

$$e_{\Gamma k} = -\frac{1}{V}\int_{A_k} \rho_k e_k (\mathbf{u}_k - \mathbf{u}_s) \cdot \mathbf{n}_k dA. \tag{6.3-16a}$$

When the interfacial tension and heat flux effects are negligible, the averaged form of the internal energy jump condition becomes:

$$\sum_{k=1}^{2}(q_{Ak} + e_{\Gamma k}) \approx 0. \tag{6.3-17}$$

If the phase k represents a true, or an analogous, viscous fluid whose total stress tensor, $\mathbf{T}_k$, can be decomposed into a compressional, or principal, tensor (such that the pressure of phase k, p_k, can be defined) and a shear stress tensor (similar to Eq. (2.2-1)), then the energy equation can be expressed as a function of enthalpy of phase k. This is represented by:

$$h_k = e_k + \frac{p_k}{\rho_k}. \tag{6.3-18}$$

The energy equation, in terms of enthalpy, is then given by:

$$\frac{\partial}{\partial t}\langle\rho_k h_k\rangle + \nabla \cdot \langle\rho_k h_k \mathbf{u}_k\rangle = \frac{\partial}{\partial t}\langle p_k\rangle + \nabla \cdot \langle p_k \mathbf{u}_k\rangle + \langle\mathbf{T}_k : \nabla\mathbf{u}_k\rangle$$
$$- \nabla \cdot \langle\mathbf{J}_{qk}\rangle + \langle J_{Ek}\rangle + q_{Ak} + h_{\Gamma k}, \tag{6.3-19}$$

where $h_{\Gamma k}$ is the enthalpy transfer due to phase change over interface, which is defined by:

$$h_{\Gamma k} = -\frac{1}{V} \int_{A_k} \rho_k h_k (\mathbf{u}_k - \mathbf{u}_s) \cdot \mathbf{n}_k dA. \tag{6.3-19a}$$

The corresponding jump condition is expressed as:

$$\sum_{k=1}^{2} (q_{Ak} + h_{\Gamma k}) \approx 0. \tag{6.3-20}$$

When the effects of wave propagation are assumed negligible, it follows that:

$$h_{\Gamma k} \approx e_{\Gamma k}. \tag{6.3-21}$$

The common form of the energy equation as a function of temperature of phase k can be readily obtained by substituting $e_k = c_{vk} T_k$ into Eq. (6.3-16). The work done, or energy dissipation by total stress, is typically negligibly small when compared to heat transfer terms, including those corresponding to interfaces, mass transfer, and thermal radiation. Assuming heat flux, $\mathbf{J}_q$, obeys Fourier's law and with contributions from total stress neglected, Eq. (6.3-16) becomes:

$$\begin{aligned}
&\frac{\partial}{\partial t} \left(\alpha_k \rho_k c_{vk}{}^i \langle T_k \rangle \right) + \nabla \cdot \left(\alpha_k \rho_k c_{vk}{}^i \langle T_k \rangle {}^i \langle \mathbf{u}_k \rangle \right) \\
&= \nabla \cdot \left(\alpha_k K_k \nabla^i \langle T_k \rangle \right) + \alpha_k{}^i \langle J_{rk} \rangle + q_{Ak} + e_{\Gamma k},
\end{aligned} \tag{6.3-22}$$

where K_k is the heat conductivity of phase k, and J_{rk} is the thermal radiation to phase k. Here, the thermal radiation is approximated to be proportional to the volume fraction of phase k although it can be physically based on the interfacial area of phase k such as in diffusive radiation.

6.4 Volume–Time-Averaged Equations

Volume-averaged equations describe spatially averaged instantaneous multiphase flows in fixed Eulerian coordinates. However, many multiphase flows are inherently stochastic, rendering the instantaneous description nondeterministic. Examples of stochastic nature include particle–fluid interactions resulting from fluid turbulence and inherently random phase particle characteristics, such as irregularity in shape, orientation, size, and collision effects. Stochastic behavior results in time-dependent fluctuations in multiphase flow transport. One common method to describe statistically averaged transport behavior is via time-averaging of the volume-averaged transport equations of multiphase flows.

6.4.1   Volume–Time Averages and Covariance

Volume–time averaging is presented in this section as an analog to Reynolds analysis of a single-phase turbulent flow: temporal decomposition into mean and high-frequency fluctuations is employed to accomplish time averaging of the already volume-averaged flow phenomena.

The time-averaging duration, τ_t, should be chosen in such a way that the following criterion is satisfied: $\tau_{HF} \ll \tau_t \ll \tau_{LF}$. The characteristic time of the high-frequency component, τ_{HF}, may be estimated from the reciprocal of the characteristic spectral frequency of the fluctuation. The characteristic time of the low-frequency component, τ_{LF}, may be determined from the time required to travel a characteristic dimension of the physical system at the local characteristic low-frequency speed. Thus, an expression for time-averaging after volume-averaging has been completed can be defined as:

$$\overline{\langle \psi_k \rangle} = \frac{1}{\tau_t} \int_0^{\tau_t} \langle \psi_k \rangle \, dt. \tag{6.4-1}$$

The instantaneous volume-averaged quantity is then expressed as a sum of two components, as:

$$\langle \psi_k \rangle = \overline{\langle \psi_k \rangle} + \left\langle \psi_k' \right\rangle, \tag{6.4-1a}$$

where the prime notation indicates the high-frequency component, and:

$$\overline{\langle \psi_k' \rangle} = 0. \tag{6.4-1b}$$

Further, it gives:

$$\overline{\langle \psi_k \rangle} = \frac{1}{\tau_t} \int_0^{\tau_t} \left\{ \frac{1}{V} \int_V \left(\overline{\psi_k} + \psi_k' \right) dV \right\} dt = \frac{1}{V} \int_V \left\{ \frac{1}{\tau_t} \int_0^{\tau_t} \left(\overline{\psi_k} + \psi_k' \right) dt \right\} dV = \overline{\langle \psi_k \rangle} \tag{6.4-2a}$$

Equation (6.4-2a) shows that superimposing time-averaging over a volume-averaged quantity is equivalent to superimposing volume-averaging over a time-averaged quantity. This equality is valid when the spatial integral and temporal integral are interchangeable due to a proper choice of T, namely: $\tau_{HF} \ll \tau_t \ll \tau_{LF}$. Considering Eq. (6.4-1) reveals that the time averages of the derivatives of a volume-averaged quantity have the following properties:

$$\overline{\frac{\partial}{\partial t} \langle \psi_k \rangle} = \frac{\partial}{\partial t} \overline{\langle \psi_k \rangle} \tag{6.4-2b}$$

$$\overline{\nabla \langle \psi_k \rangle} = \nabla \overline{\langle \psi_k \rangle} \tag{6.4-2c}$$

$$\overline{\nabla \cdot \langle \boldsymbol{\psi}_k \rangle} = \nabla \cdot \overline{\langle \boldsymbol{\psi}_k \rangle}. \tag{6.4-2d}$$

The above relations are valid because the boundaries between different phases vanish after volume-averaging. However, the time-average of a derivative generally is not equivalent to the derivative of its time average (Soo, 1989).

Averaged transport equations invariably contain averages of products, or covariance. These quantities must be further expressed as the products of averages. In a volume-time averaging approach, spatial covariance is often neglected, but the temporal covariance of high-frequency fluctuations is preserved, partly because of its significance in enhanced transport by turbulence or particle–particle interactions such as collisions. For example, applying time-averaging to a velocity-associated product, $\psi_k \mathbf{u}_k$, of phase k, it gives:

$$\overline{\langle \psi_k \mathbf{u}_k \rangle} = \overline{\langle \psi_k \rangle^i \langle \mathbf{u}_k \rangle} = \overline{\langle \psi_k \rangle^i \langle \mathbf{u}_k \rangle} + \overline{\langle \psi_k' \rangle^i \langle \mathbf{u}_k' \rangle}. \qquad (6.4\text{-}3)$$

The covariance here represents the diffusive flux for the transport of ψ_k. Traditionally, this diffusive flux is considered to be driven by the gradient of ψ_k. Hence, it can be formulated using Boussinesq's gradient approach (Boussinesq, 1877) as:

$$\overline{\langle \psi_k' \rangle^i \langle \mathbf{u}_k' \rangle} = -D_{\psi k} \nabla \overline{\langle \psi_k \rangle}, \qquad (6.4\text{-}4)$$

where $D_{\psi k}$ is the diffusive transport coefficient of ψ_k. The transport coefficients depend on the nature and mechanisms of the fluctuations causing to the transport. In single-phase turbulent flow, fluctuation is due to turbulence and the transport coefficients must be related to the eddy characteristics. This leads to a formulation that is comparatively more dependent on the characteristics of the flow than fluid properties. In a multiphase flow, such as a dense-phase gas–solid flow, the fluctuation of solids may have two distinct sources: fluid as one source and intersolids collisions as the other. In such a case, the transport coefficients must be formulated according to the physical mechanisms of the fluctuation, forming either separate or combined constitutive equations.

6.4.2   Volume-Time-Averaged Continuity Equation

Applying Eq. (6.4-2) to Eq. (6.3-7) produces the volume-time-averaged continuity equation, as:

$$\frac{\partial}{\partial t} (\overline{\alpha_k \rho_k}) + \nabla \cdot \left(\overline{\alpha_k \rho_k}^i \langle \mathbf{u}_k \rangle \right) = \overline{\Gamma}_k. \qquad (6.4\text{-}5)$$

The intrinsic density, ρ_k, represents the material density of phase k, which is regarded as time-independent for most multiphase flows. The volume-time-averaged mass flux of phase k can be further decomposed to:

$$\overline{\alpha_k \rho_k}^i \langle \mathbf{u}_k \rangle = \rho_k \overline{\alpha}_k^i \langle \overline{\mathbf{u}}_k \rangle + \rho_k \overline{\alpha_k'^i \langle \mathbf{u}_k' \rangle} = \rho_k \overline{\alpha}_k^i \langle \overline{\mathbf{u}}_k \rangle - D_{ek} \nabla (\rho_k \overline{\alpha}_k), \qquad (6.4\text{-}6)$$

where D_{ek} is the eddy diffusivity of phase k in the mixture. The volume-time-averaged continuity equation of phase k can then be written as:

$$\frac{\partial}{\partial t} (\rho_k \overline{\alpha}_k) + \nabla \cdot \left(\rho_k \overline{\alpha}_k^i \langle \overline{\mathbf{u}}_k \rangle \right) = \nabla \cdot (D_{ek} \nabla (\rho_k \overline{\alpha}_k)) + \overline{\Gamma}_k. \qquad (6.4\text{-}7)$$

6.4.3 Volume-Time-Averaged Momentum Equation

In a similar fashion to that described above, the volume-time-averaged momentum equation is obtained from Eq. (6.3-12) as:

$$\frac{\partial}{\partial t}\left(\overline{\alpha_k\,\rho_k{}^i\,\langle\mathbf{u}_k\rangle}\right) + \nabla\cdot\left(\overline{\alpha_k\,\rho_k{}^i\,\langle\mathbf{u}_k\rangle{}^i\,\langle\mathbf{u}_k\rangle}\right) = \nabla\cdot\left(\overline{\alpha_k{}^i\,\langle\mathbf{T}_k\rangle}\right) + \overline{\alpha_k\,\rho_k\,\mathbf{f}_k} + \overline{\mathbf{F}}_{Ak} + \overline{\mathbf{F}}_{\Gamma k}.$$

$$(6.4\text{-}8)$$

The volume-time average for the convective momentum flux of phase k can be further manipulated as:

$$\overline{\alpha_k\,\rho_k{}^i\,\langle\mathbf{u}_k\rangle{}^i\,\langle\mathbf{u}_k\rangle} = \rho_k\,\overline{\alpha}_k{}^i\,\langle\overline{\mathbf{u}_k}\rangle{}^i\,\langle\overline{\mathbf{u}_k}\rangle + \rho_k\,\overline{\alpha}_k{}^i\,\overline{\langle\mathbf{u}'_k\rangle{}^i\,\langle\mathbf{u}'_k\rangle}$$
$$+ \rho_k\,\overline{\alpha'_k{}^{\prime i}\,\langle\mathbf{u}'_k\rangle{}^i\,\langle\overline{\mathbf{u}_k}\rangle} + \rho_k{}^i\,\langle\overline{\mathbf{u}_k}\rangle\,\overline{\alpha'^i_k\,\langle\mathbf{u}'_k\rangle} + \rho_k\,\overline{\alpha'^i_k\,\langle\mathbf{u}'_k\mathbf{u}'_k\rangle}.$$

$$(6.4\text{-}9)$$

The second term on the right-hand side is commonly referred to as the Reynolds stress of phase k. The Reynolds stress of phase k is given by:

$$\langle\overline{\tau}_{ek}\rangle = -\rho_k\overline{\alpha}_k{}^i\,\overline{\langle\mathbf{u}'_k\rangle{}^i\,\langle\mathbf{u}'_k\rangle}.$$

$$(6.4\text{-}9a)$$

When phase k is a turbulent fluid phase, Eq. (6.4-9a) is determined directly from the turbulence model of the fluid. The last triple covariance term on the right-hand side of Eq. (6.4-9) is typically neglected to facilitate model closure. Therefore, using Eqs. (6.4-6) and (6.4-9a), the convective momentum flux in Eq. (6.4-9) is determined as:

$$\overline{\alpha_k\rho_k{}^i\,\langle\mathbf{u}_k\rangle{}^i\,\langle\mathbf{u}_k\rangle} = \rho_k\overline{\alpha}_k{}^i\,\langle\overline{\mathbf{u}_k}\rangle{}^i\,\langle\overline{\mathbf{u}_k}\rangle$$
$$- D_{ek}\left(\nabla\left(\rho_k\overline{\alpha}_k\right){}^i\,\langle\overline{\mathbf{u}_k}\rangle + {}^i\,\langle\overline{\mathbf{u}_k}\rangle\,\nabla\left(\rho_k\overline{\alpha}_k\right)\right) - \langle\overline{\tau}_{ek}\rangle.$$

$$(6.4\text{-}9b)$$

The volume–time-averaged momentum equation for phase k is then:

$$\frac{\partial}{\partial t}\left(\rho_k\overline{\alpha}_k{}^i\,\langle\overline{\mathbf{u}_k}\rangle - D_{ek}\nabla\left(\rho_k\overline{\alpha}_k\right)\right) + \nabla\cdot\left(\rho_k\overline{\alpha}_k{}^i\,\langle\overline{\mathbf{u}_k}\rangle{}^i\,\langle\overline{\mathbf{u}_k}\rangle\right) = \nabla\cdot\left(\overline{\alpha_k{}^i\,\langle\mathbf{T}_k\rangle} + \langle\overline{\tau}_{ek}\rangle\right)$$
$$+ \nabla\cdot\left(D_{ek}\left(\nabla\left(\rho_k\overline{\alpha}_k\right){}^i\,\langle\overline{\mathbf{u}_k}\rangle + {}^i\,\langle\overline{\mathbf{u}_k}\rangle\,\nabla\left(\rho_k\overline{\alpha}_k\right)\right)\right) + \rho_k\overline{\alpha}_k\mathbf{f}_k + \overline{\mathbf{F}}_{Ak} + \overline{\mathbf{F}}_{\Gamma k}.$$

$$(6.4\text{-}10)$$

6.4.4 Volume–Time-Averaged Energy Equation

The volume–time-averaged equation of energy conservation as a function of total energy, with reference to Eq. (6.3-14), can be written as:

$$\frac{\partial}{\partial t}\overline{\langle\rho_k E_k\rangle} + \nabla\cdot\overline{\langle\rho_k E_k\mathbf{u}_k\rangle} = \nabla\cdot\overline{\langle\mathbf{T}_k\cdot\mathbf{u}_k\rangle} + \overline{\langle\rho_k\mathbf{u}_k\rangle\cdot\mathbf{f}_k}$$
$$+ \nabla\cdot\overline{\langle\mathbf{J}_{qk}\rangle} + \overline{\langle\mathbf{J}_{Ek}\rangle} + \overline{q}_{Ak} + \overline{q}_{\Gamma k} + \overline{q}_{Wk}.$$

$$(6.4\text{-}11)$$

Because total energy is computed as the sum of internal energy and kinetic energy, the volume–time-averaged total energy per unit volume can be expressed as:

$$\overline{\langle\rho_k E_k\rangle} = \langle\overline{\rho}_k\rangle{}^i\,\langle\overline{e}_k\rangle + \overline{\langle\rho'_k\rangle{}^i\,\langle e'_k\rangle} + \frac{1}{2}\overline{\langle\rho_k\mathbf{u}_k\cdot\mathbf{u}_k\rangle}.$$

$$(6.4\text{-}12)$$

The second term on the right-hand side captures the energy transport due to fluctuations in density and internal energy (*e.g.*, temperature). In turbulent flow transport, this term is linked to the averaged quantities through:

$$\overline{\langle \rho_k' \rangle^i \langle e_k' \rangle} = C_{ek} \frac{k^3}{\varepsilon^2} \nabla \langle \overline{\rho}_k \rangle \cdot \nabla^i \langle \overline{e}_k \rangle, \tag{6.4-12a}$$

where k and ε are turbulence kinetic energy intensity and its dissipation rate, respectively. C_{ek} is an empirical constant associated with energy transport by fluctuations in density and temperature. If the triple covariance is neglected, the volume–time-averaged kinetic energy can be approximated as:

$$\begin{aligned}
\frac{1}{2}\overline{\langle \rho_k \mathbf{u}_k \cdot \mathbf{u}_k \rangle} &\approx \frac{1}{2}\overline{\langle \rho_k \rangle^i \langle \mathbf{u}_k \rangle \cdot^i \langle \mathbf{u}_k \rangle} \\
&= \frac{1}{2}\left(\langle \overline{\rho}_k \rangle^i \langle \overline{\mathbf{u}}_k \rangle \cdot^i \langle \overline{\mathbf{u}}_k \rangle + \langle \overline{\rho}_k \rangle^i \overline{\langle \mathbf{u}_k' \rangle \cdot^i \langle \mathbf{u}_k' \rangle} \right) - D_{ek}\nabla\langle \overline{\rho}_k \rangle \cdot^i \langle \overline{\mathbf{u}}_k \rangle.
\end{aligned} \tag{6.4-12b}$$

Using Eq. (6.4-12a) and Eq. (6.4-12b), Eq. (6.4-12) can be reformulated as:

$$\overline{\langle \rho_k E_k \rangle} = \langle \overline{\rho}_k \rangle^i \langle \overline{E}_k \rangle + C_{ek} \frac{k^3}{\varepsilon^2} \nabla \langle \overline{\rho}_k \rangle \cdot \nabla^i \langle \overline{e}_k \rangle - D_{ek}\nabla\langle \overline{\rho}_k \rangle \cdot^i \langle \overline{\mathbf{u}}_k \rangle, \tag{6.4-12c}$$

where the volume-time average of total energy is given by:

$$^i \langle \overline{E}_k \rangle =^i \langle \overline{e}_k \rangle + \frac{1}{2}^i \langle \overline{\mathbf{u}}_k \rangle \cdot^i \langle \overline{\mathbf{u}}_k \rangle + \frac{1}{2}^i \overline{\langle \mathbf{u}_k' \rangle \cdot^i \langle \mathbf{u}_k' \rangle}. \tag{6.4-12d}$$

The third term on the right-hand side of Eq. (6.4-12d) represents the turbulence kinetic energy.

Two components contribute to the total energy carried by the mass flux of phase k, as described by:

$$\overline{\langle \rho_k E_k \mathbf{u}_k \rangle} = \overline{\langle \rho_k e_k \mathbf{u}_k \rangle} + \frac{1}{2}\overline{\langle \rho_k \mathbf{u}_k (\mathbf{u}_k \cdot \mathbf{u}_k) \rangle}. \tag{6.4-13}$$

If the triple covariance is again ignored, the volume–time-averaged internal energy flux can be expressed in a similar fashion to Eq. (6.4-12) as:

$$\begin{aligned}
\overline{\langle \rho_k e_k \mathbf{u}_k \rangle} &= \langle \overline{\rho}_k \rangle^i \langle \overline{\mathbf{u}}_k \rangle^i \langle \overline{e}_k \rangle + C_{ek} \frac{k^3}{\varepsilon^2}^i \langle \overline{\mathbf{u}}_k \rangle \nabla \langle \overline{\rho}_k \rangle \cdot \nabla^i \langle \overline{e}_k \rangle \\
&\quad -^i \langle \overline{e}_k \rangle D_{ek} \nabla \langle \overline{\rho}_k \rangle - \langle \overline{\rho}_k \rangle D_{Tk}\nabla^i \langle \overline{e}_k \rangle,
\end{aligned} \tag{6.4-13a}$$

where D_{Tk} is the eddy thermal diffusivity. The second term on the right-hand side of Eq. (6.4-13) can be expanded as:

$$\begin{aligned}
\frac{1}{2}\overline{\langle \rho_k \mathbf{u}_k (\mathbf{u}_k \cdot \mathbf{u}_k) \rangle} &= \frac{1}{2}\left(\langle \overline{\rho}_k \rangle^i \langle \mathbf{u}_k \rangle^i \langle \mathbf{u}_k \rangle \cdot^i \langle \mathbf{u}_k \rangle + \overline{\langle \rho_k' \rangle^i \langle \mathbf{u}_k \rangle^i \langle \mathbf{u}_k \rangle \cdot^i \langle \mathbf{u}_k \rangle} \right) \\
&= \frac{1}{2}\langle \overline{\rho}_k \rangle^i \langle \overline{\mathbf{u}}_k \rangle \left(^i \langle \overline{\mathbf{u}}_k \rangle \cdot^i \langle \overline{\mathbf{u}}_k \rangle +^i \overline{\langle \mathbf{u}_k' \rangle \cdot^i \langle \mathbf{u}_k' \rangle} \right) +^i \langle \overline{\mathbf{u}}_k \rangle \cdot \langle \overline{\rho}_k \rangle^i \overline{\langle \mathbf{u}_k' \rangle^i \langle \mathbf{u}_k' \rangle} \\
&\quad + \frac{1}{2}\left(^i \langle \overline{\mathbf{u}}_k \rangle \cdot^i \langle \overline{\mathbf{u}}_k \rangle +^i \overline{\langle \mathbf{u}_k' \rangle \cdot^i \langle \mathbf{u}_k' \rangle} \right) \overline{\langle \rho_k' \rangle^i \langle \mathbf{u}_k' \rangle} + \overline{\langle \rho_k' \rangle^i \langle \mathbf{u}_k' \rangle} \cdot^i \langle \overline{\mathbf{u}}_k \rangle^i \langle \overline{\mathbf{u}}_k \rangle.
\end{aligned} \tag{6.4-13b}$$

Therefore, combining Eq. (6.4-13a) and (6.4-13b), the convective flux of total energy in Eq. (6.4-13) can now be written as:

$$\overline{\langle \rho_k E_k \mathbf{u}_k \rangle} = \langle \overline{\rho}_k \rangle^i \langle \overline{\mathbf{u}_k} \rangle^i \langle \overline{E}_k \rangle + C_{ek} \frac{k^3}{\varepsilon^2}{}^i \langle \overline{\mathbf{u}_k} \rangle \nabla \langle \overline{\rho}_k \rangle \cdot \nabla^i \langle \overline{e}_k \rangle - {}^i \langle \overline{E}_k \rangle D_{ek} \nabla \langle \overline{\rho}_k \rangle$$
$$- \langle \overline{\rho}_k \rangle D_{Tk} \nabla^i \langle \overline{e}_k \rangle - {}^i \langle \overline{\mathbf{u}_k} \rangle \cdot \langle \overline{\tau}_{ek} \rangle - {}^i \langle \overline{\mathbf{u}_k} \rangle^i \langle \overline{\mathbf{u}_k} \rangle \cdot D_{ek} \nabla \langle \overline{\rho}_k \rangle.$$

(6.4-13c)

The volume–time-averaged total energy equation is arrived at by replacing the corresponding terms in Eq. (6.4-11) with Eq. (6.4-12c) and (6.4-13c). In the volume-time average of the unsteady term, it is generally assumed that the transient terms of covariance are negligible when compared with the products of low-frequency components. As a result, the second and third terms in Eq. (6.4-12c) can be neglected in the unsteady term of the volume–time-averaged total energy equation. The pertinent simplification is then:

$$\frac{\partial}{\partial t} \left(\rho_k \overline{\alpha}_k{}^i \langle \overline{E}_k \rangle \right) + \nabla \cdot \left(\rho_k \overline{\alpha}_k{}^i \langle \overline{E}_k \rangle^i \langle \overline{\mathbf{u}_k} \rangle \right) = \nabla \cdot \left[{}^i \langle \overline{\mathbf{u}_k} \rangle \cdot \left(\langle \overline{\mathbf{T}}_k \rangle + \langle \overline{\tau}_{ek} \rangle \right) \right]$$
$$+ \nabla \cdot \left(\langle \overline{\mathbf{J}}_{qk} \rangle + \langle \overline{\mathbf{J}}_{eqk} \rangle \right) - \nabla \cdot \left[C_{ek} \frac{k^3}{\varepsilon^2}{}^i \langle \overline{\mathbf{u}_k} \rangle \nabla \left(\rho_k \overline{\alpha}_k \right) \cdot \nabla^i \langle \overline{e}_k \rangle \right]$$
$$+ \nabla \cdot \left[{}^i \langle \overline{E}_k \rangle D_{ek} \nabla \left(\rho_k \overline{\alpha}_k \right) \right] + \nabla \cdot \left[{}^i \langle \overline{\mathbf{u}_k} \rangle^i \langle \overline{\mathbf{u}_k} \rangle \cdot D_{ek} \nabla \left(\rho_k \overline{\alpha}_k \right) \right]$$
$$+ \left(\rho_k \overline{\alpha}_k{}^i \langle \overline{\mathbf{u}_k} \rangle - D_{ek} \nabla \left(\rho_k \overline{\alpha}_k \right) \right) \cdot \mathbf{f}_k + \langle \overline{J}_{Ek} \rangle + \overline{q}_{Ak} + \overline{q}_{\Gamma k} + \overline{q}_{Wk},$$

(6.4-14)

where $\langle \overline{\mathbf{J}}_{eqk} \rangle$ denotes the turbulent heat flux, as defined by:

$$\langle \overline{\mathbf{J}}_{eqk} \rangle = \rho_k \overline{\alpha}_k D_{Tk} \nabla^i \langle \overline{e}_k \rangle.$$

(6.4-14a)

This term must be accounted for by the turbulence modeling employed for phase k.

6.4.5   Closure of Volume–Time-Averaged Equations

Equations (6.4-7), (6.4-10), and (6.4-14) constitute the general governing transport equations of a multiphase flow: the volume–time-averaged mass, volume–time-averaged momentum, and volume–time-averaged energy equations. Closure of these equations stipulates that the number of independent variables equals the number of independent equations and constitutive relations. Constitutive relations are expressed as a function of the independent variables and include the equations of state as well as expressions to determine the transport coefficients, stresses, covariance terms, and inherent source terms.

There are several customary analogies commonly adopted in continuous modeling of multiphase flows. First, the total stress of phase k is often decomposed into a normal stress component, known as pressure, and a shear stress component, known as viscous stress. Doing so yields:

$$\langle \overline{\mathbf{T}}_k \rangle = -\alpha_k{}^i \langle \overline{p}_k \rangle \mathbf{I} + \langle \overline{\tau}_k \rangle,$$

(6.4-15)

where the intrinsic-averaged pressure only represents actual pressure when the phase k is a real fluid. A disputable point in Eq. (6.4-15) is the validation of volume fraction weighting on the pressure term, because pressure or stress is not an extensive property (rather an intensive property independent of mass) of thermodynamics in a real fluid. Next, specific internal energy, or temperature, is commonly chosen as the independent variable in the energy equation. Thus, it gives:

$$^i \langle e_k \rangle = {}^i \langle E_k \rangle - \frac{1}{2} {}^i \langle \mathbf{u}_k \rangle \cdot {}^i \langle \mathbf{u}_k \rangle - \frac{1}{2} {}^i \langle k_k \rangle \equiv c_{vk}{}^i \langle T_k \rangle, \qquad (6.4\text{-}16)$$

where k_k denotes the turbulence kinetic energy of phase k. The intrinsic-averaged temperature defined in Eq. (6.4-16) is the true thermodynamic temperature. The intrinsic-averaged density of phase k is a material property that is usually determined from its equation of state, which is expressed as a function of pressure and temperature for phase k. This is given by:

$$^i \langle \overline{\rho}_k \rangle = \rho_k \left({}^i \langle p_k \rangle, {}^i \langle \overline{T}_k \rangle \right) \qquad (6.4\text{-}17)$$

Typically, the independent quantities to be solved include four scalar variables and one vector variable for each phase: the scalars are the intrinsic-averaged volume fraction (α_k), density (ρ_k), pressure (p_k), and temperature (T_k); the vector is velocity ($\mathbf{u}_k$). Thus, for an N-phase flow, solutions of 4N scalar variables and N vector variables must be determined. Substituting Eq. (6.4-15) and Eq. (6.4-16) into the momentum and energy equations, respectively, results in 3N scalar equations (*i.e.*, Eq. (6.4-7), Eq. (6.4-14), and Eq. (6.4-17)) and N vector equation (*i.e.*, Eq. (6.4-10)). In addition, unification of volume fractions for the N phases is given in Eq. (6.3-7a); hence, N−1 relations are required to close the model. The remaining N−1 relationships often consist of additional formulations for pressure or normal stresses of phase k, which can be highly case-dependent, as Section 6.5 illustrates.

The governing equations contain terms that require additional manipulation with respect to the independent variables in order to close the model and ensure solvability. In addition to the body force, $\mathbf{f}_k$, and energy source, J_{ek}, these terms of interest include: (1) molecular fluxes or shear stresses ($\tau_k, \mathbf{J}_{qk}$); (2) turbulent fluxes or stresses ($\tau_{ek}, \mathbf{J}_{eqk}$); (3) eddy diffusivities (D_{ek}, D_{Tk}); and (4) interfacial transport terms ($\Gamma_k, F_{Ak}, F_{\Gamma k}, q_{Ak}, q_{\Gamma k}, q_{wk}$), which are formulated via constitutive equations, turbulence models, and interfacial transport models. These closure terms depend on the material properties as well as flow conditions. The interfacial transport terms for different phases are not independent because for each pair of contact phases, k and l, the terms must satisfy the following jump conditions across the interfaces:

$$\sum_{j=k,l} \Gamma_j = 0 \qquad (6.4\text{-}18)$$

$$\sum_{j=k,l} \left(\mathbf{F}_{Aj} + \mathbf{F}_{\Gamma j} \right) = \mathbf{M}_{\sigma I} \qquad (6.4\text{-}19)$$

$$\sum_{j=k,l} \left(q_{Aj} + q_{wj} + q_{\Gamma j} \right) = E_{\sigma I}, \qquad (6.4\text{-}20)$$

where $\mathbf{M}_{\sigma I}$ and $E_{\sigma I}$ are the momentum and energy source due to surface tension, respectively; their detailed forms can be obtained from Eq. (6.3-4).

Because α_k is a variable in a multiphase flow, the volume-averaged phase density likewise becomes a variable, even if the associated material density remains a constant. Thus, in a multiphase flow, the continuum of a phase is always compressible. This concept deserves special attention because most longstanding theories for single-phase fluid mechanics were developed on an assumption of constant fluid density, or incompressibility. A direct analog between single-phase incompressible flow and multiphase flow with constant material densities can be a misleading oversimplification.

For convenience, the signaling volume-averaging brackets and time-averaging bars from the preceding equations have been removed in the following equations. In particular, $\mathbf{U}_k$ is used to stand for the time-averaged velocity while $\mathbf{u}_k$ still represents the instant velocity of phase k, while both are regarded locally volume-averaged. It should be understood that all of the independent variables are intrinsic volume–time averaged in the proceeding. The volume–time-averaged equations can be summarized, in a simplified form, as:

$$\frac{\partial}{\partial t}\left(\alpha_k\rho_k\right) + \nabla\cdot\left(\alpha_k\rho_k\mathbf{U}_k\right) = \nabla\cdot\left[D_k\nabla\left(\alpha_k\rho_k\right)\right] + \Gamma_k \tag{6.4-21}$$

$$\frac{\partial}{\partial t}\left[\alpha_k\rho_k\mathbf{U}_k - D_{ek}\nabla\left(\alpha_k\rho_k\right)\right] + \nabla\cdot\left(\alpha_k\rho_k\mathbf{U}_k\mathbf{U}_k\right) = -\nabla\left(\alpha_k p_k\right) + \nabla\cdot\left[\alpha_k\left(\tau_k + \tau_{ek}\right)\right]$$
$$+ \nabla\cdot\left[D_{ek}\nabla\left(\alpha_k\rho_k\right)\mathbf{U}_k + D_{ek}\mathbf{U}_k\nabla\left(\alpha_k\rho_k\right)\right] + \alpha_k\rho_k\mathbf{f}_k + \mathbf{F}_{Ak} + \mathbf{F}_{\Gamma k} \tag{6.4-22}$$

$$\frac{\partial}{\partial t}\left(\alpha_k\rho_k E_k + C_{ek}\frac{k^3}{\varepsilon^2}\nabla\left(\alpha_k\rho_k\right)\cdot\nabla e_k - D_{ek}\nabla\left(\alpha_k\rho_k\right)\cdot\mathbf{U}_k\right) + \nabla\cdot\left(\alpha_k\rho_k E_k\mathbf{U}_k\right)$$
$$= -\nabla\cdot\left(\alpha_k p_k\mathbf{U}_k\right) + \nabla\cdot\left[\alpha_k\left(\tau_k + \tau_{ek}\right)\cdot\mathbf{U}_k\right] + \nabla\cdot\left[\alpha_k\left(\mathbf{J}_{qk} + \mathbf{J}_{eqk}\right)\right]$$
$$- \nabla\cdot\left[C_{ek}\frac{k^3}{\varepsilon^2}\mathbf{U}_k\nabla\left(\alpha_k\rho_k\right)\cdot\nabla e_k\right] + \nabla\cdot\left[E_k D_{ek}\nabla\left(\alpha_k\rho_k\right)\right]$$
$$+ \nabla\cdot\left[\mathbf{U}_k\mathbf{U}_k\cdot D_{ek}\nabla\left(\alpha_k\rho_k\right)\right] + \left[\alpha_k\rho_k\mathbf{U}_k - D_{ek}\nabla\left(\alpha_k\rho_k\right)\right]\cdot\mathbf{f}_k$$
$$+ \alpha_k J_{Ek} + q_{Ak} + q_{\Gamma k} + q_{Wk}. \tag{6.4-23}$$

In summary, the volume–time-averaged transport Eqs. (6.4-21)–(6.4-23) form the governing equations for a general multiphase flow system. In order to close the model, constitutive relations must be provided for terms that describe pressure, stress, turbulence, transport coefficients, and interfacial transport. Such relationships are highly dependent on the material property of each phase, such as compressibility, and the property of the flow, such as flow regime and volume concentration. Because different multiphase flows may exist in dramatically different forms with distinct mechanisms of phase interactions, no universal closure relations have been developed. Knowledge of the specific flow scenario is essential to establish the closure relationships, and reasonable assumptions are necessary to simply the problem.

As in the case of a single-phase flow, a closed set of governing equations must be formulated before seeking a meaningful solution. However, model closure is but a necessary condition to finding a solution; it is impossible to obtain a solution without model closure, but closure is not a sufficient condition guaranteeing a solution. Moreover, existence or uniqueness of a nontrivial solution to every problem has yet to be theoretically proven.

6.5 Constitutive Relations in Multifluid Model

Constitutive relations and closure techniques for dispersed multifluid flows are discussed in this section. These flows involve multiple non-immiscible fluid phases with interfacial boundaries between them. Examples of such flows include bubbly-liquid flow and liquid spray jets.

6.5.1 Pressure

As analyzed in Section 6.4.5, closure of the governing equations demands an additional N–1 constitutive relationships, typically for the averaged pressure, p_k. In stratified or dilute dispersed multifluid flows, it is customary to assume instantaneous pressure equilibrium between the time–volume-averaged phase pressure, p_k and the interfacial averaged pressure, $p_{k,i}$. Thus:

$$p_k = p_{k,i}, \tag{6.5-1}$$

where the interfacial averaged pressure, $p_{k,i}$, is given by (Drew, 1983):

$$p_{k,i} = \frac{\int_{Ak} p_k \mathbf{n}_k dA \cdot \int_{Ak} \mathbf{n}_k dA}{\left| \int_{Ak} \mathbf{n}_k dA \right|^2} = \frac{\int_{Ak} p_k \mathbf{n}_k dA \cdot \nabla \alpha_k}{|\nabla \alpha_k|^2}. \tag{6.5-1a}$$

This assumption is reasonable when acoustic effects, such as bubble expansion or contraction, are not significant.

The interfacial pressures, $p_{k,i}$ and $p_{l,i}$, on each side of the interface are related by:

$$p_{k,i} - p_{l,i} = \kappa \sigma_{kl}, \tag{6.5-2}$$

where σ_{kl} is the interfacial tension between the two phases, and κ is the mean curvature of the interface. For most multiphase flows suitable for modeling with the Eulerian average approach, the contribution of interfacial tension to the average momentum transport is negligible, so that:

$$p_{k,i} \approx p_{l,i}. \tag{6.5-2a}$$

With the approximation of Eq. (6.5-2a), Eq. (6.5-1) predicts that all N phases will have the same pressure, p. It then gives:

$$p_k = p \qquad (k = 1, \cdots, N). \tag{6.5-3}$$

For a dense dispersed phase, the contact between dispersed phase particles causes a collisional pressure $p_{k,c}$ in addition to the interfacial pressure p, which leads to:

$$p_k = p + p_{k,c} \qquad (k = 1, \ldots, N). \qquad (6.5\text{-}3\text{a})$$

Equation (6.5-3) or Eq. (6.5-3a) effectively establishes the required $N-1$ relations of phase pressures for the necessary condition of model closure.

6.5.2   Molecular Fluxes

Molecular transport fluxes of phase k are often assumed to be entirely dependent on the properties of phase k itself. For example, when phase k is a fluid, its stress tensor is expressed as an analog to a single-phase Newtonian fluid flow, as given by:

$$\boldsymbol{\tau}_k = \mu_k \left[\nabla \mathbf{U}_k + (\nabla \mathbf{U}_k)^{\mathrm{T}} \right] + \left(\lambda_k - \frac{2}{3} \mu_k \right) (\nabla \cdot \mathbf{U}_k) \mathbf{I}, \qquad (6.5\text{-}4)$$

where μ_k and λ_k are the shear (*i.e.*, dynamic) and bulk viscosities of phase k, respectively. The molecular heat flux is given by:

$$\mathbf{J}_{qk} = K_k \nabla T_k, \qquad (6.5\text{-}5)$$

where K_k is the thermal conductivity of phase k. Corresponding formulations for a phase of solid particles is discussed in Section 6.7.

6.5.3   Eddy Diffusivities

Eddy diffusivity, D_{ek}, signifies covariance of phase density and velocity fluctuations of phase k. It also measures the rate of mass diffusion of phase k in the mixture. Even if the material density of phase k remains constant, phase density does fluctuate in a multiphase flow because it is coupled with volume fraction. Eddy diffusivity is significant in highly dispersed systems where the characteristic dimension of the dispersed phase is less than or comparable to the mean free path of the suspending fluid phase, and it is commonly considered in mixing processes of multifluid flows where the phase distribution can be highly nonuniform. Typical application examples under the former condition include diffusion of smoke particulate (less than 0.1 micrometers) in air and Brownian motion of submicron particles in a suspending liquid. In dilute turbulent dispersed flows, eddy diffusivity also measures turbulent dispersion of the dispersed phase.

Conversely, the effect of eddy diffusivity, D_{ek}, is often neglected in multiphase flow processes where the characteristic dimension of the dispersed phase is much greater than the mean free path of the suspending fluid phase or characteristic size of eddies because turbulent dispersion is limited. In stratified gas–liquid flows, the two phases do not diffuse into each other, despite the volume fraction gradient (although diffusion of the species still takes place). Therefore, unless otherwise noted, eddy diffusivity is routinely neglected in the Eulerian formulation of the governing equations for multiphase flows.

The thermal eddy diffusivity, D_{Tk}, characterizes covariance between internal energy and velocity fluctuations, and it measures the rate of energy diffusion of phase k in the mixture. Both D_{ek} and D_{Tk} can be modeled in a similar fashion to eddy viscosity of eddy momentum diffusion in turbulence modeling of single-phase turbulent flows. Such analogy is typically premised on relationships of molecular diffusion of mass, momentum, and internal energy in accordance with molecular kinetics theory.

6.5.4   Interfacial Transport

The interfacial transport terms, introduced in Section 6.3, can be formulated by integration over the interfaces. For most multiphase flows of interest, the contributions of the interfacial stress to the momentum and the energy equations are negligible when compared with other terms. Therefore, the right-hand side of the interfacial jump conditions of Eq. (6.4-19) and Eq. (6.4-20) vanishes. Thus, it gives:

$$\mathbf{M}_{\sigma I} = \mathbf{0} \qquad\qquad E_{\sigma I} = 0. \qquad\qquad (6.5\text{-}6)$$

This relationship implies that the fluxes on both sides of the interface are continuous and consequently, can be adequately defined from only one of the two contact phases.

For a multiphase flow with N phases, the interfacial mass transfer to phase k is given by the source term, Γ_k. The source term is represented by the sum of individual interfacial mass flow, m_{kl}, per unit volume from phase k to phase l. It then yields the following:

$$\Gamma_k = -\frac{1}{V}\int_{A_k} \rho_k\,(\mathbf{U}_k - \mathbf{U}_s)\cdot\mathbf{n}_k dA$$

$$= -\frac{1}{V}\sum_{l=1,l\neq k}^{n}\int_{A_{kl}}\rho_k\,(\mathbf{U}_k - \mathbf{U}_{s,kl})\cdot\mathbf{n}_{kl}dA = -\sum_{l=1,l\neq k}^{N}(\dot{m}_{kl} - \dot{m}_{lk}). \qquad (6.5\text{-}7)$$

It is apparent that the sum of Γ_k over all phases equates to zero; the total mass of the N phases is therefore conserved. Evaluation of m_{kl} depends on the specific interfacial mass transfer mechanisms, such as phase change and chemical reaction kinetics.

The interfacial momentum transfer due to interfacial mass flux is given by:

$$\mathbf{F}_{\Gamma k} = -\frac{1}{V}\int_{A_k}\rho_k\mathbf{U}_k\,(\mathbf{U}_k - \mathbf{U}_s)\cdot\mathbf{n}_k dA = -\sum_{l=1,l\neq k}^{N}(\dot{m}_{kl}\mathbf{U}_k - \dot{m}_{lk}\mathbf{U}_l). \qquad (6.5\text{-}8)$$

Integration of the total stress on the interface reveals the hydrodynamic force between the two phases. This result is obtained by:

$$\mathbf{F}_{Ak} = \frac{1}{V}\int_{A_k}\mathbf{T}_k\cdot\mathbf{n}_k dA$$

$$= -\frac{1}{V}\int_{A_k}p_k\mathbf{n}_k dA + \frac{1}{V}\int_{A_k}\boldsymbol{\tau}_k\cdot\mathbf{n}_k dA = p_{k,i}\nabla\alpha_k + \sum_{l=1,l\neq k}^{n}\mathbf{F}_{kl}. \qquad (6.5\text{-}9)$$

The first term on the right-hand side of the expression originates from the averaged interfacial pressure, $p_{k,i}$, acting on the boundary of phase k, and is defined by

Eq. (6.5-1a). The derivation relies on the relationship of Eq. (6.2-8a). When used with the shared pressure formulation Eq. (6.5-3), the averaged interfacial pressure is further assumed to be the fluid pressure, p. The second term on the right-hand side of the equation, F_{kl}, is the generalized drag force between phase k and l. This summed term typically includes the drag force, lift force, added mass force, and others. The physical interpretation and evaluation of those forces depend on considerations of the phase properties and the flow regimes introduced in Chapters 3 and 4. Therefore, the interfacial force can be written as:

$$\mathbf{F}_{Ak} = p\nabla\alpha_k + \sum_{l=1,l\neq k}^{n} \left(\mathbf{F}_{D,kl} + \mathbf{F}_{L,kl} + \mathbf{F}_{A,kl} + \ldots\right). \tag{6.5-9a}$$

Among the interfacial forces expressed in Eq. (6.5-9a), the drag force often dominates and is typically calculated from the relative velocity of the two phases. Thus, it gives:

$$\mathbf{F}_{D,kl} = \beta_{kl}(\mathbf{U}_l - \mathbf{U}_k), \tag{6.5-9b}$$

where β_{kl} is the interfacial momentum transfer coefficient. For a phase k of dispersed particles with diameter, d_k, and density, ρ_k, in fluid phase l of viscosity μ_l, the interfacial momentum transfer coefficient can be expressed as:

$$\beta_{kl} = \frac{\rho_k}{\tau_S} \frac{C_D \mathrm{Re}_k}{24}\alpha_k, \tag{6.5-9c}$$

where τ_S is the particle Stokes relaxation time, defined by Eq. (3.2-22a); C_D is the drag coefficient and Re_k is the particle Reynolds number, respectively defined by Eqs. (3.2-2) and (3.2-2a). For particles with Stokes drag, β_{kl} simplifies to:

$$\beta_{kl} = \frac{18\mu_l}{d_k^2}\alpha_k. \tag{6.5-9d}$$

The lift force, $\mathbf{F}_{L,k}$, on the discrete phase originates from particle motion in a transverse velocity gradient or from particle rotation. It can be modeled by:

$$\mathbf{F}_{L,kl} = -C_L\alpha_k\rho_l(\mathbf{U}_l - \mathbf{U}_k) \times (\nabla \times \mathbf{U}_l), \tag{6.5-9e}$$

where C_L is the lift coefficient. The lift force on the continuous phase can be simply calculated from: $\mathbf{F}_{L,lk} = -\mathbf{F}_{L,kl}$.

The added mass force, which is due to the acceleration of the discrete phase k in a continuous phase l, is given by:

$$\mathbf{F}_{A,kl} = C_A\alpha_k\rho_l\frac{d}{dt}(\mathbf{U}_l - \mathbf{U}_k), \tag{6.5-9f}$$

where d/dt is the material derivative of relative velocity. The added mass coefficient depends on the geometry of the discrete phase particles. For a spherical particle, $C_A = 0.5$. The virtual mass force is of significance when the continuous phase density is similar to, or in excess of, the discrete phase; this was the case for the gas bubbles moving in a liquid as discussed in Chapter 1.

The interfacial convective heat transfer per unit volume is calculated from the surface integration of the heat flux, $\mathbf{J}_{qk}$, as:

$$q_{Ak} = -\frac{1}{V} \int_{A_k} \mathbf{J}_{qk} \cdot \mathbf{n}_k dA = - \sum_{l=1,l\neq k}^{n} A_{kl} h_{ckl}(T_k - T_l). \tag{6.5-10}$$

The convective heat transfer coefficient, h_{ckl}, between phase k and l is often calculated from the interfacial area per unit volume along with a Nusselt number correlation developed for different flow conditions. The term A_{kl} denotes the interfacial area concentration. It can then be written as:

$$q_{\Gamma k} = -\frac{1}{V} \int_{A_k} \rho_k E_k (\mathbf{U}_k - \mathbf{U}_s) \cdot \mathbf{n}_k dA = - \sum_{l=1,l\neq k}^{n} (\dot{m}_{kl} E_k - \dot{m}_{lk} E_l). \tag{6.5-11}$$

The work done by the interfacial stress can be expressed by separating the velocity, $\mathbf{U}_k$, into the interfacial average, $\mathbf{U}_{k,i}$, and the variation, $(\mathbf{U}_k - \mathbf{U}_{k,i})$. This process is given by:

$$\begin{aligned}
q_{Wk} &= -\frac{1}{V} \int_{A_k} (\mathbf{T}_k \cdot \mathbf{U}_k) \cdot \mathbf{n}_k dA \\
&= -\mathbf{U}_{k,i} \cdot \left[\frac{1}{V} \int_{A_k} \mathbf{T}_k \cdot \mathbf{n}_k dA \right] - \frac{1}{V} \int_{A_k} \mathbf{T}_k \cdot (\mathbf{U}_k - \mathbf{U}_{k,i}) \cdot \mathbf{n}_k dA \\
&= -\mathbf{U}_k \cdot \mathbf{F}_{Ak} + q''_{Wk},
\end{aligned} \tag{6.5-12}$$

where $\mathbf{F}_{Ak}$ in the first term on the right-hand side of the result is the interfacial momentum transfer term given by Eq. (6.5-9). The second term on the right-hand side is known as the interfacial extra work. It usually has a complex formulation (Arnold *et al.*, 1990; Drew and Lahey, 1993).

6.5.5   Turbulence Modeling

The turbulent transport terms are typically represented by the turbulent stresses in the momentum equation and need to be closed with proper constitutive correlations; equations commonly known as turbulence models. For a true continuous fluid phase, turbulent transport is caused by eddy generation, transport, and dissipation within the fluid phase, and additional phase interactions due to the turbulence. The basic formulation of the turbulence transport terms emulates the logic employed for single-phase turbulent flow in Chapter 2. Phase interactions may alter origins of turbulence for the fluid phase, thereby affecting turbulent transport of all phases. This interactive influence on turbulent transport is commonly known as turbulence modulation, which is discussed as an advanced topic in Section 6.7.2.

For a dispersed phase, such as small bubbles or droplets, turbulent transport is primarily due to eddy engulfment by the fluid phase; described by the Hinze–Tchen model (Hinze, 1975) for dilute fluid–particle flows, which is discussed in Section 6.6.2. For a dispersed phase of large fluid particles, such as large bubbles, additional mechanisms of turbulence transport or enhancement are present, such as enhanced eddy generation by wake shedding stemming from the relative motions of bubbles. These phenomena necessitate more sophisticated turbulence formulations that are beyond the scope of the utilitarian Hinze–Tchen model.

6.6   Constitutive Relations for Fluid–Solid Flows

Fluid–solid multiphase flows involve discrete solid particles that are suspended in a continuous carrying fluid. In principle, the averaging process of Sections 6.3 and 6.4 applies to such flows, provided continuum mechanics is applied to the inside of each solid particle in a way similar to the treatment of the fluid phase. In such an Eulerian–Eulerian approach, particle size is typically orders of magnitude smaller than the smallest size of control volume. As a result, internal stresses within the solid particles and the local jump condition for stresses across the individual particle surface are not considered. Therefore, the shared pressure formulation employed in fluid–fluid flow does not apply, and particle–particle interactions must be handled explicitly. After averaging, the solid phase is effectively treated as a continuous pseudo-fluid phase that mimics the collective dynamic behavior of the individual particles. New constitutive relationships for the solid-phase transport, such as stresses, need to be defined in the context of particle dynamics, rather than the intrinsic material properties of the solid itself.

6.6.1   Stresses of Solid Particles

In the transport of particle phase, the solid stress originates from two physical sources: the kinetic effect due to the particle shear motion in a viscous fluid and the collisional effect from interparticle collisions. The solid-phase stress tensor can be defined in a similar fashion to the fluid stress as:

$$\mathbf{T}_p = -p_p \mathbf{I} + \boldsymbol{\tau}_p, \tag{6.6-1}$$

where p_p is the particle pressure; τ_p is the viscous stress of the particle phase, given by:

$$\boldsymbol{\tau}_p = \mu_p \left[\nabla \mathbf{U}_p + (\nabla \mathbf{U}_p)^\mathrm{T} \right] + \left(\lambda_p - \frac{2}{3}\mu_p \right) (\nabla \cdot \mathbf{U}_p)\, \mathbf{I}. \tag{6.6-1a}$$

In the above equation, μ_p and λ_p are the shear and bulk viscosity of the particle phase, respectively. Closure of p_p, μ_p, and λ_p of solid particles is obtained either from empirical correlations or from kinetic theory of granular flows.

One way to close the constitutive relations of the particle phase is to use empirical correlations of particle-laden mixtures. It has been postulated that the mixture viscosity can be represented as a linear combination of the constituent phase viscosities. It then yields:

$$\mu_{mix} = \alpha\mu + \alpha_p\mu_p. \tag{6.6-2}$$

Once the mixture viscosity is determined, the shear viscosity of the solid phase is readily found. The shear viscosity of a fluid–particle mixture is often modeled as a function of the solid volume fraction, α_p, and the fluid viscosity, μ. For example, the viscosity of a dilute mixture with less than 3% volume fraction of particles is calculated by (e.g., Einstein, 1906; Brinkman, 1952; Roscoe, 1952):

$$\mu_{mix} = \mu\left(1 - \alpha_p\right)^{-2.5}. \tag{6.6-2a}$$

In the limiting case of dense flow with closely packed solid particles, the mixture viscosity can be modeled as (Frankel and Acrivos, 1967):

$$\mu_{mix} = \frac{9}{8}\mu\frac{\left(\alpha_p/\alpha_{p,\max}\right)^{1/3}}{1 - \left(\alpha_p/\alpha_{p,\max}\right)^{1/3}}, \tag{6.6-2b}$$

where $\alpha_{p,\max}$ is the maximum volume fraction of solid particles. The bulk viscosity, λ_p, of the solid-particle phase is often neglected in correlation. It should be noted that the above equations of mixture viscosity do not account for the particle–particle contact, and hence can only be applied to particle-containing flows in the macro-viscous regime, such as Ba < 40, as described in Section 2.5.1.

Correlations for pressure of solid particles often only consider the collisional effects. A commonly used empirical model considers the effect of the elastic mod-ulus of the solid-particle phase in such a way that the following expression is valid:

$$\nabla\left(\alpha_s p_s\right) = -E(\alpha)\nabla\alpha, \tag{6.6-3}$$

where E is the elastic modulus, modeled as an empirical function of voidage:

$$E\left(\alpha\right) = 10^{a\alpha+b}. \tag{6.6-3a}$$

E helps keep the particles separated so the calculated solid volume fraction does not exceed the maximum packing limit for a given size and shape of the particles. The parameters a and b are empirical constants, such as the values of $a = -8.76$ and $b = 5.43$, in the numerical simulation of fluidized beds (Gidaspow, 1986).

Another approach to provide closure relationships for the solid stress terms is to use kinetic theory of granular flow, which is introduced in Section 2.5. The solid-phase stress is closed using the concept of granular temperature, which reflects the magnitude of the particle velocity fluctuation. Once a solution to the transport equation of the granular temperature is obtained, the variables p_p, λ_p, and μ_p are determined as functions of the granular temperature, given in Eq. (2.5-18a, b, d).

6.6.2   Turbulent Diffusion of Particulates

For a dilute suspension of particulates in a turbulent flow, turbulent diffusion of par-ticles is effectuated by eddy transport from the turbulent fluid phase. One of the most fundamental turbulent models for particulate diffusion is the Hinze–Tchen model that considers the diffusion of discrete small particles in a homogeneous turbulent flow (Tchen, 1947; Hinze, 1975). Major assumptions of this model include the following:

(1) Fluid turbulence is homogeneous and steady.
(2) The domain of turbulence is infinite.
(3) The particles are spherical and small compared to the smallest wavelength in turbulence.
(4) Stokes' drag governs particle motion relative to the fluid.
(5) The particle is always entrapped inside the same turbulent eddy.

For brevity, only the one-dimensional case is considered. The transport rate of a scalar quantity, ψ, can be expressed in the form suggested by Boussinesq (1877) as:

$$\overline{u'\psi'} = \frac{1}{\tau_t} \int_0^{\tau_t} \psi'(t)u'(t)dt = -D_\psi \frac{d\Psi}{dx}, \qquad (6.6\text{-}4)$$

where D_ψ is the coefficient of diffusion of ψ, and Ψ is the mean value of ψ. Assuming a linear variation of Ψ with respect to x, D_ψ can be approximated by (Fan and Zhu, 1998):

$$D_\psi = \overline{u'^2} \int_0^{\tau_t} R(\tau)d\tau, \qquad (6.6\text{-}4a)$$

where $R(\tau)$ is a dimensionless autocorrelation function, also known as the Lagrangian correlation coefficient, and it is given by the expression:

$$R(\tau) = \frac{\overline{u'(t)u'(t+\tau)}}{\overline{u'^2}}. \qquad (6.6\text{-}4b)$$

The dynamic behavior of a discrete particle in a turbulent eddy is taken as similar to the slow motion of a spherical particle in a fluid at rest. Assuming the effects of the history force and carried mass force to be negligible, the motion of the particle is then only governed by the Stokes drag, so that the following is valid:

$$\frac{du_p}{dt} = \frac{u - u_p}{\tau_S}, \qquad (6.6\text{-}5)$$

where τ_S is the Stokes relaxation time. In addition, the Lagrangian correlation coefficient for the turbulent fluid motion is approximated by an exponential function, given by:

$$R(t) = \exp\left(-\frac{t}{\tau_e}\right), \qquad (6.6\text{-}6)$$

where τ_e is the characteristic time of the eddy motion, or the eddy existence time. This quantity can be estimated from the k-ε turbulence model as:

$$\tau_e = \sqrt{\frac{3}{2} C_\mu^{\frac{3}{4}} \frac{k}{\varepsilon}}. \qquad (6.6\text{-}6a)$$

Thus, with u and u_p represented by Fourier integrals and after employing inverse Fourier transform techniques, it has been shown that (Fan and Zhu, 1998):

$$\overline{u'^2}_p = \left(\frac{\tau_e}{\tau_e + \tau_S}\right) \overline{u'^2}, \qquad (6.6\text{-}7)$$

and the following is also valid:

$$R_p(t) = \frac{1}{\tau_e - \tau_S} \left\{ \tau_e \exp\left(-\frac{t}{\tau_e}\right) - \tau_S \exp\left(-\frac{t}{\tau_S}\right) \right\}. \tag{6.6-8}$$

The above expression indicates the Lagrangian correlation coefficient for discrete particle motion is not a single exponential function. From the definitions of the transport coefficients for turbulent diffusion, given in Eq. (6.6-4) and Eq. (6.6-4a), the following expressions can be written:

$$\frac{D_{\psi p}}{D_\psi} = 1 + \left\{ \left(\frac{\tau_e}{\tau_S}\right)^2 - 1 \right\}^{-1} \left\{ \frac{\exp\left(-\frac{t}{\tau_S}\right) - \exp\left(-\frac{t}{\tau_e}\right)}{1 - \exp\left(-\frac{t}{\tau_e}\right)} \right\}. \tag{6.6-9}$$

For a short diffusion time (i.e., $t \ll \tau_e$ and $t \ll \tau_S$), the well-known Hinze–Tchen equation is obtained as:

$$\frac{D_{\psi p}}{D_\psi} = \left(1 + \frac{\tau_S}{\tau_e}\right)^{-1}. \tag{6.6-9a}$$

Other turbulent transport coefficients for particles are related to the corresponding turbulent transport coefficients of fluid phase via the analog to the kinetic theory of gases. For instance, the turbulent viscosity of particle phase can be obtained using this method as:

$$\frac{\mu_p}{\mu} \frac{\rho}{\rho_p} = \frac{D_{\psi p}}{D_\psi} = \left(1 + \frac{\tau_S}{\tau_e}\right)^{-1}. \tag{6.6-9b}$$

Both D_ψ and μ in Eq. (6.6-9b) correspond to turbulent diffusivity and viscosity of fluid phase, respectively, not to the molecular ones; and the densities in the above expression are phase averaged. Hence, using Eq. (6.6-9a), the particle turbulence kinetic energy, k_p, is determined from the turbulent eddy kinetic energy, k, as:

$$\sqrt{\frac{k}{k_p}} = 1 + \frac{\tau_S}{\tau_e}. \tag{6.6-10}$$

Conversely, the effect of particles on flow turbulence is given by the relationship:

$$\frac{k}{k_0} = \left(1 + \frac{\alpha_p \rho_p}{\rho}\right)^{-1}, \tag{6.6-11}$$

where k_0 is the local flow turbulence kinetic energy, in the absence of particles. Equation (6.6-10) indicates that k_p must always be smaller than k; for the same turbulent flow, the value of k_p diminishes as particle size increases. Moreover, according to Eq. (6.6-11), the presence of particles always reduces fluid turbulence, a justifiable characterization wherever the assumptions of Hinze–Tchen model are valid.

Certain assumptions of the Hinze–Tchen model do not apply to the dispersed flow of large particles, such as those regarding Stokes drag or particles being trapped within the same eddy, or the dense-phase particle flows where interparticle collisions prevail. In addition, motion of large particles may generate additional flow disturbances, such as via wake shedding, that augment fluid turbulence rather than diminish it. More advanced models for turbulent diffusion, the transport of particle phases, and turbulence modulation between particle and fluid phases are discussed in Section 6.7.3.

6.7   Advanced Topics

This section covers some topics concerning multiphase interactions and phenomena that have been considerably difficult to describe with the Eulerian formulation. These topics include multiscaled modeling on phase interactions, modeling of flows with polydispersed discrete phase, and turbulent modulation due to the phase interaction between continuum fluid and discrete phase of particles. Applications of some of these advanced models are also illustrated in the case studies of Section 6.8.

6.7.1   Effect of Mesoscale Structures on Phase Interaction

While the averaging process yields a set of governing equations describing macroscopic behavior of a multiphase flow, some localized features of the flow are inevitably forfeited. However, the uncaptured small-scale features can have significant impact on the macroscopic (averaged) flow field because the separation of scales is not typically pronounced in most multiphase flows. Figure 6.7-1 provides a simple example of three unit volumes in a particle–fluid system. The characteristic dimension of the unit volumes is much larger than the size of individual particles, but it is far less than the entire flow domain; this condition is referred to as the meso-scale. All three unit volumes have identical particle properties, fluid properties, and overall volume fraction, but their averaged drag coefficient often varies by orders of magnitude because of different spatial phase distributions, often described as mesoscopic structures. Most current models of the drag force are founded on the assumption that the phase distribution is homogeneous within a control volume; the effects of mesoscopic structures are therefore neglected. Such formulations may yield inaccurate predictions for the macroscopic flow field. A specific example where this can occur involves gas–solid flow in a circulating fluidized bed riser. Ignoring the highly heterogonous mesoscopic structures caused by particle clustering overestimates the drag force, which can result in an unrealistically high entrainment rate of solids being carried out of the riser by the exiting gas phase. Neglecting mesoscopic structures may also result in grid dependence in the numerical solutions. As the grid resolution is increased, finer flow structures are revealed with corresponding statistical quantities, computed by flow field averaging, being noticeably refined until the grid size is on the order of ten times the particle diameter (Igci *et al.*, 2008). While this approach may yield accurate results, it relies on an extremely fine mesh and demonstrates the existence of grid-dependence on the solution. Therefore, in order to obtain an accurate grid-independent solution from a computationally efficient coarse grid, sub-grid models must be incorporated into the interfacial transport closures.

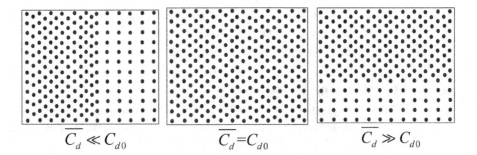

Figure 6.7-1 Dependence of drag coefficient on structure (Li and Kwauk, 2001).

There are typically three approaches employed to obtain sub-gird models for particle–fluid drag model closure. The first method is to perform direct numerical simulation (DNS) of particle–fluid flows where the particles and mesoscopic structures can be resolved directly. Interfacial transport coefficients, such as the drag force, can then be extracted from the results and provided as closure equations for averaged Eulerian models (Benyahia, 2006). The second approach is known as the filtered model that begins with a two-fluid simulation using a fine grid. Once solved, the drag force is sampled from the results in regions of different sizes. The size of the sampled region, known as the filter size, becomes a parameter in the resulting drag force closure model. When applied to the averaged Eulerian equations, the filter size corresponds to the grid size and compensates for the grid effect so that a grid-independent solution can be obtained (Igci *et al.*, 2008). The third approach for constructing the sub-grid model is known as the energy minimization multiscale (EMMS) model. This model assumes the genesis of mesoscopic structures is a balance between the tendency of the fluid to flow toward minimum resistance and of particles to suspend with the least amount of gravitational potential. Accordingly, a stability criterion is employed to qualify the mesoscopic structure (*e.g.*, cluster size) and determine the drag coefficient (Ge *et al.*, 2007). All three approaches have produced better predictions for certain particular applications. However, a general multiscale approach for comprehensive modeling of multiphase flows has yet to be developed and validated.

6.7.2   Particle Size Distribution and Interfacial Area Concentration

Area is an important factor governing the transfer rates between two phases at the interface. The interfacial area can be readily calculated from the volume fraction where the size of the discrete phase is uniform, such as in dispersed flow. In contrast, if the discrete phase is polydispersed then the interfacial area concentration in a control volume is a function of the local particle size distribution, which may evolve due to particle growth, agglomeration, breakup, or coalescence. For this class of cases, computing the interfacial area requires capturing particle size evolution and using statistical techniques, such as the population balance model. An alternative approach is to employ the interfacial area concentration transport model, which can be comparatively less computationally intensive. The latter is often used for specific applications, such as gas–liquid flow with bubbles.

6.7.2.1   Population Balance Model

The total population of particles with a particle volume, V, can be represented by the number density function $n(\mathbf{r}, V, t)$, where $\mathbf{r}$ is the position vector of the particles. The local volume fraction of particles of all sizes can be calculated by:

$$\alpha_p(\mathbf{r}, t) = \int_{\Omega_d} n(\mathbf{r}, V, t) V dV. \tag{6.7-1}$$

The governing equation for the evolution of particle size distribution, known as the population balance equation, is then expressed by:

$$\frac{\partial n(\mathbf{r}, V, t)}{\partial t} + \nabla \cdot [n(\mathbf{r}, V, t)\mathbf{U}_V] + \frac{\partial}{\partial V}\left[n(\mathbf{r}, V, t)\dot{V}\right]$$
$$= B_A(\mathbf{r}, V, t) - D_A(\mathbf{r}, V, t) + B_B(\mathbf{r}, V, t) - D_B(\mathbf{r}, V, t). \tag{6.7-2}$$

The second term on the left-hand side of Eq. (6.7-2) describes the convection of the particles by its velocity, $\mathbf{U}_V$. The third term on the left-hand side refers to the particle size change due to growth or dissolution. The four terms on the right-hand side of the model describe the change of particle size due to aggregation and breakage. These effects include:

1) The birth rate due to aggregation, as given by:

$$B_A(\mathbf{r}, V, t) = \frac{1}{2}\int_0^V a(V - V', V)n(\mathbf{r}, V - V', t)n(\mathbf{r}, V, t)dV', \tag{6.7-2a}$$

where $a(V', V)$ is the aggregation kernel that depends on the collision frequency and efficiency of aggregation of two particles of volumes V and V'.

2) The death rate due to aggregation, described by:

$$D_A(\mathbf{r}, V, t) = \frac{1}{2}\int_0^\infty a(V, V')n(\mathbf{r}, V', t)n(\mathbf{r}, V, t)dV'. \tag{6.7-2b}$$

3) The birth rate due to breakage, which is represented by:

$$B_B(\mathbf{r}, V, t) = \int_{\Omega_V} n_B g(V')\beta(V|V')n(\mathbf{r}, V', t)dV', \tag{6.7-2c}$$

where $g(V)$ is the breakage frequency of particle of volume V; $\beta(V|V')$ is the probability density function of particle breakage from volume V' to V, n_B is the number of daughter particles resulting from the said breakage.

4) The death rate due to breakage, expressed by:

$$D_B(\mathbf{r}, V, t) = g(V)n(\mathbf{r}, V, t). \tag{6.7-2d}$$

Detailed models for the rates described above are dependent on the particular physical mechanisms of aggregation and breakage in the system. More detailed information is available in the literature and in references (Jakobsen *et al.*, 2005; Chen *et al.*, 2011).

Solutions to Eq. (6.7-2) are obtained by discretizing the continuous size distribution into N classes, where N is a finite number. N_i represents the number density of particles in the ith class with size between V_i and V_{i+1}. This is represented by:

$$N_i(\mathbf{r}, t) = \int_{V_i}^{V_{i+1}} n(\mathbf{r}, V, t) dV \quad (i = 1 \ldots N). \tag{6.7-3}$$

After discretization, Eq. (6.7-2) is converted into an equation in N_i, and it is solved for each class of particles. This technique is numerically robust and directly computes the particle size distribution (PSD). However, the method is computational intensive because a large number of classes are required to resolve the distribution.

An alternate approach is to solve the size distribution indirectly by using the method of moments (MOM). The kth moment of the particle size distribution is defined as:

$$M_k(\mathbf{r}, t) = \int_0^{\infty} n(\mathbf{r}, l, t) l^k dl, \tag{6.7-4}$$

where l is the characteristic length of a single particle. The population balance equation, Eq. (6.7-2), is then transformed into N equations with respect to the N moments M_k. The properties characterizing the particle population, such as total number density, total particle volume, total surface area, and Sauter mean diameter, can be expressed via low-order moments. Extensions of the standard MOM include the quadrature method of moments (QMOM) (McGraw, 1997) and direct quadrature method of moments (DQMOM) (Fox, 2008). The advantage of the MOM class of methods is that typically only a small number of moments need to be solved instead of a large number of bins required to discretize the exact PSD, as with discrete methods.

6.7.2.2   Interfacial Area Transport Model

The interfacial area transport model accounts for various mechanisms of the change of interfacial area directly, without the need to model the particle size distribution. In a gas–liquid flow where gas bubbles represent the discrete phase, the transport of interfacial area concentration, a_I, can be modeled by the following equation (Ishii and Hibiki, 2006):

$$\frac{\partial a_I}{\partial t} + \nabla \cdot [a_I \mathbf{U}_I] = \frac{2}{3} \left(\frac{a_I}{\alpha_b} \right) \left\{ \frac{\partial \alpha_b}{\partial t} + \nabla \cdot (\alpha_b \mathbf{U}_b) - \eta \right\} + \frac{\pi}{18} \left(\frac{a_I}{\alpha_b} \right) d_v^3 \sum_j R_j + \pi d_c^2 R_\eta.$$

$$\tag{6.7-5}$$

The first term on the right-hand side of Eq. (6.7-5) captures the change of surface area due to particle volume change, and η is the rate of volume generated by nucleation. The second term models the effect of bubble interaction mechanisms, including coalescence and breakage where d_v is the bubble volume-equivalent diameter and R_j is the rate of particle number change due to jth particle interactions. The third term reflects the change in interfacial area due to phase change such as nucleation, in which d_c is the critical bubble size of nucleation, and R_η is the rate of nucleation. Mechanistic models for the rates of interfacial area change caused by bubble interaction mechanisms, nucleation, and condensation phenomena must be established as constitutive relations to close the interfacial area transport model. Details of such constitutive models are available in the literature and references (Ishii and Hibiki, 2006).

6.7.3   Turbulence Modulation

In a dispersed turbulent flow, the diffusion of particle clouds is significantly enhanced by turbulent eddy transport of the carrying fluid, typically by a factor of two or three orders of magnitude. Notably, introducing the dispersed phase can either enhance or suppress continuous phase turbulence, affecting the behavior of the mean flow field (*e.g.*, pressure drop in pipe flows). Gore and Crowe (1989) proposed a critical ratio of the particle diameter to a characteristic length scale of turbulence predicated on an extensive review of experimental results of turbulence modulation in dilute suspension pipe flows and jet flows. Their suggested relationship is given by:

$$\left(\frac{d_p}{l_e}\right)_{cr} \approx 0.1, \tag{6.7-6}$$

where l_e is the integral length scale of the most energetic eddies in the continuous phase. In the k-ε turbulence model, this quantity is expressed by:

$$l_e = C_\mu \frac{k^{3/2}}{\varepsilon}. \tag{6.7-6a}$$

Turbulence intensity is enhanced when the ratio given in Eq. (6.7-6) is greater than the critical value, and it is suppressed when the ratio is less than the critical value of 0.1. This criterion indicates turbulence is attenuated by small particles, but it is amplified by large particles.

At least six mechanisms, which are not independent of each other, contribute to turbulence modulation in fluid–solid suspension flows:

(1) Dissipation of turbulence kinetic energy by the particles.
(2) Increase in apparent viscosity due to the presence of particles.
(3) Vortex shedding or the presence of a wake behind the particles.
(4) Fluid moving with particles as carried mass of the particles.
(5) Increase in the velocity gradients between particles.
(6) Preferential concentration of particles by turbulence; particles selectively concentrated in particular structures by turbulence may cause rapid attenuation of that structure or trigger a new instability (Squires and Eaton, 1990).

Comprehensive modeling at this stage, including all of the preceding mechanisms, is not possible because of the intricate coupled relationships of turbulent interactions and incomplete information on the sources of turbulence generation. Nevertheless, simplified modeling that accounts for some predominant modulation mechanisms of turbulence is possible. For example, the Hinze–Tchen model introduced in Section 6.6.2 is a straightforward approach based on Mechanism (1) above, and is useful for small particles and decreases the continuous phase turbulence intensity.

For a single phase flow, the k and ε equations are given by Eq. (2.3-21) and Eq. (2.3-22), respectively. The k-ε turbulence model can be extended to the continuous fluid phase of a dispersed flow by including phase interactions from the dispersed phase (Elghobashi and Abou-Arab, 1983). For the turbulence kinetic energy, it can be expressed as:

$$\frac{\partial}{\partial t}(\alpha\rho k) + \nabla \cdot (\alpha\rho \mathbf{U}k) = \nabla \cdot \left(\alpha\frac{\mu_{\text{eff}}}{\sigma_k}\nabla k\right) + \alpha G - \alpha\rho\varepsilon + (1-\alpha)\rho S_k. \quad (6.7\text{-}7)$$

From its dissipation, it gives:

$$\frac{\partial}{\partial t}(\alpha\rho\varepsilon) + \nabla\cdot(\alpha\rho\mathbf{U}\varepsilon) = \nabla\cdot\left(\alpha\frac{\mu_{\text{eff}}}{\sigma_\varepsilon}\nabla\varepsilon\right) + \alpha\frac{\varepsilon}{k}(C_{1\varepsilon}G - C_{2\varepsilon}\rho\varepsilon) + (1-\alpha)\rho S_\varepsilon, \quad (6.7\text{-}8)$$

where G is the production of turbulence kinetic energy due to Reynolds stress of the continuous fluid phase. The last term on the right-hand side in Eq. (6.7-7) and Eq. (6.7-8) represents the effect of the modulation of the continuous fluid phase turbulence by the dispersed phase. Thus, the effect of the dispersed phase on the turbulence field of the continuous fluid phase can be accounted for directly by adding an extra term to the turbulent viscosity expression for the continuous phase or by introducing source terms into the turbulence transport equations for the continuous phase (Simonin and Viollet, 1990; Zhang et al., 2006).

6.7.3.1 Model of Yuan and Michaelides

The model of Yuan and Michaelides (1992) includes two predominant mechanisms of turbulence modulation in the dilute gas–solid flows: (1) turbulence reduction due to the kinetic energy dissipation from an eddy to accelerate particles; (2) turbulence enhancement due to the wake of the particles or vortex shedding.

For simplicity, the one-dimensional case is considered of a particle with velocity, U_p, entering an eddy of velocity, U, and interacting with it for a period of time, τ_i. The dominant interacting force between the particle and fluid is assumed to be the drag force, F_D. The velocity disturbance due to the particle originates from the wake behind the particle (e.g., $\text{Re}_p > 20$) and from vortex shedding (e.g., $\text{Re}_p > 400$). Hence, changes in the kinetic energy due to turbulence production are proportional to the difference between the squares of the velocities of the two phases and the volume where the velocity disturbance originates. The wake volume is taken as half of a complete ellipsoid, with a base diameter of d_p (identical to the particle diameter) and wake length of l_w. Thus, the total energy production of the gas by the particle wake or vortex shedding can be written as:

$$\Delta E_p = \frac{\pi}{12}d_p^2 l_w \rho \left(U^2 - U_p^2\right). \quad (6.7\text{-}9)$$

The rate of work done by the gas is: $F_D U$; the change of kinetic energy of the particle is: $F_D U_p$. The rate of energy dissipation, ε, is expressed by:

$$\varepsilon = F_D (U - U_p) = \frac{\pi}{8} C_D d_p^2 \rho (U - U_p)^2 |U - U_p|. \tag{6.7-10}$$

For brevity, the drag coefficient is given by the following relationship;

$$C_D = \frac{24 C}{Re_p}, \tag{6.7-10a}$$

where C, a function of Re_p in principle, is considered a constant during the interaction. During particle–eddy interaction, the equation of motion for the particles is given by:

$$\frac{\pi}{6} d_p^3 \rho_p \frac{dU_p}{dt} = \frac{\pi}{8} C_D d_p^2 \rho (U - U_p) |U - U_p|. \tag{6.7-11}$$

Combining the above expression with Eq. (6.7-10) and Eq. (6.7-11) yields an approximation for the particle velocity during the time interval, τ_i, of:

$$U_p = U_{p0} + (U - U_{p0}) \left\{ 1 - exp\left(-\frac{Ct}{\tau_S} \right) \right\} \qquad t \le \tau_i, \tag{6.7-12}$$

where τ_S is the Stokes relaxation time, and τ_i is the particle–eddy interaction time, defined by:

$$\tau_i = min\left(\frac{l_e}{|U - U_p|}, \frac{\rho l_e^2}{\mu} \right). \tag{6.7-12a}$$

The total energy dissipation from the eddy during τ_i, which is equivalent to the total work performed by the eddy on the particle during the particle–eddy interaction time, is then obtained by integrating Eq. (6.7-10) with respect to t, as:

$$\begin{aligned} \Delta E_d &= \frac{\pi}{12} d_p^3 \rho_p (U - U_{p0})^2 \left[1 - exp\left(-\frac{2C \tau_i}{\tau_S} \right) \right] \\ &\approx \frac{\pi}{12} d_p^3 \rho_p (U - U_p)^2 \left[1 - exp\left(-\frac{2C \tau_i}{\tau_S} \right) \right]. \end{aligned} \tag{6.7-13}$$

Combining the above expression with Eq. (6.7-9) and Eq. (6.7-13) yields an expression for total turbulence modulation as:

$$\begin{aligned} \Delta E_e &= \Delta E_p - \Delta E_d \\ &= \frac{\pi}{12} d_p^2 l_w \rho \left(U^2 - U_p^2 \right) - \frac{\pi}{12} d_p^3 \rho_p (U - U_p)^2 \left[1 - exp\left(-\frac{2C \tau_i}{\tau_S} \right) \right]. \end{aligned} \tag{6.7-14}$$

In the case of very small particles (*i.e.*, $\tau_i \gg \tau_S$), the particle velocity approaches the gas velocity and the wake disappears. The asymptotic expansion of Eq. (6.7-14) for fine particles is given by:

$$\Delta E_e = -\frac{\pi}{12} d_p^3 \rho_p (U - U_p)^2. \tag{6.7-14a}$$

In the case of large particles (*i.e.*, $\tau_i \ll \tau_S$), the particle velocity does not change appreciably during the particle–eddy interaction, and the production term predominates. Consequently, the asymptotic value of ΔE_t for large particles is equivalent to ΔE_p. The asymptotic expansion of Eq. (6.7-14) reveals that fine particles will cause turbulence reduction, which is proportional to the cube of the particle diameter; whereas large particles will primarily cause an increase in turbulence, which is proportional to the square of the particle diameter. This indication agrees with data compiled by Gore and Crowe (1989).

The model of Yuan and Michaelides (1992) can be incorporated into the k-ε turbulence model of the continuous phase by setting the source term to the turbulence production per unit volume and time. Thus, it yields:

$$S_k = \Delta E_e \frac{6}{\pi \rho d_p^3} \frac{1}{\tau_i} \tag{6.7-15}$$

and:

$$S_\varepsilon = C_\varepsilon \frac{\varepsilon}{k} S_k. \tag{6.7-16}$$

This model predicts a reasonable turbulence modulation in dilute gas–solid flows against experimental data (Bolio and Sinclair, 1995).

6.7.3.2 Bubble-Induced Turbulence in Liquid

In a system where gas bubbles are dispersed in a continuous liquid and bubble size is on the order of millimeters or greater, turbulence in the continuous liquid phase is enhanced due to velocity fluctuation induced by the bubble wake. Production of turbulence kinetic energy is often calculated from the work performed by the bubbles, which is equal to the product of interfacial force and local slip velocity. Consider the drag force as the only interfacial momentum transfer, the modification on source terms in the k-ε model by bubble-induced turbulence can be approximated by (Pfleger and Becker, 2001):

$$S_k = C_k \frac{3}{4} \frac{C_D}{d_b} |\mathbf{U}_b - \mathbf{U}|^3, \tag{6.7-17}$$

whereas S_ε takes the same form as Eq. (6.7-16).

A slightly different approach by Troshko and Hassan (2001) also yielded the modification in the k-equation the same as Eq. (6.7-17). The inclusion of added mass effect, however, leads to a different modification in the ε-equation, which is expressed as

$$S_\varepsilon = 0.45 \frac{3C_D |\mathbf{U}_b - \mathbf{U}|}{2C_A d_b} S_k. \tag{6.7-18}$$

The difference between Eq. (6.7-16) and Eq. (6.7-18) stems from the different time-scale assumptions made for the dissipation of bubble-induced turbulence as well as the relative significance of added mass force to the interaction. In practice, both models are capable of reproducing the dynamic bubble plume behavior (Zhang *et al.*, 2006).

6.8   Case Studies

The following case studies serve for further discussion with simple applications of the theories of Chapter 6, in particular using the continuum modeling to describe the dynamic transport of particles. The examples include mixing of suspended particles in a mechanically agitated slurry tank, bubble plume transport in a bubble column, heat transfer of tubular heat changer in a gas–solid fluidized bed, and evaporating sprays into a gas–solid flow. These examples illustrate various particle–fluid and particle–particle interactions in the Eulerian modeling for the transport of discrete phases.

6.8.1   Particle Suspension in a Stirred Tank

Problem Statement: Establish an Eulerian–Eulerian model to predict the flow and solid concentration profile in a mechanically agitated slurry tank, as shown in Figure 6.8-1. Assume that the tank is completely filled without a free surface and the agitation is isothermal and turbulent.

Analysis: The following is a simplified model based on the work of Micale *et al.* (2000). In the stirred slurry process, turbulent mixing action by the impeller encourages solid particles, which tend to settle via gravity, to remain in suspension. It is assumed that the solid particles are dispersed without interparticle collisions or agglomeration, and there is no interfacial mass transfer.

The continuity equation for both phases can be written, based on Eq. (6.4-21), as:

$$\frac{\partial}{\partial t} (\alpha\rho) + \nabla \cdot (\alpha\rho\mathbf{U}) = 0 \tag{6.8-1}$$

$$\frac{\partial}{\partial t} (\alpha_p\rho_p) + \nabla \cdot (\alpha_p\rho_p\mathbf{U}_p) = 0. \tag{6.8-2}$$

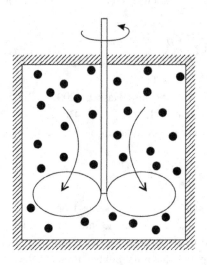

Figure 6.8-1 A mechanically agitated slurry tank.

Volume fraction conservation of Eq. (6.3-7a) gives

$$\alpha + \alpha_p = 1. \tag{6.8-3}$$

The momentum equation of liquid phase is given, based on Eq. (6.4-22), by:

$$\frac{\partial}{\partial t} (\alpha\rho\mathbf{U}) + \nabla \cdot (\alpha\rho\mathbf{U}\mathbf{U}) = -\alpha\nabla p + \nabla \cdot \left[\alpha (\mu + \mu_e) \left(\nabla\mathbf{U} + \nabla\mathbf{U}^T\right)\right] + \alpha\rho\mathbf{g} - \mathbf{F}_D, \tag{6.8-4}$$

where the turbulent viscosity, μ_e, is determined from k and ε by Eq. (2.3-18). The interphase momentum transfer is represented by the drag force, $\mathbf{F}_D$, which is expressed by:

$$\mathbf{F}_D = \frac{3}{4}\frac{C_D}{d_p}\alpha_p\rho \left|\mathbf{U}_p - \mathbf{U}\right| (\mathbf{U}_p - \mathbf{U}), \tag{6.8-4a}$$

whereas the drag coefficient C_D in the slurry is related to the drag coefficient of an isolated particle, C_{D0}, and the ratio of particle size to the Kolmogorov length scale of dissipative eddies, which is given by:

$$\frac{C_D - C_{D0}}{C_{D0}} = 8.74 \times 10^{-4} \left(\frac{\rho^3\varepsilon}{\mu^3}\right)^{\frac{3}{4}} d_p^3. \tag{6.8-4b}$$

The momentum equation of the particle phase is given, based on Eq. (6.4-22), by:

$$\frac{\partial}{\partial t} \left(\alpha_p\rho_p\mathbf{U}_p\right) + \nabla \cdot \left(\alpha_p\rho_p\mathbf{U}_p\mathbf{U}_p\right) = -\alpha_p\nabla p + \nabla \cdot \left[\mathbf{T}_p\right] + \alpha_p\rho_p\mathbf{g} + \mathbf{F}_D, \tag{6.8-5}$$

where the total stress of particle phase, $\mathbf{T}_p$, is given by Eq. (6.6-1).

A simple homogeneous k-ε model is used to handle turbulence of the mixture, in which both phases are assumed to share the same values for k and ε and the turbulent modulation due to inter-phase turbulent transfer is not considered. Thus, the k-equation is given by

$$\frac{\partial}{\partial t} (\rho_m k) + \nabla \cdot (\rho_m\mathbf{U}_m k) = \nabla \cdot \left[\left(\mu + \frac{\mu_e}{\sigma_k}\right)\nabla k\right] + S_{km}, \tag{6.8-6}$$

where

$$\rho_m = \alpha\rho + \alpha_p\rho_p \tag{6.8-6a}$$

$$\mathbf{U}_m = \alpha\mathbf{U} + \alpha_p\mathbf{U}_p \tag{6.8-6b}$$

$$\mu_e = \rho_m C_\mu \frac{k^2}{\varepsilon} \tag{6.8-6c}$$

$$S_{km} = 2\mu_e \left(\alpha\mathbf{S} : \mathbf{S} + \alpha_p\mathbf{S}_p : \mathbf{S}_p\right) - \rho_m\varepsilon, \tag{6.8-6d}$$

where $\mathbf{S}$ is the stain rate, defined by Eq. (2.2-3). Similarly, the ε-equation is given by

$$\frac{\partial}{\partial t} (\rho_m\varepsilon) + \nabla \cdot (\rho_m\mathbf{U}_m\varepsilon) = \nabla \cdot \left[\left(\mu + \frac{\mu_e}{\sigma_k}\right)\nabla\varepsilon\right] + S_{\varepsilon m}, \tag{6.8-7}$$

where

$$S_{\varepsilon m} = 2\mu_e C_1 \frac{\varepsilon}{k} \left(\alpha\mathbf{S} : \mathbf{S} + \alpha_p\mathbf{S}_p : \mathbf{S}_p\right) - C_2\rho_m\frac{\varepsilon^2}{k}. \tag{6.8-7a}$$

Thus, eight independent equations, Eqs. (6.8-1)–(6.8-7) and Eq. (6.8-6c), for eight unknowns ($\alpha, \alpha_p, \mathbf{U}, \mathbf{U}_p, p, \mu_e, k,$ and ε) are obtained. The closure of the modeling is now established. It is noted that impeller rotation is accounted for by a multiple reference frame model where the impeller zone is computed in a rotational reference frame with the rest of the tank remains stationary.

Comments: (1) The interphase momentum exchange here only includes the drag. While the drag is important in all multiphase flows, the turbulent dispersion force often has a particularly significant effect on the prediction of solids concentration profile in slurries. Ignoring the turbulent dispersion force would measurably underestimate the particle cloud height. Other interaction forces, such as lift and virtual mass, have an inconsequential impact on simulation results for this class of applications. (2) With strong agitated rotating flow, turbulence can be highly nonisotropic. Hence, the use of isotropic k-ε model may fail to reflect this nonisotropic nature of turbulence and resulted effects on phase transport. (3) The present mode does not account for the turbulent modulation due to phase interaction nor includes the particle–particle interactions such as collision or agglomeration, which can become important with increased particle concentration or larger particle size.

6.8.2   Bubble Plume Flow in Bubble Column

Problem Statement: Establish a two-fluid model to simulate the flow in a bubble column. The velocity fluctuation induced by displacement of the liquid as bubbles pass by is commonly known as the bubble-induced turbulence, whose impact on flow transport in bubbly column should be accounted for. The interphase forces include the drag, lift, and added mass forces.

Analysis: The following is a simplified model based on the work of Zhang *et al.* (2006). It is assumed that the bubbles are monodispersed without coalescence or breakup, and there is no interfacial mass transfer. Both liquid and bubbles are regarded as incompressible, with constant material densities of each phase.

The mass conservation equation for both phases can be written, based on Eq. (6.4-21), as:

$$\frac{\partial}{\partial t}(\alpha\rho) + \nabla \cdot (\alpha\rho\mathbf{U}) = 0 \tag{6.8-8}$$

$$\frac{\partial}{\partial t}(\alpha_b\rho_b) + \nabla \cdot (\alpha_b\rho_b\mathbf{U}_b) = 0. \tag{6.8-9}$$

Volume fraction conservation of Eq. (6.3-7a) gives:

$$\alpha + \alpha_b = 1. \tag{6.8-10}$$

The momentum equation of liquid phases is given, based on Eq. (6.4-22), by:

$$\frac{\partial}{\partial t}(\alpha\rho\mathbf{U}) + \nabla \cdot (\alpha\rho\mathbf{U}\mathbf{U}) = -\alpha\nabla p + \nabla \cdot \left[\alpha\mu_{eff}\left(\nabla\mathbf{U} + \nabla\mathbf{U}^T\right)\right] + \alpha\rho\mathbf{g} + \mathbf{M}, \tag{6.8-11}$$

where the effective viscosity of liquid phase has three additive contributors: the molecular viscosity, the shear-induced turbulent viscosity, and the bubble-induced turbulent dispersion term, expressed by:

$$\mu_{eff} = \mu + \mu_e + \mu_B. \tag{6.8-11a}$$

The turbulent viscosity, μ_e, is determined from k and ε by Eq. (2.3-18). The bubble-induced turbulence is given by (Sato and Sekoguchi, 1975):

$$\mu_B = \rho \alpha_b C_{\mu B} d_b |\mathbf{U}_b - \mathbf{U}|. \tag{6.8-11b}$$

The interphase momentum transfer, $\mathbf{M}$, is expressed by:

$$\begin{aligned}
\mathbf{M} &= \frac{3}{4} \frac{C_D}{d_b} \alpha_b \rho |\mathbf{U}_b - \mathbf{U}| (\mathbf{U}_b - \mathbf{U}) \\
&+ \alpha_b \rho C_L (\mathbf{U}_b - \mathbf{U}) \times \nabla \times \mathbf{U} + \alpha_b \rho C_A \left(\frac{d\mathbf{U}_b}{dt} - \frac{d\mathbf{U}}{dt} \right),
\end{aligned} \tag{6.8-11c}$$

whereas C_D, C_L, and C_A stand for the drag coefficient, lift coefficient, and added mass coefficient, respectively.

The momentum equation of the bubble phase is given, based on Eq. (6.4-22), by:

$$\begin{aligned}
\frac{\partial}{\partial t} (\alpha_b \rho_b \mathbf{U}_b) + \nabla \cdot (\alpha_b \rho_b \mathbf{U}_b \mathbf{U}_b) &= -\alpha_b \nabla p \\
+ \nabla \cdot \left[\alpha_b (\mu_b + \mu_{be}) \left(\nabla \mathbf{U}_b + \nabla \mathbf{U}_b^T \right) \right] &+ \alpha_b \rho_b \mathbf{g} - \mathbf{M},
\end{aligned} \tag{6.8-12}$$

where the turbulent viscosity of the bubble phase is related to the turbulent viscosity of the liquid phase, as suggested by Jakobsen *et al.* (1997):

$$\mu_{be} = \frac{\rho_b}{\rho} \mu_e. \tag{6.8-12a}$$

The k and ε equations are, respectively, given by

$$\frac{\partial}{\partial t} (\alpha \rho k) + \nabla \cdot (\alpha \rho \mathbf{U} k) = \nabla \cdot \left[\alpha \left(\mu + \frac{\mu_e}{\sigma_k} \right) \nabla k \right] + 2\mu_e (\alpha \mathbf{S} : \mathbf{S}) - \alpha \rho \varepsilon + S_k \tag{6.8-13}$$

$$\frac{\partial}{\partial t} (\alpha \rho \varepsilon) + \nabla \cdot (\alpha \rho \mathbf{U} \varepsilon) = \nabla \cdot \left[\alpha \left(\mu + \frac{\mu_e}{\sigma_k} \right) \nabla \varepsilon \right] + 2\mu_e C_1 \frac{\varepsilon}{k} (\alpha \mathbf{S} : \mathbf{S}) - C_2 \alpha \rho \frac{\varepsilon^2}{k} + S_\varepsilon, \tag{6.8-14}$$

where $\mathbf{S}$ is the strain rate, defined by Eq. (2.2-3), S_k and S_ε are bubble-induced turbulence terms, respectively, defined by Eq. (6.7-17) and Eq. (6.7-16).

Thus, ten independent equations, Eqs. (6.8-8)–(6.8-14) with Eqs. (2.3-18), (6.8-11b), and (6.8-12a), for ten unknowns $(\alpha, \alpha_b, \mathbf{U}, \mathbf{U}_b, p, \mu_e, \mu_{be}, \mu_B, k,$ and $\varepsilon)$ are obtained. The closure of the modeling is now established. It is noted that, due to the bubble–liquid interactions, the boundary conditions of liquid phase can be different from those of a single-phase flow, such as the logarithmic wall law (Troshko and Hassan, 2001).

Comments: (1) Different turbulent models and transport coefficients such as drag coefficients can vary, depending on the bubble flow regimes. (2) The interaction among bubbles constantly leads to coalescence and breakup, which makes the bubble size polydispersed and likely nonspherical, with different dynamic phase transport and phase interactions. (3) Case simulation examples can be found in the reference (Zhang *et al.*, 2006).

6.8.3   Heat Transfer of Immersed Tubes in Dense Gas–Solid Fluidized Bed

Problem Statement: Establish a multifluid model for the heat transfer of tubes immersed in a dense gas–solid fluidized bed. Assume that the tube surface temperature is constant, and solid particles are monodispersed without forming agglomerate or clusters. There is no interfacial mass transfer, and thermal radiations among solids or between solids and wall can also be ignored.

Analysis: The following is a simplified model based on the work of Yusuf *et al.* (2011). The porous flow distributor, immersed tubes, and other in-bed obstacles are all included in the modeling domain, represented by the given distributions of volume porosity and area porosity in the multifluid governing equations (Mathiesen *et al.*, 2000). The values of those porosities vary between zero and one, with zero for total blockage and one for total openness. The gas flow is assumed to be incompressible and without strong turbulence due to the damping of dense solid suspension.

The continuity equation for both phases can be expressed as:

$$\frac{\partial}{\partial t}(\beta_v \alpha \rho) + \nabla \cdot (\beta_a \alpha \rho \mathbf{U}) = 0 \tag{6.8-15}$$

$$\frac{\partial}{\partial t}(\beta_v \alpha_p \rho_p) + \nabla \cdot (\beta_a \alpha_p \rho_p \mathbf{U_p}) = 0, \tag{6.8-16}$$

where β_v and β_a are the volume and area porosities, respectively. The volume fraction conservation is the same as Eq. (6.8-3).

The momentum equation of gas phase is given by:

$$\frac{\partial}{\partial t}(\beta_v \alpha \rho \mathbf{U}) + \nabla \cdot (\beta_a \alpha \rho \mathbf{UU}) = -\beta_v \alpha \nabla p + \nabla \cdot \left[\beta_a \alpha \mu \left(\nabla \mathbf{U} + \nabla \mathbf{U}^T \right) \right] + \beta_v \alpha \rho \mathbf{g} + \mathbf{M}, \tag{6.8-17}$$

where the interphase momentum transfer is dominated by the gas–solid drag force in fluidization, which can be formulated by (Gidaspow, 1994):

$$\mathbf{M} = \begin{cases} \beta_v \left(150 \frac{\alpha_p^2 \mu}{\alpha d_p^2} + 1.75 \frac{\alpha_p \rho |\mathbf{U_p} - \mathbf{U}|}{d_p} \right) (\mathbf{U_p} - \mathbf{U}) & \alpha \leq 0.8 \\ \beta_v \frac{3}{4} \frac{C_D}{d_p} \alpha^{-1.65} \alpha_p \rho \left| \mathbf{U_p} - \mathbf{U} \right| (\mathbf{U_p} - \mathbf{U}) & \alpha > 0.8 \end{cases}. \tag{6.8-17a}$$

The momentum equation of particle phase is given by:

$$\frac{\partial}{\partial t}(\beta_v \alpha_p \rho_p \mathbf{U_p}) + \nabla \cdot (\beta_a \alpha_p \rho_p \mathbf{U_p U_p}) = -\beta_v \alpha_p \nabla p + \nabla \cdot \left[\beta_a \mathbf{T}_p \right] + \beta_v \alpha_p \rho_p \mathbf{g} - \mathbf{M}, \tag{6.8-18}$$

where the total stress of particle phase, $\mathbf{T}_p$, is due to interparticle collisions, which can be formulated from the kinetic theory of granular flows in Section 2.5.

The thermal energy equation of gas phase is given by

$$\frac{\partial}{\partial t}\left(\beta_v \alpha \rho c_p T\right) + \nabla \cdot \left(\beta_a \alpha \rho \mathbf{U} c_p T\right) = \nabla \cdot \left[\beta_a \alpha K \nabla T\right] + q, \tag{6.8-19}$$

where q is the interphase volumetric heat transfer, expressed by

$$q = \beta_v \frac{6\alpha_p}{d_p} h_c (T_p - T). \tag{6.8-19a}$$

The thermal energy equation of particle phase is given by

$$\frac{\partial}{\partial t}\left(\beta_v \alpha_p \rho_p c_{pp} T_p\right) + \nabla \cdot \left(\beta_a \alpha_p \rho_p \mathbf{U}_p c_{pp} T_p\right) = \nabla \cdot \left[\beta_a \alpha_p K_p \nabla T_p\right] - q, \tag{6.8-20}$$

where the thermal conductivity of gas and particle phases as well as the convective heat transfer coefficient are functions of the phases' volumetric fractions and material properties, with detailed correlations given in the reference (Yusuf et al., 2011).

So far, seven independent equations, Eqs. (6.8-15)–(6.8-20) and Eq. (6.8-3), for seven unknowns ($\alpha, \alpha_p, \mathbf{U}, \mathbf{U}_p, p, T$, and T_p) are obtained. The closure of the modeling is now established. Thus, with proper correlations of transport coefficients and boundary conditions, the temperature of phases can be solved, which leads to the local heat transfer coefficient on the tube surface, calculated by:

$$h_c = \frac{1}{(T_w - T_b)}\left(\alpha K \nabla T + \alpha_p K_p \nabla T_p\right)_w \cdot \mathbf{n}_w, \tag{6.8-21}$$

where T_w and T_b are the temperatures of the tube surface and bed bulk, respectively. The total heat transfer is therefore given from the integration over all tube surfaces, such as

$$q = \sum_{i=1}^{N} \int_{S_i} h_c (T_w - T_b)\, dS = \sum_{i=1}^{N} \int_{S_i} \left(\alpha K \nabla T + \alpha_p K_p \nabla T_p\right)_w \cdot \mathbf{n}_w dS. \tag{6.8-22}$$

Comments: (1) The gas flow in dense fluidization can still be highly turbulent. The gas turbulence can be accounted for by various turbulence models such as Sub Grid Scale (SGS) model (Mathiesen et al., 2000). (2) The total stress in particle phase is calculated in terms of granular temperature by Eq. (2.5-18) using the kinetic theory model of granular flows. The kinetic theory model is typically closed with an additional governing equation for granular temperature that is coupled with particle phase velocity and volume fraction. (3) It should be noted that β_a is a directional heterogeneous property. (4) Case simulation examples can be found in the reference (Yusuf et al., 2011).

6.8.4 Evaporating Spray in Gas–Solid Suspension Flow

Problem Statement: Consider an evaporating liquid spray, such as liquid nitrogen spray, injected into a gas–solids suspension flow. The ambient gas and solid flows are considered at an ambient temperature much higher than the evaporation temperature of liquid spray. The vapor is mixed with gas to form a gaseous mixture. Establish a model to predict this transport phenomena of the two-component three-phase flow, with coupled interphase mass, momentum, and heat transfers.

Analysis: The following is a simplified model based on the work of Wang *et al.* (2004). A hybrid Eulerian–Eulerian–Lagrangian model is adopted for this problem, with a two-fluid Eulerian modeling for gas–solid flow and a Lagrangian modeling for the evaporating droplets. It is assumed that the volumetric effect of droplets on the gas–solid flow can be neglected. In addition, the solid particles are monodispersed without agglomeration.

For the Eulerian–Eulerian model of gas–solid flow, the continuity equation for the gas mixture phase can be expressed, based on Eq. (6.4-21), as:

$$\frac{\partial}{\partial t}(\alpha\rho) + \nabla \cdot (\alpha\rho\mathbf{U}) = \Gamma, \tag{6.8-23}$$

where Γ is the volumetric vaporization rate from spray, which is calculated from the droplet Lagrangian model. The species conservation equation for vapor in the gas mixture, in terms of vapor mass fraction Y_v, is given by

$$\frac{\partial}{\partial t}(\alpha\rho Y_v) + \nabla \cdot (\alpha\rho Y_v\mathbf{U}) = \Gamma. \tag{6.8-24}$$

The continuity equation for the solid particle phase is given by

$$\frac{\partial}{\partial t}(\alpha_p\rho_p) + \nabla \cdot (\alpha_p\rho_p\mathbf{U}_p) = 0. \tag{6.8-25}$$

Volume fraction conservation gives

$$\alpha + \alpha_p = 1. \tag{6.8-26}$$

The momentum equation of gas mixture phase is given, based on Eq. (6.4-22), by:

$$\frac{\partial}{\partial t}(\alpha\rho\mathbf{U}) + \nabla \cdot (\alpha\rho\mathbf{U}\mathbf{U}) = -\alpha\nabla p + \nabla \cdot \left[\alpha\left(\mu + \mu_e\right)\left(\nabla\mathbf{U} + \nabla\mathbf{U}^T\right)\right]$$
$$+ \alpha\rho\mathbf{g} - \mathbf{M}_{gp} - \mathbf{M}_{gd} + \Gamma\mathbf{U}_d, \tag{6.8-27}$$

where the turbulent viscosity, μ_e, is determined from k and ε by Eq. (2.3-18). The gas–solid interphase momentum transfer, $\mathbf{M}_{gp}$, is given by Eq. (6.8-17a). The gas–droplet interphase momentum transfer, $\mathbf{M}_{gd}$, is represented by the volumetric droplet drag force as:

$$\mathbf{M}_{gd} = \frac{\pi}{8}C_{Dd}\,n_d\,\rho d_d^2\,|\mathbf{U}_d - \mathbf{U}|\,(\mathbf{U}_d - \mathbf{U}), \tag{6.8-27a}$$

where n_d is the number density of droplets, determined by the droplet trajectories from Lagrangian model.

The momentum equation of particle phase is given, based on Eq. (6.4-22), by:

$$\frac{\partial}{\partial t}(\alpha_p\rho_p\mathbf{U}_p) + \nabla \cdot (\alpha_p\rho_p\mathbf{U}_p\mathbf{U}_p) = -\alpha_p\nabla p + \nabla \cdot [\mathbf{T}_p] + \alpha_p\rho_p\mathbf{g} + \mathbf{M}_{gp} - \mathbf{M}_{pd}, \tag{6.8-28}$$

where the total stress of particle phase, $\mathbf{T}_p$, is given by Eq. (6.6-1) and determined from the kinetic theory of granular flows in Section 2.5. The solid-droplet interphase momentum transfer, $\mathbf{M}_{pd}$, is formulated similar to Eq. (6.8-17a) as:

$$\mathbf{M}_{pd} = \frac{\pi}{8}\alpha_p^{-1.65}C_{Dd}\,n_d\,\rho d_d^2\,|\mathbf{U}_d - \mathbf{U}_p|\,(\mathbf{U}_d - \mathbf{U}_p). \tag{6.8-28a}$$

The thermal energy equation of gas mixture phase is given by

$$\frac{\partial}{\partial t}\left(\alpha\,\rho c_p T\right) + \nabla \cdot \left(\alpha\,\rho \mathbf{U}c_p T\right) = \nabla \cdot [\alpha\,K\nabla T] - q_{gp} - q_{gd} + \Gamma c_p T_d, \tag{6.8-29}$$

where q_{gp} is the gas–particle interphase volumetric heat transfer, expressed by

$$q_{gp} = \frac{6\alpha_p}{d_p}h_c(T - T_p), \tag{6.8-29a}$$

and q_{gd} is the gas–droplet interphase volumetric heat transfer, given by

$$q_{gd} = \frac{\pi\,n_d}{8}d_d^2 h_{cd}(T - T_d). \tag{6.8-29b}$$

The thermal energy equation of solid particle phase is given by

$$\frac{\partial}{\partial t}\left(\alpha_p\rho_p c_{pp}T_p\right) + \nabla \cdot \left(\alpha_p\rho_p \mathbf{U}_p c_{pp}T_p\right) = \nabla \cdot \left[\alpha_p K_p\nabla T_p\right] + q_{gp} - q_{pd}, \tag{6.8-30}$$

where q_{pd} is the solid–droplet interphase volumetric heat transfer from solid–droplet collisions.

For the deterministic Lagrangian model of droplets, the evaporation is assumed to be diffusion-dominated. The mass conservation of the ith droplet is governed, from Eq. (3.4-13), as

$$\frac{dm_{di}}{dt} = -2\pi\,\rho D d_{pi}\ln\left(\frac{Y_{v\infty} - 1}{Y_{vd} - 1}\right). \tag{6.8-31}$$

The equation of motion of the droplet is given by:

$$\begin{aligned} m_{di}\frac{d\mathbf{U}_{di}}{dt} = \frac{\pi}{8}C_{Dd0}d_{di}^2\alpha^{-1.65}\,|\mathbf{U}(\mathbf{r}_{di}) - \mathbf{U}_{di}|\,(\mathbf{U}(\mathbf{r}_{di}) - \mathbf{U}_{di}) \\ + \frac{\pi}{8}C_{Dd0}d_{di}^2\alpha_p^{-1.65}\,|\mathbf{U}_p(\mathbf{r}_{di}) - \mathbf{U}_{di}|\,(\mathbf{U}_p(\mathbf{r}_{di}) - \mathbf{U}_{di}) + m_{di}\mathbf{g}. \end{aligned} \tag{6.8-32}$$

The equation of trajectory is given by:

$$\frac{d\mathbf{r}_{di}}{dt} = \mathbf{U}_{di}. \tag{6.8-33}$$

The thermal energy equation of the droplet is given, using the lumped thermal capacity model, as:

$$m_{di}c_{di}\frac{dT_{di}}{dt} = \pi d_{pi}^2 h_{cd}(T(\mathbf{r}_{di}) - T_{di}) + \pi d_{pi}^2 h_{pd}(T_p(\mathbf{r}_{di}) - T_{di}) + \frac{dm_{di}}{dt}L. \tag{6.8-34}$$

Thus, twelve independent equations, Eqs. (6.8-23)–(6.8-34), for twelve unknowns ($\alpha, \alpha_p, \mathbf{U}, \mathbf{U}_p, p, T, T_p, \mathbf{r}_d, \mathbf{U}_d, T_d, m_d$, and Y_v) are obtained. The closure of the modeling is now established. It is noted that the Lagrangian model is applied to each individual droplet, respectively, and the droplet terms in the Eulerian equations are contributed by all droplets found in the local domain of $(\mathbf{r}, t)$.

Comments: (1) The volumetric effect of droplets on the multifluid continuum of the Eulerian model is currently neglected for simplicity. Such an effect can be significant near the spray nozzle regime where the droplet volumetric fraction may even surpass that of the solid volumetric fraction. (2) The interphase mass, momentum, and heat transfers between droplets and solid particles are dominated by interphase collisions where sophisticated mechanistic models are yet to be established. Most of the current models are simply extended from empirical correlations in fluidization. (3) Case simulation examples can be found in the reference (Wang *et al.*, 2004).

6.9   Summary

In a continuum modeling approach of a multiphase flow, each transport phase is regarded as an individual pseudo-continuum fluid and all these "fluids" co-share the same space and time domains. This chapter delineates the volume-averaging method to construct the pseudo-continuum fluids over which the volume-averaged Eulerian modeling approach is developed. The key concepts and formula of volume-averaged continuum modeling include:

- definitions of intrinsic and phase averages and their relationship, especially the averages of velocity and energy (or temperature);
- volume-averaging theorems: Eq. (6.2-8) for averaging of a gradient, Eq. (6.2-9) for averaging of a divergent, and Eq. (6.2-10) for averaging of a time derivative;
- general form of volume-averaged transport equations (*i.e.*, Eq. (6.3-3)), and the individual equations of mass, momentum, and energy (*i.e.*, Eq. (6.3-5), Eq. (6.3-8) and Eq. (6.3-14));
- volume and mass balance conditions of all phases: Eq. (6.3-7a) and Eq. (6.3-6);
- approximated formulation (also known as constitutive relations) of the volume-averaged tensors by individual volume-averaged parameters (such as phase velocity), which has to be assumed or developed for modeling closure;
- formulation of interfacial transport between phases.

Similar to the modeling of single-phase turbulent flows given in Section 2.3, the effect of turbulence on phase transport is also handled via Reynolds decomposition and time-averaging over the volume-averaged equations (since a continuum of a transport phase is preconstructed via volume-averaging). The resulted volume–time-averaged Eulerian equations thus contain terms of covariance of fluctuating variables, which need to be closed via further modeling approaches, such as those analogous to the turbulence modeling of single-phase flows. The detailed formulation of various turbulent transport coefficients and the salient behavior of the turbulence modulation from various interactions among phases are presented.

Nomenclature

$< >$	Volume-averaging
$\overline{< >}$	Volume–time averaging
A	Area
a_I	Interfacial area concentration
B	Birth rate
C_D	Drag coefficient
c_p	Specific heat at constant pressure
c_v	Specific heat at constant volume
D	Diffusivity Death rate
d	Diameter
E	Total energy
	Elastic modulus
e	Specific internal energy
$\boldsymbol{F}$	Force
f	Body force
G	Production term of k
g	Gravity acceleration
h	Specific enthalpy
h_c	Convective heat transfer coefficient
i	ith
$\boldsymbol{J}$	Flux of ψ
K	Thermal conductivity
k	Turbulence kinetic energy
l	Characteristic length
$\mathbf{M}$	Momentum transfer
m	Mass
$\boldsymbol{n}$	Normal unit vector
n	Number density
p	Pressure
Q	Volumetric flow rate
q	Heat flow
Re	Reynolds number
$\boldsymbol{r}$	Position vector
$\boldsymbol{S}$	Spatial deformation rate
S	Source term
$\boldsymbol{T}$	Stress tensor
T	Temperature
t	Time
U	Time-averaged velocity
$\boldsymbol{u}$	Instant velocity
V	Volume
Y	Species mass fraction

Greek symbols

α	Volume fraction
β	Momentum transfer coefficient Porosity
Γ	Interfacial mass flow
ε	Dissipation rate of k
μ	Dynamic viscosity
ρ	Density
σ	Surface tension
τ	Relaxation time
τ_t	Time-averaging duration
ψ	General quantity
ϕ	Source term of ψ

Subscript

0	Isolated
∞	Ambient
A	Added mass
	Aggregation
a	Area
B	Breakage
b	Bubble
C	Collision
D	Drag
d	Droplet
E	Energy
e	Eddy
g	Gas
I	Interface
k	Phase k
	Turbulence kinetic energy
L	Lift
l	Phase l
m	Maximum
p	Particle
q	Heat transfer
r	Radiation

Superscript

$'$	Fluctuating
$-$	Time-averaging
i	Intrinsic
T	Transpose

S	Stokes
s	Interface
T	Thermal
v	Vapor
	Volume
W	Work
w	Wake
	Wall
Γ	Interphase mass transfer
ε	Dissipation rate of k

Problems

P6.1 Provide three practical examples illustrating the impact of the minimum control volume of the "continuum" concept in a multiphase flow in the context of numerical simulation or empirical measurements.

P6.2 For a dense-phase gas–solid isothermal and laminar flow, write a complete set of volume-averaged governing equations. List all independent variables and indicate the number of independent equations required to ensure a closed-form modeling.

P6.3 Consider a turbulent and dense-phase gas–solid flow with charged particles through a parallel-plate electrostatic precipitator. Develop an Eulerian–Eulerian model to calculate the particle deposition loss to the plate wall. Coupling between the gas–solid flow and the charge-induced electric field must be considered.

P6.4 Consider a turbulent gas–solid particle flow in a 90° bend. Establish an Eulerian–Eulerian model of the two-phase flow transport over the bend.

P6.5 If the particle size is much smaller than that of a control volume, show that the pressure gradient on the particles per unit volume can be approximated as:

$$< \nabla p_k >= \alpha_k \nabla p$$

P6.6 Derive the volume-average transport equation for chemical species.

P6.7 To predict the flow and solid concentration profile in a mechanically agitated slurry tank with high solid concentration, modify the Eulerian–Eulerian model in the Case Study of 6.8.1 or develop an alternative model to include the effect of interparticle collisions.

P6.8 Establish an Eulerian–Eulerian model to predict a bubbly flow in a mechanically agitated slurry tank, similar to Figure 6.8-1. Assume that the agitation is isothermal and turbulent. The bubbles have no coalescence or breakup during the agitation.

P6.9 A tank is partitioned by a removable wall into two chambers. One chamber is completely filled with a solid–liquid mixture while the other chamber contains only the clear liquid. There is no free surface in the filled tank and solids are heavier than the liquid. The partition wall is then suddenly removed out of the tank. This sudden removal triggers a gravity current driven by inertial settling of solid particles in the liquid tank. Establish an Eulerian–Eulerian model for the dispersion of solids and induced current movement.

P6.10 Consider a dense gas–solid jet injected into a stagnant gas medium. Develop a model to describe the phase distributions of gas and solids as well as trajectories of particles. Assume that the jet flow is turbulent and the interparticle effects such as collisions cannot be ignored.

P6.11 Develop an Eulerian–Eulerian model to simulate the grinding process and efficiency of a centrifugal grinder for coal grinding with steel balls (grinding by coal–steel collisions). The coarse coal is fed pneumatically into the centrifugal grinder whereas the fine coal (after grinding) is also removed pneumatically by the exiting air flow.

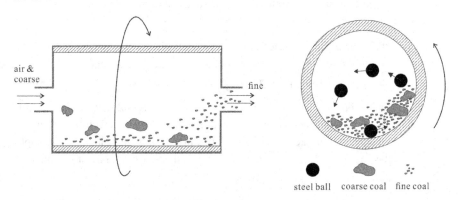

steel ball   coarse coal   fine coal

P6.12 Establish a two-fluid model for an upper bubbly flow in a vertical column with expansion. The expansion may cause a strong change in pressure reduction, flow separation, and flow shearing, leading to bubble interactions such as coalescence and breakup. The wall shearing may also drift bubbles away from the wall boundary toward mainstream of transport.

References

Arnold, G. S., Drew, D. A., and Lahey, R. T. (1990). An assessment of multiphase flow models using the second law of thermodynamics. *Int. J. Multiphase Flow*, **16**, 481–494.

Bolio, E. J., and Sinclair, J. L. (1995). Gas turbulence modulation in the pneumatic conveying of massive particles in vertical tubes. *Int. J. Multiphase Flow*, **21**, 985–1001.

Boussinesq, J. (1877). *Essai sur la théorie des eaux courantes, Mémoires présentés par divers savants à l'Académie des Sciences*, **23**, 1–680.

Brinkman, H. C. (1952). The viscosity of concentrated suspensions and solutions. *J. Chem. Phys.*, **20**, 571–573.

Cantero, M. I., Balachadar, S., and García, M. H. (2008). An Eulerian-Eulerian model for gravity currents driven by inertial particles. *Int. J. Multiphase Flow*, **34**, 484–501.

Chen, X.-Z., Luo, Z.-H., Yan, W.-C., Lu, Y.-H., and Ng, I-S. (2011), Three-dimensional CFD-PBM coupled model of the temperature fields in fluidized-bed polymerization reactors. *AIChE J.*, **57**, 3351–3366.

Delhaye, J. M., and Archard, J. L. (1976). On the averaging operators introduced in two-phase flow modeling. *OECD/NEA Specialists' Meeting on Transient Two-Phase Flow.* Toronto, Canada.

Drew, D. A. (1983). Mathematical modeling of two-phase flow. *Ann. Rev. Fluid Mech.*, **15**, 261–291.

Drew, D. A., and Lahey, R. T. Jr. (1993). Analytical modeling of multiphase flows. *Particulate Two-Phase Flow*, Ed. M. D. Roco. Boston, MA: Butterworth-Heinemann.

Einstein, A. (1906). Eine neue Bestimmung der Moleküldimensionen. *Ann. Phys.*, **19**, 289–306.

Elghobashi, S. E., and Abou-Arab, T. W. (1983). A two-equation turbulence model for two-phase flows. *Phys. Fluids*, **26**, 931–938.

Fan, L.-S., and Zhu, C. (1998). *Principles of Gas-Solid Flows.* Cambridge: Cambridge University Press.

Fox, R. O. (2008). A quadrature-based third-order moment method for dilute gas-particle flows. *J. Comput. Phys.*, **227**, 6313–6350.

Frankel, N. A., and Acrivos, A. (1967). On the viscosity of a concentrated suspension of solid spheres. *Chem. Eng. Sci.*, **22**(6), 847–853.

Ge, W., Chen, F., Gao, J., *et al.* (2007). Analytical multi-scale method for multi-phase complex systems in process engineering-Bridging reductionism and holism. *Chem. Eng. Sci.*, **62**, 3346–3377.

Gidaspow, D. (1986). Hydrodynamics of fluidization and heat transfer: supercomputer modeling. *Appl. Mech. Rev.*, **39**, 1–23.

Gidaspow, D. (1994). *Multiphase Flow and Fluidization: Continuum and Kinetic Theory Description.* Academic Press.

Gore, R. A., and Crowe, C. T. (1989). Effect of particle size on modulating turbulent intensity. *Int. J. Multiphase Flow*, **15**, 279–285.

Hinze, J. O. (1975). *Turbulence.* New York: McGraw-Hill.

Igci, Y., Andrews, A. T. IV, Sundaresan, S., Pannala, S., and O'Brien, T. (2008). Filtered two-fluid models for fluidized gas-particle suspensions. *AIChE J.*, **54**, 1431–1448.

Ishii, M. (1975). *Thermo-Fluid Dynamic Theory of Two-Phase Flow.* Paris: Eyrolles.

Ishii, M., and Hibiki, T. (2006). *Thermo-fluid Dynamics of Two-phase Flow.* New York: Springer.

Jackson, R. (2000). *The Dynamics of Fluidized Particles.* Cambridge: Cambridge University Press.

Jakobsen, H. A., Lindborg, H., and Dorao, C. A. (2005). Modeling of bubble column reactors: progress and limitations. *Ind. Eng. Chem. Res.*, **44**, 5107–5151.

Jakobsen, H. A., Sannæs, B. H., Grevskott, S., and Svendsen, H. F. (1997). Modeling of vertical bubble-driven flows. *Ind. Eng. Chem. Res.*, **36**, 4052–4074.

Li, J., and Kwauk, M. (2001). Multiscale nature of complex fluid-particle systems. *Ind. Eng. Chem. Res.*, **40**, 4227–4237.

Mathiesen, V., Solberg, T., and Hjertager, B. H. (2000). An experimental and computational study of multiphase flow in a circulating fluidized bed. *Int. J. Multiphase Flow*, **26**, 387–419.

McGraw, R. (1997). Description of aerosol dynamics by the quadrature method of moments. *Aerosol Sci. Technol.*, **27**, 255–265.

Micale, G., Montante, G., Grisafi, F., Brucato, A., and Godfrey, J. (2000). CFD simulation of particle distribution in stirred vessels. *Trans. IChemE*, **78, Part A**, 435–444.

Pfleger, D., and Becker, S. (2001). Modeling and simulation of the dynamic flow behavior in a bubble column. *Chem. Eng. Sci.*, **56**, 1737–1747.

Roscoe, R. (1952). The viscosity of suspensions of rigid spheres. *Brit. J. Appl. Phys.*, **3**, 267–269.

Sato, Y., and Sekoguchi, K. (1975). Liquid velocity distribution in two-phase bubble flow. *Int. J. Multiphase Flow*, **2**, 79–95.

Scargiali, F., D'Orazio, A., Grisafi, F., and Brucato, A. (2007). Modelling and simulation of gas-liquid hydrodynamics in mechanically stirred tanks. *Trans. IChemE, Part A, Chem. Eng. Des.*, **85(A5)**, 637–646.

Simonin, C., and Viollet, P. L. (1990). Predictions of an oxygen droplet pulverization in a compressible subsonic coflowing hydrogen flow. *Numerical Methods for Multiphase Flows*, FED'91, 65–82.

Slattery, J. C. (1967a). General balance equation for a phase interface. *I&EC Fund.*, **6**, 108–115.

Slattery, J. C. (1967b). Flow of viscoelastic fluid through porous media. *AIChE J.*, **13**, 1066–1071.

Soo, S. L. (1965). Dynamics of multiphase flow systems. *I&EC Fund.*, **4**, 426.

Soo, S. L. (1989). *Particulates and Continuum: Multiphase Fluid Dynamics*. New York: Hemisphere.

Squires, K. D., and Eaton, J. K. (1990). Particle response and turbulence modification in isotropic turbulence. *Phys. Fluids*, **A2**, 1191–1203.

Tchen, C. M. (1947). *Mean Value and Correlation Problems Connected with the Motion of Small Particles in a Turbulent Field*. Ph.D. Thesis. Delft University, Netherlands.

Troshko, A. A., and Hassan, Y. A. (2001). A two-equation turbulence model of turbulent bubbly flows. *Int. J. Multiphase Flow*, **27**, 1965–2000.

Wang, X., Zhu, C., and Ahluwalia, R. (2004). Numerical simulation of evaporating spray jets in concurrent gas-solids pipe flows. *Powder Technol.*, **140**, 56–67.

Whitaker, S. (1969). Advances in theory of fluid motion in porous media. *I&EC*, **61**, 14–28.

Yuan, Z., and Michaelides, E. E. (1992). Turbulence modulation in particulate flows – a theoretical approach. *Int. J. Multiphase Flow*, **18**, 779–785.

Yusuf, R., Halvorsen, B., and Melaaen, M. C. (2011). Eulerian-Eulerian simulation of heat transfer between a gas-solid fluidized bed and an immersed tube-bank with horizontal tubes. *Chem. Eng. Sci.*, **66**, 1550–1564.

Zhang, D., Deen, N. G., and Kuipers, J. A. M. (2006). Numerical simulation of the dynamic flow behavior in a bubble column: A study of closures for turbulence and interface forces. *Chem. Eng. Sci.*, **61**, 7593–7608.

7 Numerical Modeling and Simulation

7.1 Introduction

In Chapters 3 to 6, general governing equations for multiphase flows that incorporate microscopic descriptions of individual particle dynamics and macroscopic descriptions of averaged behavior of phase transport were introduced. The additional formulation of constitutive relations and properly assigned boundary and initial conditions provide the necessary problem closure so that the multiphase system modeling equations become theoretically solvable. The solution provides the sought-after insight into the dynamic transport behavior of the multiphase flows. However, the analytical solutions to closed-form models are rarely attainable. This conundrum can be due to the inherent nature of nonlinear coupled partial-differential equations, intricate interphase coupling, and complex domain geometries of flow systems. Therefore, solutions to closed governing equations of multiphase flows often necessitate numerical methods.

A multiphase flow modeling coupled with numerical simulation yields valued results on multiphase flows. For example, numerical simulation is capable of yielding useful information on a transient flow field, which is difficult to obtain from direct experimental measurements. Further, adverse conditions and accessibility challenges also often render empirical measurements difficult to perform. Modeling and simulation of the flow field often requires less time to retrieve the necessary information compared to experimentation. With the current computing power, numerical simulation has become the de facto standard in the study of multiphase flows. Also, due to the advances in computational methods, simulation has become one of most prominent approaches to describing multiphase flow behavior as evidenced by the availability of many commercial computational fluid dynamics (CFD) software packages used for academic and industrial research and development. However, it should be noted that the current knowledge of multiphase flows is still far from complete to allow the results of modeling and numerical simulation to be directly used with full confidence. Thus, experimental verification of the simulations is deemed necessary. Experimental measurement methods for obtaining data for model verification are given in Chapter 8.

This chapter introduces common numerical methods used to solve multiphase flow model equations. The information contained herein also provides general guidelines for developing and implementing new numerical techniques for solving model equations as well as for selecting models and solution techniques when using existing CFD software packages, such as Fluent, StarCCM+ and OpenFOAM. The typical workflow for a numerical simulation of multiphase flows is introduced in Section 7.2. In this section, an overview of the preprocessing, solution, and postprocessing steps is given. In Section 7.3, numerical algorithms for solving the partial differential equations for general transport problems and single phase flows are presented. They serve as a basis for solving more complex multiphase flow models. Section 7.4 covers numerical techniques for simulating the motion of particles, droplets, and bubbles. The simulation is capable of resolving the flow of the nonrigid surface of the discrete phase, and is suitable to account for the "microscopic" problems discussed in Chapters 3 and 4. Sections 7.5 and 7.6 are focused on methods for "macroscopic" flows, which describe the averaged flow behavior on a scale much larger than the individual particles. Section 7.5 illustrates the Eulerian–Lagrangian algorithm, which is the numerical adaptation of the continuum-discrete model discussed in Chapter 5. Section 7.6 presents the numerical version of the Eulerian–Eulerian model where all phases are treated as a domain-shared continuum, which corresponds to the model presented in Chapter 6. Finally, Section 7.7 discusses the Lattice Boltzmann Method (LBM) that is based on the solution of the discrete Boltzmann equation. It has unique advantages in modeling flows with both rigid and deformable particles.

7.2   General Procedure of Numerical Modeling and Simulation

Numerical simulation can be apportioned into three main stages: preprocessing by the preprocessor, solution by the solver, and postprocessing by the postprocessor. The corresponding computational tools are shown in Figure 7.2-1. The three required components used to be offered as independent software programs that needed to be purchased separately. Communication is facilitated via standardized kernel formats and by importing or exporting compatible data structures. However, it is becoming more common for commercially available CFD and Multiphysics platforms to bundle all three component engines into a single software suite.

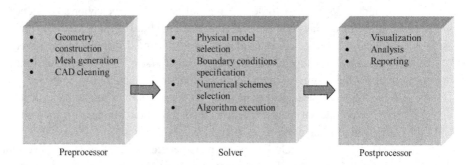

Figure 7.2-1 Workflow of numerical simulation of a flow problem.

At the *preprocessing* stage, a geometric model is constructed to represent the flow domain, and it is discretized into a computational mesh. The flow domain can be constructed in the geometric modeler or imported from another Computer Aided Design (CAD) software package. Generating the element mesh involves an intricate process that divides the flow geometry into a finite number of computational cells where information about the flow field is computed and stored. The computational domain may be discretized and resulted in a structured or unstructured mesh; the differences are illustrated in Figure 7.2-2. In a structured mesh, the position of a grid point, or node, is uniquely defined by a set of two or three indices, in the two-dimensional and three-dimensional domains, respectively. Thus, nodes can logically be mapped to a grid system in Cartesian coordinates. The cell shape in such mesh is typically quadrilateral or hexahedron for two- and three-dimensional domains, respectively. Using a structured mesh often is conducive to efficient solution algorithms, and its simple structure makes implementation of the simulation code straightforward. Conversely, an unstructured mesh contains a theoretically infinite variety of element shapes, and the relative position of the elements cannot be easily identified via standard indexing. This makes algorithms developed for an unstructured mesh more complicated. However, an unstructured mesh is sometimes advantageous for

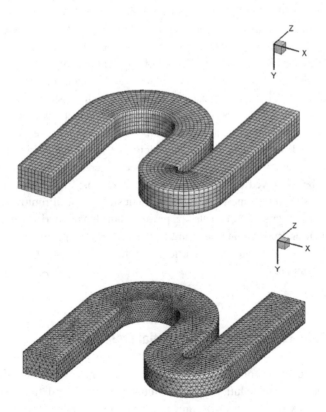

Figure 7.2-2 Example of a structured mesh (top) and an unstructured mesh (bottom).

complex domain boundaries, and it can be more flexible in general although sometimes structured meshes just cannot handle certain geometries without compromising the mesh quality and subsequently the simulation results. The choice between the two mesh paradigms often requires balancing efficiency with flexibility. Mesh generation may also include the CAD cleaning and geometry simplification. Because industrial applications of multiphase flows routinely involve complex flow domain geometries, preprocessing is often the most time-consuming and labor-intensive stage.

After preprocessing is complete, the second stage, known as *problem solving*, can begin. Here, once the computational domain is meshed, material properties, boundary conditions, and initial conditions must be determined and assigned. Next, solution techniques are specified, which include the numerical algorithms, schemes, and parameters. This is often where turbulence models are selected and solver features are enabled, such as the energy equation. Numerical parameters may also be specified or adjusted to ensure solution success. Once all aspects of the simulation are defined, the algorithms iteratively execute to obtain a numerical solution over the discretized volumes of the mesh. Some current commercially available CFD packages allow a user to define and solve a flow problem without the need for detailed deliberation of the solution algorithms or their implementation. However, general knowledge of multiphase flows and numerical methods will help the user to understand the principles of the code and inform their choices so that appropriate models and solution techniques are employed for their applications. Such knowledge will also help to detect and correct the situation where the CFD software produces a so-called solution that is unrealistic or devoid of physical meaning.

At the *postprocessing* stage, results are analyzed and visualized. A CFD postprocessing engine usually includes robust graphical features for visualizing the solved fields via an intuitive environment. Available graphical depictions of results typically include contour plots of scalar distributions, such as temperature, pressure, and concentrations; and, vector fields, such as velocity; isosurfaces capturing interfaces between phases; streamline plots; and trajectories of discrete particles. All of these quantities can be displayed for selected solution iterations, including steady-state convergence, or shown for time-dependent transient simulations. Primitive variables and various fluxes, or gradients, can also be presented in alphanumeric form for interrogation by the user. Result quantities can be interrogated at a specific point in space or time, averaged, or integrated on a particular surface or volume of interest in the computational domain.

7.3   Numerical Solutions of Partial Differential Equations

The governing equations of multiphase flows are commonly partial differential equations (PDEs) whose solutions are continuous in space and time. However, numerical solutions for this class of equations can only be computed for discrete locations of interest in space and time. To be solved efficiently, continuous PDEs are

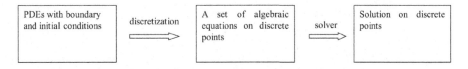

| PDEs with boundary and initial conditions | discretization $\Longrightarrow$ | A set of algebraic equations on discrete points | solver $\Longrightarrow$ | Solution on discrete points |

Figure 7.3-1 General procedure for obtaining numerical solutions to PDEs.

approximated by a system of discrete algebraic equations. The typical procedure for numerically solving a PDE is illustrated in Figure 7.3-1.

7.3.1   Numerical Solution of General Transport Equation

The general transport equation of a variable ϕ can be expressed by:

$$\frac{\partial}{\partial t}(\rho\phi) + \nabla \cdot (\rho\mathbf{U}\phi) = \nabla \cdot \left(\Gamma_\phi \nabla\phi\right) + S_\phi. \tag{7.3-1}$$

To solve this equation, the convection and diffusion terms need to be discretized in spatial coordinates; conversely, the transient terms require discretization in the temporal coordinate.

The most common spatial discretization techniques in CFD include the finite difference method, finite element method, and finite volume method. In the *finite difference* method, partial derivatives governing a grid point, or node, are approximated by the finite difference between adjacent grid points. This technique is most effective when using a structured mesh with simple domain geometry; accuracy and efficiency are hindered for complex geometries. In the *finite element* method, the governing equations are multiplied by a weight function and integrated over the domain. The approximate solution is arrived at by minimizing the weighted residues over successive solution iterations. The advantage of the finite element method is evidenced in its capacity to accommodate with complex computational domain geometries; however, application to coupled, nonlinear equations leads to reduced efficiency. The *finite volume* method is based on the integral form of the conservation equations that are applied to each cell volume of the mesh. It is suitable for complex domain geometries, and has a superior ability to ensure conservation of quantities, which is important to obtain accurate solutions of transport phenomena models. As such, the remaining topics in this chapter are based on the finite volume method.

Integrating Eq. (7.3-1) over a control volume and applying the Gauss theorem yields the general transport equation. This can be expressed in integral form as:

$$\frac{\partial}{\partial t}\iiint_V (\rho\phi)dV + \oiint_S (\rho\mathbf{U}\phi) \cdot ndS = \oiint_S \left(\Gamma_\phi \nabla\phi\right) \cdot \mathbf{n}dS + \iiint_V S_\phi dV, \tag{7.3-2}$$

where V and S represent the volume and surface area of control volume, respectively. Equation (7.3-2) is the foundation of the finite volume method.

Consider the example of a two-dimensional domain mapped onto a uniform Cartesian mesh, as shown in Figure 7.3-2. The variable ϕ is defined at the center of a control volume, which is a rectangular cell centered at point P and bounded by the

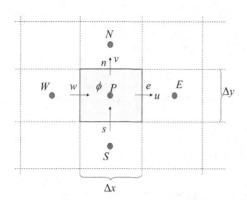

Figure 7.3-2 A uniform rectangular mesh in a 2D domain.

solid line. In Figure 7.3-2, the capital letters P, E, W, N, and S refer to the cell centers of the control volume and of its four neighboring control volumes; the lowercase letters e, w, n, and s refer to the cell boundaries of the control volume of interest. Equation (7.3-2) is applied to the control volume of Figure 7.3-2; the condition $\Delta V = \Delta x \Delta y$ is utilized; the discretized values at the cell center are employed; and fluxes across cell surfaces are considered. Thus, Eq. (7.3-2) becomes:

$$\frac{\partial}{\partial t}(\rho\phi)_P + \frac{1}{\Delta x}\left[(\rho U\phi)_e - (\rho U\phi)_w\right] + \frac{1}{\Delta y}\left[(\rho V\phi)_n - (\rho V\phi)_s\right]$$

$$= \frac{1}{\Delta x}\left[(\Gamma_\phi\nabla\phi)_e - (\Gamma_\phi\nabla\phi)_w\right] + \frac{1}{\Delta y}\left[(\Gamma_\phi\nabla\phi)_n - (\Gamma_\phi\nabla\phi)_s\right] + (S_\phi)_P$$

$$(7.3\text{-}3)$$

The convective fluxes on the left-hand side of the above equation require the evaluation of ϕ at the control volume surfaces. Since ϕ is defined at the cell center, its value at the cell boundaries must be determined prior to applying Eq. (7.3-3). Different techniques are used to specify the value of ϕ and result in unique discretization schemes. For example, if the *central differencing scheme* is used to specify ϕ, it gives:

$$\phi_w = \frac{\phi_W + \phi_P}{2}. \tag{7.3-4}$$

Thus, the cell face value, ϕ_w, simply computes the average of the two adjacent cell center values, ϕ_W and ϕ_P. Alternatively, if the *first-order upwind scheme* is employed, it yields:

$$\phi_w = \begin{cases} \phi_W & \text{if } U_w \geq 0 \\ \phi_P & \text{if } U_w < 0 \end{cases}. \tag{7.3-5}$$

In Eq. (7.3-5), the cell face value, ϕ_w, is a function of the flow direction. Other commonly used differencing schemes for the convective terms include the *power law scheme*; the *QUICK scheme*; and higher-order differencing schemes, such as the total variation diminishing (TVD) schemes and essentially nonoscillatory (ENO) schemes. Selecting a differencing scheme often necessitates consideration of inherent numerical properties, such as accuracy, degree of conservation, boundedness, and transportiveness (Versteeg and Malalasekera, 2007).

Different discretization schemes are also available for the diffusive fluxes on the right-hand side of Eq. (7.3-3), which are used to evaluate the gradient of ϕ at the cell boundaries. For example, the central differencing scheme is expressed by:

$$(\nabla\phi)_w = \frac{\phi_P - \phi_W}{\Delta x}. \tag{7.3-6}$$

Temporal discretization schemes are used to discretize the transient term in Eq. (7.3-3), and they specify how the subsequent solution iteration for the variable, ϕ^{n+1}, at a new time step, t^{n+1}, is obtained from the previous solution for the variable, ϕ^n, at the past time step, t^n. For example, in the *first-order Euler forward scheme*, the transient term becomes:

$$\frac{(\rho\phi)^{n+1} - (\rho\phi)^n}{\Delta t} = -[\nabla \cdot (\rho U\phi)]^n + [\nabla \cdot (\Gamma_\phi \nabla\phi)]^n + S_\phi^n. \tag{7.3-7}$$

Here, the convective and diffusive flux terms are written in shorthand form and any spatial discretization schemes described above are applicable. Since all the terms at the time step, t^n, are already solved, the value of the subsequent iteration for the variable, ϕ^{n+1}, is readily obtained from Eq. (7.3-7). The Euler forward scheme is thus referred to as an *explicit* scheme.

Conversely, when the right-hand side of Eq. (7.3-3) is represented by the values at the new time step t^{n+1}, its value is unknown. This result characterizes *implicit* schemes, such as the *first-order Euler backward* scheme, represented by:

$$\frac{(\rho\phi)^{n+1} - (\rho\phi)^n}{\Delta t} = -[\nabla \cdot (\rho U\phi)]^{n+1} + [\nabla \cdot (\Gamma_\phi \nabla\phi)]^{n+1} + S_\phi^{n+1}. \tag{7.3-8}$$

Obtaining ϕ^{n+1} from implicit schemes, such as by Eq. (7.3-8) above, typically requires matrix inversion. The need to invert matrices often renders implicit schemes more computationally intensive when compared to explicit schemes. The advantage of implicit schemes is their superior numerical stability, which enables larger time steps to be selected.

After spatial and temporal discretization, Eq. (7.3-1) is converted into a set of algebraic equations describing the transport of ϕ in each individual cell for the entire domain. This process can be expressed in the general form:

$$a_P \phi_P^{n+1} = \sum_{nb} a_{nb} \phi_{nb}^{n+1} + s, \tag{7.3-9}$$

where a is the coefficient and s is the source term, all determined by the discretization schemes that are employed for each term. For explicit temporal discretization schemes, all the terms on the right–hand side of the expression and at t^{n+1} are neglected so that ϕ_P^{n+1} can be evaluated directly. Conversely, implicit temporal discretization schemes, or steady-state problems, require the solution of Eq. (7.3-9) to undergo matrix inversion. Due to the large number of equations embodied in

Eq. (7.3-9), obtaining a solution can be an intensive and lengthy process. For nonlinear governing equations (*e.g.*, for the momentum equation where $\phi = U$), the coefficient matrix of Eq. (7.3-9) contains functions of the variable ϕ. In this case, iterative solutions are necessary. Specialized algorithms that take advantage of the large sparse matrices present are often employed to solve such systems. Some of these techniques include direct methods, such as the tridiagonal matrix algorithm (TDMA), and iterative methods, such as the line successive over-relaxation (LSOR) method and the conjugate gradient methods.

7.3.2   Numerical Methods for Single-Phase Flow

As discussed in Section 2.2.3, the solution for a nonreacting single-phase flow is given by the solutions for the variables ρ, U, p, and T. They are obtained from four independent governing equations, three of which share a similar form with the general transport equation Eq. (7.3-1) and describe the transport of mass, momentum, and energy. These equations can be written, respectively, as:

$$\frac{\partial \rho}{\partial t} + \nabla \cdot (\rho U) = 0 \tag{7.3-10}$$

$$\frac{\partial \rho U}{\partial t} + \nabla \cdot (\rho UU) = \nabla \cdot (-p\mathbf{I} + \tau) + \rho \mathbf{g} \tag{7.3-11}$$

$$\frac{\partial \rho e}{\partial t} + \nabla \cdot (\rho Ue) = -\nabla \cdot \mathbf{J}_q + J_e + \tau : \nabla U - p\nabla \cdot U. \tag{7.3-12}$$

Solving the above equations requires properly formulated constitutive equations for e, τ, $\mathbf{J}_q$, and J_e. In addition, a mechanism to compute the pressure, p, must be provided. One source for this is the equation of state that characterizes the thermodynamic properties of the fluid by expressing pressure as a function of density and temperature, as given in Eq. (7.3-13) below:

$$p = p(\rho, T) \tag{7.3-13}$$

In conjunction with Eq. (7.3-13), the transport equations, Eqs. (7.3-10)–(7.3-12) can be solved for ρ, p, U, and T via the numerical techniques described in Section 7.3.1. When the transport coefficients are dependent on temperature or flow status, such as in the case of turbulence, the equation solution procedure becomes very complicated because of inherently strong coupling, the nonlinear nature of PDEs, and the need for additional coupled transport equations (*e.g.*, the k–ε model or RSM).

For an *incompressible and isothermal* Newtonian fluid flow, the governing transport equations are simplified because they are decoupled from the energy equation, Eq. (7.3-12) and the equation of state (7.3-13), in addition to the fact that density remains constant. Equation (7.3-10) is therefore reduced to:

$$\nabla \cdot U = 0 \tag{7.3-14}$$

The above is used in conjunction with Eq. (7.3-11) to obtain p and U. The most common technique used to solve the pressure-velocity coupling indicated above is known as the semi-implicit pressure linked equation (SIMPLE) algorithm (Patankar and Spalding, 1972; Patankar, 1980), which is based on the iterative solution for pressure-correction. The SIMPLE algorithm relies on a trial pressure, p^*, in the momentum equation to obtain a provisional velocity field, $\mathbf{U}^*$. Because p^* may not represent the actual pressure field, $\mathbf{U}^*$ does not necessarily satisfy the incompressibility condition. Thus, a pressure correction, p', is calculated from the incompressible constraint that is used to find corrections to the velocity field. The corrected velocity and pressure field are then applied as initial conditions for the next solution iteration. The process continues successively as the flow field is expected to approach the final solution, and the algorithm continues until the velocity field satisfies the incompressibility condition to a specified accuracy.

The SIMPLE algorithm is often carried out on a staggered mesh with the pressure variable situated at the cell center, while the individual velocity components are positioned on corresponding cell faces, as shown in Figure 7.3-3. The staggered mesh is a preferable grid arrangement for many incompressible flow solvers because it avoids a numerical artifact, a result referred to the checkerboard effect. An alternative arrangement where all variables are located at the same position, often the cell center, is known as a collocated mesh (Rhie and Chow, 1983). The collocation approach is more conveniently implemented for unstructured mesh of complex

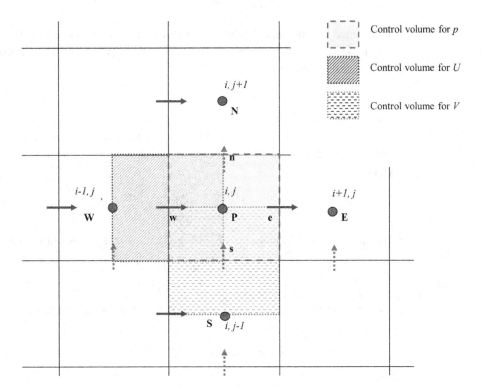

Figure 7.3-3 Definition of pressure and velocity components on a staggered mesh.

computational domain geometries. However, intricate numerical manipulation and expansive computer random access memory are required to avoid the checkerboard effect (Prakash and Patankar, 1985).

To illustrate the SIMPLE algorithm, it is again considered a two-dimensional steady flow problem. The discretized momentum equation can be written as a set of algebraic equations, represented by:

$$a_w U_w = \sum_{nb} a_{nb} U_{nb} + (\rho g_x)_w - (p_P - p_W) A_w$$

$$a_s V_s = \sum_{nb} a_{nb} V_{nb} + (\rho g_y)_s - (p_P - p_S) A_s.$$

(7.3-15)

Summations in the above equations include velocity components on adjacent velocity nodes. The coefficients a_w, a_s, and a_{nb} depend on the specific spatial discretization schemes as discussed in the previous section, and they are functions of the velocity field.

With an initial estimate of pressure p^*, eqs. (7.3-15) yield intermediate velocity components U^* and V^*. This is written as:

$$a_w U_w^* = \sum_{nb} a_{nb} U_{nb}^* + (\rho g_x)_w - (p_P^* - p_W^*) A_w$$

$$a_s V_s^* = \sum_{nb} a_{nb} V_{nb}^* + (\rho g_y)_s - (p_P^* - p_S^*) A_s.$$

(7.3-16)

It is assumed that the actual solutions of p, U, and V can be obtained by adding corrections p', U', and V', respectively, to the estimated values. Thus, it gives the following expressions:

$$p = p^* + p' \qquad U = U^* + U' \qquad V = V^* + V'.$$

(7.3-17)

Thus, the governing equations for the correction terms can be gleaned from the difference between Eqs. (7.3-15) and (7.3-16). The result is further simplified by neglecting the summation with respect to subscript nb on the right-hand side. Thus, it yields:

$$a_w U_w' = - (p_P' - p_W') A_w$$

$$a_s V_s' = - (p_P' - p_S') A_s.$$

(7.3-18)

Substituting Eq. (7.3-18) into Eq. (7.3-17) gives rise to the calculated velocity, as:

$$U_w = U_w^* - (p_P' - p_W') A_w/a_w$$

$$V_s = V_s^* - (p_P' - p_S') A_s/a_s.$$

(7.3-19)

Substituting U from Eq. (7.3-19) into Eq. (7.3-14) at the cell center leads to the pressure-correction equation of:

$$a_P p_P' = a_e p_E' + a_w p_W' + a_n p_N' + a_s p_S' + s,$$

(7.3-20)

where

$$a_P = a_e + a_w + a_n + a_s$$

(7.3-20a)

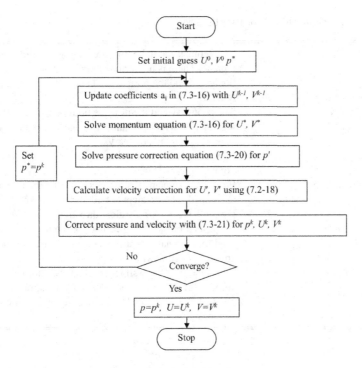

Figure 7.3-4 Flowchart of SIMPLE algorithm.

$$s = -\left[\left(\rho U^*\right)_e - \left(\rho U^*\right)_w\right]\Delta y - \left[\left(\rho V^*\right)_n - \left(\rho V^*\right)_s\right]\Delta x. \qquad (7.3\text{-}20\text{b})$$

If $\mathbf{U}^*$ is the true velocity, the term s should vanish. Therefore, the magnitude of s can serve as a convergence criterion, and the iterative calculation should lead to a diminishing value of this term as the variables p, U, and V approach their actual values.

In practice, direct application of Eq. (7.3-17) often results in divergence. To avoid this result, underrelaxation factors ω are introduced into the iteration, as given by:

$$p^k = p^* + \omega_p p'$$
$$U^k = \omega_U U + (1 - \omega_U)U^{k-1} \qquad (7.3\text{-}21)$$
$$V^k = \omega_V V + (1 - \omega_V)V^{k-1},$$

where U^{k-1} and V^{k-1} are the velocities computed in the previous iteration; and U and V are the corrected velocity components according to (7.3-19). ω_p, ω_U, and ω_V are the under-relaxation coefficients for p, U, and V, respectively. The under-relaxation coefficients are chosen between 0 and 1 in an effort to promote iteration stability.

A flowchart of the SIMPLE algorithm, as applied to this two-dimensional example, is summarized in Figure 7.3-4.

In addition to the SIMPLE algorithm, similar approaches for pressure-velocity decoupling exist, including SIMPLER, SIMPLEC, and PISO algorithms

Table 7.3-1 Typical Types of Boundary Conditions.

Symmetric condition	The geometry and the flow are symmetric about an axis or a plane. There is no normal velocity or flux across the boundary, and the normal gradient to the boundary equates to zero.
Periodic condition	The geometry and the flow are both periodic, so that fluxes leaving through one boundary reenter the domain from the opposite periodic boundary.
Inlet condition	The velocity and fluxes are specified for the flow into the domain.
Outflow condition	Placed far from any disturbances, so that the flow is fully developed. The gradients of all variables, except the pressure, are zero in the flow direction.
Constant pressure condition	A constant pressure is specified for the flow. Often this is used for external flow around obstacles, free surface flows, and internal flow with multiple outlets.
Wall condition	The velocity at the wall is specified. The normal velocity component is zero and no orthogonal penetration is allowed. The tangential velocity is set based on no-slip, slip, or partial slip conditions.

(Versteeg and Malalasekera, 2007). For unsteady problems, the SIMPLE category of algorithms require iterations within each time step, while PISO employs a non-iterative transient computation procedure. Another viable algorithm for unsteady problems is the projection method (Chorin, 1968).

7.3.3   Boundary Conditions

In order to solve the governing PDEs of a flow model, boundary conditions must be supplied. Formulation of conditions relies on the overall physical nature of the flow, such as compressibility and turbulence, and the flow characteristics at the boundaries of the domain. Typical types of boundary conditions are included in Table 7.3-1.

A flow problem can only be well defined if the boundary conditions are specified in a complete and consistent manner. More details on selecting and assigning boundary conditions can be found in multiple CFD texts, in the general literature, and the references (Versteeg and Malalasekera, 2007).

7.4   Resolved Interface Approach for Dispersed Phase Objects

The resolved interface approaches are most suitable for modeling a small number of dispersed phase objects as particles, bubbles, and droplets in multiphase flows, and can be applied to address the problems described in Chapters 3 and 4. The boundary for these objects is resolved during the simulation, which requires the

Table 7.4-1 Resolved Interface Approaches for Dispersed Phase Objects.

Method	Mesh	Application	Advantage	Disadvantage
Boundary-fitted mesh	conformal	Particles, bubbles	Very accurate surface representation	Complicated mesh generation for multiple objects
Arbitrary Lagrangian Eulerian	conformal	Particles	Can handle multiple objects	Expensive mesh regeneration
Immersed boundary	nonconformal	Particles	Relatively simple mesh generation	Less accurate than conformal mesh methods
Front-tracking	nonconformal	Droplets, bubbles	Accurate surface motion	Difficult to handle breakup/coalescence
Volume of fluid	nonconformal	Droplets, bubbles	Volume conservation guaranteed	Difficult to reconstruct interface from volume fraction
Level-set	nonconformal	Droplets, bubbles	Easy representation of interface	Volume conservation not guaranteed
Diffuse interface	nonconformal	Droplets, bubbles	Especially suited for near-critical fluid or surface wetting	Relatively thick interface

mesh size to be much smaller than the phase object's characteristic dimension. The flow field surrounding the objects is directly solved from the flow equations; the interfacial transport of mass, momentum, and energy is directly accounted for with proper interface conditions, without any model closure relations. These methods are consequently referred to as direct numerical simulations (DNS) of multiphase flow.

Motion and deformation of the dispersed phase objects often elicits the moving boundary problem, which can be solved using a conformal or nonconformal mesh. In conformal mesh approaches, the computation domain boundary coincides with the object surface, and no mesh is generated for the volume within the object. In nonconformal mesh strategies, the object surface is not aligned with the mesh, and the volumes on both sides of the surface are discretized. The commonly employed numerical methods using both conformal and nonconformal mesh are listed in Table 7.4-1.

7.4.1 Conformal Mesh Methods

As mentioned in Table 7.4-1, the most popular conformal mesh methods include boundary-fitted mesh technique and arbitrary Lagrangian–Eulerian (ALE) technique.

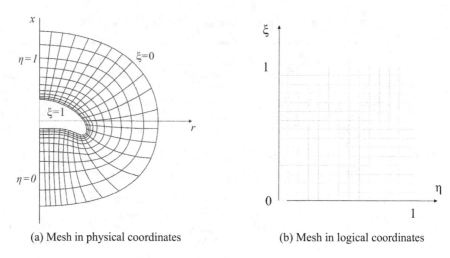

(a) Mesh in physical coordinates   (b) Mesh in logical coordinates

Figure 7.4-1 Computational mesh for a bubble rising in an axis-symmetric domain.

7.4.1.1   Boundary-Fitted Mesh Technique

The boundary-fitted mesh technique solves the flow equations using a curvilinear coordinate system that follows the shape of the object surface. Only the continuous phase domain is meshed and solved; no mesh is generated within the objects. Therefore, this technique is useful for flow around solid particles or gaseous bubbles where only the external flow is considered important. The advantage of this method is straightforward control of mesh size and quality near the object surface, enhancing the accuracy of boundary treatment. The drawback of this technique is a complex mesh generation scheme, especially for unsteady three-dimensional problems that involve moving interfaces or multiple particles.

Figure 7.4-1 illustrates an example of a mesh for an axis-symmetric flow around a bubble. Mapping is necessary to establish the transformation between the curvilinear coordinates (ξ, η) and the cylindrical (x, r) coordinates of the physical domain. The mapping is found by solving for $x(\xi, \eta)$ and $r(\xi, \eta)$ from the following Laplace equation:

$$\frac{\partial}{\partial \xi}\left(f\frac{\partial x}{\partial \xi}\right) + \frac{\partial}{\partial \eta}\left(\frac{1}{f}\frac{\partial x}{\partial \eta}\right) = 0$$

$$\frac{\partial}{\partial \xi}\left(f\frac{\partial r}{\partial \xi}\right) + \frac{\partial}{\partial \eta}\left(\frac{1}{f}\frac{\partial r}{\partial \eta}\right) = 0$$

(7.4-1)

where $f(\xi, \eta)$ is the distortion function that can be freely specified to control mesh density. The solution to Eqs. (7.4-1) requires proper boundary conditions. In the example shown in Figure 7.4-1, the boundaries $\eta = 0$ and $\eta = 1$ correspond to the symmetrical axis; $\xi = 0$ corresponds to the far field; $\xi = 1$ represents the bubble surface.

The governing flow equations are solved in the (ξ, η) coordinates using either the stream-function/vorticity formulation (Ryskin and Leal, 1984) or the primitive velocity and pressure variables from the Navier–Stokes equation (Magnaudet *et al.*, 1995). The governing equations for general curvilinear coordinates have an intricate form when compared to Cartesian coordinates. For deformable surfaces, an iterative process is necessary to ascertain the surface shape. For flow around a bubble, the normal stress balance is verified on the bubble surface, and the unbalanced stress is used as a criterion to adjust the shape of the interface. A new boundary-fitted mesh is then generated, and the new flow field is solved. This iteration process continues until the normal stress is balanced on the bubble surface, indicating the converged bubble shape and a valid flow field description.

7.4.1.2 Arbitrary Lagrangian–Eulerian (ALE) Technique

The arbitrary Lagrangian–Eulerian (ALE) technique permits the mesh nodes within the continuous phase domain to move arbitrarily during the simulation in an effort to optimize the shape of the mesh elements as the discrete particles move within the suspending medium. The ALE technique is distinguished from the Lagrangian formulation, in which mesh is affixed to the discrete phase, and from the Eulerian formulation where mesh is static in space. Mesh motion on the particle surface is defined by the particle's linear and rotational velocity, and minor adjustments are made to the mesh nodes within the continuous phase domain. The quality of the adjusted mesh is then checked. If unacceptable distortion is detected, the entire mesh is regenerated to ensure suitable shapes of the mesh elements. Interpolation is required to project the flow field that was computed for the prior mesh onto the new mesh.

Figure 7.4-2 shows an example of the unstructured mesh used in the ALE approach for a particle suspension. The effects on the flow field due to the presence of solid particles are treated by enforcing the no-slip condition on the particle

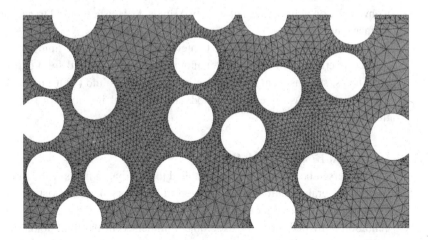

Figure 7.4-2 ALE method with unstructured mesh for a particle suspension flow.

surface. The hydrodynamic force exerted on the particle surface is used to calculate the particle's motion. Numerical instability can result from the method's technique of direct integration of the fluid–particle interaction force due to the fluid stress acting on the particle boundary and explicitly applying the result to the equation for particle motion. As a remedy, a weak formulation can be derived via the finite element method to incorporate the coupled fluid and particle motions. The fluid–solid interacting force does not explicitly appear in the equation; therefore, instability is circumvented (Hu *et al.*, 2001).

7.4.2 Nonconformal Mesh Methods

In nonconformal mesh methods, the computation domain boundary does not overlap with the discrete phase boundary. The interface between different phases is immersed within a fixed mesh that does not move or deform with interfacial motion. The interface is represented explicitly by Lagrangian markers or implicitly by contours of a scalar field. The flow fields on both sides of the interface are solved from a unified set of governing equation; proper boundary conditions on the interface are enforced by additional source terms in the conservation of mass, momentum, and energy equations, which represent the transport of such quantities across the interface.

In general, for an incompressible two-phase flow, the continuity and momentum equations for both phases can be written as:

$$\nabla \cdot \mathbf{U} = 0$$

$$\rho(\frac{\partial \mathbf{U}}{\partial t} + \mathbf{U} \cdot \nabla \mathbf{U}) = -\nabla p + \nabla \cdot (2\mu \mathbf{S}) + \rho \mathbf{g} + \mathbf{f}_\sigma, \qquad (7.4\text{-}2)$$

where $\mathbf{S}$ is the strain rate tensor defined by Eq. (2.2-3); and $\mathbf{f}_\sigma$ is the interfacial force density used to enforce accurate interfacial boundary conditions. For example, on a particle surface, $\mathbf{f}_\sigma$ is generally specified such that no-slip condition is realized. Conversely, on an interface between two fluids, $\mathbf{f}_\sigma$ is usually used to account for the surface tension effect.

Among the different immersed interface techniques, the immersed boundary method is employed for flows containing deformable or rigid solid particles; other techniques, such as front tracking method, volume of fluid (VOF) method, level-set method, and diffuse interface methods are leveraged for flows of two immiscible fluids.

7.4.2.1 Immersed Boundary Method

The immersed boundary method is applied to multiphase flows with immersed solid objects. Several variations of the immersed boundary method exist whose applicability depends on whether the solid object is elastic or rigid and on how the interface condition is treated. For an elastic object immersed in the fluid domain, Ω, shown in Figure 7.4-3, the boundary is represented by a collection of massless markers described in Lagrangian coordinates, $\mathbf{R}(s, t)$, along the surface, Γ.

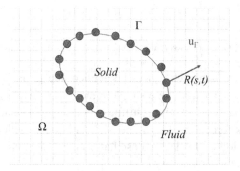

Figure 7.4-3 Illustration of a mesh setup in immersed boundary method.

The massless markers are connected by mathematically elastic fibers that obey the constitutive equation of the solid material. The motion and deformation of the object is represented by the motion of the Lagrangian markers, which have the same velocity as the local flow field. Thus, the position of a marker can be found by integrating its velocity function, $\mathbf{u}(\mathbf{R}(s,t),t)$, that is extracted from the local fluid velocity, $\mathbf{U}(\mathbf{r},t)$. This process yields:

$$\frac{\partial \mathbf{R}(s,t)}{\partial t} = \mathbf{u}(\mathbf{R}(s,t),t) = \int_{\Omega} \mathbf{U}(\mathbf{r},t)\delta(\mathbf{r} - \mathbf{R}(s,t))d\mathbf{r}. \qquad (7.4\text{-}3)$$

In the numerical implementation, the Dirac delta function, δ, is usually smoothed via a small set of Eulerian grid nodes around the Lagrangian marker. The Lagrangian markers $\mathbf{R}(s,t)$ do not typically overlap with the Eulerian nodes $\mathbf{r}$ (Mittal and Iaccarino, 2005).

At the elastic solid and fluid interface, the boundary condition demands the solid stresses due to the elastic modulus and bending rigidity be appropriately transmitted to the fluid. The elastic force density, $\mathbf{F}(s,t)$, is calculated from the variation of the elastic tension σ_E along a segment of the immersed boundary. This can be expressed as:

$$\mathbf{F}(s,t) = \frac{\partial(\sigma_E \mathbf{t})}{\partial s} = \frac{\partial \sigma_E}{\partial s}\mathbf{t} + \sigma_E \left|\frac{\partial \mathbf{R}}{\partial s}\right| \kappa \mathbf{n}. \qquad (7.4\text{-}4)$$

The elastic tension, σ_E, along an elastic fiber that connects two adjacent Lagrangian markers is calculated from Hooke's law, as:

$$\sigma_E = k\left\{\left|\frac{\partial \mathbf{R}}{\partial s}\right| - 1\right\}, \qquad (7.4\text{-}4a)$$

where k is the stiffness coefficient. The tangential unit vector $\mathbf{t}$, normal unit vector $\mathbf{n}$, and the curvature κ of the boundary in Eq. (7.4-4) are given by:

$$\mathbf{t} = \frac{\partial \mathbf{R}}{\partial s} \bigg/ \left|\frac{\partial \mathbf{R}}{\partial s}\right| \qquad \mathbf{n} = \frac{\partial \mathbf{t}}{\partial s} \bigg/ \left|\frac{\partial \mathbf{t}}{\partial s}\right| \qquad \kappa = \frac{\partial \mathbf{t}}{\partial s} \bigg/ \left|\frac{\partial \mathbf{R}}{\partial s}\right|. \qquad (7.4\text{-}4b)$$

The force density exerted by the fluid on the immersed boundary must be balanced by the solid stress within the elastic body. Therefore, the interfacial force density, $\mathbf{f}_\sigma$, in Eq. (7.4-2) is expressed by:

$$\mathbf{f}_\sigma(\mathbf{r}, t) = \int_\Gamma \mathbf{F}(s, t)\delta(\mathbf{r} - \mathbf{R}(s, t))ds. \qquad (7.4-5)$$

For a rigid body, the above approach cannot be applied due to an absent constitutive equation for the solid material. One simple remedy is to approximate the rigid surface as an elastic boundary with high stiffness. Another viable approach is to apply a fictitious feedback force that restores the boundary nodes to their equilibrium positions. From an analog to the damped oscillator model, the restoring force can be represented by (Goldstein *et al.*, 1993):

$$\mathbf{F}(t) = \alpha \int_0^t \mathbf{u}(\tau)d\tau + \beta\mathbf{u}(t). \qquad (7.4-6)$$

In the above expression, α and β are empirical coefficients used to determine the required magnitude of the restoring and damping forces, respectively. This approach has demonstrated encouraging results for low-Reynolds-number flows. However, enforcing the rigid boundary condition requires large values of α and β, which can result in numerical instability for highly unsteady flows.

Another class of immersed boundary techniques for rigid objects is known as the direct forcing approach. Instead of using a restoring force, direct corrections to the velocity field at the immersed boundary are utilized to impose the desired velocity condition (Mohd-Yusof, 1997). While this approach circumvents the need for the user-specified parameters of α and β of the feedback forcing approach discussed above, it still uses the delta function in Eq. (7.4-5) to distribute the interfacial force near the boundary. Spreading the force into the fluid region may affect the accuracy of the method in the boundary layer for high-Reynolds-number flows, and its effect is dependent upon the details of the discretization method.

A third class of immersed boundary method also uses the direct forcing approach, but maintains a sharp interface at the solid surface by using interpolation techniques such as "ghost cells" in finite differencing models or "cut-cell" techniques for finite volume methods (Fadlun *et al.*, 2000; Udaykumar *et al.*, 2001). This results in improved local accuracy near the immersed boundary, but is more complex and usually requires a special discretization scheme at the immersed boundary.

7.4.2.2   Front Tracking Method

The front-tracking method (Unverdi and Tryggvason, 1992; Tryggvason *et al.*, 2001) is a numerical technique used to solve for the flow of two immiscible fluid phases separated by a moving interface. It is commonly used to investigate the dynamics of bubbles and droplets. Similar to the immersed boundary method, it employs two separate meshes for two- and three-dimensional simulations, as shown in Figure 7.4-4. A fixed Eulerian mesh is used to solve the flow equations; a separate surface mesh, represented by a set of interconnected markers, is used to track the motion of the discrete phase interface.

Throughout the simulation, each surface marker is convected individually by the local flow field, which is interpolated from the velocity as defined on the Eulerian

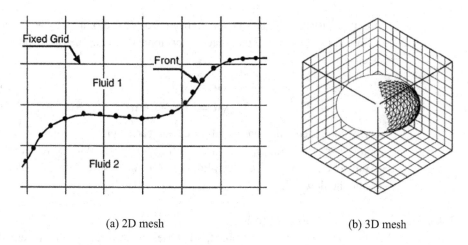

(a) 2D mesh   (b) 3D mesh

Figure 7.4-4 Meshes in the front-tracking method (Tryggvason *et al.*, 2001).

Figure 7.4-5 Reconstruction for surface mesh elements in 3D front-tracking method.

mesh. If the surface undergoes severe deformation, the distribution of the markers on the surface may become unfavorable. Dynamic reconstruction of the interface is often necessary during the simulation to avoid mesh degradation. Figure 7.4-5 illustrates the three basic operations in surface reconstruction for three-dimensional computations.

In general, the size of the surface elements should be similar to the background Eulerian mesh. Thus, an excessively large surface element can be further divided by adding new marker points to improve resolution. Markers are removed from disproportionately small elements to avoid small "wiggles." Reconnection of markers is performed to avoid elements with large aspect ratio.

With the interface location identified, density ρ and viscosity μ can be determined for each node on the Eulerian mesh. Although the interface is represented by distinct markers, distributions of density and viscosity are mathematically smoothed across the interface within a narrow band on the order of several grid widths to facilitate computation. The interfacial force density is calculated from the surface tension, which acts in the direction normal to the interface. The force density is given by:

$$\mathbf{f}_\sigma(\mathbf{r}, t) = \sigma\kappa(\mathbf{r})\mathbf{n}(\mathbf{r})\delta(\mathbf{r}).$$

(7.4-7)

In practice, the interfacial force is first calculated on each surface mesh element, Δs, by:

$$\delta\mathbf{f}_\sigma = \int_{\Delta s} \sigma\kappa\mathbf{n}\,ds,$$

(7.4-8)

where the curvature κ is calculated from Eq. (7.4-8a) in the two-dimensional domain and Eq. (7.4-8b) in the three-dimensional domain. The two expressions are, respectively:

$$\kappa \mathbf{n} = \partial \mathbf{s}/\partial s, \tag{7.4-8a}$$

$$\kappa \mathbf{n} = (\mathbf{n} \times \nabla) \times \mathbf{n}. \tag{7.4-8b}$$

Once computed, the interfacial force on each surface element is distributed smoothly to the Eulerian mesh via weighting functions. With ρ, μ, and $\mathbf{f}_\sigma$ identified, the governing equations (e.g., Eq. (7.4-2)) can be solved on the Eulerian mesh to obtain the flow field.

7.4.2.3   Volume of Fluid Method

The distribution of two immiscible fluid phases in the volume of fluid method (VOF) is predicated on the volume fraction of a particular phase within each computational cell or control volume. Specifically, the volume fraction function in a cell, $\alpha(\mathbf{r}, t)$, centered at spatial coordinate $\mathbf{r}$ is defined as:

$$\begin{cases} \alpha(\mathbf{r}, t) = 1 & \text{if the cell is fully filled with fluid 1} \\ 0 < \alpha(\mathbf{r}, t) < 1 & \text{if the cell is partially filled with fluid 1} . \\ \alpha(\mathbf{r}, t) = 0 & \text{if the cell is empty of fluid 1} \end{cases} \tag{7.4-9}$$

From this definition, the interface must be located in the cells where the volume fraction function is between 0 and 1.

The interface is assumed to be passively transported by the flow field, and the evolution of the volume fraction function is therefore described by the scalar convection equation, as:

$$\frac{\partial \alpha}{\partial t} + \mathbf{U} \cdot \nabla \alpha = 0, \tag{7.4-10}$$

where $\mathbf{U}$ is the velocity field, obtained from the momentum equation.

In order to solve Eq. (7.4-10), the mass flux across the cell boundaries must be evaluated, requiring the precise interface location to be known. The volume fraction, α, is insufficient to signal the interface position. Thus, reconstruction of the interface from the volume fraction, $\alpha(\mathbf{r}, t)$, is required in VOF methods. Earlier incarnations of VOF employed the donor-acceptor scheme (Hirt and Nichols, 1981), which uses a horizontal or vertical line to approximate the interface inside an interface cell, as shown in Figure 7.4-6(b). A preferred approach is the geometrical reconstruction scheme that approximates the interface with piecewise linear elements in each cell (Youngs, 1982), as shown in Figure 7.4-6(c). Here, the gradient of the volume fraction function is first used to compute the orientation of the interface, and the value of α is then used to determine the exact location of interface inside the cell.

The fluid properties in Eq. (7.4-2) at any location are calculated as the average properties of the individual phases, weighted by their local volume fraction. For example, the density of the fluid can be expressed by the relationship:

$$\rho(\alpha) = \alpha \rho_1 + (1 - \alpha)\rho_2. \tag{7.4-11}$$

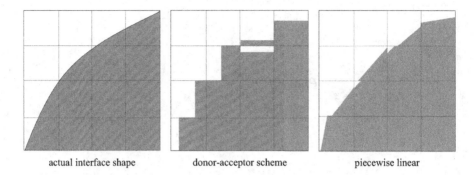

actual interface shape   donor-acceptor scheme   piecewise linear

Figure 7.4-6 Geometric reconstruction of interface location in VOF method.

Such an expression reduces to the intrinsic phase density within the bulk region, while simultaneously providing a smooth transition at the interface.

The interfacial force density $\mathbf{f}_\sigma$ due to the surface tension is calculated via Eq. (7.4-7). Using the continuum surface force (CSF) model (Brackbill *et al.*, 1992), $\mathbf{f}_\sigma$ can be further expressed by the gradient of the volume fraction function, as:

$$\mathbf{f}_\sigma = \sigma \kappa \delta \mathbf{n} = \sigma \kappa \frac{2\rho \nabla \alpha}{\rho_1 + \rho_2}, \tag{7.4-12}$$

where the curvature of the interface can be obtained from the expression:

$$\kappa = -\nabla \cdot \left(\frac{\nabla \alpha}{|\nabla \alpha|} \right). \tag{7.4-12a}$$

7.4.2.4 Level-Set Method

The level-set method (Chang *et al.*, 1996; Sussman *et al.*, 1998; Desjardins *et al.*, 2008) implicitly expresses the interface between two immiscible fluid phases via the contour of a scalar function ϕ that is defined throughout the domain. In practice, it is preferred to set the level-set function $\phi(\mathbf{r}, t)$ as the assigned distance from point $\mathbf{r}$ to the interface. Thus, the value of ϕ has a different sign on different sides of the interface, and the interface is represented by the contour where $\phi = 0$, as shown in Figure 7.4-7. It then yields the following:

$$\begin{cases} \phi(\mathbf{r}, t) < 0 & \text{for } \mathbf{r} \text{ in phase 1} \\ \phi(\mathbf{r}, t) > 0 & \text{for } \mathbf{r} \text{ in phase 2} \\ \phi(\mathbf{r}, t) = 0 & \text{for } \mathbf{r} \text{ at interface} \end{cases} \tag{7.4-13}$$

Interface evolution is controlled by the local interface velocity, which transports the interface according to the convection equation. This can be expressed using the relationship:

$$\frac{\partial \phi}{\partial t} + \mathbf{u}_\Gamma \cdot \nabla \phi = 0, \tag{7.4-14}$$

where $\mathbf{u}_\Gamma$ is the interface velocity interpolated from the velocity field $\mathbf{U}$.

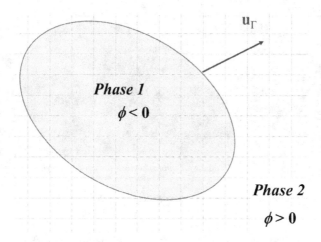

Figure 7.4-7 Schematic illustration of the level-set method.

Using Eq. (7.4-14) to update the position of the interface does not guarantee that ϕ is kept as the signed distance to the interface. In order to guarantee this condition, the level-set function needs to be re-initialized as the computation proceeds. A simple way to accomplish this is to successively solve the following partial differential equation until a steady state is obtained:

$$\frac{\partial d}{\partial \tau} = \text{sign}(\phi)(1 - |\nabla d|).$$ (7.4-15)

Here, τ is the pseudo time step; the initial condition is given by:

$$d(\mathbf{r}, 0) = \phi(\mathbf{r}).$$ (7.4-15a)

The sign function is expressed by:

$$\text{sign}(\phi) = \begin{cases} -1 & \text{if } \phi < 0 \\ 0 & \text{if } \phi = 0 \\ 1 & \text{if } \phi > 0 \end{cases}.$$ (7.4-15b)

Because the value of the sign function is zero at the interface, $d(\mathbf{r}, \tau)$ must have the same zero-level set as ϕ and the location of the interface is therefore unchanged during re-initialization. When the steady-state condition is reached, the function $d(\mathbf{r}, \tau)$ will have the exact property of the distance function, or $|\nabla d| = 1$. Therefore, its steady-state value is assigned to ϕ, effectively maintaining the distance function.

With the level-set function defined, material properties, such as density and viscosity, can be assigned to each region for the intrinsic quantities of the corresponding phase. However, a sharp change of the properties across the interface can strain numerical computation. To alleviate this possibility, a small finite thickness is generally assumed for the transitional region across the interface. With an interface thickness of 2λ, the density and viscosity at any location within the domain are expressed by:

$$\rho(\phi) = \rho_1 + (\rho_2 - \rho_1)H_\lambda(\phi)$$
$$\mu(\phi) = \mu_1 + (\mu_2 - \mu_1)H_\lambda(\phi),$$

(7.4-16)

where H_λ is the smoothed Heaviside function, expressed as:

$$H_\lambda(\phi) = \begin{cases} 0 & \text{if } \phi \leq -\lambda \\ \dfrac{\phi + \lambda}{2\lambda} + \dfrac{\sin(\pi\phi/\lambda)}{2\pi} & \text{if } |\phi| \leq \lambda. \\ 1 & \text{if } \phi \geq \lambda \end{cases}$$

(7.4-16a)

Similar to the VOF method, the interfacial force density, $\mathbf{f}_\sigma$, can be expressed by the continuum surface force (CSF) model, as:

$$\mathbf{f}_\sigma = \sigma\kappa\delta\mathbf{n}.$$

(7.4-17)

In the level-set approach, the normal vector, $\mathbf{n}(\phi)$, and the local interface curvature, $\kappa(\phi)$, are conveniently obtained from the level-set function via the following:

$$\mathbf{n}(\phi) = \frac{\nabla\phi}{|\nabla\phi|} \qquad \kappa(\phi) = \nabla \cdot \left(\frac{\nabla\phi}{|\nabla\phi|}\right).$$

(7.4-17a)

The smoothed delta function $\delta(\phi)$, which ensures that the surface tension only appears at the interface, is calculated by differentiating the smoothed Heaviside function given in Eq. (7.4-16a). Thus, it gives:

$$\delta(\phi) = \frac{dH_\lambda}{d\phi}.$$

(7.4-17b)

7.4.2.5   Diffuse Interface Method

In stark contrast to the VOF and level-set methods that adopt the concept of a physically sharp interface with infinitesimal thickness, the diffuse interface method registers the finite thickness of the interfacial region. Physical properties, such as density and viscosity, change rapidly albeit smoothly from one phase to another across the interface. This renders the model applicable to physical phenomena with length scales commensurate with the thickness of the interface, such as the near-critical fluids and moving contact lines. The diffuse interface method has been developed for single-component dual-phase flows as well as flows of binary fluids (Anderson et al., 1998). The binary fluid extension is spotlighted in the following discussion (Jacqmin, 1999).

In the binary fluid model, the two fluid phases are identified by a composition variable $C(\mathbf{r}, t)$, known as the phase field function or the order parameter, which represents a mass fraction of the binary fluid mixture. Based on the phase field function, the free energy density of a fluid mixture, f, can be defined as:

$$f = \beta\Psi(C) + \frac{1}{2}\alpha|\nabla C|^2.$$

(7.4-18)

The first term on the right-hand side of the free energy density expression defines the bulk energy density, with the coefficient β, which determines the immiscibility between the two fluid components; the second term captures the gradient energy at the interface between the two phases, with the gradient energy coefficient α. One simple model of the bulk free energy density is to use the double well potential in a form as:

$$\Psi(C) = (C + 1/2)^2 (C - 1/2)^2. \tag{7.4-18a}$$

The equilibrium condition requires minimization of the free energy, and the values $C = 1/2$ and $C = -1/2$ correspond to the quantity of the phase field function in the two stable bulk phases.

The phase field function, C, is transported by the modified Cahn-Hilliard equation, originally developed to model spinodal decomposition (Cahn, 1961; Hilliard, 1970) and now modified with an advection, which is given as:

$$\frac{\partial C}{\partial t} + \mathbf{U} \cdot \nabla C = k \nabla^2 \Phi. \tag{7.4-19}$$

The diffusion term above is given by the mobility k and Laplacian of the chemical potential Φ. The Laplacian of the chemical potential is given by:

$$\Phi = \beta \frac{d\Psi}{dC} - \alpha \nabla^2 C. \tag{7.4-19a}$$

At equilibrium, the chemical potential satisfies the requirement that $\Phi = $ constant. Otherwise, its Laplacian in Eq. (7.4-19) tends to drive the segregation of the two phases and maintain a thin but finite interface.

The interfacial force density in Eq. (7.4-2) is expressed by:

$$\mathbf{f}_\sigma = -C \nabla \Phi. \tag{7.4-20}$$

With the above force density, the momentum equation of Eq. (7.4-2) can also be written in the following form:

$$\rho \left(\frac{\partial \mathbf{U}}{\partial t} + \mathbf{U} \cdot \nabla \mathbf{U} \right) = \nabla \cdot \mathbf{P} + \nabla \cdot (2\mu \mathbf{S}) + \rho \mathbf{g}, \tag{7.4-21}$$

where the modified pressure tensor $\mathbf{P}$, which incorporates the interfacial force term, is written as:

$$\mathbf{P} = -\left(p - C\Phi + \beta\Psi - \frac{\alpha}{2} |\nabla C|^2 \right) \mathbf{I} - \alpha \nabla C \nabla C. \tag{7.4-21a}$$

The phase field function, C, across a planar interface at equilibrium can be analytically derived from the constant chemical potential condition. For example, for the bulk free energy density given in Eq. (7.4-18a), C across a planar interface normal to the x direction can be expressed by:

$$C = \frac{1}{2} \tanh(\xi), \tag{7.4-22}$$

where $\xi = \sqrt{\beta/2\alpha x}$. Thus, if the interfacial region is defined between $C = -0.45$ and $C = 0.45$, the width of the interface is approximated by $4.16\sqrt{\alpha/\beta}$.

According to the mechanical definition, surface tension is given by the integration of the stress balance across the interface as:

$$\sigma = \int_{-\infty}^{\infty} \left(P_{xx} - P_{yy}\right)dx. \tag{7.4-23}$$

Using the stress tensor given in Eq. (7.4-21a), the surface tension can also be expressed as:

$$\sigma = \alpha \int_{-\infty}^{\infty} \left(\frac{dC}{dx}\right)^2 dx. \tag{7.4-23a}$$

From the interfacial profile in Eq. (7.4-22), the above integration yields a surface tension value of:

$$\sigma = \sqrt{\alpha\beta/18}, \tag{7.4-23b}$$

for the bulk free energy density given in Eq. (7.4-18a).

7.5   Eulerian–Lagrangian Algorithms for Multiphase Flows

This section examines numerical implementations of the Eulerian–Lagrangian model introduced in Chapter 5. Recall, the Eulerian–Lagrangian model describes a coupled flow involving one continuous fluid phase and M discrete particle phases. For brevity, the discussion herein is focused on momentum transport, without considering mass or energy coupling. No provision for chemical reactions, interphase mass transfer, or heat transfer is made in the algorithm for solving momentum transport.

An overall flow diagram of the Eulerian–Lagrangian algorithm is provided by Figure 7.5-1. The algorithm can be essentially divided into three subroutines: calculating fluid–particle forces, computing particle–particle forces, and solving the particle and fluid phase governing equations. Each of these stages is treated separately in the subsequent subsections.

The general algorithm shown in Figure 7.5-1 can be simplified if the discrete phase concentration is sufficiently low. For very dilute flows, the influence of discrete phase on continuous phase flow can be neglected; particle collisions are also rare. Thus, only the fluid force acting on particles is required to calculate the particle trajectory. Such a simplified approach embodies *one-way coupling*. For moderately dilute flows, particle–fluid interaction affects both the continuous and discrete phase motions. Here, as in the case of very dilute flows, particle–particle interaction remains negligible. However, the solution algorithm must be constructed to iteratively compute particle–fluid interactions between both the continuous and discrete phase momentum equations simultaneously. This type of approach is known as *two-way coupling*. For relatively dense flows, both particle–fluid and particle–particle interactions must be considered when solving for the momentum transport of particles and fluid phase. A *four-way coupling* iterative algorithm is required to capture the effects of particle–fluid interactions between the continuous and discrete phase momentum equations as well as particle–particle interactions within the discrete

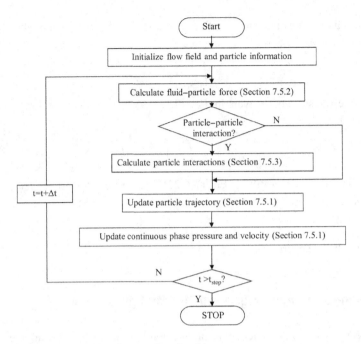

Figure 7.5-1 Eulerian–Lagrangian algorithm.

phase momentum equations. The degree of model sophistication for particle–fluid and particle–particle coupling depends on the nature of the problem and the specified accuracy tolerance. One simple criterion to signal the nature of coupling is the particle volumetric concentration. For example, one-way coupling is appropriate for cases with $\alpha_p < 10^{-6}$, whereas four-way coupling is likely necessary when $\alpha_p > 10^{-3}$. A more complete classification map of phase coupling can be found in Figure 5.1-1 (Elghobashi, 2006).

7.5.1   Governing Equations

When the volumetric effect of particles on the fluid transport must be considered, the fluid phase continuity and momentum equations are volume-averaged, as given by Eq. (6.3-7) and Eq. (6.3-12). If phase change and energy coupling can be neglected, the equations are simplified to the following:

$$\frac{\partial(\alpha\rho)}{\partial t} + \nabla \cdot (\alpha\rho\mathbf{U}) = 0 \tag{7.5-1}$$

$$\frac{\partial(\alpha\rho\mathbf{U})}{\partial t} + \nabla \cdot (\alpha\rho\mathbf{U}\mathbf{U}) = -\alpha\nabla p + \alpha\nabla \cdot \boldsymbol{\tau} + \alpha\rho\mathbf{g} - \sum_{m=1}^{M} \mathbf{M}_m. \tag{7.5-2}$$

The two equations above are similar for single-phase flows, except for the addition of fluid volume fraction term, α, for phase averaging and the interaction term, $\mathbf{M}_m$, that accounts for momentum transport between the fluid phase and the mth particle

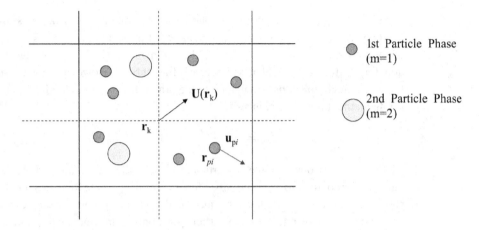

Figure 7.5-2 Schematic illustration of the Eulerian–Lagrangian model.

phase. Once the particle positions are obtained using the discrete phase equations, the volume fraction of the fluid phase, α, is readily calculated for any location. The fluid–particle interaction terms, $\mathbf{M}_m$, are determined via the methodology of Section 7.5.2. Thus, the fluid phase governing equations, Eq. (7.5-1) and Eq. (7.5-2), can be solved using the procedure illustrated for single-phase flows.

The discrete phase governing equations for position and momentum of the ith particle are given in the Lagrangian reference frame, as:

$$\frac{d\mathbf{r}_{pi}}{dt} = \mathbf{u}_{pi}(t) \tag{7.5-3}$$

$$m_{pi}\frac{d\mathbf{u}_{pi}}{dt} = \mathbf{F}_{pfi} + \mathbf{F}_{ppi} + m_{pi}\mathbf{g}. \tag{7.5-4}$$

The terms, $\mathbf{F}_{pf}$ and $\mathbf{F}_{pp}$, on the right-hand side of the momentum equation represent the fluid–particle force and the particle–particle force, respectively; these forces are individually described in Sections 7.5.2 and 7.5.3. If the particle rotation is considered, equations capturing particle orientation and angular velocity must also be prescribed. Equations (7.5-3) and (7.5-4) are first-order ordinary differential equations (ODEs) that can be solved using typical numerical integration techniques, such as the first-order explicit method or the fourth-order Runge–Kutta algorithm.

7.5.2   Continuous-Discrete Phase Coupling

The fluid–particle interaction force is present in the momentum equations of both the continuous phase and the discrete phase. When the Eulerian–Lagrangian algorithm is formulated with the finite volume method, $\mathbf{F}_{pfi}$ in Eq. (7.5-4) represents the force exerted by the local fluid on a single particle i, and $\mathbf{M}_m$ in Eq. (7.5-2) captures the combined effect of all particles that belong to the mth discrete phase on the fluid volume in the Eulerian cell $\mathbf{r}_k$, as shown in Figure 7.5-2.

The force, $\mathbf{F}_{pfi}$, acting on the ith particle is a combination of the pressure gradient force and fluid stress gradient force, and it is commonly expressed as a sum of simple-mode hydrodynamic forces. The summation may include the steady-state drag force, lift force, virtual mass force, and Basset force. However, if the constituent forces are negligible when compared to the drag force, the total fluid–particle force acting on the ith particle reduces to:

$$\mathbf{F}_{pfi} = -\frac{V_{pi}}{(1 - \alpha_{pi})}\beta_{ik}(\mathbf{U}(\mathbf{r}_k) - \mathbf{u}_{pi}), \tag{7.5-5}$$

where V_{pi} is the volume of the ith particle, and β_{ik} is the local fluid–particle momentum transfer coefficient. β_{ik} is usually a function of the particle Reynolds number and particle concentration, and it can be quantified from the appropriate drag correlation introduced in Chapter 4. Since the particle position, $\mathbf{r}_{pi}$, may not be positioned at the exact cell center $\mathbf{r}_k$, the gradient terms and velocity difference in Eq. (7.5-5) are often represented using interpolated or cell-averaged values.

Conversely, the momentum transport from the mth particle phase to the fluid phase in the kth Eulerian cell is found from the summation of the fluid–particle force, $\mathbf{F}_{pfi}$, acting on all the particles of the mth phase. Thus, it gives:

$$\mathbf{M}_{mk} = \frac{1}{V_k}\sum_{i=1}^{N_m}\mathbf{F}_{pfi}K\left(\mathbf{r}_{pi}, \mathbf{r}_k\right). \tag{7.5-6}$$

In Eq. (7.5-6), V_k is the volume of the kth cell; $K(\mathbf{r}_{pi}, \mathbf{r}_k)$ is a generic kernel for properly distributing $\mathbf{F}_{pfi}$, to the fluid cells, $\mathbf{r}_k$, near the particle located at $\mathbf{r}_{pi}$ such that $\mathbf{M}_{mk}$ does not abruptly change when the particle moves across the cell boundary. The total momentum coupling term in Eq. (7.5-2) is given as:

$$\sum_{m=1}^{M}\mathbf{M}_m = \frac{1}{V_k}\sum_{m=1}^{M}\sum_{i=1}^{N_m}\frac{V_{pi}}{(1 - \alpha_{pi})}\beta_{ik}(\mathbf{U} - \mathbf{u}_{pi})K\left(\mathbf{r}_{pi}, \mathbf{r}_k\right). \tag{7.5-7}$$

7.5.3   Particle–Particle Interactions

Theories and models for particle–particle interactions are given in Chapters 4 and 5. In general, the particle–particle interactions in Eq. (7.5-4) include both noncontact force (*e.g.*, electrostatic force) and contact forces; the latter can be modeled with either the hard-sphere model or the soft-sphere model. In the hard-sphere model, particle collisions are assumed to be binary and instantaneous. Therefore, it is required to determine the first pair of colliding particles. Given the large number of particles in a typical simulation, locating the colliding pair of particles is a laborious task demanding optimized search algorithms. In the soft-sphere model, multiparticle and enduring contact is permitted; detection of the first colliding pair of particles is therefore not required. However, any modeling of particle–particle interactions still requires efficient search algorithm to identify the contacting particles.

For particle contact and other short-range interaction forces, it is often useful to store the sorted index of particles in a data structure that reflects their relative positions to facilitate expedited location of potential collision pairs. Such data structures can be constructed either with or without a background Eulerian grid. The two-dimensional quadtree and three-dimensional octree algorithms are examples of grid-free approaches that recursively divide the entire space into a hierarchy of respective quads or octants; within each quad or octant no more than 4 or 8 particles are stored. Adjacent particles are readily identified by traversing the quadtree or octree structure (Lohner, 1998). A cell-linked list search algorithm can also be employed using the background Eulerian grid (Garg *et al.*, 2010). Only particles in the same Eulerian cell, or the immediate adjacent cells, are considered potential neighbors. In the numerical adaptation of these search algorithms, the adjacent particle list is only constructed after a specific series of time steps execute because the displacements, and corresponding change in relative particle positions, are expected to be minimal at each iteration.

7.6   Eulerian–Eulerian Algorithms for Multiphase Flows

In the Eulerian–Eulerian approach discussed in Chapter 6, all phases are treated as interpenetrating continuous fluids, and the governing equations for each phase are of the same form for a single-phase flow. Thus, the solution procedure outlined in Section 7.3 for a single-phase CFD algorithm can be extended to solve multiphase flow problems in the Eulerian–Eulerian framework. In particular, pressure correction schemes remain effective techniques for incompressible and even weakly compressible multiphase flows. However, there are a few notable differences between the single-phase algorithms and their multiphase counterparts, namely (Spalding, 1980):

(1) Additional variables and equations are required for multiphase flows because each phase has its own separate volume fraction, velocity, and governing equations.
(2) Derivation of the pressure correction equation for a multiphase flow situation is comparatively less obvious. Because pressure is common to constituent phases, each of which has a distinct continuity equation, the condition necessary to derive the pressure correction equation is not unique.
(3) Strong coupling exist between different phases via interphase transport terms. Implicit phase coupling must be used to update all phases simultaneously; iteratively updating distinct phases with explicit coupling leads to unacceptably sluggish solution convergence.

The general framework of the Eulerian–Eulerian algorithm is shown in Figure 7.6-1. The velocity and pressure fields are obtained using the pressure correction method. Two techniques, including the semi-implicitly partial elimination algorithm (PEA) (Spalding, 1980) and the fully implicitly phase-coupled SIMPLE (PC-SIMPLE) algorithm (Vasquez and Ivanov, 2000), are illustrated in Section 7.6.1. Techniques for calculating the volume fraction are described in Section 7.6.2. For multiphase

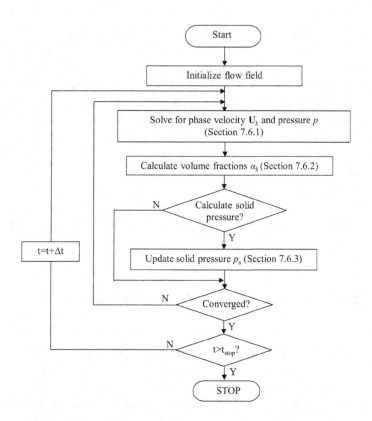

Figure 7.6-1 General framework of Eulerian–Eulerian algorithm.

flows with high solid concentrations, special treatment of the solid pressure is required, which are presented in Section 7.6.3.

7.6.1 Calculation of Velocity and Pressure Field

The phase velocity, U_k, and pressure, p, in the Eulerian–Eulerian algorithm are obtained using pressure correction approaches similar to those described in Section 7.3 for single-phase flows. For simplicity, this section demonstrates calculating U_k and p using an example of a two-phase flow in two dimensions; coupling between the two phases is emphasized.

Similar to Eq. (7.3-15) for a single-phase flow, the discretized momentum equations in the x-direction for a two-phase flow system can be written as:

$$a_{1w}(U_1)_w = \sum_{nb} a_{1nb}(U_1)_{nb} + b_1 - A_w(\alpha_1)_w (p_P - p_W) + \Delta VF_{12}\left[(U_2)_w - (U_1)_w\right]$$

$$a_{2w}(U_2)_w = \sum_{nb} a_{2nb}(U_2)_{nb} + b_2 - A_w(\alpha_2)_w (p_P - p_W) + \Delta VF_{12}\left[(U_1)_w - (U_2)_w\right].$$

$$(7.6\text{-}1)$$

The subscripts 1 and 2 signify the two flow phases. The other subscripts are as used in Eq. (7.3-15) and correspond to the locations on the finite volume grid shown in Figure 7.3-3. The last term on the right-hand side represents the interphase transport, in which ΔV is the control volume and F_{12} is related to the interfacial transport coefficient. In the momentum equation, this term represents the drag force.

The pressure correction approach first assumes a tentative pressure field, p^*, and the tentative velocity fields, $\mathbf{U}_1^*$ and $\mathbf{U}_2^*$, are computed from Eq. (7.6-1). The differences between the tentative and actual pressures and velocities are given by their respective correction expressions as:

$$p = p^* + p' \qquad \mathbf{U}_1 = \mathbf{U}_1^* + \mathbf{U}_1'' \qquad \mathbf{U}_2 = \mathbf{U}_2^* + \mathbf{U}_2''. \tag{7.6-2}$$

Following the same procedure used to derive Eq. (7.3-18), the relationship between the velocity and the pressure corrections is obtained as:

$$\begin{aligned}
a_{1w}\left(U_1'\right)_w &= -A_w(\alpha_1)_w\left(p_P' - p_W'\right) + \Delta V F_{12}\left[\left(U_2'\right)_w - \left(U_1'\right)_w\right] \\
a_{2w}\left(U_2'\right)_w &= -A_w(\alpha_2)_w\left(p_P' - p_W'\right) + \Delta V F_{12}\left[\left(U_1'\right)_w - \left(U_2'\right)_w\right].
\end{aligned} \tag{7.6-3}$$

Substituting Eq. (7.6-2) and Eq. (7.6-3) into the continuity equation results in a linear expression from which the pressure correction, p', can be solved. The corrected p and velocities, $\mathbf{U}_1$ and $\mathbf{U}_2$ from Eq. (7.6-2), are then used as the tentative values for the next solution iteration. The iterative process continues until the pressure and velocities converge to their actual values that satisfy both the momentum and the continuity equations.

The procedure for computing velocities from Eq. (7.6-1) and Eq. (7.6-3) and selecting the continuity equation to formulate the pressure correction equation depends on a choice of algorithms. Two common options are the partial elimination algorithm (PEA) and the phase-coupled SIMPLE (PC-SIMPLE) algorithm.

In PEA, velocities for each phase are solved explicitly from Eq. (7.6-1) by partial elimination. For example, by eliminating $(U_2)_w$ in Eq. (7.6-1), $(U_1)_w$ can be explicitly solved as:

$$\left\{a_{1w} + \frac{\Delta V F_{12}a_{2w}}{a_{2w} + \Delta V F_{12}}\right\}(U_1)_w = \sum_{nb} a_{1nb}(U_1)_{nb} + b_1$$

$$+ \frac{\Delta V F_{12}}{a_{2w} + \Delta V F_{12}}\left(\sum_{nb} a_{2nb}(U_1)_{nb} + b_2\right) - A_w\left\{a_{1w} + \frac{\Delta V F_{12}a_{2w}}{a_{2w} + \Delta V F_{12}}\right\}(p_P - p_W). \tag{7.6-4}$$

When more than two phases are present, the partial elimination method requires a matrix inversion. Thus, similar expressions for $(U_2)_w$ and other velocity components can also be derived. When these equations are used to calculate $\mathbf{U}_1^*$ and $\mathbf{U}_2^*$, the terms $(U_1)_{nb}$ and $(U_2)_{nb}$ are evaluated using the values obtained in the previous solution iteration. Therefore, the velocity update in Eq. (7.6-4) is semi-implicit.

The same partial elimination procedure can be applied to Eq. (7.6-3), which yields:

$$\left(U_1'\right)_w = -d_{1w}\left(p_P' - p_W'\right), \tag{7.6-5}$$

where the coefficient d_{1w} is given by the expression:

$$d_{1w} = A_w \left[(\alpha_1^*)_w + \frac{(\alpha_2^*)_w \Delta V F_{12}}{a_{2w} + \Delta V F_{12}} \right] \Big/ \left[a_{1w} + \frac{a_{2w} \Delta V F_{12}}{a_{2w} + \Delta V F_{12}} \right]. \qquad (7.6\text{-}5a)$$

Substituting the above relationships into Eq. (7.6-2) gives the corrected velocity as:

$$(U_1')_w = (U_1^*)_w - d_{1w} \left(p_P' - p_W' \right). \qquad (7.6\text{-}6a)$$

Corrected predictions for other velocity components for the phase on other faces of the control volume can be obtained in a similar way. Thus, it gives:

$$\begin{aligned}
(U_1)_e &= (U_1^*)_e - d_{1e} \left(p_E' - p_P' \right) \\
(V_1)_n &= (V_1^*)_n - d_{1n} \left(p_P' - p_N' \right) \\
(V_1)_s &= (V_1^*)_s - d_{1s} \left(p_S' - p_P' \right).
\end{aligned} \qquad (7.6\text{-}6b)$$

When the velocities in Eq. (7.6-6) are substituted into the discretized version of the continuity equation of phase 1, it yields:

$$\frac{\partial}{\partial t} (\alpha_1 \rho_1) + \nabla \cdot (\alpha_1 \rho_1 \mathbf{U}_1) = \Gamma_1. \qquad (7.6\text{-}7)$$

A linear equation results for the pressure corrections of the form:

$$a_P p_P' = \sum_{nb} a_{nb} p_{nb}' + b. \qquad (7.6\text{-}8)$$

The PC-SIMPLE algorithm solves the phase-coupled discretized momentum equation (Eq. (7.6-1)) in a fully implicit fashion, without partial elimination. In order to accomplish this, the equation must first be written in matrix form as:

$$(\mathbf{A}_w - \mathbf{R}_w)\mathbf{U}_w^i = \sum_{nb}^{n} \mathbf{A}_{nb}\mathbf{U}_{nb}^i - \mathbf{\Omega}_w(p_P - p_W) + \mathbf{B}_w^i, \qquad (7.6\text{-}9)$$

where:

$$\mathbf{U}^i = \begin{pmatrix} U_1 \\ U_2 \end{pmatrix} \qquad \mathbf{A}_w = \begin{bmatrix} a_{1w} & 0 \\ 0 & a_{2w} \end{bmatrix} \qquad \mathbf{R}_w = \Delta V F_{12} \begin{bmatrix} -1 & 1 \\ 1 & -1 \end{bmatrix}$$

$$\mathbf{A}_{nb} = \begin{bmatrix} a_{1nb} & 0 \\ 0 & a_{2nb} \end{bmatrix} \qquad \mathbf{\Omega}_w = \mathbf{A}_w \begin{pmatrix} \alpha_1 \\ \alpha_2 \end{pmatrix}_w \qquad \mathbf{B}_w^i = \begin{pmatrix} b_1 \\ b_2 \end{pmatrix} \qquad (7.6\text{-}9a)$$

To determine $\mathbf{U}_w^i$ and $\mathbf{U}_{nb}^i$ simultaneously, Eq. (7.6-9) must be incorporated into an overall matrix equation that contains all control volumes. Solving this master linear equation can be protracted, and expediting techniques, such as the block algebraic multigrid method, are often required (Vasquez and Ivanov, 2000).

In the PC-SIMPLE algorithm, a tentative pressure field, p^*, from the previous iteration is used in Eq. (7.6-9) to obtain tentative velocity field, $\mathbf{U}_k^*$. The velocity, $\mathbf{U}_k$, is then expressed as the combination of $\mathbf{U}_k^*$ and $\mathbf{U}_k'$, and substituted into the overall continuity equation for all phases. Thus, it gives:

$$\sum_{k=1}^{N} \left\{ \frac{\partial}{\partial t} (\alpha_k \rho_k) + \nabla \cdot (\alpha_k \rho_k \mathbf{U}_k) \right\} = 0. \qquad (7.6\text{-}10)$$

Notably, all interfacial mass transfer terms are canceled by overall mass balance. For a given control volume, the discretized form of Eq. (7.6-10) can be expressed as:

$$\sum_{k=1}^{N} \left(\sum_{f=1}^{n} a_{k,f} \left(\mathbf{U}_{k,f}^* + \mathbf{U}_{k,f}' \right) + b_k \right) = 0, \tag{7.6-10a}$$

where the inner summation is performed for all faces, f, of the control volume. After converting the velocity corrections, $\mathbf{U}_{k,f}'$, into pressure corrections, p', using Eq. (7.6-3), Eq. (7.6-10a) becomes a linear system in p' with a form similar to that of Eq. (7.6-8).

7.6.2   Volume Fraction

The volume fraction, α_k, can be obtained from the volume-averaged continuity equation of phase k presented in Eq. (6.3-7). This is expressed as:

$$\frac{\partial}{\partial t} (\alpha_k \rho_k) + \nabla \cdot (\alpha_k \rho_k \mathbf{U}_k) = \Gamma_k. \tag{7.6-11}$$

The term on the right-hand side, Γ_k, denotes the interphase mass transfer. This equation can be solved through discretization using the standard techniques for a control volume.

In practice, of the N phases, $N - 1$ volume fractions are actually solved from Eq. (7.6-11). The volume fraction of the omitted phase is obtained using the fact that the sum of α_k over all phases should satisfy the relationship:

$$\sum_{k=1}^{N} \alpha_k = 1. \tag{7.6-12}$$

When implementing the numerical algorithm, it is often helpful to employ so-called traps that confine the value of α_k between zero and one. Under-relaxation can also be utilized during iterations to allow sufficient time for the velocity field to be adjusted in relation to the continuity equation, before α_k differs significantly from initial values.

7.6.3   Pressure and Volume Fraction for Dense Solid Phase

In a multiphase flow with a sufficiently high solid volume fraction, an additional solid pressure term exists as part of solid stresses of the momentum equations because of solid particle collision or contact. In collisional flow with a moderate solid volume fraction, the solid pressure and solid shear stress can be obtained from the kinetic theory for granular material. In a flow with a high solid volume fraction, the plastic regime model from Section 2.5 can be useful.

For closely packed solids, the solid volume fraction cannot exceed the maximum packing limit, which is determined by the particle shape and size distribution. If the solid phase is assumed to be compressible, an equation of state can be imposed on the solid phase. Solid pressure increases exponentially as the solid volume fraction approaches the maximum packing limit and serves to restrict further solids compaction (Syamlal, 1998; Garg *et al.*, 2010).

7.7   Lattice Boltzmann Method

The lattice Boltzmann method (LBM) is a specialized numerical technique for flow simulation. Its theoretical basis is gas kinetics theory, which describes the propagation and collision of fluid molecules. The particle evolution distribution function, $f(\mathbf{r}, \mathbf{v}, t)$, represents the probability density of the molecules, or particles, with a specified velocity, $\mathbf{v}$, at spatial coordinates, $\mathbf{r}$, and time, t. The particle evolution distribution is governed by the well-known Boltzmann equation, as given by:

$$\frac{\partial f}{\partial t} + \mathbf{v} \cdot \nabla f + \frac{\mathbf{F}}{m} \cdot \nabla_{\mathbf{v}} f = Q. \tag{7.7-1}$$

The three terms on the left-hand side of Eq. (7.7-1) are the unsteady, propagation, and external force terms, respectively. The right-hand side of the expression represents the particle collision. The macroscopic hydrodynamic variables, including the mass density, momentum, and energy, can be directly obtained from the particle distribution function. The Navier–Stokes equation governing macroscopic flow behavior can be derived systematically from the Boltzmann equation (Wolf-Gladrow, 2000). Thus, the Boltzman equation accords with macroscopic fluid mechanics theory, while it simultaneously preserves pertinent intricacies of microscopic particle dynamics. The LBM has many desirable features that make it a suitable mesoscopic numerical approach for solving fluid dynamics problems, particularly for multiphase flow applications.

Application of the LBM to single-phase flows is introduced in the next subsection. Techniques are further developed to illustrate the use of LBM to fluid–solid flows and flows involving two fluids. The latter two extensions are individually discussed in Sections 7.7.2 and 7.7.3, respectively.

7.7.1   LBM for Single-Phase Flows

The governing equation in the LBM is the lattice Boltzmann equation (LBE). This is derived from the Boltzmann equation, Eq. (7.7-1), through trio discretization in space, time, and velocity space (Wolf-Gladrow, 2000). Using a finite set of velocities, $\{\mathbf{c}_i\}$, to represent the entire continuous distribution of the particle velocities, the particle distribution function, $f(\mathbf{r}, \mathbf{v}, t)$, can be discretized in the velocity space as $\{f_i(\mathbf{r}, t)\}$. Discretization of the velocity space corresponds to the lattice structure used for the discretization of the spatial coordinates. The most commonly used lattice structures for two- and three-dimensional simulations are shown in Figure 7.7-1; these structures are the D2Q9 and D3Q19 lattice, respectively. The discrete velocities, $\{\mathbf{c}_i\}$, in

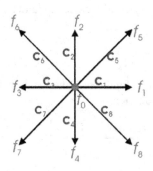

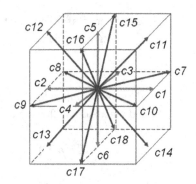

(a) D2Q9 lattice for 2D simulation   (b) D3Q19 lattice for 3D simulation

Figure 7.7-1 Typical lattice structures in LBM.

the D2Q9 lattice can be represented as:

$$
\mathbf{c}_i = \begin{cases} (0,0) & i = 0 \\ (c,0),\ (0,c),\ (-c,0),\ (0,-c) & i = 1,2,3,4 \\ (c,c),\ (-c,c),\ (-c,-c),\ (c,-c) & i = 5,6,7,8 \end{cases}
\tag{7.7-2a}
$$

Similarly, the discrete velocities in the D3Q19 lattice are given by:

$$
\mathbf{c}_i = \begin{cases} (0,0,0) & i = 0 \\ (\pm c,0,0), (0,\pm c,0), (0,0,\pm c) & i = 1 \sim 6 \\ (\pm c,\pm c,0), (\pm c,0,\pm c), (0,\pm c,\pm c) & i = 7 \sim 19 \end{cases}
\tag{7.7-2b}
$$

The constant c is known as the lattice speed, and it is defined by:

$$
c = \Delta r / \Delta t
\tag{7.7-2c}
$$

Representing the continuous velocity space with a small number of discrete velocities remarkably simplifies the computation procedure for the model. However, the computational benefit has a detrimental consequence of lost fidelity with respect to particle velocity fluctuations. A sufficiently large velocity set is required to adequately describe particle velocity fluctuations, and the simplified velocity space diminishes capacity in this regard. Sacrificing particle velocity fluctuation data leads to a corresponding diminished capacity for the model to consistently represent temperature. Thus, the following discussions are narrowed to isothermal flows only. Flows with heat transfer can be treated via more sophisticated LBM models (Wolf-Gladrow, 2000), which is beyond the scope of this section.

In the absence of external force and after discretization, the LBE can be written in a general form as:

$$
f_i(\mathbf{r} + \mathbf{c}_i \Delta t, t + \Delta t) - f_i(\mathbf{r}, t) = -\frac{1}{\tau}\left(f_i(\mathbf{r}, t) - f_i^{eq}(\rho, \mathbf{U})\right).
\tag{7.7-3}
$$

The right-hand side of Eq. (7.7-3) represents the change in particle distribution due to particle collisions as approximated through a relaxation process. The relaxation parameter, τ, is related to the fluid viscosity, μ, and the pseudo lattice sound speed, c_s, through the relationship:

$$\tau = \frac{1}{2} + \frac{\mu}{\Delta t \rho c_s^2}. \tag{7.7-3a}$$

The equilibrium velocity distribution, f^{eq}, is a function of the local fluid density, ρ, and macroscopic flow velocity, **U**. Thus, it yields:

$$f_i^{eq}(\rho, \mathbf{U}) = \rho w_i \left\{ 1 + \frac{\mathbf{c}_i \cdot \mathbf{U}}{c_s^2} + \frac{(\mathbf{c}_i \cdot \mathbf{U})^2}{2c_s^4} - \frac{\mathbf{U} \cdot \mathbf{U}}{2c_s^2} \right\}. \tag{7.7-3b}$$

where the pseudo lattice sound speed is given by: $c_s = c/\sqrt{3}$; w_i is the weight associated with each lattice velocity, which can be determined by:

$$w_i = \begin{cases} 4/9 & i = 0 \\ 1/9 & i = 2 \sim 4 \\ 1/36 & i = 5 \sim 8 \end{cases} \quad \text{for D2Q9 lattice} \tag{7.7-3c}$$

or:

$$w_i = \begin{cases} 1/3 & i = 0 \\ 1/18 & i = 2 \sim 4 \\ 1/36 & i = 5 \sim 8 \end{cases} \quad \text{for D3Q19 lattice.} \tag{7.7-3d}$$

For low-speed conditions characterized by low Mach numbers, Eq. (7.7-3b) approximates the Boltzmann–Maxwell distribution as:

$$f^M(\mathbf{v}) = \frac{\rho}{(2\pi RT)^{D/2}} \exp\left(-\frac{(\mathbf{v} - \mathbf{U})^2}{2RT}\right). \tag{7.7-3e}$$

In the presence of an external force, an additional forcing term must be added to the right-hand side of Eq. (7.7-3) (Guo et al., 2002).

Equation (7.7-3) is explicit with respect to time. In practice, updating the particle distribution function is performed in two sequential steps. First, the collision step is executed to calculate an intermediate distribution, given by:

$$f_i^* = \left(1 - \frac{1}{\tau}\right) f_i - \frac{1}{\tau} f_i^{eq}. \tag{7.7-4}$$

During the second step, known as streaming (or propagation), the post-collision distribution, f_i^*, moves along the ith lattice direction to the adjacent lattice node. Thus, it gives:

$$f_i(\mathbf{r} + \mathbf{c}_i \Delta t, t + \Delta t) = f_i^*(\mathbf{r}, t). \tag{7.7-5}$$

The macroscopic hydrodynamic variables in the LBM are obtained from the particle distribution functions, which can be written as:

$$\rho(\mathbf{r}, t) = \sum_i f_i(\mathbf{r}, t), \tag{7.7-6}$$

$$\rho(\mathbf{r}, t)\mathbf{U}(\mathbf{r}, t) = \sum_i \mathbf{c}_i f_i(\mathbf{r}, t). \tag{7.7-7}$$

Following the kinetic theory of gases, the pressure is calculated from:

$$p = \rho c_s^2. \tag{7.7-8}$$

The above has the form of the equation of state (EOS) for an ideal gas. Despite the ideal gas EOS, the LB algorithm outlined above cannot adequately model compressible flow because the velocity set, $\{c_i\}$, is not sufficiently large enough to consistently represent the temperature. Additionally, approximating the Maxwell distribution using Eq. (7.7-3b) is only valid for flows at low Mach numbers. Thus, from a numerical standpoint, the LBM is generally considered a pseudo-compressible algorithm useful for modeling incompressible flows of liquid or gaseous phase.

7.7.2 LBM for Particle Suspensions

The LBM has extensive application for modeling particle–fluid systems, particularly for simulations where the particles are treated as macroscopic immersed objects with resolved surfaces. Compared to the immersed boundary modeling techniques described in Section 7.4, the LBM has the advantage of being an efficient flow solver that benefits from its inherent kinetic description of the flow, facilitating proficient treatment of boundary conditions on the particle surface. Therefore, the LBM is especially suitable for simulations that contain large numbers of particles.

A unique approach, known as the "bounce-back" scheme, is available that imposes a no-slip condition on a solid surface in the LBM (Ladd, 1994; 2001). The particle population moving toward the solid boundary during the propagation (or streaming) step is simply bounced back at the solid surface. There are two possible implementations of the bounce-back scheme: the solid boundary can be placed either at the lattice node (*i.e.*, "node bounce-back") or midway between the fluid node and solid node (*i.e.*, "link bounce-back"). For a flat boundary, the node bounce-back scheme offers only first-order accuracy with respect to velocity, but the link bounce-back yields second-order precision. Because the LBM has second-order accuracy of spatial discretization within the domain, it is often preferred to use the link bounce-back scheme to maintain equivalent accuracy on the boundaries.

Figure 7.7-2 illustrates the link bounce-back scheme on the D2Q9 lattice for a stationary vertical wall.

After the collision step, the fluid node, $\mathbf{r}$, has the post-collision distribution, $f_{i'}^*$ ($i' = 1, 5, 8$), that indicates imminent propagation toward the solid wall located midway between $\mathbf{r}$ and $\mathbf{r} + \mathbf{c}_1$. During propagation, $f_{i'}^*$ contacts the wall and reverses direction, such that $f_3 = f_1^*$, $f_6 = f_5^*$, and $f_7 = f_8^*$, before propagating back toward the fluid node, $\mathbf{r}$. The bounce-back scheme can also be applied to a moving boundary with velocity, $\mathbf{u}_b$. In such a case, the distribution on the fluid node after the bounce-back can be written in a general form as:

$$f_i(\mathbf{r}, t) = f_{i'}^*(\mathbf{r}, t) + \frac{2\rho w_i}{c_s^2} (\mathbf{u}_b \cdot \mathbf{c}_i). \tag{7.7-9}$$

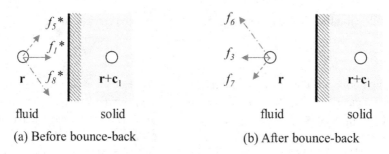

(a) Before bounce-back   (b) After bounce-back

Figure 7.7-2 Schematic illustration of the bounce-back condition.

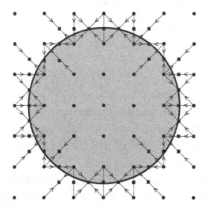

Figure 7.7-3 Bounce-back condition on a curved boundary.

Figure 7.7-3 depicts a curved solid boundary immersed in a regular lattice. In this case, the accuracy of the link bounce-back scheme degrades, and the exact location of the boundary becomes ambiguous. For a spherical particle, the effective radius (*i.e.*, hydrodynamic radius) is determined by comparing its steady drag force to the standard drag curve of Section 3.2. The effective radius is usually distinct from the value specified at input; the discrepancy is caused by factors such as fluid viscosity and particle Reynolds number (Ladd, 1994). Therefore, a priori simulation must be conducted to calibrate the hydrodynamic radius for each condition. This is often regarded as a drawback, but it can be circumvented via higher-order bounce-back schemes that interpolate the particle distribution functions based on the distance to the exact boundary location (Bouzidi *et al.*, 2001; Lallemand & Luo, 2003).

The immersed boundary method described in Section 7.4.2.1 can be incorporated into the LBM flow solver to accommodate fluid–particle flows. With this technique, the flow field is solved by the LBM, and particle motion and deformation are tracked by a set of Lagrangian markers. A force density is introduced on the particle surface to ensure a no-slip condition of the fluid phase and to control the deformation of the particle surface. The immersed boundary lattice Boltzmann method (IB-LBM) is capable of solving for motion of deformable objects (Wu and Aidun, 2010) and rigid

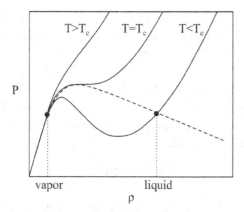

Figure 7.7-4 Pressure-density relation of a nonideal gas.

objects alike (Feng and Michaelides, 2004). High-order schemes such as the Runge–Kutta scheme have been shown to improve the numerical accuracy and stability of IB-LBM (Zhou and Fan, 2014).

7.7.3   LBM with Two Fluid Phases

Flows consisting of two fluid phases can be simulated using several approaches within the LBM framework, including interaction potential (Shan and Chen, 1993), free energy (Swift, 1996; van der Graaf, 2006), and index function (He *et al.*, 1999; Lee and Lin, 2005). Among these options, the interaction potential approach leverages the microscopic physics of the LBM and employs a straightforward and efficient algorithm. As a result, this method is introduced below as a representative class of LBM techniques for gas–liquid and liquid–liquid systems.

The collision-propagation dynamics of particles in Eq. (7.7-3) mimics the behavior of an ideal gas as suggested by the equation of state, Eq. (7.7-8). The interactive forces between the particles are neglected with this method. However, these interactions are the genesis of phase segregation in a fluid or a fluid mixture. In a nonideal fluid, intermolecular forces warrant a nonideal gas equation of state, as shown in Figure 7.7-4.

In a subcritical condition (*i.e.*, $T < T_c$), the nonideal fluid spontaneously separates into a lean vapor phase and a dense liquid phase that share an equilibrium pressure. Conversely, in a mixture of two immiscible fluids, the degree of immiscibility is determined by the interactions between the two types of fluid molecules responsible for repulsion. Therefore, by incorporating the interparticle force into the LBM, a two-phase flow can be simulated without needing to track or capture the interface. Depending on the role of the interaction force, the interaction potential-based two-phase LBM can be categorized into two types: one designed for a single component nonideal fluid and the other suited for two immiscible fluids (Shan and Chen, 1994). In both types, the LBM algorithms can be further combined with adaptive mesh refinement (AMR) technique to enhance the interface resolution, thereby improving the capability in simulating flows with complex topology change at the interface (Yu and Fan, 2009).

7.7.3.1 Interaction Potential Model for a Nonideal Fluid

For a single component nonideal fluid, the interacting force on the particle population at location $\mathbf{r}$ is defined as the summation of the pairwise interactions between particles at $\mathbf{r}$ and those at neighboring sites, $\mathbf{r} + \mathbf{c}_i$ $(i = 1 \ldots q)$. Thus, it gives:

$$\mathbf{F}_{\text{int}}(\mathbf{r}) = -\psi(\mathbf{r})G \sum_i w_i \psi(\mathbf{r} + \mathbf{c}_i)\mathbf{c}_i, \tag{7.7-10}$$

where Ψ is the interaction potential, and the negative scalar, G, indicates the strength of the interaction. Expanding the terms $\Psi(\mathbf{r} + \mathbf{c}_i)$ at $\mathbf{r}$, the interaction force can also be expressed as:

$$\mathbf{F}_{\text{int}} = -c_s^2 G\psi\nabla\psi + o(\Delta r^2). \tag{7.7-10a}$$

On the macroscopic level, this force can be combined with the pressure gradient term in the Navier–Stokes equation to yield to a modified pressure, given by:

$$p(\rho) = \rho c_s^2 + \frac{1}{2}Gc_s^2\psi^2(\rho). \tag{7.7-11}$$

The above is essentially a nonideal equation of state that controls the phase equilibrium behavior of the fluid. The interaction potential can be defined, for example, by the expression:

$$\psi(\rho) = 1 - \exp(-\rho). \tag{7.7-12}$$

Other forms of the interaction potential are available, but they may affect the density ratio and numerical stability of the algorithm (Yuan and Schaefer, 2006).

In order to incorporate the interaction force, $\mathbf{F}_{\text{int}}$, into the LB algorithm, the velocity used to calculate the equilibrium distribution from Eq. (7.7-3b) in the collision step is shifted by an amount proportional to the force. This is expressed by:

$$f_i^{eq} = f_i^{eq}\left(\rho, \mathbf{U}^{eq}\right) = f_i^{eq}\left(\rho, \tilde{\mathbf{U}} + \frac{\tau}{\rho}\mathbf{F}\right), \tag{7.7-13}$$

where the intermediate velocity is calculated by the relationship:

$$\tilde{\mathbf{U}} = \frac{1}{\rho}\sum_i \mathbf{c}_i f_i \tag{7.7-13a}$$

Because Eq. (7.7-13) results in the addition of momentum, $\Delta t\mathbf{F}_{\text{int}}$, during the collision, the macroscopic velocity of the fluid must be calculated from the average momentum, both before and after the collision. The following expression is employed to calculate the fluid velocity in lieu of Eq. (7.7-7):

$$\rho(\mathbf{r},t)\mathbf{U}(\mathbf{r},t) = \sum_i \mathbf{c}_i f_i(\mathbf{r},t) + \frac{1}{2}\Delta t\mathbf{F}_{\text{int}}. \tag{7.7-14}$$

7.7.3.2 Interaction Potential Model for Immiscible Fluids

The interaction potential model for a binary mixture employs two separate particle distribution functions, f_i^σ ($\sigma = 1$ or 2): one for each individual component. The interparticle forces thus include two types of interactions: the interactions between particles of the same component and the interactions between particles of different components. The interaction force on component σ is written as:

$$\mathbf{F}_\sigma = -\psi_\sigma(\mathbf{r}) \sum_{\bar{\sigma}} G_{\sigma\bar{\sigma}} \sum_i w_i \psi_{\bar{\sigma}}(\mathbf{r} + \mathbf{c}_i)\mathbf{c}_i. \qquad (7.7\text{-}15)$$

The interaction potential of each component is specified based on its individual local density, such that: $\psi_\sigma = \psi_\sigma(\rho_\sigma(\mathbf{r}))$. A positive value of the interaction strength, $G_{\sigma\bar{\sigma}}$, indicates reactions tending to repel components σ and $\bar{\sigma}$, while a negative $G_{\sigma\bar{\sigma}}$ signals attractive interactions. For example, models for two immiscible liquids usually incorporate the condition $G_{12} = G_{21}$; positive values introduce a repelling force between different types of particles, and thus facilitate the segregation of the two immiscible phases. At the same time, both G_{11} and G_{22} can be set to zero in an effort to simplify the computation. However, when modeling gas–liquid systems, G_{11} is often assigned a negative value so that the first component is nonideal and produces a high density contrast between the two phases; G_{22} is often designated as zero so that the second component is an ideal gas that preferably remains in the gaseous phase. A positive value for $G_{12} = G_{21}$ is also applied in this case to promote a sharply defined interface between the two phases.

The two distinct particle distribution functions, f_i^σ, both follow the LBE, Eq. (7.7-3), and evolve through the collision-propagation steps. During the collision, the shifted velocity is utilized to introduce the interaction force to the LBE, as in the single-component model. The equilibrium velocity of component, σ, is related to the interaction force, $\mathbf{F}_\sigma$. The following condition can therefore be obtained:

$$\mathbf{U}_\sigma^{eq} = \tilde{\mathbf{U}} + \frac{\tau_\sigma}{\rho_\sigma}\mathbf{F}_\sigma, \qquad (7.7\text{-}16)$$

where the intermediate velocity, $\tilde{\mathbf{U}}$, is calculated by the expression:

$$\tilde{\mathbf{U}} = \frac{1}{\rho}\left(\sum_{\sigma=1}^{2}\frac{1}{\tau_\sigma}\sum_{i=1}^{q}\mathbf{c}_i f_i^\rho\right). \qquad (7.7\text{-}16\text{a})$$

Finally, the macroscopic density and velocity are calculated from the particle distribution functions of both components, respectively, as:

$$\rho = \sum_\sigma \rho_\sigma = \sum_\sigma \sum_i f_i^\sigma \qquad (7.7\text{-}17)$$

$$\mathbf{U} = \frac{1}{\rho}\sum_\sigma\left[\sum_i \mathbf{c}_i f_i^\rho + \frac{1}{2}\Delta t \mathbf{F}_\sigma\right]. \qquad (7.7\text{-}18)$$

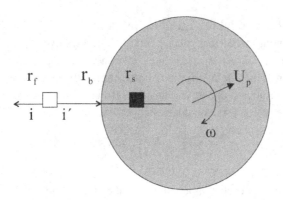

Figure 7.8-1 Link bounce-back condition on particle surface.

7.8   Case Studies

The following case studies serve for further discussion with simple applications of various numerical methods of Chapter 7. The examples include the formulation of particle–fluid force by LBM method, modeling selections for aerosol transport from an inhaler, VOF method for air entrainment in a hydraulic jump, and modeling logics for design of a gas sparger in a bubble column and design of a cyclone.

7.8.1   Particle–Fluid Force in LBM

Problem Statement: A LBM model is used to study the dynamics of a solid particle moving in a continuous flow field. The link bounce-back rule is chosen as the boundary condition on the particle surface. Derive the expression for the hydrodynamic force on the particle.

Analysis: The link bounce-back condition is illustrated in Figure 7.8-1. A particle with its center at location $\mathbf{R}$ has a linear velocity of $\mathbf{U}_p$, and an angular velocity of ω. The velocity at $\mathbf{r}_b$ on the particle surface is

$$\mathbf{u}_b = \mathbf{U}_p + \omega \times (\mathbf{r}_b - \mathbf{R}). \tag{7.8-1}$$

Assume $\mathbf{r}_f$ is a lattice node adjacent to the boundary on the fluid side, and $\mathbf{r}_s$ is a lattice node adjacent to the boundary within the particle. As a common treatment in LBM, the interior of the particle is also filled with fluid and participates in all LBM operations. In order to realize the no-slip condition on the particle boundary, the following bounce-back rule is applied to $\mathbf{r}_f$:

$$f_i(\mathbf{r}_f, t) = f_{i'}^*(\mathbf{r}_f, t) + \frac{2\rho w_i}{c_s^2} (\mathbf{u}_b \cdot \mathbf{c}_i). \tag{7.8-2}$$

The momentum exchange at the boundary location $\mathbf{r}_b$ due to the above bounce-back condition is given by

$$\mathbf{F}_i(\mathbf{r}_f, t) = \mathbf{c}_i f_i(\mathbf{r}_f, t) - \mathbf{c}_{i'} f_{i'}^*(\mathbf{r}_f, t) = 2\mathbf{c}_i f_{i'}^*(\mathbf{r}_f, t) \frac{2\rho \omega_i}{c_s^2} (\mathbf{u}_b \cdot \mathbf{c}_i) \mathbf{c}_i. \tag{7.8-3}$$

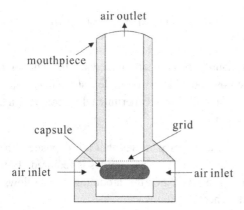

Figure 7.8-2 Schematic diagram of a dry powder inhaler.

Thus, the total fluid–particle interaction force on the particle can be expressed as

$$\mathbf{F}(t) = \sum_{\mathbf{r}_f} \sum_{i} \mathbf{F}_i(\mathbf{r}_f, t). \tag{7.8-4}$$

The summation is over all fluid lattice nodes $\mathbf{r}_f$ adjacent to the particle surface, and along all lattice directions i that point to the solid boundary. Similarly, the torque due to the hydrodynamic interaction on the particle can be expressed as

$$\mathbf{T}(t) = \sum_{\mathbf{r}_f} \sum_{i} (\mathbf{r}_b - \mathbf{R}) \times \mathbf{F}_i(\mathbf{r}_f, t). \tag{7.8-5}$$

Comments: These LBM algorithms for the hydrodynamic force and torque on a particle surface are simple to calculate as compared to those used in typically finite volume methods, where derivatives of the fluid velocity field need to be calculated based on the geometry of the boundary. This shows the advantage of LBM in simulation of complex boundary shapes (such as porous medium or packed bed) and multiphase flows with a large number of suspending particles. However, for curved surfaces, determining the exact location of the boundary in the link bounce-back condition is not straightforward, and often needs a priori calibrations (Ladd and Verberg, 2001).

7.8.2   Modeling of Aerosol Delivery by a Powder Inhaler

The dry powder inhaler is a device commonly used to deliver dry powder drug formulations to the lung. As shown in Figure 7.8-2, air flow enters the capsule chamber at the bottom of the inhaler, carries the drug powder released from the capsule upward through the grid into the mouthpiece, and enters the mouth of the patient.

One of the performance criteria is to minimize the fraction of powders that are retained in the device by particle–wall collisions, which is typically evaluated from the numerical simulation. A typical design and operation of a dry powder inhaler is as follows: inhaler height of 80 mm; mouthpiece internal diameter of 10 mm; particles of 3 μm in diameter and density of 1.5 g/cm^3; characteristic particle flow rate of 0.2 g/s, with air flow rate of 1.0 l/s.

Problem Statement: (1) To quantify the powder retained and the inhaling flow characteristics, which multiphase flow model and sub-models should be used? (2) What are the proper boundary conditions? (3) What experimental technique can be used to validate the numerical simulation?

Analysis: In order to find the fraction of particles that are deposited on the wall, tracking the trajectory of representative particles samples is required. Therefore the Eulerian–Lagrangian model is preferred. For the modeling of gas flow, it is also necessary to evaluate the turbulence conditions.

For the ambient air, based on the typical case of inhaler operation, the air velocity in the mouthpiece is about 13 m/s, which leads to Re around 4,350, indicating a weak turbulent jet flow. Thus the RANS model (such as k–ε) can be applied due to its low computational cost. Low Re correction to the standard RANS model can be considered. The drug particles form the discrete phase and are treated by tracking their motion in the Lagrangian approach. Due to the dilute nature of the flow, one-way coupling between the fluid phase and solid phase can be used. To simplify the problem, the flow is assumed to be in pseudo-steady state. The continuous phase flow can be simulated first, and the particle trajectory tracking can be performed as a post-process step using the steady-state continuous flow field.

Now consider the boundary conditions: at the air inlet, the air velocity components and turbulence model parameter are specified, with no particle injection. At the inhaler outlet, a fully developed air jet flow may be assumed, with all airborne particles escaped. On the inner wall, there is no-slip for air flow while assuming all particles are deposited after colliding with wall.

In order to validate the air flow field, the velocity field can be measured using experimental techniques such as LDV. To validate the prediction from the discrete phase model, it is necessary to determine the fraction of particles deposited in the inhaler device versus that exit from the outlet. Therefore the particle mass emitted from the inhaler needs to be collected (for example, with an aerosol impactor), and the mass of both retained and emitted particles will be measured and compared to the total mass loaded into the capsule.

Comments: In a very dilute particle–fluid flow, it is often relatively easier to predict the continuous phase flow field. It is expected that the predicted air velocity field will be in good agreement with the measure velocity profile. However, accurate prediction of the fate of particles is often more challenging. The model proposed above was shown to significantly overpredict the particle retention in the device, likely due to the zero restitution coefficient condition at the walls (Coates *et al.*, 2007). Other factors, such as particle size distribution, particle morphology, and particle–particle interactions may also contribute to the discrepancy. However, the model was able to qualitatively predict the effect of the inhale internal geometry on the retention, and the effect can be explained by the simulated flow field of the continuous phase.

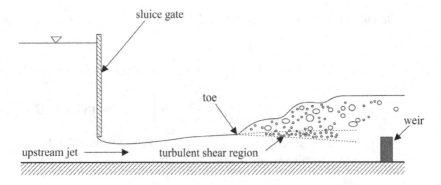

Figure 7.8-3 Air entrainment in a hydraulic jump downstream of a dam.

Thus, although not exactly accurate, the model still provides value to a practical engineering problem.

7.8.3 Air Entrainment in a Hydraulic Jump

Problem Statement: A hydraulic jump downstream of a dam can be used to dissipate energy and improve aeration. Air is entrained at the free surface and sheared by turbulence, producing bubbles that either rise to the surface or travel downstream with the bulk flow, as shown in Figure 7.8-3. Develop a model to calculate the void fraction in the flow as the result of the entrainment.

Analysis: The following is a simplified model based on the work of Witt *et al.* (2015). Since the flow field involves a dynamic free surface and breaking/coalescing bubbles, an interface-capturing model such as VOF is needed. In the VOF approach, the two phases are treated as a single continuum medium. The volume of fluid is governed by its own transport equation, such as Eq. (7.4.10).

For an incompressible two-phase flow, the continuity and momentum equations can be expressed, based on Eq. (7.4-2), as:

$$\nabla \cdot \mathbf{U} = 0 \tag{7.8-6}$$

and

$$\frac{\partial}{\partial t}(\rho \mathbf{U}) + \nabla \cdot (\rho \mathbf{U}\mathbf{U}) = -\nabla p + \nabla \cdot \left[(\mu + \mu_e)\left(\nabla \mathbf{U} + \nabla \mathbf{U}^T\right) \right] + \rho \mathbf{g} + \sigma \kappa \nabla \alpha, \tag{7.8-7}$$

where the last term on the right-hand side of Eq. (7.8-7) is the interfacial force determined by Eq. (7.4-12), and the turbulent eddy viscosity is related to k and ε by Eq. (2.3-18). The fluid properties are calculated by the weighted contribution of volume fractions of each phase, such as

$$\rho = \alpha \rho_l + (1 - \alpha)\rho_g \tag{7.8-7a}$$

$$\mu = \alpha \mu_l + (1 - \alpha)\mu_g. \tag{7.8-7b}$$

Based on Eq. (2.3-21) and Eq. (2.3-22), the transport of k and ε can be governed, respectively, by

$$\frac{\partial}{\partial t} (\rho k) + \nabla \cdot (\rho \mathbf{U} k) = \nabla \cdot \left[\left(\mu + \frac{\mu_e}{\sigma_k} \right) \nabla k \right] + 2\mu_e \, (\mathbf{S} : \mathbf{S}) - \rho \varepsilon \qquad (7.8\text{-}8)$$

$$\frac{\partial}{\partial t} (\rho \varepsilon) + \nabla \cdot (\rho \mathbf{U} \varepsilon) = \nabla \cdot \left[\left(\mu + \frac{\mu_e}{\sigma_k} \right) \nabla \varepsilon \right] + 2\rho C_1 \frac{\varepsilon}{k} \, (\mathbf{S} : \mathbf{S}) - C_2 \rho \frac{\varepsilon^2}{k}, \qquad (7.8\text{-}9)$$

where the rate of the deformation tensor, $\mathbf{S}$, is defined by Eq. (2.2-1).

The VOF equation is given by

$$\frac{\partial \alpha}{\partial t} + \nabla \cdot \alpha \mathbf{U} = 0. \qquad (7.8\text{-}10)$$

Thus, five independent equations, Eqs. (7.8-6)–(7.8-10), for five unknowns (α, $\mathbf{U}$, p, k, and ε) can be obtained. The closure of the modeling is now established. Once the simulation completes, the bubble size and the gas volume fraction in the flow can be calculated from the VOF function. Detailed simulation examples can be found in the reference (Witt *et al.*, 2015).

7.8.4 Evaluation of Sparger in Bubble Column

Problem Statement: In a bubble column, gas is introduced from a sparger, as shown in Figure 7.8-4. The sparger is often designed as a perforated plate between the column and the gas chamber. Establish a modeling approach to study the effect of sparger design on the gas holdup and distribution.

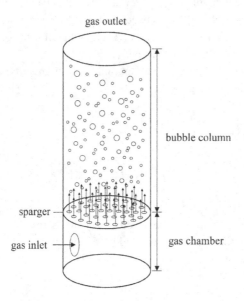

Figure 7.8-4 Illustration of a bubble column with sparger.

Analysis: The performance of the sparger can be affected by both the gas chamber and the bubble column, so both regions should be considered in the model. The two zones can be simulated individually with different models, but need to be coupled via the gas flow through the sparger. The following modeling approach is adapted from the work of Dhotre and Joshi (2007).

In the gas chamber, the single-phase continuity, momentum, and turbulence model (*e.g.*, k–ε model) equations need to be solved. The gas velocity and turbulence intensity are specified at the inlet, and no-slip condition for gas velocity is specified at the chamber wall. The orifices on the sparger plate are designated as pressure outlet. The specified pressure for each orifice is obtained through coupling with the bubble column zone.

In the bubble column, an Eulerian–Eulerian model is typically used, with liquid as the continuous phase and gas as the dispersed phase. The momentum exchange between the two phases is modeled as drag, virtual mass, and lift forces. The two-phase k–ε model is used to account for the turbulence. No-slip velocity condition is assumed at the wall. Since the freeboard region is not included in the computational domain, the top boundary represents the liquid surface and a degassing condition is assumed, where the liquid phase has zero normal velocity while the gas phase is allowed to escape. On the sparger surface, the liquid velocity and volume fraction are assumed to be zero in the orifice. The gas velocity from the orifices on the sparger to the bubble column can be calculated based on the volumetric inlet gas flow to the gas chamber, assuming an initial uniform pressure for all orifices. Thus, the coupling between the gas chamber and bubble column is established through continuity of gas velocity at the sparger boundary between the two zones. The pressure distribution at the sparger is then calculated from the bubble column model and is fed back to the gas chamber model, and a new gas velocity distribution from the sparger is obtained. It takes several iterations before the solutions in both chambers become stable (*i.e.*, the change between two consecutive iterations is less than a predetermined criterion, for example, 5%).

Once the simulation is completed, the results can be processed to plot the gas holdup along radial direction at different heights. Generally, a more uniform gas distribution is preferred. Therefore, the model can be used to study the effect of sparger design parameters, including orifice size, distribution, location, and total open area, on the uniformity of the gas distribution and total gas holdup in the column.

Comments: Bubble size is an important parameter in the bubble column model, as it affects the interphase forces. If the bubble column operates under the homogeneous regime or with a non-coalescing medium, the bubble size can be relatively uniform and a constant bubble diameter can be used for the model. If there is strong coalescing and a wide range in bubble size distribution, it might be necessary to model the size distribution with tools such as the population balance model. Thus, it is critical to analyze the flow regime in order to determine the proper model to apply. In practice, the initial estimate of the bubble size often comes from literature data, empirical correlations, or experimental observations.

7.9   Summary

This chapter introduces the basic algorithms used to solve the governing equations of multiphase flows. The algorithms for incompressible, isothermal single-phase Newtonian fluid flow, represented by the SIMPLE algorithm (Fig. 7.3-4), form the basis for more complex multiphase flow algorithms. Numerical techniques for the *microscopic* descriptions of fluid–particle interactions are focused on the discrete particle phase with rigid or nonrigid surfaces. Such methods are associated with the direct numerical simulation and can be categorized into two classes:

- the conformal mesh technique, where the surface of the object is represented as the boundary of the computational domain, and the flow equation is only solved in the continuous phase domain but not inside the particle phase. Typical conformal mesh methods include the boundary-fitted method and the arbitrary Lagrangian–Eulerian method.
- the nonconformal mesh technique, which either use a set of Lagrangian makers to track the location of the interfaces (as in immersed boundary method and front tracking method), or capture the interface from an evolving scalar field (as in volume of fluid, level-set, and diffuse interface methods).

Numerical techniques for the *macroscopic* descriptions of multiphase flow include the Eulerian–Lagrangian algorithm (Fig. 7.5-1) for continuum-discrete modeling described in Chapter 5, and the Eulerian–Eulerian algorithm (Fig. 7.6-1) for continuum modeling described in Chapter 6. The lattice Boltzmann method is a unique numerical technique for flow simulation. It is based on the discrete Boltzmann equation, rather than the typical Navier–Stokes equation in other CFD techniques. Its computational efficiency and some special treatment for multiphase models make it a suitable tool for flows with complex phase interactions.

Nomenclature

A Area
a Coefficient of ϕ in discretized scheme
C Phase field function
$\mathbf{c}$ Discrete or lattice velocity
c_s Speed of sound
e Specific internal energy
$\mathbf{F}$ Force
$\mathbf{f}_\sigma$ Interfacial force density
f Free energy density
        Particle evolution distribution function
G Interaction strength parameter
g Gravity acceleration
H Heaviside function

$\mathbf{I}$	Identity tensor
$\mathbf{J}_q$	Heat flux vector
J_e	Volumetric rate of heat generation
K	Kernel function
k	Turbulence kinetic energy
	Stiffness coefficient
	Mobility
$\mathbf{M}$	Momentum transfer
m	Mass
$\mathbf{n}$	Normal unit vector
$\mathbf{P}$	Pressure tensor
p	Pressure
$\mathbf{R}$	Surface position vector
$\mathbf{r}$	Position vector
$\mathbf{S}$	Spatial deformation rate
S	Source term in governing equation
	Surface area
s	Source term in discretized scheme
	Surface mesh element
T	Temperature
t	Time
$\mathbf{U}$	Time-averaged velocity
$\mathbf{u}$	Instant velocity
U	x-component of $\mathbf{U}$
V	Volume
	y-component of $\mathbf{U}$
w	Weighting factor

Greek symbols

α	Volume fraction
β	Momentum transfer coefficient
Γ	Interfacial mass flow
	Transport coefficient
δ	Dirac delta function
ε	Dissipation rate of k
κ	Curvature
λ	Half thickness of interface
μ	Dynamic viscosity
ρ	Density
σ	Surface tension
σ_E	Elastic tension
$\boldsymbol{\tau}$	Shear stress tensor
τ	Relaxation parameter

ϕ General quantity

ψ Interaction potential

ω Under-relaxation coefficient

Subscript

E Center of east-side cell

e Cell boundary on east

i ith

k Phase k

N   Center of north-side cell

n   Cell boundary on north

P Cell center point

p Particle

q   Heat transfer

S Center of south-side cell

s Cell boundary on south

W Center of west-side cell

w Cell boundary on west

ϕ Property of ϕ

Superscript

$'$ Correction

*   Estimate

n nth iteration

Problems

P7.1 In an interfacial cell in VOF method, the volume fraction of each phase is known. How to determine the location of interface within the cell? Propose an algorithm for a 2D case.

P7.2 Using a 2D level-set function, how to determine the volume of a phase inside an interface cell? Assume that the uniform Cartesian mesh is adopted.

P7.3 Similar to the Chapman–Enskog expansion in the derivation from the Boltzmann equation to the Navier–Stokes equation in statistical mechanics, it can be shown that the LBM can also lead to the Navier–Stokes equation. Using the D2Q9 LBM model with BGK collision operator as an example, show the derivation. What is the viscosity of the macroscopic fluid?

P7.4 Consider a rotating cylindrical tank partially filled with dense particles and water. As the tank rotates, both the air–liquid interface and the particle bed will deform. A numerical simulation will be used to study the effect of liquid motion

on the mixing of solid particles. The tank diameter is 0.2 m, the particle diameter is 2 mm, and density is 2,500 kg/m^3.
(1) Discuss the proper model for each phase.
(2) What would be the considerations for mesh size and time resolution in the numerical simulation?

P7.5 A spray injector generates a polydispersed liquid spray in a swirling gas flow field. The droplet size ranges from 1 to 100 μm. It is known that the turbulent coherent structure has a significant effect on the droplet dynamics and dispersion.
(1) What numerical models should be used in order to simulate the dynamics of the droplets?
(2) Discuss the approaches to model the effect of the gas phase turbulence on the droplet dispersion.

P7.6 In conventional CFD for single-phase flows, the numerical solution is expected to converge to the exact solution as mesh resolution is increased. However, it has been observed that in the stochastic Eulerian–Lagrangian model for dispersed two-phase flows, the numerical error could actually increase when the mesh is refined. Discuss the effect of mesh resolution on the two-way coupling between the Eulerian and Lagrangian phases, and what could be the reason for the unusual convergence behavior in the Eulerian–Lagrangian model.

P7.7 A single-phase, incompressible, transient flow expressed by

$$\frac{\partial(\rho)}{\partial t} + \nabla \cdot (\rho U) = 0$$

$$\frac{\partial(\rho U)}{\partial t} + \nabla \cdot (\rho UU) = -\nabla p + \nabla \cdot \tau + \rho g$$

can be solved by an algorithm known as the projection method in the following three steps (superscript n denotes the time step n):
(a) Calculate intermediate velocity U^*: $U^* = U^n - \frac{\Delta t}{\rho}[-\nabla \cdot (\rho UU) + \nabla \cdot \tau + \rho g]^n$
(b) Solve the pressure Poisson equation: $\nabla^2 p^{n+1} = \frac{\rho}{\Delta t}\nabla \cdot U^*$
(c) Obtain the velocity for time step $n + 1$: $U^{n+1} = U^* - \frac{\Delta t}{\rho}\nabla p^{n+1}$

(1) Derive the pressure Poisson equation in step b).
(2) Propose an extension of the above projection method to solve Eqs. (7.5-1) and (7.5-2) in an Eulerian–Lagrangian method.

P7.8 A stirred tank is used to mix a liquid with a kinematic viscosity of $\nu = 10^{-6}$ m^2/s. The impeller diameter is $d = 0.3$ m and impeller rotation speed is 6.0 rpm. In order to find out the proper grid resolution for a LES simulation of the stirred tank, a RANS simulation is first performed. From the results for the turbulence kinetic energy and dissipation, the integral length scale of the problem is found to be 0.01 m. Calculate the Kolmogorov length scale. Is it appropriate to use $\Delta x = 0.001$ m for the LES simulation?

P7.9 Why is the CFD algorithm of finite volume methods (FVM) normally computationally quicker than that of finite element method (FEM) when solving governing equations (*e.g.*, N-S equation and k–ε equations) for a turbulent fluid flow?

P7.10 The relaxation time for a solid or fluid particle moving toward a solid wall in a fluid can be estimated by:

$$\tau = \frac{(\rho_p + C_{M\infty}\rho)d^2}{18\mu(1 + 0.15\text{Re}^{0.687})}, \tag{P7.10-1}$$

where $C_{M\infty} = 0.73$ is the added mass coefficient of a spherical object moving toward a wall near contact. The contact time of a droplet on a surface can be estimated by its first-order vibration period as:

$$t_c = \frac{\pi}{4}\sqrt{\frac{(\rho_d + C_{M\infty}\rho)\,d^3}{\sigma}}. \tag{P7.10-2}$$

The solid particle contact time can be found by Eq. (4.3-7h) from the Hertzian theory.
(a) Calculate the ratio of the contact time to the relaxation time under the following conditions: (1) 1 mm glass bead in water; (2) 1 mm toluene drop in water ($\rho_p = 860$ kg/m^3, $\sigma = 0.026$N/m); and (3) 2 mm water drop in air in a film-boiling regime.
(b) Discuss the impact of this ratio on the selection of numerical integral steps in the Lagrangian particle tracking algorithms.

P7.11 Dense medium cyclones are used in coal industry to separate coal particles with different densities. A mixture of water, coal particles, and small magnetite particles enters tangentially into the cyclone and forms a swirling flow. Under the centrifugal force, heavier particles move toward the wall and travel downward. Lighter particles near the center are carried by the upward vortex and exit from the top. Develop a numerical model to calculate the percentage of coal particles exiting from the bottom of the cyclone.

References

Anderson, D. M., McFadden, G. B., and Weber, A. A. (1998). Diffuse-interface methods in fluid mechanics. *Ann. Rev. Fluid Mech.*, **30**, 139–165.

Bouzidi, M., Firdaouss, M., and Lellemand, P. (2001). Momentum transfer of a Boltzmann-lattice fluid with boundaries. *Phys. Fluids*, **13**, 3452–3459.

Brackbill, J. U., Kothe, D. B., and Zemach, C. (1992). A continuum method for modelling surface tension. *J. Comput. Phys.*, **100**, 335–354.

Cahn, J. W. (1961). On spinodal decomposition. *Acta Metall.*, **9**, 795–801.

Chang, Y. C., Hou, T. Y., Berriman, B., and Osher, S. (1996). A level set formulation of Eulerian interface capturing methods for incompressible fluid flows. *J. Comput. Phys.*, **124**, 449–464.

Chorin, A. J. (1968). Numerical solution of the Navier-Stokes equations. *Math. Comp.*, **22**, 745–762.

Coates, M. S., Chan, H. K., Fletcher, D. F., and Chiou, H. (2007). Influence of mouthpiece geometry on the aerosol delivery performance of a dry powder inhaler. *Pharm. Res.*, **24**, 1450–1456.

Desjardins, O., Moureau, V., and Pitsch, H. (2008). An accurate conservative level set/ghost fluid method for simulating turbulent atomization. *J. Comp. Phys.*, **227**, 8395–8416.

Dhotre, M. T., and Joshi, J. B. (2007). Design of a gas distributor: Three-dimensional CFD simulation of a coupled system consisting of a gas chamber and a bubble column. *Chem. Eng. J.*, **125**, 149–163.

Elghobashi, S. (2006). An updated classification map of particle-laden turbulent flows. *IUTAM Symposium on Computational Approaches to Multiphase Flow*, 3–10.

Fadlun, E. A., Verzicco, R., Orlandi, P., and Mohd-Yusof, J. (2000). Combined immersed-boundary finite-difference methods for three-dimensional complex flow simulations. *J. Comput. Phys.*, **161**, 35–60.

Feng, Z. G., and Michaelides, E. E. (2004). The immersed boundary-lattice Boltzmann method for solving fluid-particle interaction. *J. Comput. Phys.*, **195**, 602–628.

Garg, R., Galvin, J., Li, T., and Pannala, S. (2010). Documentation of open-source MFIX-DEM software for gas-solid flows. From URL https://m?x.netl.doe.gov/documentation/dem doc 2010.pdf

Goldstein, D., Handler, R., and Sirovich, L. (1993). Modeling a no-slip flow boundary with an external force field. *J. Comput. Phys.*, **105**, 354–366.

Guo, Z., Zheng, C., and Shi, B. (2002). Discrete lattice effects on the forcing terms in the lattice Boltzmann method. *Phys. Rev. E.*, **65**, 046308.

He, X., Chen, S., and Zhang, R. (1999). A lattice Boltzmann scheme for incompressible multiphase flow and its application in simulation of Rayleigh-Taylor instability. *J. Comput. Phys.*, **152**, 642–663.

Hilliard, J. E. (1970). Spinodal decomposition. *Phase Transformations*, ed. H. I. Aaronson. Metals Park, OH: Am. Soc. Metals.

Hirt, C. W., and Nichols, B. D. (1981). Volume of fluid (VOF) method for the dynamics of free boundaries. *J. Comput. Phys.*, **39**, 201–225.

Hu, H. H., Patankar, N. A., and Zhu, M. Y. (2001). Direct numerical simulation of fluid-solid systems using the arbitrary Lagrangian-Eulerian technique. *J. Comp. Phys.*, **169**, 427–462.

Jacqmin, D. (1999). Calculation of two-phase Navier-Stokes flows using phase-field modeling. *J. Compt. Phys.*, **155**, 96–127.

Ladd, A. J. C. (1994). Numerical simulations of particulate suspensions via a discretized Boltzmann equation. Part 1. Theoretical foundation. *J. Fluid Mech.*, **271**, 285–309.

Ladd, A. J. C., and Verberg, R. (2001). Lattice-Boltzmann simulation of particle-fluid suspensions. *J. Stat. Phys.*, **104**, 1191–1251.

Lallemand, P., and Luo, L-S. (2003). Lattice Boltzmann method for moving boundaries. *J. Comput. Phys.*, **184**, 406–421.

Lee, T., and Lin, C. (2005). A stable discretization of the lattice Boltzman equation for simulation of incompressible two-phase flows at high density ratio. *J. Comput. Phys.*, **206**, 14–47.

Lohner, R. (1998). Some useful data structures for the generation of unstructured grids. *Comm. Appl. Numerical Methods*, **4**, 123–135.

Magnaudet, J., Riverot, M., and Fabre, J. (1995). Accelerated flows past a rigid sphere or a spherical bubble. Part 1. Steady straining flow. *J. Fluid Mech.*, **284**, 97–135.

Mittal, I. R., and Iaccarino, G. (2005). Immersed boundary methods. *Ann. Rev. Fluid Mech.*, **37**, 239–261.

Mohd-Yusof, J. (1997). Combined immersed boundaries/B-splines methods for simulations of flows in complex geometries. *CTR Annual Research Briefs*, NASA Ames/Stanford University.

Patankar, S. V. (1980). *Numerical Heat Transfer and Fluid Flow*. Taylor & Francis.

Patankar, S. V., and Spalding, D. B. (1972). A calculation procedure for heat, mass and momentum transfer in three-dimensional parabolic flows. *Int. J. Heat Mass Transfer*, **15**, 1787–1806.

Prakash, C., and Patankar, S. V. (1985). A control-volume-based finite-element method for solving the Navier-Stokes equations using equal-order velocity-pressure interpolation. *Num. Heat Transfer*, **8**, 259–280.

Rhie, C. M., and Chow, W. L. (1983). Numerical study of the turbulent flow past an airfoil with trailing edge separation. *AIAA J.*, **21**(10), 1525–1532.

Ryskin, G., and Leal, L. G. (1984). Numerical solution of free-boundary problems in fluid mechanics. Part 1, the finite-difference technique. *J. Fluid Mech.*, **148**, 1–17.

Shan, X., and Chen, H. (1993). Lattice Boltzmann model for simulating flows with multiple phases and components. *Phys. Rev. E*, **47**, 1815–1819.

Shan, X., and Chen, H. (1994). Simulation of nonideal gases and liquid-gas phase transitions by the lattice Boltzmann equation. *Phys. Rev. E*, **49**, 2941–2948.

Spalding, D. B. (1980). Numerical computation of multi-phase fluid flow and heat transfer. *Recent Adv. Numerical Methods Fluids*, **1**, 139–167.

Sussman, M., Fatemi, E., Smereka, P., and Osher, S. (1998). An improved level set method for incompressible two-phase flows. *Comput. Fluids*, **27**, 663–680.

Swift, M. R., Orlandini, E., Osborn, W. R., and Yeomans, J. M. (1996). Lattice Boltzmann simulations of liquid-gas and binary fluid systems, *Phys. Rev. E*, **54**, 5041–5052.

Syamlal, M. (1998). MFIX documentation: Numerical technique. *Tech. Rep. DOE/MC31346-5824 (DE98002029)*, MET Center, Morgantown, WV.

Tryggvason, G., Bunne, B., Esmaeeli, A., *et al.* (2001). A front-tracking method for the computations of multiphase flow, *J. Comput. Phys.* **169**, 708–759.

Udaykumar, H. S, Mittal, R, Rampunggoon, P., and Khanna, A. (2001). A sharp interface Cartesian grid method for simulating flows with complex moving boundaries. *J. Comput. Phys.*, **174**, 345–380.

Unverdi, S. O. and Tryggvason, G. (1992). A front-tracking method for viscous, incompressible multi-fluid flows. *J. Comput. Phys.*, **100**, 25–37.

van der Graaf, S., Nisisako, T., SchroeÈn, C. G. P. H., van der Sman, R. G. M., and Boom, R. M. (2006). Lattice Boltzmann simulations of droplet formation in a T-shaped microchannel. *Langmuir*, **22**, 4144–4152.

Vasquez, S. A., and Ivanov, V. A. (2000). A phase coupled method for solving multiphase problems on unstructured meshes. *Proceedings of ASME FEDSM 2000*, 743–748.

Versteeg, H. K., and Malalasekera, W. (2007). *An Introduction to Computational Fluid Dynamics: the Finite Volume Method*. Prentice Hall.

Witt, A., Gulliver, J., and Shen, L. (2015). Simulating sir entrainment and vortex dynamics in a hydraulic jump. *Int. J. Multiphase Flow*, **72**, 165–180.

Wolf-Gladrow, D. A. (2000). *Lattice-Gas Cellular Automata and Lattice Boltzmann Models: An Introduction (Lecture Notes in Mathematics).* Springer.

Wu, J., and Aidun, C. K. (2010). Simulating 3D deformable particle suspensions using lattice Boltzmann method with discrete external boundary force. *Int. J. Num. Methods Fluids,* **62,** 765–783.

Youngs, D. L. (1982). Time-dependent multi-material flow with large fluid distortion. *Numerical Methods for Fluid Dynamics.* Ed. K. W. Morton and M. J. Baines, Academic Press.

Yu, Z., and Fan, L.-S. (2009). An interaction potential based lattice Boltzmann method with adaptive mesh refinement (AMR) for two–phase flow simulation. *J. Comput. Phys.,* **228,** 6456–6478.

Yuan, P., and Schaefer, L. (2006). Equation of state in a lattice Boltzmann model. *Phys. Fluids,* **18,** 042101.

Zhou, Q., and Fan, L.-S. (2014). A second-order accurate immersed boundary-lattice Boltzmann method for particle-laden flows. *J. Comput. Phys.,* **268,** 269–301.

8   Measurement Techniques

Measurement techniques are of significance in monitoring, controlling, and diagnosing multiphase processes, and they are essential components of multiphase flow research seeking to discover novel flow phenomena, validate experimental results, and support modeling and numerical simulations. Measurements of multiphase flows are more complicated than those of single-phase flows because a multiphase flow system is far more complex and involves a much larger set of flow variables, and a variety of flow patterns and regimes. Although some of the techniques for single-phase flow measurement can be extended to multiphase measurement, many specialized techniques have been developed on measuring dispersed phase properties in multiphase flows. As excellent reference material for measurements of single-phase flows is available in abundance (*e.g.*, Holman, 2012), this chapter is scoped to introducing techniques that are uniquely applicable to multiphase flows.

8.1   Introduction

A complete measurement consists of two stages: sampling and data analysis. Sampling in situ refers to data acquisition in place, often on-site, without isolation from conditions that may influence the flow. The in situ measurement category can be further subdivided into two groups. In the first group, a specimen is removed from the primary process flow and analyzed. This group of approaches is often employed to measure the particle size distribution and particle morphology. The second group is an online method, where measurements are made within the actual process flow directly or via a representative bypass sample stream. Online approaches can be further classified into the integral measurements, local-point measurements, and full-field measurements; full-field techniques are derived from the spatial representation of the measurements and are often referred to as such. The integral method measures the average, or the integral, of a quantity over a specified cross section or sampling path. Local measurements provide transient information of a variable at a single point in space. A full-field measurement produces a complete spatial distribution of a variable in a defined cross section or volume. In addition to spatial representation methods, other criteria exist that further classify the measurement techniques. For example, measurement techniques may be organized by the physical principles employed, including mechanical, electrical, optical, acoustical, and other mechanisms. Measurement techniques can even be divided into invasive methods and noninvasive methods, depending on how the technique interacts with the flow to be studied.

The objective of measurement is to obtain information about the different variables that characterize the multiphase flows. These variables are typically present in the governing equations, such as the transport of mass, momentum, energy, species,

Table 8.1-1 Typical Measurement Techniques for Multiphase Flows.

Property	Off-line methods	Online methods		
		Integral	Local point	Full field
Particle size and morphology	Microscopy (SEM, TEM) Sieving Sedimentation	N/A	Optical visualization Laser scattering Isokinetic sampling with cascade impactors	N/A
Volume fraction		Optical beam attenuation γ-ray densitometry	Optical fiber probe	Magnetic resonance imaging
		Capacitance transducer		Ultrasonic or electrical tomography
		Microwave transducer		Positron emission tomography
Position			Tracer tracking	Particle image velocimetry
Mass flux			Isokinetic sampling	
Velocity		Dual-beam cross-correlations Venturimeter	Optical fiber probe Laser doppler anemometry	Particle image velocimetry
Temperature	N/A	N/A	Thermocouple, Optical pyrometer Rainbow thermometry	
Pressure	N/A	N/A	Manometer Pressure transducer	
Species			Gas chromatography	Planar laser-induced fluorescence
Electric charge	Faraday cage		Electrostatic probe	

and electric charge. For each phase in a multiphase flow, the descriptive variables may include the volume fraction, velocity, pressure, temperature, or species concentration. For a dispersed phase, the variables to be measured may also include the particle size distribution and particle morphology. Table 8.1-1 organizes a list of typical measurement techniques for the major transport variables of interest in multiphase flows. Other less common measurement techniques such as nuclear magnetic resonance (NMR) imaging (Gladden and Sederman, 2013) and confocal microscopy (Hemminger *et al.*, 2007) have also gained attention. In this chapter, typical measurement methods will be introduced for the quantifiable transport variables in a multiphase flow.

8.2   Particle Size and Morphology Measurement

Particles in a multiphase flow are dispersed in the carrying fluid flow and can take the form of solid particles, droplets, or bubbles. How the size of a particle is defined depends on the shape, also known as morphology, of the particle under consideration. Depending on the measurement technique employed, several unique size values may result for an isolated particle; an exception is perfect spheres, which have only one size description. The measurement techniques available to characterize particles by size can be grouped into mechanical methods (*e.g.*, sieving), optical and electronic methods (*e.g.*, optical image, microscopy, diffraction, laser scattering, Doppler phase shift, and Coulter counting), and dynamic methods (*e.g.*, sedimentation). The particle size measurement procedure can be performed either dynamically on particles in flight or off-line on particle samples collected prior to, during, or after the experiment concludes. Table 8.2-1 delineates typical techniques used to measure the size of solid particles and provides an applicable size range for each (Fan and Zhu, 1998).

Table 8.2-1 Typical Size Measurement Methods for Solid Particles.

Method	Size range (μm)
Sieving	
Woven wire	37–5,660
Electroformed	5–120
Punched plate	50–125,000
Microscopy	
Optical	0.8–150
Electron	0.001–5
Sedimentation	
Gravitational	5–100
Centrifugal	0.001–1,000
Fraunhofer diffraction	0.1–1,000
Doppler phase shift	1–10,000

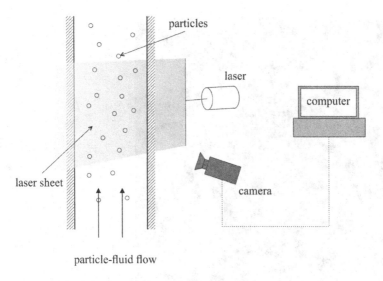

Figure 8.2-1 Schematic diagram of optical visualization system.

8.2.1   Optical Visualization Methods

One of the popular methods to examine particle shape and size is optical visualization. This technique is often performed as an in situ method, especially common for multiphase flows with constituent droplets and bubbles that challenge the effectiveness of off-line sizing approaches. The optical visualization method is usually based on a two-dimensional image projected on the collection plane. Figure 8.2-1 depicts a schematic representation of the arrangement.

Optical visualization methods are typically used to analyze large particles and low velocity flows, but its application range is more precisely defined by available image resolution and exposure time. The inherent requirement of optical access also tends to limit the applicability of this method to dilute multiphase flows. Figure 8.2-2 shows a photographic image of rising bubbles in a bubble column (under a high pressure) obtained using a visualization system. Optical visualization methods are prolific in two dimensions, but there are also three-dimensional extensions available, such as holography.

8.2.2   Microscopy Methods

Microscopy methods determine particle size distribution via image amplification that facilitates direct visualization of individual particles. This practice is often employed in an off-line setting, and it is suitable to cases of much smaller particles than optical visualization methods. Particle images are commonly obtained using one of three techniques: optical microscopy, transmission electron microscopy (TEM), and scanning electron microscopy (SEM). If using TEM, an electron beam is transmitted through an ultrathin film containing a particle sample; an image results from the interaction of the electrons with the specimen. SEM uses a fine beam of electrons to

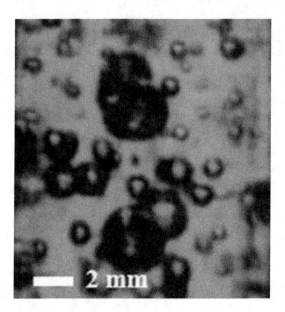

Figure 8.2-2 Image of rising bubbles in a bubble column (under a high pressure).

scan the sample surface. With microscopy methods, diffraction determines the lower limit of particle size resolution, which is proportional to the wavelength of the light or the de Broglie wavelength of the electron beam. Optical microscopy is typically utilized to measure particle size in the micrometer range, while SEM and TEM are commonly employed for expected particle sizes in the nanometer range. TEM technically has a higher resolution when compared to SEM, but the advantage of SEM is its superior ability to examine a larger area or even bulk materials. In addition to particle sizing, SEM and TEM also provide descriptive information such as particle pore structure and particle shape, or morphology (Kay, 1965; Hay and Sandberg, 1967). Figure 8.2-3 illustrates representative images obtained using SEM and TEM.

8.2.3   Sieving Analysis

Sieving is another popular off-line method for determining the size distribution of granular materials. In a sieving analysis, the particle sample is encouraged to pass through a series of sieves with successively decreasing mesh size; the particle size distribution is computed from the weight of the particles collected in each sieve bin. The sieves, or screens, are constructed of woven wire mesh with a standardized free area. There are currently two dominant standards used to govern mesh size: the Tyler Equivalent and the U.S. Sieve Series ASTM Standard. Table 8.2-2 illustrates the parameters for each. Generally, the mesh number of a sieve is defined as the number of wires per linear inch.

The weight of particles accumulated during sieving is affected by the sieving time. Excessive sieving will cause particle attrition and wear. It introduces error in the

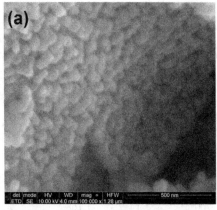 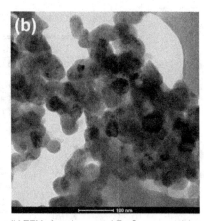

(a) SEM of agglomerated Fe_2O_3 nanoparticles   (b) TEM of agglomerated Fe_2O_3 nanoparticles

Figure 8.2-3 Typical images of SEM and TEM size sampling.

particle size distribution measurement, particularly at the limits of the predicted distribution. There are currently no viable theories or reliable empirical formulae to estimate the optimum sieving time. Dry powders, with small particles in the range of 40 to 71 μm, generally require 20 to 30 minutes of hand or machine sieving. However, larger particles, over 160 μm, often require only 5 to 10 minutes of sieving (Crowe *et al.*, 1998). In addition to hand and machine sieving, air-jet sieving is frequently applied to cohesive and large wet powders; wet sieving with micro-mesh sieves is also available and is useful for small wet powders.

8.2.4   Sedimentation Methods

The sedimentation method relies on the physical fact that particles with different sizes, or densities, have unique settling velocities in a gaseous medium or a liquid, when subjected to either gravity or a centrifugal force. Thus, the characteristic particle "size" determined by sedimentation is a dynamic diameter corresponding to a sphere of identical material density that settles at an equivalent terminal velocity. There are several ways to accomplish sedimentation. Two sedimentation methodologies are jetting particle samples into a stagnant and immiscible fluid, as shown in Figure 8.2-4, and introducing particle samples into a traverse flow to measure the various penetration depths.

The particle sample can also be introduced into a liquid by placing a thin layer of particles at the top of the liquid column to study their time-dependent settling characteristics, as shown in Figure 8.2-5. To uniformly distribute the particle in the whole column or to avoid agglomeration of the particles, additional provisions may be required, including manual or mechanical mixing, ultrasonic agitation, or the addition of a dispersing agent.

For a uniformly dispersed particle suspension, two approaches can be used to obtain the particle sedimentation rate. The first, known as the incremental method,

Table 8.2-2 Tyler Standard and U.S. ASTM Sieve Series.

Tyler Standard			U.S. Series ASTM Standard		
Mesh no.	Size (μm)	Wire diameter (μm)	Mesh no.	Size (μm)	Wire diameter (μm)
3½	5,660	1,280–1,900	3½	5,613	1,650
4	4,760	1,140–1,680	4	4,699	1,650
5	4,000	1,000–1,470	5	3,962	1,120
6	3,360	870–1,320	6	3,327	914
7	2,830	800–1,200	7	2,794	833
8	2,380	740–1,100	8	2,362	813
10	2,000	680–1,000	9	1,981	838
12	1,680	620–900	10	1,651	889
14	1,410	560–800	12	1,397	711
16	1,190	500–700	14	1,168	635
18	1,000	430–620	16	991	597
20	840	380–550	20	833	437
25	710	330–480	24	701	358
30	590	290–420	28	589	318
35	500	260–370	32	495	300
40	420	230–330	35	417	310
45	350	200–290	42	351	254
50	297	170–253	48	295	234
60	250	149–220	60	246	179
70	210	130–187	65	208	183
80	177	114–154	80	175	142
100	149	96–125	100	147	107
120	125	79–103	115	124	97
140	105	63–87	150	104	66
170	88	54–73	170	88	61
200	74	45–61	200	74	53
230	62	39–52	250	61	41
270	53	35–46	270	53	41
325	44	31–40	325	43	36
400	37	23–35	400	38	25

determines the particle concentration at a depth, h, below the surface through the light attenuation method, or X-ray attenuation. Figure 8.2-5(a) provides a conceptual representation. Assume that all particles settle at their respective terminal velocities. Hence, all particles of size larger than d_{pi}, with the terminal velocity $U_{pti} > h/t_i$,

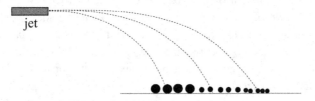

Figure 8.2-4 Sedimentation of a horizontal spray jet.

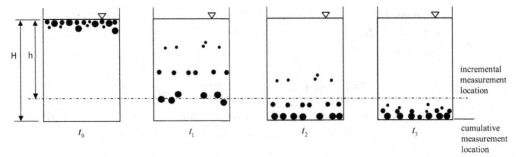

(a) incremental or cumulative measurement locations

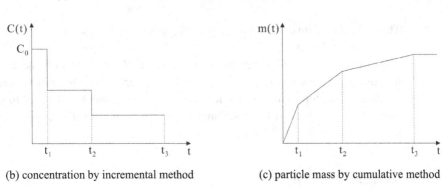

(b) concentration by incremental method   (c) particle mass by cumulative method

Figure 8.2-5 Time-dependent sedimentation measurements.

will have passed the measurement location at time $t > t_i$. Therefore, the measured change of the particle mass concentration, $c(t)$, as a function of time corresponds to the cumulative particle size distribution $F_m(d_p)$, which is defined as the fraction of the particles less than d_p. Figure 8.2-5(b) depicts the pertinent measurement data. In general, the cumulative particle size distribution is given by:

$$F_m(t) = 1 - \frac{c(t)}{c_0},$$

(8.2-1)

where c_0 is the initial particle mass concentration. Considering the balance between gravity, buoyancy, and the drag force and if Stokes drag is assumed for all particles tested, the particle size is uniquely determined from the settling time, t, by the relationship:

$$d_p = \sqrt{\frac{18\,\mu h}{(\rho_p - \rho_l)gt}}.$$

(8.2-2)

Using Eq. (8.2-1) with Eq. (8.2-2) and neglecting interactions among particles and the volumetric replacement effect of particles on the fluid during the settling, the cumulative particle size distribution, $F_m(d_p)$, can be estimated from the incremental method.

The second approach measures the cumulative mass of particles at the bottom of the test column. The pertinent data is illustrated in Figure 8.2-5(c). At time t, the mass accumulated on the column bottom, $m(t)$, corresponds to the large particles that have completely settled from the liquid and the small particles that have partially settled. Thus, it yields:

$$m(t) = m_0 \left(1 - F_m(d_p)\right) + m_0 \int_0^{d_p} \frac{t U_{pt}(x)}{H} f_m(x) dx, \qquad (8.2\text{-}3)$$

where f_m is the mass-based size density distribution function. Taking the derivative of the above expression with respect to time yields the following:

$$F_m(d_p) = t \frac{d}{dt}\left(\frac{m}{m_0}\right) - \frac{m}{m_0} + 1. \qquad (8.2\text{-}4)$$

Thus, the cumulative size distribution, F_m, can be calculated from the measured curve of $m(t)$ (Bishop, 1934).

Accuracy of the sedimentation method is affected by convection and Brownian motion during particle settling. The lower limit of particle size feasibly measured by gravitational sedimentation is about 5 μm. This method can be extended to smaller particles by using centrifugal sedimentation, which is normally applicable for a particle size range of 0.001 μm to 1 mm.

8.2.5   Cascade Impaction

When particles are sufficiently small, the sedimentation method becomes inefficient, and likely unfeasible, because of inherently protracted settling times. To mitigate this potential drawback, an inertial technique is employed in a class of devices known as cascade impactors. This apparatus samples and classifies particles via inertia from jet impingement. A single-stage inertial impactor separates impinging particulates into two groups: Particles larger than the cutoff aerodynamic size of the impactor are collected on the impaction plate, and effectively removed, from the particulate-laden stream, and those smaller than that cutoff size remain airborne and continue with the flow path. To improve the functionality, efficiency, and accuracy of the aerodynamic size classification process, cascade impactors have been augmented to include several series-connected single-stage impactors with different cutoff sizes, as shown in Figure 8.2-6.

When sampling a polydispersed particulate flow having a wide particle size distribution, it is a common practice to employ a simple particle presizing device, such as a settling chamber or cyclone, to remove very coarse particles before they enter the cascade impactor. Similarly, the final stage of the impactor often includes a subsequent backup filter to capture any remaining fines. Thus, a multistage cascade impactor yields particle mass distribution using the principle of aerodynamic particle sizing.

Cascade impactors can be operated with in situ sample devices, such as isokinetic sampling probes, or as an off-line technique. In the latter case, a particle resuspension device, such as a fluidized bed, may be necessary to preprocess the specimen prior to jetting. Cascade impactors are typically used for particle-laden flows that have a large density ratio of particles to the carrying fluid media, such as gas–solid flows

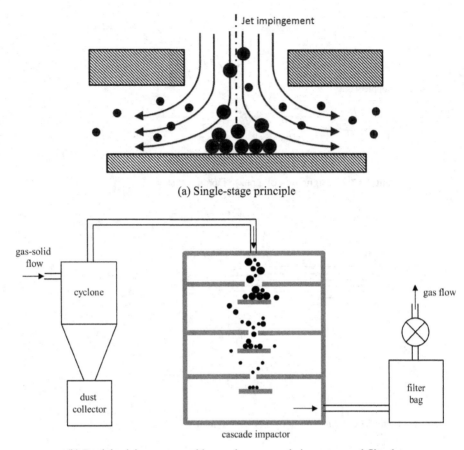

(a) Single-stage principle

(b) Particle sizing system with a cyclone, cascade impactor, and filter bag

Figure 8.2-6 Cascade impactor system.

and liquid–gas atomized sprays. Measurements obtained from cascade impactors can be employed to determine the chemical compositions of particles in addition to the aerodynamic size distribution of particles. On these merits, various designs of cascade impactors with unique nozzle jet configurations and collection surfaces have been developed (Zhu, 1999).

8.2.6   Phase Doppler Method

When a monochromatic wave, such as a laser beam, is reflected from a moving spherical particle, a frequency and phase shift results between the incident wave and reflected wave; this phenomenon is known as the Doppler effect. Notably, frequency shift can be related to translation velocity of a particle, and phase shift is a function of particle size (Bachalo, 1980). Therefore, the Doppler effect is a useful indicator

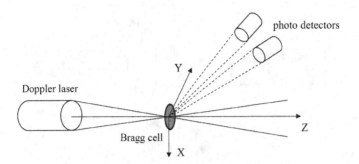

Figure 8.2-7 Configuration of a phase Doppler system.

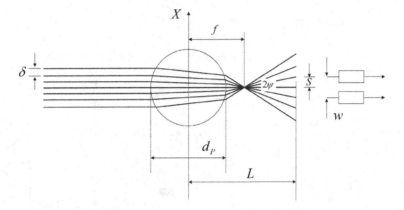

Figure 8.2-8 Fringe model for phase Doppler size measurement.

of particle velocity and size; this discovery is embodied in the phase Doppler particle analyzer (Sankar and Bachalo, 1991). Determining particle size through Doppler
phase is introduced in this section, and the particle velocity measurement from the
Doppler frequency is discussed in Section 8.5.3 on velocity measurements.

In the measurement of frequency shift or phase shift, traditionally a dual-beam
configuration by splitting the same source beam is adapted. Figure 8.2-7 shows such
a configuration in which a laser beam is introduced into the Bragg cell and split into
two interfering beams, which are then focused to the control volume of measurement.
The light refracted by a particle inside the measuring volume is received by two
photodetectors.

The principle of particle size measurement by phase Doppler can be explained
using the fringe mode, as shown in Figure 8.2-8 for a transparent particle (Bachalo
and Sankar, 1998).

Due to interference, the two intersecting incident beams produce a fringe pattern with spacing in the measuring volume. The spacing, δ, can be related to the
wavelength, λ, and the half angle of beam intersection, θ, by the expression:

$$\delta = \frac{\lambda}{2 \sin \theta}.$$

(8.2-5)

The parallel fringes are focused by the transparent particle, which acts as a lens with a focal length, l. The projected spacing, s, between the fringes as perceived by the photodetectors at a distance L from the particle can be approximated by:

$$s \approx (L - l)\frac{\delta}{l}. \tag{8.2-6a}$$

The focal length of a spherical particle is given by the relationship:

$$l = \frac{\gamma}{\gamma - 1}\left(\frac{d_p}{4}\right), \tag{8.2-6b}$$

where γ is the ratio of relative refraction indices of the particle and the surrounding medium. Therefore, under the imposed condition $L \gg l$, the projected fringe spacing, s, on the detection plane is approximated by:

$$s \approx \frac{4(\gamma - 1)}{\gamma}\frac{L\delta}{d_p}. \tag{8.2-6}$$

As the particle moves through the measuring volume, the fringe pattern similarly moves across the two photodetectors, which are separated by a distance w. The signals perceived by the two photodetectors thus have a phase difference of:

$$\Delta\phi = \frac{2\pi w}{s} = \frac{4\pi L}{s}\sin\psi \tag{8.2-7}$$

where ψ is the diverging angle defined in Figure 8.2-8. Using Eq. (8.2-5) and Eq. (8.2-6), Eq. (8.2-7) yields the relationship between the phase difference and particle size as:

$$\Delta\phi = \frac{2\pi\gamma}{(\gamma - 1)}\frac{d_p}{\lambda}\sin\theta\sin\psi \tag{8.2-8}$$

The analysis described above, and its result, are valid only for small-angle refraction. In practice, the phase Doppler method can be applied for light scattering of particles by refraction as well as by reflection. A generalized relationship between particle size and phase difference can be expressed for a two-photodetector arrangement as:

$$d_p = \frac{\lambda}{2\pi n_m}\frac{\Delta\phi}{\Phi}, \tag{8.2-9}$$

where n_m is the relative refraction index of surrounding medium, and Φ is determined by the scattering mode. In the reflection mode, Φ is given by:

$$\Phi = \sqrt{2}(1 + \sin\theta\sin\psi - \cos\theta\cos\psi\cos\varphi)^{1/2}$$
$$- \sqrt{2}(1 - \sin\theta\sin\psi - \cos\theta\cos\psi\cos\varphi)^{1/2}, \tag{8.2-9a}$$

where φ is the scattering angle. In the refraction mode, Φ becomes:

$$\Phi = 2\left[1 + \gamma^2 - \sqrt{2}\gamma(1 + \sin\theta\sin\psi - \cos\theta\cos\psi\cos\varphi)^{1/2}\right]^{1/2}$$
$$- 2\left[1 + \gamma^2 - \sqrt{2}\gamma(1 - \sin\theta\sin\psi - \cos\theta\cos\psi\cos\varphi)^{1/2}\right]^{1/2}. \tag{8.2-9b}$$

More references on laser-based optical measurement techniques of particle sizing can be found in many reviewed articles (*e.g.*, Gouesbet and Gréhan, 2015).

8.2.7   Particle Size Distribution and Averaged Size

Particle size distribution is represented by the particle size density function (pdf), which is defined as the fraction of particles within a given size range. The pdf can be formulated either as a function of the quantities or mass of sample particles. The measured category of pdf depends on the nature of measurement techniques employed, such as number-based pdf from optical methods or mass-based pdf from sieving and sedimentation methods.

The number density function is interconvertible with its corresponding mass density function by the expression:

$$f_m(x) = \frac{N_T m}{M_T} f_n(x), \qquad (8.2\text{-}10)$$

where N_T and M_T represent the total particle quantity and mass of the sample, respectively; m is the mass of a single particle of size, x. The relationship between m and x depends on the density and morphology of the particles. For a spherical particle, m can be written as:

$$m = \frac{\pi}{6} \rho_p x^3 . \qquad (8.2\text{-}10a)$$

Both the number density function and mass density function are subjected to a normalized condition. Thus, it gives:

$$\int_0^\infty f_n(x)dx = 1 \qquad \int_0^\infty f_m(x)dx = 1. \qquad (8.2\text{-}10b)$$

Particle size distribution can be obtained from measurement techniques such as an optimized cascade impactor, which consists of a series of stages with each stage collecting particles larger than a specific size while smaller particles should pass to subsequent collectors. In this manner, the distribution of sampled mass on each collection stage, along with the backup filter for the final stage, should directly represent the true mass distribution of particles. In practice though, very few stages are capable of such idealized cutoff characteristics. The realistic collection efficiency curve of a given stage, with respect to the particle size, typically traces an "S" shape, as shown in Figure 8.2-9.

As the figure illustrates, there is an overlap of particle size among prior and subsequent cascading stages, which means that the same-sized particles can be found on multiple collectors. Due to this inherent sorting inefficiency, the actual mass distribution needs to be obtained via deconvolution of the measurements. This process is often accomplished with the aid of modeling, as exemplified in the following.

If the single-stage collection efficiency of particle size x on the ith stage is denoted as $\eta_i(x)$, the actual collection of particle size x on the jth stage in a series of collectors

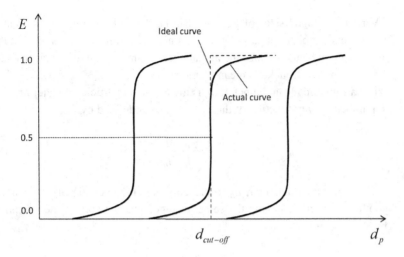

Figure 8.2-9 Collection efficiency distributions of a multistage collector.

may be expressed as:

$$\kappa_j(x) = \eta_j(x) \prod_{i=1}^{j-1} (1 - \eta_i(x)). \tag{8.2-11}$$

The predicted mass collected on the ith stage, m_{pi}, can then be computed using:

$$m_{pi} = \frac{m_T \int_0^\infty f_m(x)\kappa_i(x)\,h_i(x)dx}{\int_0^\infty f_m(x)[1 - \eta_0(x)]dx}, \tag{8.2-12}$$

where $f_m(x)$ is the particle mass density distribution function; m_T is the measured total mass concentration; the denominator represents the mass removed by the cascade impactor inlet (*i.e.*, stage $i = 0$); $h(x)$ accounts for the effect of losses at the wall between stages, which is given by:

$$h_1(x) = 1 - WL_1(x)$$
$$h_i(x) = h_{i-1}[1 - \eta_{i-1}(x)] - WL_i(x), \tag{8.2-13}$$

where $WL_i(x)$ is the wall loss factor of particle size x between the $(i - 1)$th stage and the ith stage. In practice, $\eta_i(x)$ and $WL_i(x)$ are predetermined from a calibration of the ith stage. Thus, the particle mass distribution, $f_m(x)$, can be determined by use of deconvolution on matching the predicted mass, m_{pi}, to the measured mass collected at the stage of interest, m_i. Deconvolution, in general, does not yield a unique solution. Instead, there are an infinite number of possible solutions that can fit the same set of measurements. Hence, for most engineering applications, a paradigmatic particle size distribution should be intuitively assumed to achieve a unique solution. An example of using deconvolution, specifically the chi-squared method, to extract the particle size distribution from cascade impactor data is provided in Section 8.7, where a multimodal lognormal size distribution is assumed (Dzubay and Hasan, 1990).

Various averaged sizes of particles can be defined for a given size distribution. For systems of spherical particles, commonly used averaged sizes include the arithmetic mean diameter, surface mean diameter, volume mean diameter, Sauter's mean diameter, and DeBroucker's mean diameter.

The arithmetic mean diameter, d_l, is the averaged particle diameter predicated on the number density function of the sample; this is defined by:

$$d_1 = \frac{\int_0^\infty x f_n(x) dx}{\int_0^\infty f_n(x) dx}. \tag{8.2-14a}$$

The surface mean diameter, d_S, represents the characteristic diameter of a single hypothetical particle having the same averaged surface area as the overall sample. This quantity is computed as:

$$d_S = \left(\frac{\int_0^\infty x^2 f_n(x) dx}{\int_0^\infty f_n(x) dx} \right)^{\frac{1}{2}}. \tag{8.2-14b}$$

The volume mean diameter, d_V, represents the equivalent diameter of a single fictitious particle with an identical averaged volume as the specimen; this is determined from:

$$d_V = \left(\frac{\int_0^\infty x^3 f_n(x) dx}{\int_0^\infty f_n(x) dx} \right)^{\frac{1}{3}}. \tag{8.2-14c}$$

Sauter's mean diameter, d_{32}, is represented as the diameter of a theoretical single particle with the same averaged specific surface area per unit volume as the sample being studied; this is obtained from the relationship:

$$d_{32} = \frac{\int_0^\infty x^3 f_n(x) dx}{\int_0^\infty x^2 f_n(x) dx}. \tag{8.2-14d}$$

DeBroucker's mean diameter, d_{43}, is the averaged diameter based on the mass density function of the sample, and it can be evaluated by:

$$d_{43} = \frac{\int_0^\infty x^4 f_n(x) dx}{\int_0^\infty x^3 f_n(x) dx} = \frac{\int_0^\infty x f_m(x) dx}{\int_0^\infty f_m(x) dx}. \tag{8.2-14e}$$

Selecting an appropriate averaged diameter for a particle system depends on the specific application. For instance, a pulverized coal combustion process benefits from the surface area per unit volume. In this case, Sauter's mean diameter is a strong candidate. A similar nuanced approach is required for other scenarios. It should also be noted that both Sauter's and DeBroucker's mean diameters in Eq. (8.2-14) are defined for a range of particle size; this presentation may differ from Sauter's or DeBroucker's mean diameter as constructed for a nonspherical single particle.

8.3   Volume Fraction Measurement

Measuring the volume fraction of each phase is an essential part of multiphase flow measurements. Volume fraction measurement relies on the fact that different phases have contrasting physical properties that respond predictably to the transmission of light, electromagnetic waves, or acoustic waves. In an effort to balance fidelity and brevity, two types of volume fraction measurement techniques are discussed in this section. The first group consists of the beam attenuation method and the permittivity measurement process, which are both used to measure spatially averaged volume fractions over a measurement path or a cross section. These techniques are introduced in Sections 8.3.1 and 8.3.2, respectively. The second group is represented by transmission tomography and electrical impedance tomography, both of which excel at yielding a detailed spatial distribution of volume fractions for a cross section of flow. These two types of tomography techniques are introduced in Sections 8.3.3 and 8.3.4, respectively.

Tomography entails acquiring a set of integral measurements over a consistent cross section and applying an inverse method to generate the cross-sectional distribution of a physical property; the result is then manipulated to obtain the volume fraction distribution. The cross-sectional distribution of physical properties includes density in X-ray tomography, electrical permittivity in electrical impedance tomography, radiation by nuclear spins in magnetic resonance imaging tomography, or radioactive particles in positron emission tomography. Depending on the degree of resolution, tomographic techniques can also be employed to determine the particle shapes or even cluster morphology.

Selecting a tomography technique depends on the required spatial and temporal resolutions. Figure 8.3-1 depicts the spatial and temporal resolution associated with the various techniques.

Electrical tomography has the highest temporal resolution, and it is an ideal strategy for transient flow measurements. X-ray and γ-ray tomography are particularly useful where high spatial resolution is required. Another important factor to consider when choosing a tomography technique is suitability of the measurement principle for the specific application. For example, electrical capacitance tomography is applicable for dielectric materials, but electrical resistance tomography is appropriate when studying conductive materials.

8.3.1   Beam Attenuation Method

The beam attenuation method is based on the Lambert–Beer law, which is a simplified form of the general radiation transport equation discussed in Section 4.2.3. When a beam of radiation passes through a scattering medium of monodispersed spherical particles, its intensity predictably varies due to absorption, emission, and scattering by the particles. Neglecting emission from the beam itself and ignoring the scattering contribution from adjacent particles to the beam allows the radiation transport equation, Eq. (4.2-12), to be modified to capture the change in monochromatic

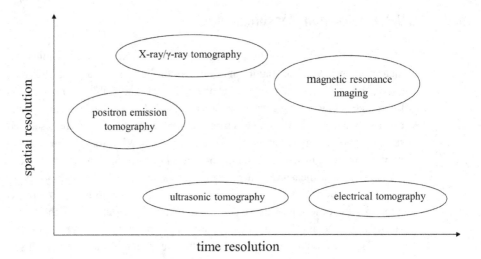

Figure 8.3-1 Time and spatial resolution of different tomography methods.

radiant intensity along the beam path. The simplification leads to:

$$\frac{dI(s)}{ds} = -\sigma_e I(s). \tag{8.3-1}$$

The extinction coefficient, σ_e, accounts for the scattering and absorption effects of the beam by the particles. For dilute particle suspensions where all interparticle effects on the beam transport can be ignored, σ_e is proportional to the particle scattering cross-sectional area and number density. This is expressed as:

$$\sigma_e = C_e \frac{\pi}{4} d_p^2 n_p, \tag{8.3-1a}$$

where C_e is the modified extinction coefficient, d_p is the particle diameter, and n_p is the particle number density. Thus, by integrating Eq. (8.3-1) over a specified beam path, the resulting attenuation in beam intensity can be used to determine the quantity of particles present, or the path-averaged volume fraction, in a particle-laden flow. It therefore yields:

$$\alpha_p = \frac{\pi}{6} d_p^3 n_p = \frac{2 d_p}{3 C_e L} \ln\left(\frac{I}{I_0}\right), \tag{8.3-2}$$

where L is the path length, and I_0 is the beam source intensity.

Equation (8.3-2) can be applied to various beam resources, including both optical beams, such as laser or collimated lights as illustrated in Figure 8.3-2, and invisible beams, such as X-ray or γ-ray, as illustrated in Figure 8.3-3.

Implementing optical beam methods requires optical access to the particle-laden flow, such as through optical windows on the pipe wall or the insertion of optical probes into the flow. Conversely, short-wavelength invisible beams, such as γ-rays, do not require optical accessibility, and they can therefore be applied to opaque flow systems. However, the superior penetrating ability of such invisible beams

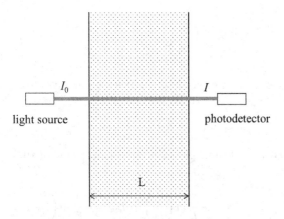

Figure 8.3-2 Measurement of particle concentration by light attenuation.

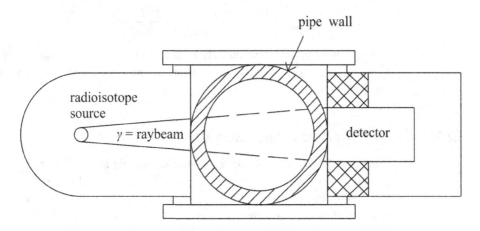

Figure 8.3-3 Single-beam γ-ray densitometer over a pipe flow (Kumara *et al.*, 2010).

consistently leads to comparatively less absorption and scattering by the particles; correspondingly reduced signal resolutions are therefore typical.

The modified extinction coefficient, C_e, in Eq. (8.3-1a) depends on fluid absorption and scattering relative to the beam as well as to other particle properties; notable properties include refraction index, size distribution, and morphology. Therefore, C_e must be calibrated for the specific flow condition, particle size distribution, and type of beam involved. For polydispersed particles, the dependence of the modified extinction coefficient on particle size may introduce a sizeable error into particle concentration computation.

The extinction coefficient, σ_e, is nearly proportional to the density of the material. For multiphase systems, the extinction coefficient of the mixture is expressed as the volume-weighted average of the extinction coefficients for each individual phase.

Thus, it gives:

$$\sigma_{em} = \sum_i \alpha_i \sigma_{ei},\tag{8.3-3}$$

where α_i is the volume fraction of phase i. For example, in a two-phase system, it can be written as:

$$\sigma_{em} = \sigma_{e1} + \alpha_2(\sigma_{e2} - \sigma_{e1}).\tag{8.3-3a}$$

Applying Eq. (8.3-1) to each distinct phase, as well as the overall mixture, yields the following expressions:

$$\sigma_{e1} = -\frac{1}{L}\ln\left(\frac{I_1}{I_0}\right) \quad \sigma_{e2} = -\frac{1}{L}\ln\left(\frac{I_2}{I_0}\right) \quad \sigma_{em} = -\frac{1}{L}\ln\left(\frac{I_m}{I_0}\right).\tag{8.3-4}$$

Thus, using Eq. (8.3-3a) together with Eq. (8.3-4), the volume fraction of phase 2 can be determined as:

$$\alpha_2 = \frac{\sigma_{em} - \sigma_{e1}}{\sigma_{e2} - \sigma_{e1}} = \frac{\ln(I_m/I_1)}{\ln(I_2/I_1)}.\tag{8.3-5}$$

A companion expression can similarly be derived for the volume fraction of phase 1. From Eq. (8.3-5), it is evident that calibration measurements of σ_{e1} and σ_{e2} are necessary.

8.3.2   Permittivity Measurement Method

Mixture permittivity and volume fractions are relatable in a well-mixed two-phase flow with sufficiently large differences in conductivity, σ, and electric permittivity, ε, between the two phases, such that $\sigma_1 \ll \sigma_2$ and $\varepsilon_1 \ll \varepsilon_2$. When phase 1 denotes continuous phase, the mixture permittivity can be related to the volume fraction of phase 2, α_2, by the simple expression:

$$\varepsilon_m = \varepsilon_1 \frac{1 + 2\alpha_2}{1 - \alpha_2}.\tag{8.3-6}$$

Conversely, when phase 2 signifies the continuous phase, the mixture permittivity is then given by:

$$\varepsilon_m = \varepsilon_2 \frac{2\alpha_2}{3 - \alpha_2}.\tag{8.3-7}$$

Thus, the volume fractions of each phase can be determined by measuring mixture permittivity, but the procedure depends on which phase forms the continuous phase. This method is often employed for gas–solid, gas–liquid, and even liquid–liquid flows, such as oil and water suspension flows.

One common technique to measure mixture permittivity is via capacitance transducers, as shown in Figure 8.3-4.

The relationship between the capacitance and the medium's permittivity depends on the structure of the capacitance sensor. For example, in a sensor comprising two

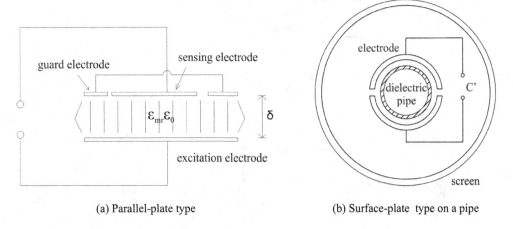

(a) Parallel-plate type   (b) Surface-plate type on a pipe

Figure 8.3-4 Schematic arrangements of capacitance transducers.

parallel plates, as shown in Figure 8.3-4(a), the measured capacitance, C, is related to the permittivity of the material between the plates through the following:

$$C = \frac{\varepsilon_{mr}\varepsilon_0 A}{\delta}, \tag{8.3-8}$$

where ε_{mr} is the relative permittivity of the mixture, ε_0 is the permittivity of free space, A is the area of the plate, and δ is the distance between the plates. In practice, the sensor is often equipped with screening and guarding components to protect it from undesired disturbances that would adversely influence the accuracy of the measurement. Figure 8.3-4(b) illustrates an example of this via a surface plate capacitance sensor installed on a circular pipe. For nonuniformly dispersed two-phase flows, the mixture permittivity parameter, ε_{mr}, of Eq. (8.3-8) depends on the spatial distribution of the phases in addition to the volume fractions of each phase (Irons and Chang, 1983).

The electric permittivity of a mixture can also be determined using microwave transducers that measure the traveling speed of a microwave through a flow medium. The traveling speed, U, of a microwave propagating in a two-phase mixture is a function of the electric permittivity, ε_{mr}, of the mixture. This is shown by:

$$U = c/\sqrt{\varepsilon_{mr}\mu_r}, \tag{8.3-9}$$

where c is the speed of light, and μ_r is the relative permeability. For liquids, such as water and oil, μ_r has a value of unity.

Figure 8.3-5 shows a cross-sectional view of a microwave transducer installed on a pipe containing a two-phase flow. The transducer consists of one microwave transmitter and two receivers placed at prescribed distances from the transmitter. The propagation of the microwave follows the wave equation:

$$w(x, t) = A_0 \sin\left[2\pi f(t - x/U)\right], \tag{8.3-10}$$

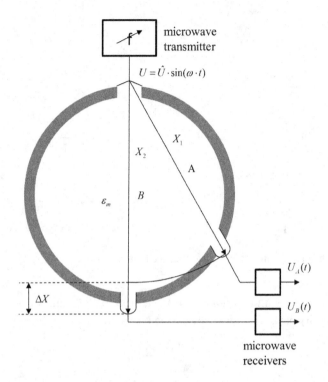

Figure 8.3-5 Schematic representation of microwave transducers.

where A_0 and f are the amplitude and frequency of the microwave, respectively. The superposition of signal received by the two receivers will always be zero, provided the following condition holds:

$$f = \frac{(1 + 2k)c}{2\Delta x\sqrt{\varepsilon_m}}, \tag{8.3-11}$$

where k is an integer. The mixture permittivity can be found from the lowest frequency, f_0, that yields zero voltage at the detector frequency. Thus, mixture permittivity is given by:

$$\varepsilon_m = \left(\frac{c}{2\Delta x f_0}\right)^2. \tag{8.3-11a}$$

Once the mixture permittivity is known, the volume fraction of the two phases can be readily determined. Flow regime and spatial distribution of the phases determine the degree of mixture permittivity dependence on volume fraction.

8.3.3   Transmission Tomography

Transmission-based tomography in volume fraction measurements and visualization of multiphase flows includes X-ray, γ-ray, and ultrasound varieties (Kumar

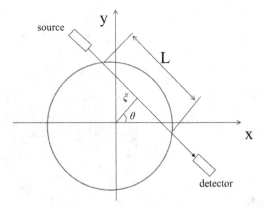

Figure 8.3-6 Principle of transmission tomography using transmission beam.

et al., 1995; Warsito *et al.*, 1999; Heindel *et al.*, 2008). The attenuation of a transmission beam passing through a scattering and absorbing medium is given by Eq. (8.3-1). The attenuation is represented as the line integral of the mass density distribution along the path, as given by:

$$P(\xi, \theta) = \int_L f(x, y)dl, \tag{8.3-12}$$

where $P(\xi, \theta)$ is the attenuation of the signal received by the detector; ξ is the distance from the origin of the coordinate, θ is the angle of projected direction, $f(x, y)$ is a function of the phase volume fractions or attenuation coefficient distribution, as illustrated in Figure 8.3-6.

By varying the distance ξ, and angular position θ, a set of measurements can be obtained for $P(\xi, \theta)$. This result is achieved by both translating and rotating the transmission source and detector pair. Hence, the spatial distribution of phases $f(x, y)$ and associated volume fractions can be computed from the obtained set of measurements $P(\xi, \theta)$ combined with solving the inverse problem of Eq. (8.3-12). Figure 8.3-7 shows the configurations of four generations of scanning modes, including: (a) translate-rotate parallel beam; (b) translate-rotate multiple-source parallel beam; (c) rotating fan; and (d) fixed detector rotating source.

The principle of ultrasonic tomography is founded on the interaction between an acoustic wave and an obstruction; the obstacle could be a dispersed phase particle, such as a solid particle or a gaseous bubble. Such interactions affect the speed and amplitude of the emitted sound wave in a predictable fashion. The change in sound wave speed is attributed to particle oscillation, sound transmission within particles, and microbubble pulsation. Attenuation of acoustic energy is the result of scattering on a bubble or particle surface, absorption by bubble pulsation, and dissipation by particle oscillation. In general, changes in the amplitude and speed of the acoustic wave depend on dispersed phase volume fraction, particle size distribution, and properties of the continuous phase. Wave amplitude and speed variations are detected

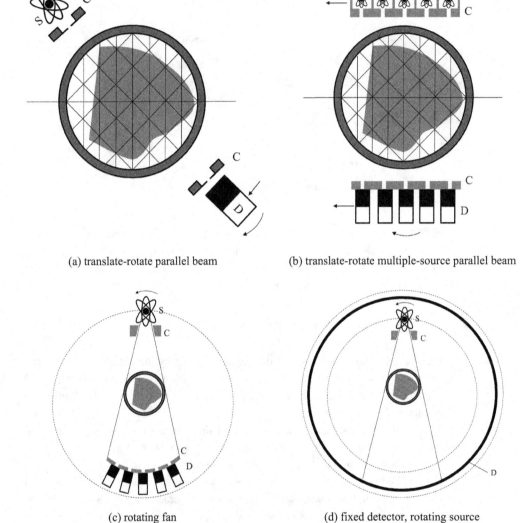

(a) translate-rotate parallel beam

(b) translate-rotate multiple-source parallel beam

(c) rotating fan

(d) fixed detector, rotating source

Figure 8.3-7 Typical scanning modes in transmission tomography.

by an ultrasonic sensor and subsequently converted to the volume fractions of the disperse phases.

Ultrasonic tomography is capable of simultaneous measurements of the gas and solid volume fractions in a slurry-bubble column (Warsito *et al.*, 1999). The difference in sound speed between a three-phase mixture and a pure liquid is expressed by the change in the transmission time. This is given by:

$$\Delta t = (K_s \alpha_s + K_g \alpha_g)L. \tag{8.3-13}$$

The energy attenuation rate in a three-phase mixture is expressed by:

$$\Omega = -\ln(I/I_0) = (X_s\alpha_s + X_g\alpha_g)L. \tag{8.3-14}$$

In the above equations, α_s and α_g are the path-averaged volume fraction of the solid and gas phases, respectively; L is the distance between the transmitter and the receiver; I_0 and I are the intensity of the incident and received acoustic waves, respectively; K_s, K_g, X_s, and X_g are coefficients that depend on the wave frequency, particle diameter, and bubble diameter, which can be calibrated in a system with known solid and gas volume fractions.

At a sufficiently high frequency, the effect of gas bubbles on wave velocity can be assumed to be negligible, which leads to $K_g \to 0$. Thus, the path integral of the volume-fraction distributions of the gas and solid phases can be derived from Eq. (8.3-13) and Eq. (8.3-14) as:

$$\int \alpha_g(x,y)dl = (\Omega - X_s\Delta t/K_s) \Big/ X_g = P_g(\xi,\theta) \tag{8.3-15}$$

$$\int \alpha_s(x,y)dl = \Delta t/K_s = P_s(\xi,\theta). \tag{8.3-16}$$

For ultrasonic tomography, Δt and Ω are calculated from the signal measured by the ultrasonic transducers; the phase distributions $\alpha_s(x,y)$ and $\alpha_g(x,y)$ are obtained by solving the inverse problem defined by Eq. (8.3-15) and Eq. (8.3-16), respectively. Reconstruction of $\alpha_s(x,y)$ and $\alpha_g(x,y)$ from the projections $P_s(s,\theta)$ and $P_g(s,\theta)$ is achieved via filtered back-projection techniques that are based on Fourier transform algorithm.

Ultrasonic tomography must be conducted in a medium through which the acoustic wave can propagate. Thus, as distinguished from other tomographic techniques, the ultrasonic variant is appropriate for media flows that are poorly penetrated by light or other electromagnetic waves. Additional advantages include comparatively reduced energy requirements, cost-effectiveness, and straightforward implementation. However, there are three major limiting factors to consider. First, the total volume fraction of the gas and solid phase needs to be less than 20% to ensure sufficient free area for the acoustic wave to penetrate the flow medium along a straight path. Second, a compound acoustic field is created by the overlap of multiple reflected waves, which can impede accuracy. To mitigate this effect, only the first transmission signal corresponding to a straight path can be used; alternatively, a high-frequency signal can be selected to reduce multiple scattering. Third, there can be reduced temporal resolution as compared to electrical methods because the speed of sound is a fraction of electromagnetic wave speed (about 1/2,000).

8.3.4   Electrical Impedance Tomography

Electrical impedance tomography is a noninvasive technique used to measure the distribution of electrical properties. Two prominent techniques are electrical capacitance tomography (ECT) and electrical resistance tomography (ERT). The former

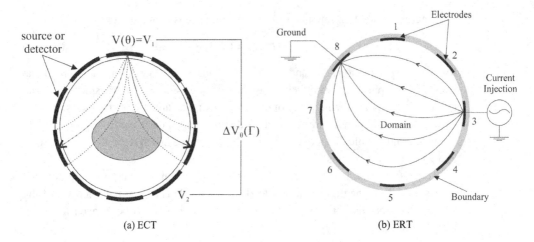

source or detector

$V(\theta)=V_1$

$\Delta V_\theta(\Gamma)$

V_2

(a) ECT

Ground

Electrodes

1

8

2

7

Current Injection

Domain

3

6

5

4

Boundary

(b) ERT

Figure 8.3-8 Arrangement of electrodes in electrical impedance tomography.

is suitable for electrically insulating, or dielectric materials; the latter is appropriate for electrically conductive materials. In both adaptations, electrodes are mounted peripherally along the circumference of the selected measuring plane. The key differentiating factor is the electrode arrangement: ECT positions the electrodes on the outer surface of the insulating column wall, and the electrodes in ERT are mounted on the inner surface in direct contact with the flow medium. Figure 8.3-8 illustrates the two variations.

In ECT measurement, as a voltage is applied sequentially to the electrodes, electrical charges accumulate on the remaining electrodes because of the capacitance between electrode pairs. The amount of charge on each electrode is measured as an electrical current. The electric field in the vessel can be described by the Poisson equation from electrostatic theory as:

$$\nabla \cdot (\varepsilon(\mathbf{r})\nabla\Phi(\mathbf{r})) = -\rho_e(\mathbf{r}). \tag{8.3-17}$$

In the above equation, $\Phi(\mathbf{r})$ is the electrical potential, $\varepsilon(\mathbf{r})$ is the permittivity, and $\rho_e(\mathbf{r})$ is the electrical charge density; all of these functions are functions of spatial coordinates. For applications where the charge density is absent in the vessel, the right-hand side of Eq. (8.3-17) is zero. The charge accumulated on the ith electrode, q_{ei}, and the applied electrical voltage across a pair of electrodes, $\Delta\Phi_i$, are related by the capacitance, which can be expressed by:

$$C_i = \frac{q_{ei}}{\Delta\Phi_i} = \frac{\iint_A \varepsilon(\mathbf{r})\nabla\Phi(\mathbf{r}) \cdot d\mathbf{A}}{\Delta\Phi_i}. \tag{8.3-18}$$

Here, the charge is integrated on the electrode surface. By measuring capacitance, the distribution of the permittivity can be obtained by solving the inverse problem defined by Eq. (8.3-18). In the sensitivity model, the relationship between the measured capacitance and the permittivity distribution can be written in the form:

$$\mathbf{C} = \mathbf{SG}, \tag{8.3-19}$$

where **C** is a vector consisting of capacitance measurements; **G** is the vector containing the permittivity distribution; **S** is the sensitivity matrix obtained by solving Eq. (8.3-18). This model assumes that each permittivity pixel contributes linearly to the overall capacitance. The sensitive matrix **S** is usually constructed by computational methods or through experimental measurement. The inverse problem, which seeks the solution of **G** when **C** is known, is often ill-posed. The solution must be obtained via reconstruction algorithms, such as iterative linear back projection or artificial neural network (Wang *et al.*, 2009).

ERT measurement employs an alternating electrical current that is injected through the electrodes into the fluid medium; the voltage signal generated on the electrodes is subsequently measured. In order for the resistive effect to dominate capacitance, the conductivity of the medium, σ, must satisfy the condition:

$$\sigma \gg 2\pi f \varepsilon \varepsilon_0, \tag{8.3-20}$$

where f is the excitation frequency of the AC current input. To satisfy the quasi-static requirement, the product of the vessel diameter, D, and the excitation frequency must be significantly smaller than the speed of light c. Thus, it yields:

$$fD \ll c. \tag{8.3-21}$$

Under the conditions outlined above, the electric field can be described by Ohm's law, as:

$$\nabla \cdot (\sigma(\mathbf{r})\nabla\Phi(\mathbf{r})) = 0, \tag{8.3-22}$$

with the boundary condition:

$$\mathbf{n} \cdot \sigma\nabla\Phi + \mathbf{J} = 0. \tag{8.3-23}$$

Equations (8.3-22) and (8.3-23) can be represented in matrix form as:

$$\mathbf{I} = \mathbf{YV}, \tag{8.3-24}$$

where **I** is the vector of the applied boundary current, **V** is the vector for measured voltage, and **Y** is the global admittance matrix. Obtaining a meaningful solution to Eq. (8.3-24) necessitates the use of image reconstruction algorithms.

Solving the inverse problems in ECT and ERT is more complicated than in the so-called hard-field techniques, such as the X-ray tomography. Equations (8.3-17) and (8.3-22) are nonlinear in terms of the permittivity and conductance; additionally, the electric field is dependent on the entire domain, rather than on only a small part of it. Thus, more robust reconstruction algorithms are required for ECT and ERT as compared with the other tomographic implementations discussed. In addition to being economical and energy efficient, the main advantage of ECT and ERT lies in their ability to capture transient behavior of the flow. The drawback of such techniques is their poor spatial resolution. A quasi-three-dimensional image can be generated by stacking a series of two-dimensional images along the axial direction in the tier sequence that captures the images. However, in order to achieve a true three-dimensional image of the field, multiple layers of electrodes must be installed in the sensor to detect the field variation in both radial and axial directions, a technique used in electrical capacitance volume tomography (ECVT) (Warsito *et al.*, 2007; Wang *et al.*, 2010).

8.4   Mass Flow Measurement

Determining mass flow is one of the most important transport quantities for many multiphase flow applications. Local mass flux is defined as the product of material density, local volume fraction, and velocity of the transport phase. Realistically, a measurement is often obtained from a convenient control volume over a given sampling period; resulting measured dynamic transport properties for multiphase flows are volume–time-averaged values. Thus, for a phase k, the local averaged mass flux can be expressed using Eq. (6.4-6) as:

$$\mathbf{J}_{mk} = \rho_k \alpha_k \mathbf{U}_k + \overline{\rho_k \alpha''^i_k \langle \mathbf{u}'_k \rangle}, \tag{8.4-1}$$

where the last term on the right-hand side is the covariance of phase density, or alternatively volume fraction, and fluctuation velocity; for turbulent transport, this covariance term is referred to as diffusive mass flux. Equation (8.4-1) reveals that mass flux cannot be determined from the measurements of volume fraction and phase velocity, or vice versa, unless the covariance can be independently measured or simply ignored. Available strategies to measure mass flux and overall mass flow are introduced in this section, and phase velocity measurement techniques are introduced in the subsequent section. Isokinetic sampling and the electric ball-probe method represent two viable approaches to quantify local mass flux.

8.4.1   Overall Mass Flow Measurement

The overall mass flow of a phase k at a cross section of phase transport can be obtained through direct integration of the local mass flux over the cross section. This is expressed as:

$$J_{TMK} = \iint_A J_{mk} \, dA. \tag{8.4-2}$$

For a flow-through multiphase system without phase change, the overall mass flow during the transport process of each phase is conserved. Hence, in many cases, the overall mass flow can conveniently be determined either before phase mixing or after phase separation. These two options are respectively exemplified in the ejector spray scenario of Figure 8.4-1 and the pneumatic pipeline conveying solids in Figure 8.4-2, where the gas flow rate is measured by the venturi and the solid mass flow rate is measured from the weight change on the scale.

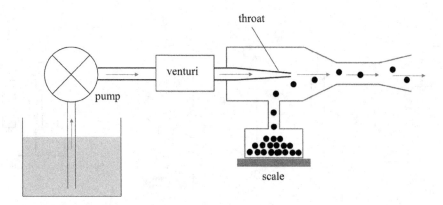

Figure 8.4-1 Ejector spray system.

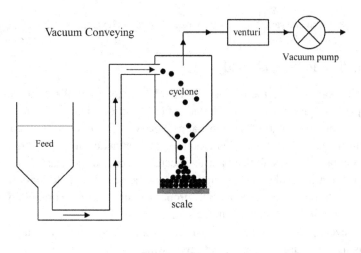

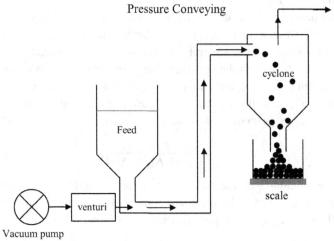

Figure 8.4-2 Pneumatic pipeline conveying.

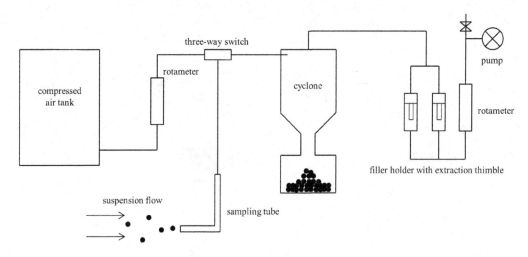

Figure 8.4-3 Isokinetic sampling with a two-stage collector (cyclone and thimble).

8.4.2   Isokinetic Sampling Method

Isokinetic sampling is the de facto standard technique for direct measurement of local particle mass flux in a particle-laden flow. A probe aligned with the direction of flow is employed to extract a sample at a velocity identical to the local fluid velocity prior to inserting the probe; this is known as isokinetic sampling. After an initial transition period, the stabilized particle accumulation rate as measured at a filter, or an equivalent particulate collector, represents the actual particle mass flux for the flow. In addition to mass flux, particle properties such as size, morphology, and bulk density can also be obtained from the sample.

To prevent pre-accumulation of particles inside the sampling system, pressurized flow is used to purge the sampling probe when the probe is inserted into the process flow but not in use for the sampling. Hence, when transitioning from the purge process to actual measurement during a sampling, there is an initial period that, if not accommodated, can cause significant measurement error because of the transient unsteady flow effects and the influence due to the so-called holdup volume in the system. To mitigate these undesirable effects, two samples over different sampling durations should be collected (Soo *et al.*, 1969). Hence, the particle mass flux can then be determined from the expression:

$$J_{mp} = \frac{m_{p1} - m_{p2}}{A(\Delta t_1 - \Delta t_2)},\tag{8.4-3}$$

where each sampling duration must be longer than the initial transition period. Equation (8.4-3) indicates that the measured particle mass flux is chiefly time-averaged and partly volume-averaged, governed by the effective flow area of the sampling probe, A.

An isokinetic sampling apparatus, suitable for both dilute and dense particle-laden flows, is illustrated in Figure 8.4-3.

The cyclone separator functions as a first-stage particle collector, and final particle accumulation occurs at the filter holder with the extraction thimble. In addition, the extraction flow rate is modulated, via a rotameter control valve, to achieve the isokinetic requirement. A pressurized clean air hose is employed to prevent the particles from entering the sampling tube when not in use. Once again, two samples are extracted over different durations to mitigate transient unsteady influences and the effects due to the holdup volume caused by the switching from probe purge to sampling modes. The difference between the two measurements yields the actual local mass flux. To minimize error due to entry effects, the sampling probe should be sharp edged with thin-well structures. The isokinetic condition is attained by equalizing the internal and external static pressure with respect to the sampling probe; two pressure differential pressure taps provide the monitoring provision. Isokinetic sampling can also be applied to flows containing liquid droplets. In these cases, the droplet sample is usually collected by an immiscible liquid.

In many multiphase flows, isokinetic sampling is difficult to implement for myriad reasons. As illustrated in Figure 8.4-4, it is almost impossible to attain a precise velocity match between the specimen and the local fluid velocity in the process of aerosol sampling in an exhaust stack or in nearly stagnant air. Additionally, with some confined three-dimensional flows, correct sample probe orientation is infeasible to realize due to space constraints. For certain flows with rapid or erratic change in velocity, instantaneous matching of sampling velocity is all but impossible. Moreover, the insertion of a sampling probe inevitably disturbs the original flow field. When sampling large particles, heavy particles, or with dense suspensions, effectuating and sustaining the required isokinetic condition consistently fails to provide

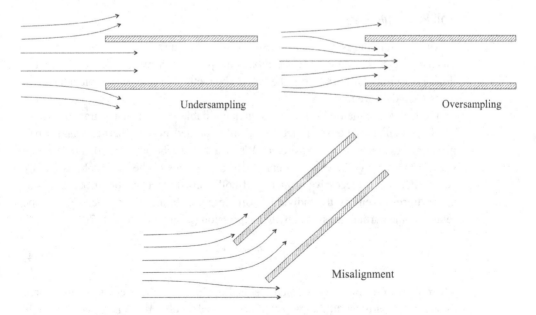

Undersampling   Oversampling

Misalignment

Figure 8.4-4 Anisokinetic sampling conditions.

sufficient aerodynamic lift to convey such particles through the sampling system. In practice, these challenges result in the more likely case of an anisokinetic condition when tracking particle mass flux. Anisokinetic sampling results from a mismatch of specimen and local fluid velocities or misalignment of sampling probe orientation; possibly both. Hence, further investigation of anisokinetic sampling results is essential to estimate the deviation from the ideal isokinetic prediction.

It is worth exploring the limiting cases of sampling very small or very large particles. Fine particles routinely mimic the motion of a fluid stream. Therefore, sampled particle mass flux of very small particles obtained through anisokinetic sampling is proportional to sampling velocity. For very large particles, the inertia of the particles enables them to maintain their trajectory and withstand the sudden change in direction and regime of fluid stream near the sampling probe inlet. Therefore, the sampled particle mass flux is independent of the variance from an isokinetic condition. Between these extremes, lies sampling of medium-sized particles, for which Lagrangian trajectory modelling can be enlisted to correct measurements obtained by anisokinetic sampling (Zhu *et al.*, 1997).

Numerous mechanisms contribute to the margin of error for particle mass flux measurements acquired through isokinetic sampling. Some examples include gravitational sedimentation, impaction on the wall or at the tube bends, wall deposition due to diffusion of small particles, flow turbulence, surface drag, agglomeration of fine particles, electrostatic charge, tendency of particles to stick the wall, and flow disturbance caused by insertion of the sample probe; the potential anisokinetic conditions discussed previously also are factors to consider.

8.4.3   Ball Probe Method

Isokinetic sampling is an inconvenient method to acquire instantaneous or continuous measurements of particle mass flux. If the particles carry an electrostatic charge during the multiphase transport, the ball probe method is a more suitable alternative when such data are required.

The ball probe method is premised on quantifiable particle charge transfer resulting from collisions between electrostatically charged moving particles and a ball probe inserted into the flow stream. When a ball probe is inserted into a dilute particle-laden flow, the charging current from particles to the ball probe is mainly contributed via a successive sequence of collisions that can be described as a single scattering event by the individual particles. Hence, the charging current can be related to the particle mass flux by the expression (Cheng and Soo, 1970):

$$I = \frac{\eta_{bp} A_b q_{bp}}{m} J_{mp}, \tag{8.4-4}$$

where η_{bp} is the fraction impacted, defined as the ratio of the cross-sectional area of stream of particles to the projected area of ball probe A_b; m is mass of a single particle; q_{bp} is the charge transfer per impact between a single particle and the ball

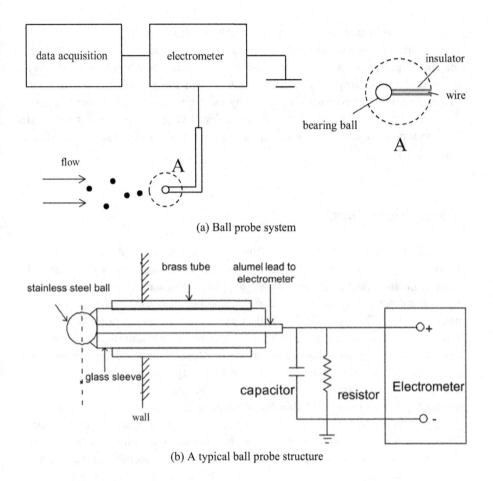

(a) Ball probe system

(b) A typical ball probe structure

Figure 8.4-5 Ball probe system.

probe, which can be expressed by Eq. (4.3-19b). Notably, q_{bp} is a function of particle velocity, charge properties of particles, and the ball probe features.

When applied to a dense suspension, effects of multiple scattering and charge saturation of the ball probe become significant, and they can no longer be ignored. To account for the consequences, Eq. (8.4-4) can be modified to a more general form as (Zhu and Soo, 1992):

$$I = C_{b1} U_p^{3/5} J_{mp} \exp\left(-C_{b2}\frac{J_{mp}}{U_p^2}\right), \tag{8.4-5}$$

where C_{b1} and C_{b2} are semiempirical coefficients, and U_p is the particle velocity. Due to the coupled nature of particle mass flux and velocity measurements when using a ball probe, this method is routinely employed in a joint measurement of particle velocity or volume fraction.

Figure 8.4-5 shows a typical ball probe system that consists of an electrically conductive metal ball connected to an electric-current measurement device, such as electrometer.

The probe is insulated except for the actual ball part and the internal wiring, which both function to promote electric current sampling and charge transfer. In practice, the charging current measured by a ball probe includes both the charges directly transferred as the suspended particles collide with the probe surface and those induced by the charged particles passing by the probe; charges from direct impact dominate though. The contribution of the induced charge can be either included into the semiempirical coefficient, C_{b1}, of Eq. (8.4-5) or modeled separately (Park *et al.*, 2002).

8.5   Velocity Measurement

Dynamic transport characteristics of different phases in a multiphase flow are typically distinguished by their different velocities. There are some well-developed velocity measurement techniques for single-phase flows, such as Pitot tube and hot-wire anemometry, which have been well introduced in many textbooks of fluid mechanics (*e.g.*, Holman, 2012). While these methods may be extendable to the velocity measurement of continuum fluid phase in some very dilute multiphase flows, most of them become invalid in applications of dense-phase multiphase flows. The following introduction is thus focused on the velocity measurement techniques of multiphase flows, especially on those of discrete phases.

The type of velocity available for measurement in multiphase flows can be grouped into the following: local-time-averaged velocity of a phase, phase velocity field at a given moment, and volume–time-averaged velocity of a phase. The class of velocity selected depends on the measurement principle adopted. Methods for volume–time-averaged velocity are embodied in the cross-correlation and venturi-meter methods. Techniques for local-time-averaged velocity measurement include laser Doppler velocimetry and the corona-charging ball probe method. Field imaging techniques, such as particle tracking velocimetry (PTV) and particle image velocimetry (PIV), yield an overall representation of the velocity field.

8.5.1   Cross-correlation Method

The cross-correlation method is founded on the principle that a disturbance, represented by a physical quantity of interest with time history $x(t)$, will propagate through a nondispersive linear path and combine with statistically independent noise, $n(t)$, to produce a response, $y(t)$. For simplicity, let the frequency response function of the propagation path be a constant: $H(f) = H = $ constant; this approximation is reasonable for unsteady motion with very low frequency. Thus, for propagation distance, L, and propagation velocity, U, it gives:

$$y(t) = H \left[x \left(t - \frac{L}{U} \right) + n(t) \right]. \tag{8.5-1}$$

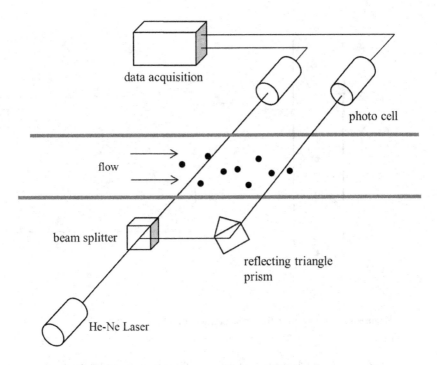

data acquisition

photo cell

flow

beam splitter

reflecting triangle
prism

He-Ne Laser

Figure 8.5-1 Cross-correlation measurement of a gas–solid pipe flow.

The cross-correlation function, $R_{xy}(\tau)$, of $x(t)$ and $y(t)$, for a time delay, τ, is given by the relationship:

$$R_{xy}(\tau) = \lim_{T \to \infty} \frac{1}{T} \int_0^T x(t)y(t + \tau)d\tau = HR_{xx}\left(\tau - \frac{L}{U}\right). \qquad (8.5\text{-}2)$$

From the condition $|R_{xx}(\tau)| < R_{xx}(0)$, the first peak after time zero gives τ_1. It then gives:

$$U = L/\tau_1. \qquad (8.5\text{-}3)$$

Therefore, the propagation velocity can be determined from the known propagation distance and time delay where the peak value of the cross-correlation occurs (Bendat and Piersol, 1986).

The cross-correlation technique can be applied to any nonuniform and unsteady transport of multiphase flow where some slowly dispersed phase properties can either be directly identified or measured by two sensors located at a given spacing along the phase transport. Figure 8.5-1 depicts dual-beam cross-correlation of unsteady pneumatic transport of solids (Zhu *et al.*, 1991), and Figure 8.5-2 illustrates the cross-correlation strategy for a bubbly column flow using a pair of optical or capacitance sensors or a dual-sensor hot-film probe (Gurau *et al.*, 2004).

Various types of sensors can be employed for cross-correlation measurements. For example, the cross-sectional dispersed phase volume fraction can be acquired by a capacitance sensor, microwave sensor, ultrasonic sensor, or γ-ray densitometer. If

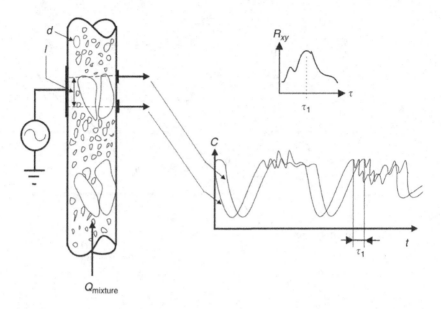

Figure 8.5-2 Cross-correlation measurement of a bubbly column flow.

the distance between the two sensors is moderate, the transit dispersed phase flow structure will not significantly vary as it passes the two sensors; two similar signals will be generated within a time interval that can be used to obtain a meaningful correlated velocity. Hence, the velocity measured by cross-correlation is volume-averaged over the control volumes of sensors and time-averaged over the integral time, T, during which a series of phase transport signals can be correlated.

Under realistic flow conditions, sensor locations need to be arranged such that the measurement correctly represents the averaged velocity of desired phase. For example, in the case represented by Figure 8.5-2, the optical or capacitance sensor may register all bubble sizes present. Hence, the predicted averaged velocity by cross-correlation of the direct output from sensors will similarly represent an average of all bubble sizes. To obtain a velocity value for a specified bubble size, the output signal should be scrutinized via signal filtration or sensors must be selected that can only recognize the target bubble size. As a final note, cross-correlation methods cannot be used for uniformly dispersed flows nor for rapidly dispersing flows.

8.5.2   Venturimeter

The venturimeter is a commonly used and straightforward option to measure the overall volumetric flow rate for a single-phase pipe flow. The device relies on the Bernoulli principle, which is valid only for flows that are steady-state, incompressible, inviscid, and one-dimensional. A venturimeter may be applicable for multiphase flow measurements if the constituent phases of interest can be approximated as a mixture; the flow must be uniformly dispersed and devoid of any relative phase movement so all effects on phase interaction, such as drag forces, can be neglected. In such cases, the multiphase flow is regarded as a single-phase flow with a specified mixture density. The volumetric flow rate, Q_m, can then be determined by the expression:

$$Q_m = \frac{A}{\sqrt{1 - \beta^4}} \sqrt{\frac{2\Delta p}{\rho_m}}, \tag{8.5-4}$$

where ρ_m is the mixture density; Δp is the differential pressure due to a change in device cross-sectional area; A is the cross-sectional area of the venturi throat; β is the throat-to-pipe diameter ratio.

A scenario that obeys the mixture approximation is a gas–liquid two-phase flow where the gas–liquid phase interaction can be ignored. In a gas–liquid mixture, the liquid density is orders of magnitude larger than the gas density. Therefore, it yields:

$$Q_m \rho_m = Q_m(1 - \alpha_g)\rho_l = Q_l\rho_l, \tag{8.5-5}$$

where α_g is the gas volume fraction. From Eq. (8.5-4) and Eq. (8.5-5), the volumetric flow rate of the liquid can be approximated by the following relationship:

$$Q_l = \frac{A}{\sqrt{1 - \beta^4}} \sqrt{\frac{2(1 - \alpha_g)\Delta p}{\rho_l}}. \tag{8.5-6}$$

Hence, the liquid flow rate for a gas–liquid flow in a pipe can be captured using a combination apparatus consisting of a venturimeter and a volume-fraction measurement device, such as a densitometer. The relevant equipment and arrangement are shown in Figure 8.5-3 (Hammer *et al.*, 2005).

Figure 8.5-3 specifically illustrates a measurement arrangement that includes both a venturimeter and a γ-ray densitometer to deduce the liquid-phase flow rate of a

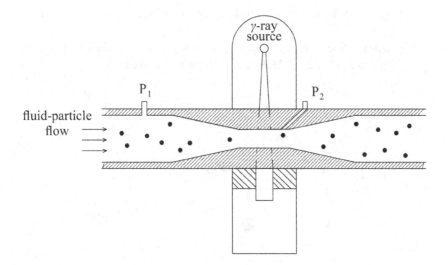

Figure 8.5-3 Combination of a venturimeter and a γ-ray densitometer.

gas–liquid dual-phase pipe flow. The differential pressure quantity, Δp, is measured from the two pressure ports, p_1 and p_2. The gas volume fraction, α, is monitored via the γ-ray densitometer, which is installed at the throat of the venturimeter. With the addition of the geometric dimensions for the physical venturimeter, the volumetric liquid flow rate, q_l, can be calculated from Eq. (8.5-6).

8.5.3   Laser Doppler Velocimetry

Laser Doppler velocimetry (LDV), also referred to as laser Doppler anemometry (LDA), measures the local point velocity by detecting the Doppler frequency shift of the light scattered by a moving particle that is suspended in a carrier fluid flow. Figure 8.5-4 depicts the typical LDV arrangement whereby a laser beam is emitted toward a moving particle that scatters a predictable fraction of light to a receiving photodetector.

The frequency perceived by the observer, f_r, is related to the frequency emitted by the source, f_0, by the expression:

$$f_r = \left(\frac{c - \mathbf{u}_p \cdot \hat{\mathbf{r}}}{c - \mathbf{u}_p \cdot \hat{\mathbf{s}}} \right) f_0, \tag{8.5-7}$$

where c is the speed of light in the medium; $\mathbf{u}_p$ is the particle velocity; $\mathbf{s}$ is a unit vector pointing from the source to the particle; $\mathbf{r}$ is a unit vector pointing from the particle to the observer. The particle velocity is found from the frequency shift observed by the stationary photodetector.

In practice, the relatively high frequency in the formulation of Eq. (8.5-7) is often prohibitive to permit direct detection; a more realistic LDV measurement employs a much reduced frequency that is modulated by the interference of two light beams. Figure 8.5-5 conceptually illustrates two popular configurations for the double-beam approach.

According to Eq. (8.5-7), the frequency shift in a dual-beam arrangement, as it appears in Figure 8.5-5(a), is given by the expression:

$$f_D = f_{r1} - f_{r2} \approx \frac{\mathbf{u}_p \cdot (\mathbf{s}_2 - \mathbf{s}_1)}{\lambda} = \frac{2 u_{py} \sin \theta}{\lambda}. \tag{8.5-8a}$$

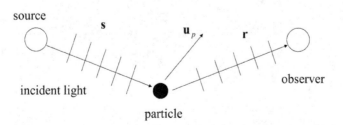

Figure 8.5-4 Doppler frequency shift in scattered wave from a moving particle.

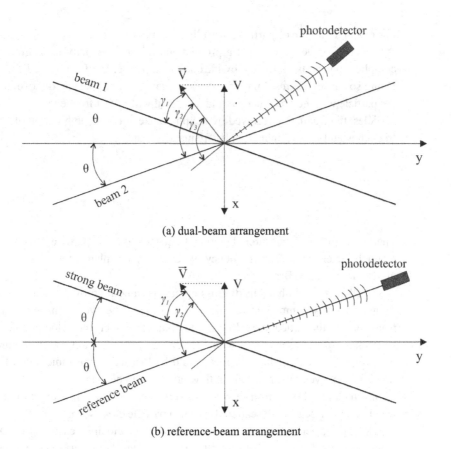

(a) dual-beam arrangement

(b) reference-beam arrangement

Figure 8.5-5 Modes of two-beam arrangement in LDV.

Alternatively, the frequency shift in a reference-beam arrangement, as appearing in Figure 8.5-5(b), is determined by:

$$f_D = f_r - f_0 \approx \frac{\mathbf{u}_p \cdot (\mathbf{r} - \mathbf{s})}{\lambda} = \frac{2u_{py} \sin \theta}{\lambda},\tag{8.5-8b}$$

where λ is the wavelength, and θ is the half angle between the two beams. Equation (8.5-8) can alternatively be formulated by the fringe model. The spacing, δ, in the fringe pattern due to the interference between two beams can be obtained via Eq. (8.2-5). Because the Doppler frequency detected by the photodetector is the reciprocal of the time interval required for the particle to move across the fringe spacing, the particle velocity in the direction that is perpendicular to the fringes can be calculated from:

$$u_{py} = \delta f_D = \frac{\lambda}{2 \sin \theta} f_D,\tag{8.5-9}$$

which has the same form as Eq. (8.5-8).

The Doppler frequency in Eq. (8.5-8) provides a mechanism to measure the magnitude of the velocity component, u_{py}, but it fails to recognize its direction.

For example, a pair of particles traveling in opposite directions would generate an equivalent Doppler frequency; a problematic result in velocity measurement. This complication can be mitigated by introducing a frequency offset to the LDV system. Doing so causes the resulting fringe of the scattered light beams to move; a stationary particle will then return a continuous signal with offset frequency, f_s. For moving particles, the frequency received by the photodetector is a combination of the offset frequency and the original Doppler frequency. It then can give:

$$f_D = f_s + \frac{2u_{py} \sin \theta}{\lambda}, \qquad (8.5\text{-}10)$$

Therefore, given a measured Doppler frequency f_D, the particle velocity can be uniquely determined. The frequency offset is often implemented by employing an optical unit called a Bragg cell.

A LDV system with a pair of crossing beams can only provide measurements for a single velocity component that lies in the plane of the two beams and is also perpendicular to the optical axis. In order to obtain the other two velocity components, additional pairs of laser beams, with different frequencies or colors, are required. The resulting three pairs of laser beams should be focused to the same control volume where particle velocities of a given flow phase are to be measured.

In principle, LDV is restricted to measuring velocities of moving particles upon which the laser beams are scattered. When a particle closely tracks the fluid medium with negligible velocity slip between the phases, the velocity measured by LDV actually represents the local fluid velocity. Such particles, known as tracer particles, are typically very small and often commensurate with the micron scale, ensuring negligible impact on the characteristics of the local fluid medium. When applied to a dispersed multiphase flow, Doppler frequencies measured by a dual-beam LDV can encompass a wide range, partly due to its encumbrance of merely capturing one component of velocity and the inability to differentiate signals reflecting from disparately sized particles in the flow. Hence, to yield a statistically meaningful result, LDV measurement is typically time-averaged and scoped to a strategically selected frequency screening window. Another limitation with LDV measurements is that only one particle a time is permitted in the control volume. When multiple particles coexist within the control volume simultaneously, the Doppler signal of one particle is indistinguishable from others in the control volume, confining the application of LDV to very dilute multiphase flows. More introduction of LDV can be found in many references such as Fingerson and Menon (1998).

Phase Doppler anemometry (PDA), also known as the phase Doppler particle analyzer (PDPA), is a hybrid technique that combines LDV and phase Doppler measurement to simultaneously measure particle size and velocity in a given flow. This method is regarded as an extension of conventional LDA methodologies. PDPA is also capable of discerning the particle size and phase velocity distributions (Bachalo and Sankar, 1998).

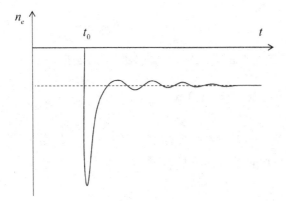

Figure 8.5-6 Schematic time variation in ion density from a corona discharger.

8.5.4   Corona Discharge Method

For a dilute suspension flow, the local flow velocity can be measured using LDV or cross-correlation techniques. However, these techniques routinely require optical access to the flow, which limits their application to dilute suspensions. In addition, the preponderance of techniques cannot differentiate signals generated from different sized particles, which is problematic for flows with polydispersed particles. PDPA possesses the ability to distinguish velocity measurements from different-sized particles via phase shift, but it has the same hindrances as LDV when extended to multiphase flow phenomena.

For multiphase flows, especially dense suspensions, simultaneous measurement of both the fluid phase and the particle phase velocities is of vital importance. However, few currently available measurement techniques can provide such functionality. For a gas–solid flow in a simple translational motion (*e.g.*, without flow rotation and devoid of recirculation), local velocities of both gas and particles along the transport direction can be concurrently determined via the corona discharge method (Soo *et al.*, 1989). This measurement method relies on ions generated from local gas ionization to trace the phase transport. A predictable peak in ion density that is generated by corona discharge results for a local gas that is initially neutral but suddenly becomes ionized when subjected to a strong DC electric field. This outcome is reflected by the plot in Figure 8.5-6.

The excessive quantity of transient ions instantaneously attach to the local gas molecules as well as nearby particles passing by. The local gas and particles carry a distinctively excessive charge and become tracers that travel to a downstream charge sensor. The sensor, such as ball probe, records the characteristics of the charged particles and local gas, which is ultimately used to determine the gas velocity and particle velocity via the different phase transport behavior.

As shown in Figure 8.5-7(a), a negative corona discharge in a particle-laden gas at a high voltage produces ions in air through attachment, and it modifies electrostatic charges on solid particles suspended in the flow. With the aid of a high-speed

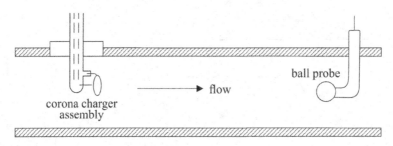

(a) Corona-discharging & ball-probe system

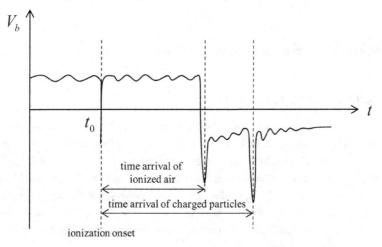

(b) Typical phase velocity measurements

Figure 8.5-7 Corona-discharging method.

data acquisition system and an electrostatic probe stationed downstream, the corona discharge probe arrangement utilizes the transient behavior of a corona discharge to earmark the arrival of ions and solid particles with modified charges. This is illustrated in Figure 8.5-7(b). Using the location of the electric probe to the corona source, the arrival time intervals correspond to the local, instant, and one-dimensional air velocity, and particle velocity. In a dense solid suspension flow, the probe arrangement of Figure 8.5-7 does not suffer the shortcomings that other available probe alternatives do. Notably, negative corona is normally preferred over positive corona because of its inherent ionization stability.

8.5.5   Particle Image Velocimetry

Planar field visualization for a section of multiphase flow can be achieved using laser-based optical imaging techniques, such as the laser-sheet from a laser beam expansion method or the fast-sweeping laser beam from a parabolic reflector approach. Intense or large field visualization via laser beam expansion often

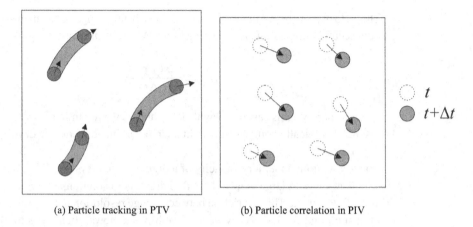

(a) Particle tracking in PTV   (b) Particle correlation in PIV

Figure 8.5-8 Methods for identifying particle position shift in PTV/PIV.

demands noncontinuous lasers be employed that create the field illumination effect by laser pulses. Using this method requires synchronization provisions between the camera and the pulse laser to capture a meaningful image of illuminated field. A series of images of a tracked particle in transport can be recorded if a high-speed camera, or a double-exposure cross-correlation camera, is used (Adrian, 2005).

Particle tracking velocimetry (PTV) and particle image velocimetry (PIV) are two paradigmatic techniques used for field velocity measurement of multiphase flows. PTV typically employs straightforward particle identification and track-ing methods, such as streaking using long exposure times or a series of images of an individual particle in transport. These techniques are shown in Figure 8.5-8(a). PIV often relies on two images taken using a double-exposure cross-correlation camera, which is capable of identifying the same particle via spatial cross-correlation and the corresponding pair of position shifts. This method is shown in Figure 8.5-8(b). The number density of the particles in an image is typically modest when a PTV strategy is utilized. Thus, pairs of particle images can be readily identified, and the motion of individual particles can be studied accordingly.

PIV technique employs a much higher particle concentration when compared to PTV; therefore, recognizing and analyzing each individual particle motion is not considered practical. Instead, the image is often divided into multiple interrogation regions where the flow velocity and particle velocity are assumed to be uniform. For each subregion being scrutinized, the two consecutive images are compared and the closest match to the averaged displacement that represents the overall motion of the totality of the particles is selected. The selection process is often accomplished via cross-correlation between the two images. The procedure is repeated for all of the interrogation regions, and an overall velocity map is ultimately generated for the entire flow field.

The velocity of the particles can be obtained from their displacements between two consecutive images taken at time t and $t + \Delta t$. Thus, it gives:

$$\mathbf{u}_p\left(\mathbf{r}, t\right) = \frac{\Delta \mathbf{r}_p(\mathbf{r}, t)}{\Delta t}. \qquad (8.5\text{-}11)$$

The measuring systems leveraged with PTV and PIV are similar to the system in Figure 8.2-1, typically comprising a light source, an image recording device, and an image processor.

Due to the limited memory capacity of a digital camera system, it is often impossible to warrant both large view field and high image resolution are within the same frame of flow-image. This restriction between image resolution necessary for particle identification and the large view field of multiphase flows often leads to measurements using trackable particles much larger than tracers of the continuous phase. Thus, PTV and PIV are typically relegated to velocity measurement of particle phase only. However, when the field view becomes sufficiently small, velocities of both the fluid and particle phases can be gleaned using PIV or PTV. In a two-phase flow, PTV and PIV techniques can also be combined to simultaneously measure the velocity field of both the continuous phase and the disperse phase flow (Bröder and Sommerfeld, 2007). During such image processing, the tracers representing the continuous phase and bubbles or particles of the disperse phase can be discerned either based on inherent size difference or by introducing fluorescent tracer particles excited by the incident laser.

8.6   Charge Measurement

Electrostatic charge carried by particles in a multiphase flows can be generated or altered by various charging or discharging mechanisms, such as triboelectric charging by contact and field charging by external electric fields (Fan and Zhu, 1998). Measuring electrostatic charge should be executed under the strict condition that no alteration of the existing charge on sampled particles will be tolerated during the measurement process; the total electrostatic charge of particles sampled must be registered by the measurement. Therefore, contact-based online measurement techniques are typically not suitable. For example, the ball probe method measures electrostatic charge transfer, which merely reflects a partial, not complete, discharge of particles colliding with the probe; additional triboelectric charges can also be generated by the probe–particle collisions that may occur during sampling, which are not evinced. There are two predominant methods of charge measurement: direct and indirect. Direct measurement techniques are embodied by isokinetic sampling with a complete discharge measurement device, such as Faraday cups. Indirect measurement approaches include noncontact charge measurement probes, which are based on changes in capacitance or induction by surrounding charged particles.

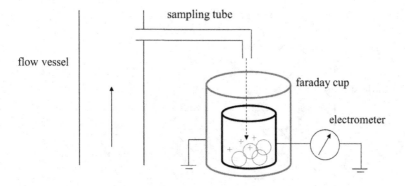

Figure 8.6-1 Faraday cup charge density measurement system.

8.6.1 Sampling with Faraday Cup

The sign and density of charges on particles can be measured using the Faraday cup concept illustrated in Figure 8.6-1. Particles are removed from the flow vessel by an isokinetic sampling tube or scooper.

The sampled particles are subsequently collected in a Faraday cup, which is placed into, but insulated from, a grounded cup; the Faraday cup itself is also grounded. Charges collected in the Faraday cup are neutralized via an induced electric current, which is measured and regulated by the electrometer. The total charge, q_e, collected in a time period, Δt, is related to the current, I, by the expression:

$$I = \frac{q_e}{\Delta t}. \tag{8.6-1}$$

When the mass of the particles is obtained, the charge density per unit mass of particles can be calculated.

8.6.2 Induction Probe

The operating principle of an induction probe is similar to a ball probe. However, the induction probe head is shielded by an insulating medium, and it is not directly exposed to the flow. A schematic representation for a typical induction probe arrangement is shown in Figure 8.6-2. The insulation mitigates charge transfer due to collisions between the particles and the probe; thus, only the desired induced charge is measured.

The main disadvantage of the induction probe is spotlighted when charges on the vessel walls interfere with measurement signals. This drawback can occur often and is particularly prominent when the walls are fashioned of nonconductive materials, such as Plexiglas. The electric capacitance tomography (ECT) strategy, introduced in Section 8.3.4, can also be used for measuring electrostatic charge distributions.

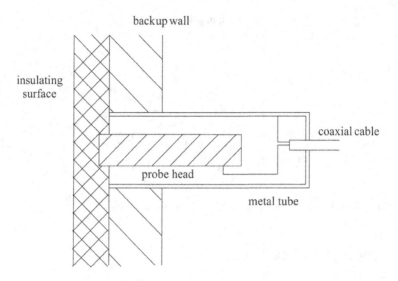

Figure 8.6-2 Schematic design of an induction probe (Boland and Geldart, 1971).

8.7   Case Studies

The following case studies serve for further discussion with simple applications or extensions of measurement methods of this chapter. The examples include an application of the deconvolution method to obtain the particle size distribution from a set of cascade-impactor measurements, an optical method to measure the microbubble sizes or droplet sizes, and a pressure-measurement method to determine volume fraction in slurry-bubble column.

8.7.1   Particle Size Distribution by Deconvolution Method

Sampling for particle size distribution is commonly done using a set of collectors in series, such as cascade impactors or a series of sieves. There is a size overlapping of particles collected of neighboring stages. To obtain an estimate that is close to the true size distribution, a deconvolution is typically preferred.

Problem Statement: Consider a particle size distribution that can be represented by an overlapping of trimodal lognormal distributions from a set of measurements. The data is obtained from several collectors in series. Assume that the collection characteristics of each individual collector can be predetermined. Develop a method of deconvolution to obtain the particle size distribution from the measurements.

Analysis: The following is a simplified model from the work of Dzubay and Hasan (1990) that proposed a method of fitting multimodal lognormal size distributions to cascade impactor data. The collection efficiency of each stage can be expressed as a two-parameter function, such as for the collector of the jth stage,

$$\eta_j(x) = \left(1 + \left(\frac{x_{cj}}{x}\right)^{s_j}\right)^{-1},$$ (8.7-1)

where x_c is the cutoff size where collection efficiency is 50%, and s is the steepness of the efficiency curve. Both x_c and s are assumed to be known or can be predetermined for each individual collector. If the last stage ($j = N$) is a backup filter, then $\eta_N = 1$.

Assume that there is no wall loss from the sampling (i.e., $h(x) = 1$ for all stages). Based on Eqs. (8.2-11) and (8.2-12), the predicted mass collected on the collector of the ith stage is calculated by

$$m_{pi} = \frac{m_T \int_0^\infty f_m(x)\eta_j(x) \prod_{i=1}^{j-1} (1 - \eta_i(x))\,dx}{\int_0^\infty f_m(x)[1 - \eta_0(x)]dx},$$ (8.7-2)

where m_T is the measured total mass concentration, and the denominator accounts for the mass removed by the sampler inlet, which is the initial stage ($i = 0$), and thus is excluded from size analysis. The mass density distribution function is assumed to be a trimodal lognormal function, which is given by

$$f_m(x) = \sum_{k=1}^{3} C_k f_{mk}(x),$$ (8.7-3)

where, for modes $k = 1, 2$, and 3, C_k is the partition factor of each modal, with

$$\sum_{k=1}^{3} C_k = 1,$$ (8.7-3a)

and $f_{mk}(x)$ of modal k follows the lognormal distribution:

$$f_{mk}(x) = \frac{1}{\sqrt{2\pi}\,(\ln \sigma_k)\,x} \exp\left\{-\frac{1}{2}\left(\frac{\ln x - \ln x_k}{\ln \sigma_k}\right)^2\right\},$$ (8.7-3b)

where x_k is the mass-averaged size and σ_k is the geometric standard deviation of stage k, respectively.

Size distribution parameters C_k, x_k, and σ_k are determined by a least squares method that minimizes χ^2, defined by

$$\chi^2 = \frac{1}{N - N_p} \sum_{i=1}^{N} \left(\frac{m_{pi} - m_i}{\delta m_i}\right)^2,$$ (8.7-4)

where m_{pi} is the predicted mass from Eq. (8.7-2); m_i is the mass measured on stage i; N is the total number of measurements; N_p is the number of fitted parameters, for the trimodal distribution and with Eq. (8.7-3a), $N_p = 8$; δm_i is the random measurement error in m_i, which can be further estimated by

$$\delta m_i = \sqrt{\varepsilon_1^2 + (\varepsilon_2 m_i)^2},$$ (8.7-4a)

where ε_1 is the random error in concentration near blank values, and ε_2 is a fraction that represents random error for large mass concentration.

An iterative search is used to find the χ^2 minimum, at which the best-fit parameters of C_k, x_k, and σ_k are determined. Detailed examples of such applications and iterative algorithm can be found in the reference (Dzubay and Hasan, 1990).

8.7.2   Optical Measurement of Microbubbles and Droplets

Optical methods are constantly applied to study the flows with microbubbles or droplets, in which sizing out-of-focus bubbles or droplets can be a major source in biased errors in the associated size or volume fraction measurements. For instance, a bubble or droplet measurement by a simple backlighting or shadowgraphy is to illuminate the bubble-carrying flow field by a diffuse light source and record the associated shadow image for bubble sizing by a front-facing camera or camcorder. The images constantly contain out-of-focus bubbles that can cause large errors in sizing, especially for cases with small bubbles or droplets.

Problem Statement: How to eliminate or significantly reduce the biased errors from the out-of-focus bubbles or droplets in the optical images of sizing measurements?

Analysis: The following is a brief introduction of a method from the work of Dehaeck *et al.* (2009), in which an optical planar technique, known as the laser marked shadowgraphy, is developed to avoid sizing the out-of-focus bubbles or droplets.

A combination of backlighting and glare-point velocimetry is implemented for measurement of microbubbles or droplets, as shown in Figure 8.7-1. Specifically, bubbles are illuminated from two sides (with a laser beam or thin-sheet and a total reflection mirror on the opposite side) and the image recording device such as a camera is placed in front of the measurement plane at a scattering angle of 90° with

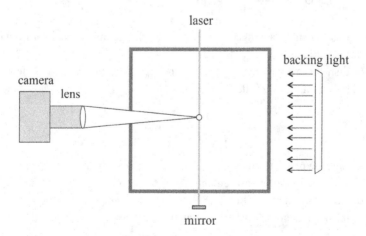

Figure 8.7-1 Schematic system of laser marked shadowgraphy.

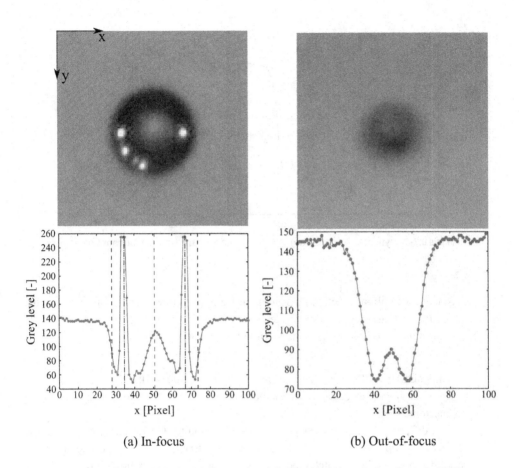

(a) In-focus (b) Out-of-focus

Figure 8.7-2 Bubbles and their gray-level distributions (from Dehaeck *et al.*, 2009).

respect to the laser beam or sheet, while a diffuse backlighting is provided to enhance the image of bubbles or droplets.

An example of single-bubble images is given in Figure 8.7-2, with (a) for an in-focus bubble (identified with glare-point markers) and (b) for an out-of-focus bubble (without any glare points). A sample of gray-level distribution of each bubble is also given, respectively, under the corresponding optical image.

A digitized filtration of gray-level images can be performed to filter out the out-of-focus bubbles first and then the glare-points to get a sharper image of the in-focus bubble, from which the size and volume can be calculated. Detailed application examples of bubble size distribution can be found in the reference (Dehaeck *et al.*, 2009).

8.7.3 Volume Fraction in a Pressurized Slurry-Bubble Column

Problem Statement: In slurry-bubble columns at high pressure, the gas holdup may increase with an increase in pressure, especially at high slurry concentration.

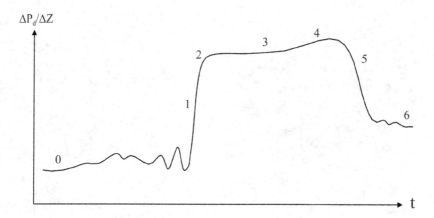

$\Delta P_d/\Delta Z$

Figure 8.7-3 Dynamic pressure drop in a disengagement measurement (from Luo *et al.*, 1999).

Since most volume fraction measurement techniques have various limitations when applied to high-pressure flow systems, is it possible to use pressure signal for the determination of bubble size and volume fraction at high pressure?

Analysis: The following is a simplified method from the work of Luo *et al.* (1999), in which a dynamic disengagement method is developed for bubble size and gas holdup determination.

A differential pressure transducer is used to monitor the variation of the dynamic pressure drop between two fixed locations in a slurry-bubble column. The liquid and solid phases are in the batch mode, while the gas continuously flows through the column. After the flow reaches steady state, the gas inlet and outlet are simultaneously shut off to maintain a constant system pressure.

The variation of the dynamic pressure drop signal with time, as shown in Figure 8.7-3, can be decomposed into several stages. Initially at stage 0, the slurry-bubble column is at steady state. Then, there is a sharp increase in pressure drop after the shutoff of gas flow corresponds to the disengagement of bubbles, which is marked as stage 1. After the disengagement of all large bubbles, there is a more gradual pressure drop increase corresponding to the slower disengagement of small bubbles, marked as stage 2. In stage 3, the nearly constant pressure drop indicates that, after all bubbles are disengaged, the solid particles remain fully suspended. During stage 4, particles start to settle down, and increased solids concentration in the region between the two pressure ports leads to higher pressure drop. Then, the solid bed surface falls between the two pressure ports and continues to move down, leading to sharp decrease of the differential pressure, which is marked as stage 5. Finally, in stage 6, the solid bed surface moves below the lower pressure port, and the dynamic pressure drop approaches zero.

Assume the densities of each phase are known. The volume fractions of each phase, α_g, α_l, and α_s, can be related to the dynamic pressure drop, Δp_d. At stage 0, the dynamic pressure drop can be expressed by:

$$\left.\frac{\Delta p_d}{\Delta z}\right|_0 = \left(\alpha_g \rho_g + \alpha_l \rho_l + \alpha_s \rho_s - \rho_l\right) g, \tag{8.7-5}$$

with

$$\alpha_g + \alpha_l + \alpha_s = 1. \tag{8.7-6}$$

At stage 3, only liquid and solid phases are present, and therefore:

$$\left.\frac{\Delta p_d}{\Delta z}\right|_3 = \left(\alpha'_l \rho_l + \alpha'_s \rho_s - \rho_l\right) g, \tag{8.7-7}$$

with

$$\alpha'_l + \alpha'_s = 1. \tag{8.7-8}$$

Since both liquid and solid phases are in batch mode, the ratio of the two phases is a constant:

$$K = \frac{\alpha_s}{\alpha_l} = \frac{\alpha'_s}{\alpha'_l}. \tag{8.7-9}$$

Using Eqs. (8.7-7)–(8.7-9), the ratio K can be expressed as:

$$K = \frac{|\Delta p_d / \Delta z|_3}{(\rho_s - \rho_l) g - |\Delta p_d / \Delta z|_3}. \tag{8.7-10}$$

Combining Eqs. (8.7-5), (8.7-6), (8.7-9), and (8.7-10), the gas holdup can be obtained by:

$$\alpha_g = \frac{|\Delta p_d / \Delta z|_3 - |\Delta p_d / \Delta z|_0}{(\rho_s - \rho_l) g + |\Delta p_d / \Delta z|_3}. \tag{8.7-11}$$

Then the liquid and solid holdups can be calculated using Eq. (8.7-6), (8.7-9), and (8.7-11):

$$\alpha_l = \frac{1 - \alpha_g}{1 + K} \tag{8.7-12}$$

$$\alpha_s = \frac{1 - \alpha_g}{1 + 1/K}. \tag{8.7-13}$$

8.8 Summary

This chapter introduces the principles of experimental methods to measure or determine various transport properties in multiphase flows. Typical properties include geometric characteristics of dispersed phase (such as size and shape of solid particles, droplets, and bubbles), phase volume fractions, mass fluxes or flow rates, velocities, and electrostatic charges. It discusses the basic mechanisms and applicability of the measurement techniques involved. Basic methods based on the primary measurement objectives include:

- particle size and morphology: optical image method, sieving method, sedimentation method, cascade impaction method, and laser-scattering method;

- volume fraction: beam-attenuation method, permittivity method (via capacitance transducer), and tomography principles;
- mass flow rate: isokinetic sampling method, and ball probe method;
- velocity: cross-correlation method, LDV method, and PIV method;
- charge: Faraday cup method, and induction probe method.

The chapter also discusses the data analysis methods describing the particle size distribution from overlapped size-sampling, such as the deconvolution method. It is also important to identify the equivalent diameter of nonspherical particles that a size measurement reveals. The commonly used equivalent diameters include the arithmetic mean diameter, surface mean diameter, volume mean diameter, Sauter's mean diameter, and DeBroucker's mean diameter.

Nomenclature

A Amplitude
     Area
C Coefficient
     Capacitance
c Speed of light
     Mass concentration
D Diameter
d Particle diameter
F Cumulative size distribution
f Size density distribution function
     Frequency
g Gravity acceleration
H Chamber height
h Height
$\mathbf{I}$ Current vector
I Beam intensity
     Current
J_m Mass flux
K Time difference coefficient
     Phase volume fraction ratio
k Wave number
L Characteristic length
l Focusing length
m Mass
$\mathbf{n}$ Normal unit vector
n Relative refraction index
     Number density
P Attenuation function

p	Pressure
Q	Flow rate
q	Electric charge
$\boldsymbol{r}$	Position vector
s	Projected fringe spacing
	Steepness
T	Time period
t	Time
U	Velocity
$\boldsymbol{u}$	Instant velocity
$\boldsymbol{V}$	Voltage vector
w	Distance between photodetectors
	Wave amplitude
WL	Wall loss factor
X	Energy attenuation rate coefficient
x	Size
$\boldsymbol{Y}$	Admittance vector

Greek symbols

α	Volume fraction
γ	Ratio of refraction indices
δ	Spacing distance
	Fringe spacing
ε	Permittivity
η	Single-stage collection efficiency
Φ	Electric potential
κ	Collection efficiency
λ	Wave length
μ	Dynamic viscosity
	Permeability
ρ	Density
σ	Conductivity
σ_e	Extinction coefficient
τ	Time shift
ψ	Diverging angle
Ω	Energy attenuation rate
ω	Angular frequency

Subscript

0	Initial
b	Ball probe
c	Cutoff

d	Dynamic
D	Doppler
e	Charge
g	Gas
i	ith
l	Liquid
m	Mass
	Mixture
n	Number
p	Particle
	Predicted
pt	Particle terminal
r	Relative
S	Surface
s	Solid
T	Total
V	Volume

Problems

P8.1 Particle image velocimetry is a nonintrusive measurement technique of the flow field, as introduced in Section 8.5.5. Information of the local velocity field is obtained by image analysis for the trajectories of small tracer particles suspended in the fluid. Therefore the tracer particles need to be able to closely follow the flow. Determine if the following particles can be used as the tracers in their intended applications:

(1) Polymer microspheres with density of 1,050 kg/m^3, and diameter of 7 μm, to be used in deionized water with density of 995 kg/m^3 and viscosity of 0.95×10^{-3} Pa-s. The characteristic length and velocity of the flow are 0.03 m and 0.1 m/s, respectively.

(2) TiO$_2$ particles with density of 3,500 kg/m^3, and diameter of 3 μm, to be used in air with density of 1.2 kg/m^3 and viscosity of 1.8×10^{-5} Pa-s. The characteristic length and velocity of the flow are 0.2 m and 5 m/s, respectively.

P8.2 A capacitance transducer is used to measure the void fraction in gas–liquid flow in a pipe. Two ring electrodes ae installed at a given separation distance, and the capacitance between the two electrodes is measured. The geometric dimensions such as pipe diameter and the wall thickness are also given. Derive the equation that can be used to calculate the gas volume fraction from the measured capacitance for a core-annular flow.

P8.3 Analytical ultracentrifuge is a technique to measure the size of macromolecules such as proteins. It measures the concentration profile of a sample under high-speed rotation (*e.g.*, about 10^4 rpm), and calculate the mass of the macromolecule from the concentration profile. The ratio of the particle velocity u to the centrifugal acceleration $(\omega^2 r)$ is defined as the sedimentation coefficient.

(1) Using force balance, express the sedimentation coefficient as a function of molecule weight M, partial volume v, solvent density ρ, diffusion coefficient D, gas constant R, and absolute temperature T.

(2) In the sedimentation equilibrium experiment, the sample concentration reaches equilibrium due to the balance between diffusion and sedimentation. Derive the equation for concentration profile at equilibrium.

(3) Suggest a data processing technique to obtain the molecule weight from the measured equilibrium concentration profile $c(r)$.

P8.4 Assume that there are spherical particles having a lognormal number density distribution as follows:

$$f_n(x)\frac{2}{\sqrt{2\pi}x}\exp(-2(lnx)^2).$$

Plot the volume density function with the number density function and calculate the surface, volume, and Sauter's and DeBroucker's mean diameter.

P8.5 Calculate the coefficients, C_{b1} and C_{b2} of Equation (8.4-5) based on the following ball probe current calibration results.

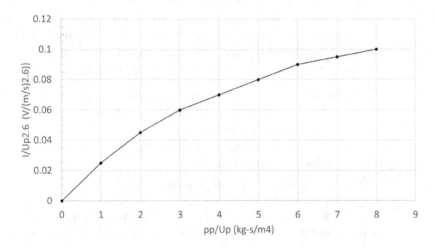

Figure P8.5 Ball probe calibration curve data.

P8.6 Based on Eq. (8.5-9), determine the velocity (u_{py}) of the particles that gives the following LDV measurement results.

$$V(t) = \cos 2\pi\, at\, e^{-bt^2},$$

where $a = 10$ kHz and $b = 10^6$ sec^{-2}. The laser beam angle, the initial frequency, and the wavelength are $30°$, 20 kHz, and 550 nm, respectively.

P8.7 Consider a gas–liquid–solid fluidized bed with the measured distribution of the total pressure along the column as follows:

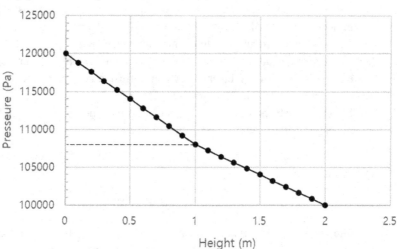

Figure P8.7 Total pressure variation along gas–liquid–solid fluidized bed column.

This pressure distribution includes regions both in the bed and the free board. Find the dynamic pressure variation along the column based on the total pressure plot given above and calculate the solid volume fraction in the bed. Densities of solid, liquid, and gas phases are 2,500, 1,000, and 1 kg/m^3, respectively. Assume that the gas volume fraction is constant along the entire column.

P8.8 A lab-scaled experimental system needs to be established to help the design of a pneumatic drug inhaler (*e.g.*, intermittent sprayer of drug powders). The designed system must meet the following objectives:
(1) Variable pressure of feeding gas (from a pressurized gas container);
(2) Variable time duration of the feeding;
(3) Measureable jetting velocity of drug powder from inhaler;
(4) Measurable drug dose per feeding.
    (a) Provide a schematic design of the experimental system;
    (b) Identify the measurement devices needed (one for each measurand);
    (c) Explain (by providing quantifiable equation or formula) how to determine the jetting velocity and drug dose from your suggested measurement methods.

P8.9 Viscosity (or stress and strain-rate relationship) of general rheological fluids can be determined using a co-cylinder viscometer, in which the rotation speed and corresponding torque can be measured. Develop a relationship between the fluid viscosity (or the slope of shear stress to shear strain) and measured rotation speed and torque.

P8.10 Viscosity (or stress and strain-rate relationship) of very viscous fluids can be determined using a falling-body viscometer, in which the settling velocity of the falling object (typically a solid sphere) under gravity can be measured. Develop a relationship between the fluid viscosity (or the slope of shear stress to shear strain) and the measured settling velocity, assuming the falling is in the creeping flow regime and the wall effect can be ignored.

P8.11 Explain the principle and limitations of Laser Doppler Velocimetry (LDV) with basic equations for discussion. Can LDV be used to measure the droplets' velocity of an atomized spray jet that has a wide size distribution and a distribution of injection angles? Explain or justify your comments.

P8.12 For optical measurements of fluid flow velocity, the actual measured is the velocity of tracer(s). Explain why such a measurement is only an approximate of the true fluid velocity. How to quantify the difference?

P8.13 The convective heat transfer coefficient of a sphere can be determined via temperature measurement in transient heat transfer to the sphere in a convective flow environment. Assume that the lumped heat capacity model can be applied to the sphere heat transfer, and the convective heat transfer coefficient remains unchanged during the heat transfer process. All material and geometric properties can be regarded as constants.

(a) Derive an expression of convective heat transfer coefficient as a function of time, initial temperature (T_0) and ambient temperature (T_∞) (note that $T = T_\infty$ as $t = \infty$).

(b) Based on the least squares method and a set of measurements $\{T_i, t_i\}$ ($i = 1, \ldots, N$), derive equation(s) to best estimate the convective heat transfer coefficient.

P8.14 Consider the choices of flow velocity measurement methods: (1) LDV; (2) PDPA; (3) PIV; (4) corona-charger ball probe; (5) dual-beam cross-correlation; and (6) radioactive particle tracking.

(a) List the methods of intrusive measurement;

(b) List the methods of "instant" velocity measurement (e.g., suitable for turbulent characteristics study);

(c) List the methods of time-averaged velocity measurement;

(d) List the methods of space-averaged velocity measurement;

(e) List the methods that do not require "tracers."

P8.15 A lab-scaled experimental system needs to be established to help the design of a spray atomizer based on the principle of induction of an ejector. The designed system must meet the following objectives: (1) Controllable and measurable pressure of jetting gas; (2) Measureable flowrates of gas and induced liquid; (3) Measureable jetting velocities of gas and droplets (note: velocities are different between gas and droplets); (4) Measurable droplet size distributions.

(a) Provide a schematic design of the experimental system;

(b) Identify the measurement devices needed.

References

Adrian, R. J. (2005). Twenty years of particle image velocimetry. *Exp. Fluids*, **39**, 159–169.

Bachalo, W. D. (1980). Method for measuring the size and velocity of spheres by dual-beam light-scatter interferometry. *Appl. Opt.*, **19**, 363–370.

Bachalo, W. D., and Sankar, S. V. (1998). Phase Doppler particle analyzer. *The Handbook of Fluid Dynamics*, ed. R. W. Johnson, Boca Raton, FL: CRC Press.

Bendat, J. S., and Piersol, A. G. (1986). *Random Data Analysis and Measurement Procedures*. 2nd Ed., New York: John Wiley & Sons.

Bishop, D. L. (1934). A sedimentation method for the determination of the particle size of finely divided materials (such as hydrated lime). *Bur. of Stand. J. Res.*, **12**, 173–183.

Boland, D., and Geldart, D. (1971). Electrostatic charging in gas fluidized bed. *Powder Technol.*, **5**, 289–297.

Bröder, D., and Sommerfeld, M. (2007). Planar shadow image velocimetry for the analysis of the hydrodynamics in bubbly flows. *Meas. Sci. Tech.*, **18**, 2513–2528.

Cheng, L., and Soo, S. L. (1970). Charging of dust particles by impact. *J. Appl. Phys.*, **41**, 585–591.

Crowe, C. T., Sommerfeld, M., and Tsuji, Y. (1998). *Multiphase Flow with Droplets and Particles*. Boca Raton, FL: CRC Press.

Dehaeck, S., Van Parys, H., and Hubin, A. (2009). Laser marked shadowgraphy: a novel optical planar technique for the study of microbubbles and droplets. *Exp. Fluids*, **47**, 333–341.

Dzubay, T. G., and Hasan, H. (1990). Fitting multimodal lognormal size distributions to cascade impactor data. *Aerosol Sci. Technol.*, **13**, 144–150.

Fan, L.-S., and Zhu, C. (1998). *Principles of Gas-Solid Flows*. Cambridge: Cambridge University Press.

Fingerson, L. M., and Menon, R. K. (1998). Laser Doppler velocimetry. *The Handbook of Fluid Dynamics*, ed. R. W. Johnson, Boca Raton, FL: CRC Press.

Gladden, L. F., and Sederman A. J. (2013). Recent advances in flow MRI. *J. of Magnetic Resonance*, 229, 2–11.

Gouesbet, G., and Gréhan, G. (2015). Laser-based optical measurement techniques of discrete particles: A review [invited keynote]. *Int. J. Multiphase Flow*, **72**, 288–297.

Gurau, B., Vassallo, P., and Keller, K. (2004). Measurement of gas and liquid velocities in an air–water two-phase flow using cross-correlation of signals from a double sensor hot-film probe. *Exp. Therm. Fluid Sci.*, **28**, 495–504.

Hammer, E. A., Johansen, G. A., Dyakowski, T., Roberts, E. P. L., Cullivan, J. C., Williams, R. A., Hassan, Y. A., and Claiborn, C. S. (2005). Advanced experimental techniques. *Multiphase Flow Handbook*, ed. C. T. Crowe, Boca Raton, FL: CRC Press.

Hay, W., and Sandberg, P. (1967). The scanning electron microscope, a major breakthrough for micropaleontology. *Micropaleontology*, **13**, 407.

Heindel, T. J., Gray, J. N., and Jensen, T. C. (2008). An X-ray system for visualizing fluid flows. *Flow Meas. Instrum.*, **19**, 67–78.

Hemminger, O., Yu, Z., Lee, J., and Fan, L.-S. (2007). Microparticle flow in liquid medium: 3-D confocal microscopic imaging and computation of flow in microchannels, in *Fluidization-XII* ed. Berruti, F., Bi, X., and Pugdley, T., Engineering Conference Institute, 489–496.

Holman, J. P. (2012). *Experimental Methods for Engineers*, 8th ed., McGraw-Hill.

Irons, G. A., and Chang, J. S. (1983). Particle fraction and velocity measurement in gas-powder streams by capacitance transducers. *Int. J. Multiphase Flow*, **9**, 289–297.

Kay, D. H. (1965). *Techniques for Electron Microscopy*, 2nd ed. Oxford: Blackwell Scientific.

Kumar, S. B., Moslemian, D., and Dudukovic, M. P. (1995). A γ-ray tomographic scanner for imaging voidage distribution in two-phase flow systems. *Flow Meas. Inst.*, **6**, 61–73.

Kumara, W. A. S., Halvorsen, B. M., and Melaaen, M. C. (2010). Single-beam gamma densitometry measurements of oil-water flow in horizontal and slightly inclined pipes. *Int. J. Multiphase Flow*, **36**, 467–480.

Luo, X., Lee. D. J., Lau, R., Yang, G., and Fan, L.-S. (1999). Maximum stable bubble size and gas holdup in high-pressure slurry bubble column. *AIChE J.*, **45**, 665–680.

Park, A.-H., Bi, H. T., Grace, J. R., and Chen, A. (2002). Modeling charge transfer and induction in gas–solid fluidized beds. *J. Electrostatics*, **55**, 135–158.

Sankar, S. V., and Bachalo, W. D. (1991). Response characteristics of the phase Doppler particle analyzer for sizing spherical particles larger than the light wavelength. *Appl. Opt.*, **30**, 1487–1496.

Soo, S. L., Baker, D. A., Lucht, T. R., and Zhu, C. (1989). A corona discharge probe system for measuring phase velocities in a dense suspension. *Rev. Sci. Instrum.*, **60**, 3475.

Soo, S. L., Stukel, S. S., and Hughes, J. M. (1969). Measurement of mass flow and density of aerosols in transfer. *J. Environ. Sci. Tech. (Ind. Eng. Chem.)*, **3**, 386–393.

Wang, F., Marashdeh, Q., Fan, L.-S., and Warsito, W. (2010). Electrical capacitance volume tomography: design and applications. *Sensors*, **10**, 1890–1917.

Wang, F., Marashdeh, Q., Fan, L.-S., and Williams, R. A. (2009). Electrical capacitance, electrical resistance, and positron emission tomography techniques and their applications in multi-phase flow systems. *Adv. in Chem. Eng.*, **37**, 179–222.

Warsito, W., Ohkawa, M., Kawata, N., and Uchida, S. (1999). Cross-sectional distributions of gas and solid holdups in slurry bubble column investigated by ultrasonic computed tomography. *Chem. Eng. Sci.*, **54**, 4711–4728.

Warsito, W., Marashdeh, Q., and Fan, L.-S. (2007). Electrical capacitance volume tomography. *IEEE Sensors J.*, **7**(4), 525–535.

Zhu, C. (1999). Isokinetic sampling and cascade samplers. *Instrumentation for Fluid Particle Flow*, ed. S. L. Soo, New York: Noyes Publications/William Andrew.

Zhu, C., Slaughter, M. C., and Soo, S. L. (1991). Measurement of velocity of particles in a dense suspension by cross correlation of dual laser beams. *Rev. Sci. Instrum.*, **62**, 2036–2037.

Zhu, C., and Soo, S. L. (1992). A modified theory for electrostatic probe measurements of particle mass flows in dense gas-solid suspensions. *J. Appl. Phys.*, **72**(5), 2060–2065.

Zhu, C., Yu, T., and Huang, D. (1997). Numerical study of effect of velocity slip on isokinetic/anisokinetic sampling of gas-solid flows. Int. Symp. on Multiphase Fluid, Non-Newtonian Fluid and Physicochemical Fluid Flows, Oct. 7–9, Beijing, China.

Part II

Application-Based Analysis of Multiphase Flows

9   Separation of Multiphase Flows

9.1   Introduction

Separation processes associated with particle removal, sampling, or recirculation are central to many multiphase flow systems. The particles of interest can be solids, droplets, or bubbles. The principles of separation are conceptually based on the notion that different phases exhibit distinctly dissimilar dynamic responses or transport characteristics. Any physical mechanism that causes, or enhances, disparity can be selected as a controlling identifier for phase separation. Typical examples of governing indicators include a field force (*e.g.*, gravity or electrostatic force), flow-induced inertia (*e.g.*, centrifugation or jet impaction), and interception (*e.g.*, sieving or filtration). These mechanisms provide the fundamental design principles for industry separation devices, such as gravity settling chambers, cyclones, impactors, and filters. To realize high separation efficiencies, multistage separators are commonly employed that leverage several phenomena at each stage or in the process.

This chapter introduces the predominant separation mechanisms currently in use for various multiphase flows. Theoretical considerations required for modeling each mechanism are presented. Section 9.2 is focused on inertia-based separation methods induced by gravity, centrifugal force, and jet-enhanced impaction. Section 9.3 presents filtration methods with collection mechanisms based on a combination of diffusion, interception, and inertia impaction. Section 9.4 discusses the separation mechanisms that leverage an external electric field. The separated particles can be charged or composed of dielectric materials with charge neutrality. Section 9.5 provides case studies with selected modeling examples.

9.2   Separation by Phase Inertia

When subjected to gravity or a centrifugal force, inertia causes suspended particles to depart from the carrying fluid phase if their densities are different. The resulting trajectory deviation provides a unique and convenient method to achieve predictable particle separation that does not require moving parts for the majority of installations. When considering the available separation systems, the phase inertia technique is the most straightforward and cost-effective option in the context of operation and maintenance.

Section 9.2.1 introduces the simplest and most popular inertia-based separation methods, which are focused on the structure and operation principles of key components or subsystems. Section 9.2.2 continues with various basic models representing the inertia-based separation principles. Most of these models yield analytical solutions, obtained using governing equations from Chapters 2 to 6 and with much simplified assumptions or geometric conditions that are of particular use to practical design of industrial systems due to their simplicity in parametric analysis.

9.2.1   Phase-Inertia Separation Methods

Typical inertia-based separation systems include gravitation settling chambers, cyclones, and cascade impactors. Phase separation can also be augmented by combining with other multiphase flow separation techniques, such as by electrostatic precipitation or by filtration.

9.2.1.1   Gravitational Sedimentation

Gravitational sedimentation, also known as gravity settling, refers to the separation of flowing particles from a carrying medium by the varying gravitational force acting in proportion to distinct material densities of different particles. Applicable particle-laden multiphase flows include gas–solid flows, solid–liquid flows, immiscible liquid–liquid flows, and gas–solid–liquid flows. In gravity settling separation, the particle-laden fluid flow is injected, either horizontally or at an oblique angle, against gravity into a large expansion chamber or open space, as shown in Figure 9.2-1. Progressive enlargement of the cross-sectional chamber area normal to the flow stream is required to significantly reduce the carrying fluid velocity and minimize flow-assisted penetration of solids.

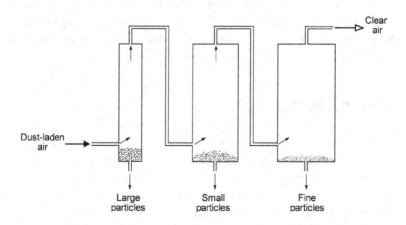

Figure 9.2-1 Cascade gravity settlers (elutriators) of a dust-laden air flow.

There are two basic types of gravity settlers: the simple expansion chamber, or horizontal flow settling chamber, as shown in Fig. 9.2-2(a), and the multiple-tray settling chamber, or Howard settling chamber, as illustrated in Fig. 9.2-2(b).

In a multiple-tray settling chamber, the presence of thin collection plates results in a marginally increased linear gas velocity and a significantly reduced vertical settling distance, when compared with the simple expansion chamber. As a result, the overall collection efficiency for a multiple-tray settling chamber is comparatively greater than that of the simple expansion chamber. Thus, for a specified collection efficiency, the required physical size of a multiple-tray settler will be less than that of a simple expansion settler.

Gravitational settling is often employed to encourage solid–liquid separation. When the primary goal is to obtain a clarified liquid, the process is called clarification. Conversely, if the intended purpose is to produce a concentrated slurry, the process is called thickening. A typical sedimentation thickener is schematically illustrated in Figure 9.2-3.

Slurry is continuously fed into the center of the tank at a depth several feet below the surface of the liquid in the tank. Liquid radially disperses while particulate begins settling to the bottom of the tank. This sedimentation process results in a concentration gradient that increases with tank depth and is stratified into three characteristic zones: the clarification zone, the settling zone, and the compression zone. Clear liquid flows out of the system over the top edge of the tank. A slowly revolving rake scrapes the sludge from the tank bottom and directs it to the central outlet, where the sludge is removed.

A sedimentation classifier typically operates using either the sink-and-float method or differential settling. The sink-and-float method is useful when two different constituent particle types are to be separated from a liquid phase of intermediate density. In this approach, the particles that have a greater density than the liquid medium settle from the liquid, while the particles with less density float. The differential settling method is applicable when the liquid density is less than either of the two constituent types of particles. Here, separation is a function of the difference in the terminal velocity of the particles. However, if the mixture includes a range of

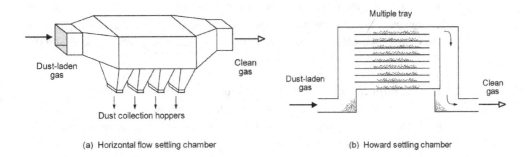

(a) Horizontal flow settling chamber   (b) Howard settling chamber

Figure 9.2-2 Examples of gravity settlers.

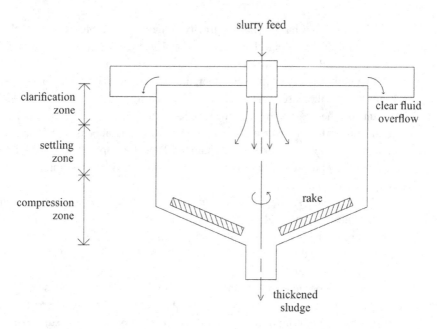

Figure 9.2-3 Schematic illustration of a circular basin thickener.

particle sizes, the potential exists for large-light particles and small-heavy particles to settle at an identical rate; in this case, no effective separation occurs.

The gravitational settling principle can also be utilized to separate a gas–liquid mixture, two immiscible liquids, or three-phase flows (*e.g.*, as gas–oil–water). Conceptual examples of gas–liquid separators and three-phase separators are shown, respectively, in Figure 9.2-4. When the gas–liquid mixture stream enters the separator given in the figure, a preponderance of liquid rapidly descends into the liquid pool; fine liquid droplets remain suspended in the gas phase. These residual droplets are carried by the upward gas flow, but ultimately settle under the influence of gravity. For the separation initiative to be considered a success, all droplets must be retained in the vessel; they cannot be carried out by the gas flow.

The main advantages of gravity settlers are their elementary structure and straightforward operation. The primary disadvantages include a relatively weak driving force for separation and the requirement for a collection chamber that is comparatively larger than other alternatives such as cyclones, fabric filters, and electrostatic precipitators. Gravity settlers are often used in industrial processes as precleaners with the aim of removing relatively large particles.

9.2.1.2   Inertia Enhancement by Flow Rotation

Inertia of flowing phases can be enhanced by flow controls. For techniques founded on rotating flow principles, multiple phases are subjected to various centrifugal forces, a common practice in multiphase flow separation. One embodiment of a rotation-based flow separator is the cyclone, which establishes a vortex using a tangential inlet or guide vanes, as shown in Figure 9.2-5.

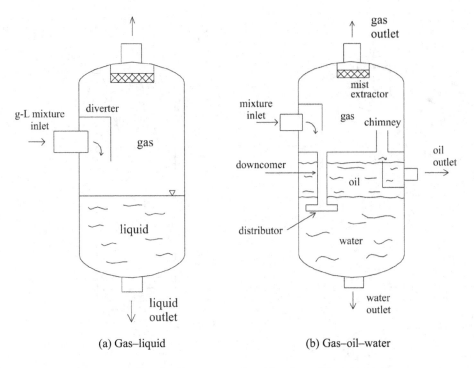

Figure 9.2-4 Gas–liquid separators (from Stewart and Arnold, 2008).

The rotating flow subjects suspended particles to a horizontal centrifugal force, which usually exceeds the vertical gravitational force by at least two orders of magnitude. Thus, even lightweight particles are readily directed toward the wall. Because of their relatively high inertia, particles vigorously collide with the wall, and boundary layer interaction causes either a dramatic change in particle momentum or a significant loss of kinetic energy. Post-collision particles are either deposited on the wall surface or driven by gravity and boundary-layer flow toward the particle collection bunker. Meanwhile, the carrying fluid flow, with its relatively low inertia, diverges from the wall region and moves toward the cyclone outlet.

The cyclone collector is one of the most elegant particle collectors: It has no moving parts and is straightforward to maintain. Despite its simple arrangement, centrifugal acceleration in a cyclone collector can potentially exceed gravitational acceleration by a factor of 300 to 2,000. Consequently, cyclones have a unique ability to achieve a large disparity in phase inertia, which results in high separation efficiencies. Notably, the separation efficiency attainable for a cyclone with guide vanes may be diminished from that of the tangential inlet variant. This is because guide vanes impart a relatively slow rotation effect to the fluid. However, guide vane cyclones feature a coaxial arrangement of the fluid and solid outlets to prevent re-entrainment of particles, which may occur when using a tangential inlet cyclone. Additional advantages of cyclones include economical operation, reliability, and suitability for high-temperature operations.

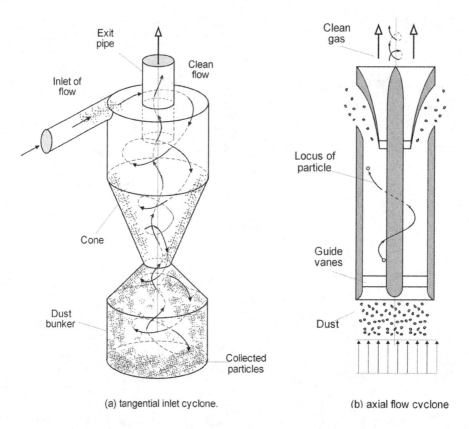

(a) tangential inlet cyclone.   (b) axial flow cyclone

Figure 9.2-5 Typical cyclones.

Detailed information about the local rotating flow velocity and resulting pressure distribution is of vital importance to predict particle separation efficiency or describe cyclone characteristics. For example, flow rotation generates centrifugal acceleration that enhances the particle separation. However, flow rotation also creates a low-pressure zone, or pressure gradient force, that tends to counterbalance the centrifugal force, which can function to reduce separation efficiency. In addition, when the carrying fluid velocity in a particle separator is sufficiently high, particles deposited on the collection plates can be picked up through the shear lifting effect, such as the Saffman force, that countervails settling and other retaining forces, such as gravity and van der Waals forces. Re-entrainment of already-collected particles can also substantially reduce overall collection efficiency.

9.2.1.3   Separation by Impaction

Particles can be separated by flow-assisted impaction on collection surfaces, such as the cascade impactors of Section 8.2.5. In a particle jet impingement, as exemplified by the May cascade impaction system shown in Figure 9.2-6, particles with high inertia do not flow with the fluid under an abrupt turn of the stream. Such particles

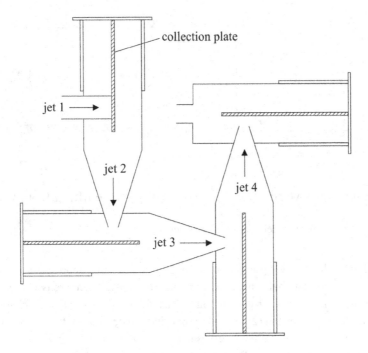

Figure 9.2-6 May cascade impactor (from May, 1945).

are trapped by surface collisions, but the fluid flow is permitted to diverge from the impacting surface.

A cascade impactor, consisting of a series of impaction stages arranged successively so the fluid impact velocity within each stage is progressively increased, can be employed as either a particle separator or a size analyzer. A high-efficiency filter is typically leveraged as a final stage collector in an effort to capture any fines that escaped the previous stages.

9.2.2   Modeling Approaches

Process objectives must be specified in order to select an appropriate separation approach. For example, if an inertia-based separator, such as a cyclone, is to be employed in a multiphase flow, the collection efficiency, minimum diameter of particles to be collected, and the tolerable effects on flow operations must all be evaluated. The collection efficiency is typically a function of the following: particle size, particle density, carrying fluid characteristics, inlet tangential velocity, vortex swirl intensity, cyclone geometry, cyclone material, and cyclone wall surface properties that impact particle–wall collisions.

Modeling predictions of particle separation efficiency for an inertia-based separator are principally obtained by solving Eulerian–Lagrangian equations; the Eulerian method describes the fluid flow and the Lagrangian method predicts particle flow and trajectories. Depending on the primary modeling objectives in the context of imposed

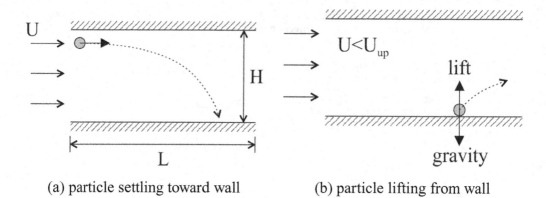

(a) particle settling toward wall   (b) particle lifting from wall

Figure 9.2-7 A simple model of horizontal settler.

constraints, the following must be considered: degrees of approximation, solution accuracy, computational time, and computing capacity. There are two general categories of modeling that can accommodate such requirements: (1) the mechanistic and parametric approach; and (2) the comprehensive and detailed method. The former is a starkly simplified approach that serves to yield a quick assessment of primary mechanisms and key parametric effects in an effort to semi-quantify an optimized system design or operating mode. The latter is usually scoped to provide the most detailed and realistic assessment possible, which requires pushing the envelope of the existing state of the art in terms of modeling techniques and available computational power.

9.2.2.1   Mechanistic and Parametric Modeling

The main objective of mechanistic and parametric modeling is to glean the most fundamental characteristics of a process, or system, through simplified assumptions. The method typically relies on straightforward algebraic expressions or ordinary differential equations to assess primary mechanisms and key parametric effects conducive to an optimized design or operating strategy. A few such models for gravitational settlers and cyclones are provided below that demonstrate the general procedure.

Figure 9.2-7(a) shows a basic model that can be used to estimate the minimum diameter of particles collected in a horizontal settling chamber of a given length, L, and height, H. Consider the limiting case where a particle enters the chamber at height, H, such that the settling time for the particular particle is at a maximum with respect to all identically sized entering particles. It is further assumed that (1) the fluid velocity, U, is uniform throughout the chamber; (2) there is no velocity slip between the particle and fluid in the horizontal direction; and (3) the particle settles at its terminal velocity.

To ensure all particles of the same size are collected, the maximum settling time of the particle, t_{sm}, should be less than the particle residence time in the chamber. Thus, it yields:

$$t_{sm} = \frac{H}{U_{pt}} \leq \frac{L}{U},\qquad\text{(9.2-1)}$$

where U_{pt} is the particle terminal velocity. For a small spherical particle setting at its terminal velocity that is balanced by the Stokes drag and the gravitational force, the diameter of collected particles can be estimated by the expression:

$$d_p^2 \geq \frac{18\mu UH}{L(\rho_p - \rho)g}. \tag{9.2-2}$$

Pickup velocity is a constraint to the fluid flow velocity worthy of consideration. The pickup velocity is the onset criterion for particle re-entrainment that occurs when the particle lift forces, such as the Saffman force due to flow shearing near the wall, begin to exceed the restraining forces, such as the gravity or van der Waals forces, which hold particles to the wall or form agglomerates. As shown in Figure 9.2-7(b), the carrying fluid flow rate should be controlled so the average transport velocity in the settling chamber is far less than the pickup velocity. Neglecting interparticle friction, the pickup velocity (defined in Section 11.3.2), U_{pu}, may be estimated from the relationship (Zenz and Othmer, 1960):

$$U_{pu} = \sqrt{\frac{4gd_p}{3}\left(\frac{\rho_p}{\rho} - 1\right)}. \tag{9.2-3}$$

For a vertical settler, the condition that all particulates remain inside the vessel implies the terminal velocity for all particles must be greater than the superficial fluid velocity. This condition is guaranteed by:

$$U_{pt} > U. \tag{9.2-4}$$

The predominant mechanism in cyclone separation is the centrifugal force that causes particles to separate from the carrying fluid flow. However, a rotational flow also creates a low-pressure zone, as in the center of a tornado, which can lift deposited particles by suction and reduce the overall separation efficiency. The flow rotation and associated pressure distribution can be approximated with a single-phase flow model because the particle concentration in a typical cyclone application is very dilute; the particle–fluid interaction can therefore be regarded as one-way coupled.

One popular model, known as Burgers' viscous vortex model (Burgers, 1948), considers a simple rotating flow in a cylinder. The flow is assumed to be steady-state, incompressible, axisymmetric, devoid of body force, and laminar. Thus, the continuity and momentum conservation equations of the flow are governed by Eq. (2.2-10) and Eq. (2.2-12), which can be expressed in cylindrical coordinates as:

$$\frac{\partial U}{\partial r} + \frac{\partial W}{\partial z} + \frac{U}{r} = 0 \tag{9.2-5}$$

$$U\frac{\partial U}{\partial r} + W\frac{\partial U}{\partial z} - \frac{V^2}{r} = -\frac{1}{\rho}\frac{\partial p}{\partial r} + \frac{\mu}{\rho}\left(\frac{\partial^2 U}{\partial r^2} + \frac{\partial^2 U}{\partial z^2} + \frac{1}{r}\frac{\partial U}{\partial r} - \frac{U}{r^2}\right)$$

$$U\frac{\partial V}{\partial r} + W\frac{\partial V}{\partial z} + \frac{UV}{r} = \frac{\mu}{\rho}\left(\frac{\partial^2 V}{\partial r^2} + \frac{\partial^2 V}{\partial z^2} + \frac{1}{r}\frac{\partial V}{\partial r} - \frac{V}{r^2}\right) \tag{9.2-6}$$

$$U\frac{\partial W}{\partial r} + W\frac{\partial W}{\partial z} = -\frac{1}{\rho}\frac{\partial p}{\partial z} + \frac{\mu}{\rho}\left(\frac{\partial^2 W}{\partial r^2} + \frac{\partial^2 W}{\partial z^2} + \frac{1}{r}\frac{\partial W}{\partial r}\right),$$

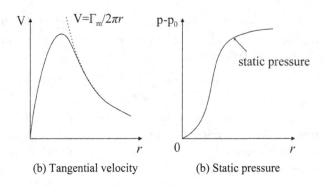

(b) Tangential velocity   (b) Static pressure

Figure 9.2-8 Typical vortex flow distribution.

where U, V, and W are the velocity components in the respective cylindrical coordinates (r, θ, z). A typical solution of velocity and pressure in the following form can be verified to satisfy the above equations:

$$U = -Ar \qquad V = V(r) \qquad W = 2Az \tag{9.2-7a}$$

$$p = -\frac{1}{2}\rho A^2(r^2 + 4z^2) + \rho \int \frac{V^2}{r}dr, \tag{9.2-7b}$$

where A is a constant, and $V(r)$ needs to be solved from the tangential-component momentum equation of Eq. (9.2-6b). Substituting Eq. (9.2-7a) and Eq. (9.2-7b) into Eq. (9.2-6b) yields:

$$-Ar\frac{dV}{dr} - AV = \frac{\mu}{\rho}\frac{d^2V}{dr^2} + \frac{1}{r}\frac{dV}{dr} - \frac{V}{r^2}, \tag{9.2-7c}$$

which has a solution, for the tangential velocity, as:

$$V = \frac{\Gamma_m}{2\pi r}\left\{1 - \exp\left(-\frac{A\rho r^2}{2\mu}\right)\right\}, \tag{9.2-7d}$$

where Γ_m is a constant representing the maximum circulation in the system. Equation (9.2-7d) further shows that, as the viscosity vanishes, the tangential velocity of a free vortex is given by

$$V = \frac{\Gamma_m}{2\pi r}. \tag{9.2-7e}$$

Thus, the flow pattern in a cyclone can be roughly represented as a free vortex core of nearly-solid body rotation with a viscous vortex flow in the outer-wall region, as illustrated in Figure 9.2-8(a). The corresponding pressure distribution can be solved from Eq. (9.2-7b) with Eq. (9.2-7d), as schematically represented by Figure 9.2-8(b).

Based on the analysis above, the tangential flow velocity in a tangential-inlet cyclone can be approximated through the relationship (Alexander, 1949):

$$Vr^\beta = \text{const}, \tag{9.2-8a}$$

where the power index, β, is an empirical constant; other velocity components are assumed to be negligible. When a cyclone separator is operated at a high flow rate,

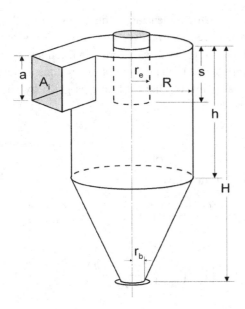

Figure 9.2-9 Geometric configuration of a tangential-inlet cyclone.

the particle-laden flow is often highly turbulent. However, turbulence effects on the flow velocity distribution and static pressure in the preceding vortex flow models are similarly ignored. Detailed account of turbulent flows and associated particle–fluid interactions in cyclone separations typically require a numerical modeling approach.

Because of the nature of the flow in a cyclone, as predicted by the rotational flow model above, particle trajectories and hence collection potential can be determined using the Lagrangian approach with an assumption of one-way coupling between fluid and particle flows. Such an approach was exemplified in a parametric model developed by Leith and Licht (1972), which estimates the collection efficiency of the tangential-inlet cyclone shown in Figure 9.2-9.

In the model, fluid flow is approximated by Eq. (9.2-8a). Other key assumptions include (1) a constant radial velocity of particles; (2) Stokes drag force is considered; and (3) no slip in tangential velocity between the fluid and particles. The collection efficiency can then be estimated with the expression:

$$\eta_c = 1 - \exp\left\{-2\left(\frac{2\tau_S V_i^2}{R^2}(1+\beta)t_m\right)^{\frac{1}{2\beta+2}}\right\}, \qquad (9.2\text{-}8b)$$

where τ_S is the Stokes relaxation time, defined by Eq. (3.2-22a); t_m is mean particle residence time, defined as the average of the minimum and maximum residence times. The minimum residence time is the time required for the flow to descend from the mid-level of the entrance to the exit at the bottom of the cyclone, which is estimated by:

$$t_{m1} = \frac{\pi R^2(s - a/2)}{V_i A_i}\left\{1 - \left(\frac{r_e}{R}\right)^2\right\}, \qquad (9.2\text{-}8c)$$

where s, a, r_e, R, and A_i are geometric dimensions defined in Figure 9.2-9. In order to estimate the maximum residence time, t_{m2}, the lowest point to which the flow descends must be identified. Due to the effects of vortex-induced low pressure, this point does not necessarily correspond to the bottom of the cyclone as described by the full depth, H. The deepest point, of chord length l, is given by an empirical equation (Alexander, 1949) of the form:

$$\frac{l}{r_e} = 5\left(\frac{\pi R^2}{A_i}\right)^{\frac{1}{3}}. \tag{9.2-8d}$$

Hence, the mean residence time is then given by:

$$t_{m2} = t_{m1} + \frac{\pi R^2}{V_i A_i}\left\{h - s + \frac{(l+s-h)}{3}\left(1 + \xi + \xi^2\right) - \left(\frac{r_e}{R}\right)^2 l\right\}, \tag{9.2-8e}$$

where:

$$\xi = 1 - \left\{1 - \left(\frac{r_b}{R}\right)\right\}\left(\frac{l+s-h}{H-h}\right). \tag{9.2-8f}$$

The above equations of separation efficiency are obtained for monodispersed or identically sized spherical particles with a uniform inlet flow condition, or equal probability inlet position. However, a size distribution of flowing particles at the inlet is more realistic. An important performance index for cyclone separation is the cutoff size of a cyclone, which is defined as the minimum particle size that yields a collection efficiency of at least 50%. According to Eq. (9.2-8b), the cutoff size of a tangential-inlet cyclone can be estimated by the relationship:

$$d_{pc} = \frac{3R}{V_i}\sqrt{\frac{\mu}{\rho_p(1+\beta)t_m}}\left(\frac{\ln 2}{2}\right)^{1+\beta}. \tag{9.2-9}$$

The inertial separation effect due to abrupt shift in direction of a particle-laden flow stream is evinced by the particle stopping distance, which is defined as the travelling distance of a particle in its forward direction prior to coming to rest with respect to the surrounding fluid. For small particles, assuming Stokes drag during impingement, the particle stopping distance for a spherical particle, with initial velocity U_0, can be calculated as:

$$\delta = U_0 \tau_S. \tag{9.2-10a}$$

For large or heavy particles with an initial Re ranging from 1 to 400, the stopping distance may be estimated using an empirical formula proposed by Mercer (1973) as:

$$\delta = \frac{\rho_p d_p}{\rho}\left\{Re^{\frac{1}{3}} - \sqrt{6}\tan^{-1}\left(\frac{Re^{\frac{1}{3}}}{\sqrt{6}}\right)\right\}, \tag{9.2-10b}$$

where the particle Reynolds number is based on maximum terminal velocity in the stagnant fluid. Notably, if the flow experiences right-angle impaction on a collecting plate, only particles with deviated trajectories that are capable of actually reaching the plate are collected. Hence, to estimate particle separation efficiency, it is necessary to evaluate the complete departure trajectory of a particle and its tendency toward the impaction plate. One simple, but very crude, method to estimate the total departure is by equating it to the stopping distance parameter (Reist, 1993).

9.2.2.2 Comprehensive and Detailed Modeling

Mechanistic and parametric models are suitable for characteristic and interdependence studies, yielding primarily trends or qualitative value rather than quantitative results. Actual systems operating in industrial processes rarely can benefit from such simple models that are idealized and struggle to yield realistic, reliable, and quantifiable predictions. For example, the following conditions are indicative of actual cyclone separators: cyclones have inherently complex geometries, even with internals; the fluid flows are rarely uniform and are often turbulent; the particles are polydispersed; inlet conditions of particle flows are randomly distributed; and the relative motion between fluid and particles is beyond the Stokes regime. In this class of cases, the particle settling behavior – hence the collection efficiency – can be potentially described using the basic Eulerian–Lagrangian formulation introduced in Chapter 5, which needs to be solved numerically.

The dominant technique is to employ the Eulerian method to represent the fluid flow and the Lagrangian method to characterize the particle trajectories. Most industrial applications involve turbulent flows. For example, cyclone separation processes generate vigorous swirling turbulence that is erratic near the wall boundary regions due to both powerful tangential movement and viscous damping of the flow. Careful and deliberate intention is required to properly select an adequate turbulence model due to the irregular nature of separation, which will directly influence the validity of the overall flow modeling. Anisotropic turbulence provisions, such as the Reynolds stress approach, should be favored over isotropic models, such as the $k-\varepsilon$ model.

Particle transport using Lagrangian modeling can be affected by the local and instantaneous turbulent behavior of the fluid flow. Random variations in the fluid velocity due to turbulent fluctuations, the initial entrance locations, and particle size can be accommodated with a Monte Carlo simulation method or with probability density functions (Theodore and Buonicore, 1976).

For a typical single-stage impactor, the flow will not make a collective right-angle turn. The collection efficiency with respect to particle size, or a characteristic efficiency curve, may be obtained if the individual efficiencies for a series of monodispersed spherical particles are known. Determination of the individual efficiency for a particle of specified size can be accomplished either through empirical measurements or theoretical predictions. Employing theory to make the efficiency prediction relies on particle trajectory modeling in an impingement flow. In this case, it is traditionally assumed that all the particles impacting the collection surface are successfully captured and remain attached to that surface. The flow field of the jets for a particular nozzle and collection plate can be calculated by solving the Navier–Stokes equations together with the corresponding impactor geometry.

9.3   Filtration

Solid particles can be separated from a carrying fluid if the solid-laden suspension flow passes through a single-layer screen, a multiple-layer screen, or a permeable medium that is designed to retain particulate of a specified size. This separation method, known as filtration, is straightforward and is highly effective at collecting particles of all sizes.

Filtration is a physical separation technique whereby particles are removed from the fluid and retained by the filter media. Three basic collection mechanisms that employ fibers exist: inertial impaction, interception, and diffusion. In collection by inertial impaction, the particles with large inertia depart from the fluid streamlines in the vicinity of the fiber collector and collide with it. In collection by interception, particles with relatively little inertia track the streamline around the fiber collector closely and are partially, or completely, immersed in the boundary region. The particle velocity is thereby diminished, the particles graze the barrier, and ultimately stop on the surface of the collector. Collection by diffusion is particularly applicable for fine particles. In this collection mechanism, particles with zigzag Brownian motion in the immediate area of the collector accumulate on its surface. The efficiency of collection by diffusion increases with decreasing particle size and suspension flow rate. There are several other collection mechanisms to consider: gravitational sedimentation, induced electrostatic precipitation, and van der Waals deposition. The impact of these techniques on filtration effectiveness may be important in some processes, but these topics are beyond the scope of this text.

In the context of particle collection, filtration techniques can be classified into either cake filtration or depth filtration. With cake filtration, particles are deposited on the front surface of a collecting filter, as shown in Fig. 9.3-1(a). In the case of depth filtration, particles flow through the filter and are collected during their passage, as conceptually illustrated in Fig. 9.3-1(b). Examples of filters employing cake filtration include fabric filters, sieves, nuclepore filters, and membrane filters. Filters relying on depth filtration include packed beds, meshes, fluidized beds, and glass-fiber filters.

Modeling of multiphase flow separation by filtration is typically accomplished with the mechanistic and parametric approach. Newtonian fluid flow in a permeable medium is described by Darcy's law or Ergun's equation of Section 2.4. Lagrangian trajectory modeling is embraced for conducting in-depth penetration analyses of fine particles. For filtration flows, the most meaningful results include the filtration efficiency and associated pressure drop across the filtration media. The following section is centered on fundamental models for collection mechanisms and resulting separation efficiencies.

9.3.1   Collection Efficiency of a Single Fiber or Granular Particle

The filtration efficiency of a filter depends on the effectiveness of particle capture for the basic elements that comprise the device. Typical constituent filter elements include fibers and granular particles. Recall, the basic three collection mechanisms

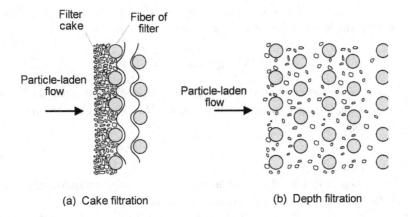

Figure 9.3-1 Types of filtration.

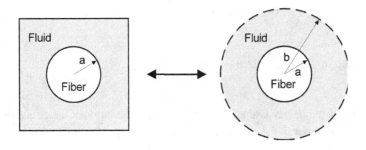

Figure 9.3-2 Happel cell model of flow around a single fiber.

are diffusion, interception, and inertia. The total collection efficiency of a single element is intuitively given by:

$$\eta = \eta_d + \eta_{int} + \eta_{imp}. \tag{9.3-1}$$

9.3.1.1 Collection Efficiency of a Single Fiber

The flow around a single fiber inside a fibrous filter can be analyzed using the Happel cell model. A conceptual representation of the model is provided as Figure 9.3-2.

The fiber collection efficiency resulting from Brownian diffusion can be expressed as:

$$\eta_d = 2.9\left(\frac{1 - \alpha_f}{\mathrm{Ku}}\right)^{\frac{1}{3}} \mathrm{Pe}_f^{-\frac{2}{3}}, \tag{9.3-2}$$

where α_f is the fiber volumetric fraction or packing density; Ku is the Kuwabara factor, which can be further defined by the expression:

$$Ku = \alpha_f - \frac{\alpha_f^2}{4} - \frac{3}{4} - \frac{1}{2}\ln\alpha_f. \qquad (9.3\text{-}2a)$$

Pe_f is the Peclet number of fiber. With the Stokes–Einstein equation for the diffusion of spherical particles through a fluid with low Re, Pe_f can be expressed by the following:

$$Pe_f = \frac{2Ud_f}{D} = \frac{6\pi\mu Ud_p d_f}{C_c k_B T}, \qquad (9.3\text{-}2b)$$

where D is diffusivity, C_c is the Cunningham slip correction factor, μ is viscosity, k_B is Boltzmann constant, and d_f is the fiber diameter.

The collection mechanisms that employ interception and inertia impaction are generally coupled by the same flow convection through filter media. Hence, the efficiency of filtration by interception and impaction is typically represented in a combined fashion as proposed by Flagan and Seinfeld (1988). The combined efficiency can be expressed as an implicit and coupling function of parameter λ as:

$$\eta_{imp+int} = \frac{1+\lambda}{2Ku}\left(2\ln(1+\lambda) - 1 + \alpha_f + \frac{1-\alpha_f/2}{(1+\lambda)^2} - \frac{\alpha_f}{2}(1+\lambda)^2\right)$$

$$\lambda = \frac{d_p}{d_f} + Stk\sqrt{\alpha_f}\left(1 + \frac{\eta_{imp+int}}{\lambda}\right)(1+\lambda-\eta_{imp+int})\left\{1 - \exp\left(-\frac{1}{Stk\sqrt{\alpha_f}\left(1+\frac{\eta_{imp+int}}{\lambda}\right)}\right)\right\},$$
$$(9.3\text{-}3)$$

where the parameter λ stands for the ratio of the deviation of the critical streamline at the top of a fiber to the fiber radius; Stk is Stokes number, given by:

$$Stk = \frac{C_c \rho_p d_p^2 U}{18\mu d_f} = \frac{C_c U\tau_s}{d_f}. \qquad (9.3\text{-}3a)$$

The Cunningham slip correction factor, C_c, may be estimated from the following relationship (Ounis *et al.*, 1990):

$$C_c = 1 + \left\{2.514 + 0.8\exp\left(-\frac{0.55d_p}{\lambda_g}\right)\right\}\left(\frac{\lambda_g}{d_p}\right), \qquad (9.3\text{-}3b)$$

where λ_g is the mean free path of the gas. It is noted that when $d_p/d_f \ll 1$ and $Stk\alpha_f^{1/2} \ll 1$, Equation (9.3-3) is simplified into a decoupled expression for collection efficiencies of interception and impaction. These can be written respectively as:

$$\eta_{int} = \frac{1-\alpha_f}{Ku}\left(\frac{d_p}{d_f}\right)^2 \qquad (9.3\text{-}4a)$$

$$\eta_{imp} = \frac{2(1-\alpha_f)\sqrt{\alpha_f}}{Ku}Stk\left(\frac{d_p}{d_f}\right) + \frac{(1-\alpha_f)\alpha_f}{Ku}Stk^2. \qquad (9.3\text{-}4b)$$

From Equations (9.3-2) and (9.3-3), or equivalently Eq. (9.3-4a) and Eq. (9.3-4b), the total collection efficiency can be calculated as a function of particle size.

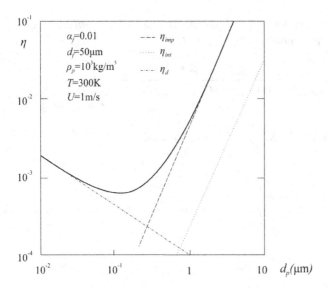

Figure 9.3-3 Typical collection efficiency of a single fiber.

Figure 9.3-3 illustrates an example of the computed individual-mode and combined collection efficiencies of a fiber over a range of particle sizes of suspended aerosols. For the specific filter and flow conditions considered in the model, a minimum of the total collection exists for fibrous filtration of suspended aerosols, which is typically in the submicron range of 0.1 to 1.0 μm.

9.3.1.2 Collection Efficiency of a Single Granular Particle

In a similar fashion to the derivation for fiber collection, the collection efficiency of a single granular particle by Brownian diffusion can be expressed by (Pfeffer and Happel, 1964):

$$\eta_d = 4(1-\alpha)^{\frac{2}{3}} A_s^{\frac{1}{3}} \mathrm{Pe_s}^{-\frac{2}{3}},$$
(9.3-5)

where the subscript s stands for the single granular particle. A_s is a function of bed porosity or volume fraction of fluid, given by:

$$A_s = \frac{2\left(1-(1-\alpha)^{\frac{5}{3}}\right)}{2-3(1-\alpha)^{\frac{1}{3}}+3(1-\alpha)^{\frac{5}{3}}-2(1-\alpha)^2}.$$
(9.3-5a)

$\mathrm{Pe_s}$ is the granular Peclet number, expressed by the relationship:

$$\mathrm{Pe_s} = \frac{d_s U}{D} = \frac{3\pi \mu U d_p d_s}{C_c k_B T}.$$
(9.3-5b)

The collection efficiency by impaction for an isolated granular particle may be estimated by the following (Beizaie, 1977):

$$\eta_{imp} = \begin{cases} \dfrac{0.2453\,(\text{Stk} - 1.213)^{0.955}}{1 + 0.2453\,(\text{Stk} - 1.213)^{0.955}} & \text{Stk} > 1.213 \\ 0 & \text{Stk} < 1.213 \end{cases}, \qquad (9.3\text{-}6)$$

where the granular Stokes number is defined as:

$$\text{Stk} = \frac{C_c \rho_p d_p^2 U}{9\,\mu d_s}. \qquad (9.3\text{-}6a)$$

The collection efficiency by interception may be predicted by the expression (Rajagopalan and Tien, 1976):

$$\eta_{\text{int}} = 1.5 A_s (1 - \alpha)^{\frac{2}{3}} \left(\frac{d_p}{d_s}\right)^2. \qquad (9.3\text{-}7)$$

In granular filtration, the collision between flowing particles and granular particles may result in a rebounding or re-entrainment of collected particles. The probability of capture is often represented by the so-called adhesion probability, σ, which may be estimated by (Tien, 1989):

$$\sigma = \frac{\eta}{\eta_0} = \begin{cases} 0.00318\ \text{Stk}^{-1.248} & \text{Stk} > 0.01 \\ 1 & \text{Stk} < 0.01 \end{cases}, \qquad (9.3\text{-}8)$$

where η_0 denotes collection efficiency without rebounding or re-entrainment effects.

9.3.2   Collection Efficiency of a Filter

To this point, the discussion has been focused on the collection of a single element, be it a fiber or granular particle, of an overall filter assembly. To determine the collection efficiency of a fibrous filter without sacrificing generality, a scenario involving a uniform and transverse flow with respect to cylindrical fibers that comprise fibrous filter media is analyzed.

Consider transverse flow over a cylindrical fiber as depicted in Figure 9.3-4.

A simple model of the collection efficiency of a fiber can be directly given by:

$$\eta_f = \frac{2y}{d_f}, \qquad (9.3\text{-}9a)$$

where y is the limiting width of the particle stream to be collected. Now consider a rectangular bag filter of thickness L, height H, and width W. The number of fibers, dN_f, situated within an elemental thickness dl, is given by the expression:

$$dN_f = \frac{4\alpha_f H dl}{\pi d_f^2}. \qquad (9.3\text{-}9b)$$

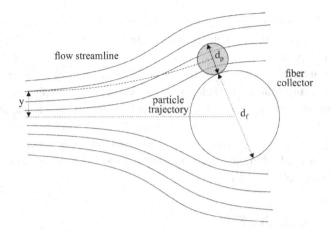

Figure 9.3-4 Particle capture by a single fiber.

Within a time interval, dt, the total number of particles collected within the element is given by:

$$- dN_p = n_p \frac{2yUdt}{(1 - \alpha_f)} WdN_f = n_p \frac{2yUdt}{(1 - \alpha_f)} W \left(\frac{4\alpha_f Hdl}{\pi d_f^2} \right), \qquad (9.3\text{-}9c)$$

where U is the superficial velocity of the dust-laden gas, and n_p is the particle number density. The total number of particles entering the system within time, dt, is expressed by:

$$N_p = n_p UdtHW. \qquad (9.3\text{-}9d)$$

The relative number of particles collected can be predicted by the relationship:

$$- \frac{dN_p}{N_p} = \frac{8y}{\pi d_f^2} \frac{\alpha_f dl}{(1 - \alpha_f)} = \frac{4\eta_f}{\pi d_f} \frac{\alpha_f}{(1 - \alpha_f)} dl. \qquad (9.3\text{-}9e)$$

Thus, for a depth fibrous filter with a filter thickness of L, the overall filter efficiency of monosized aerosols, η_F, is computed using:

$$\eta_F = 1 - \exp \left(- \int_0^L \frac{4\eta_f}{\pi d_f} \frac{\alpha_f}{(1 - \alpha_f)} dl \right). \qquad (9.3\text{-}10)$$

For a suspension flow of polydispersed particles, the averaged collection efficiency of a single fiber may be calculated from:

$$\eta_{f,ave} = \int_0^\infty f_M \eta_{f,x} dx, \qquad (9.3\text{-}11)$$

where f_M is the mass-based size distribution of suspended particles. Thus, as a logical extension of Eq. (9.3-10), the overall filter efficiency of polydispersed particles can be estimated as:

$$\eta_F = 1 - \exp \left(- \int_0^L \frac{4\eta_{f,ave}}{\pi d_f} \frac{\alpha_f}{(1 - \alpha_f)} dl \right). \qquad (9.3\text{-}12)$$

For a depth granular filter, the overall collection efficiency may be estimated using the above equations by replacing the collection efficiency of a single fiber by that of a single granular particle.

The accumulation of collected particles inside the filter media can have a profound impact on the collection efficiency of the device. For example, the collected particles themselves can function as additional collection elements inside the original filter media; if this occurs, the formulation for the collection efficiency of single elements may change. If the particles function as collection elements in the companion paradigm of a cake filtration, the filter thickness and the filter-bed distribution will change in the collection process.

9.3.3   Pressure Drop through a Filter

Separation by filtration near peak efficiency or high flow rate usually requires a relatively high static pressure head to overcome the inherent flow resistance of the porous media and obstruction due to particle accumulation as the filter loads. The resulting pressure drop leads to increased energy consumption for the system to overcome the resistance, and necessitates robust filter material to withstand operation. The flow pattern in a filter is complex as a result of the intricate structure of the fiber, which is further complicated by the process of particle deposition. Consequently, the pressure drop across a particulate filter is challenging to accurately predict. Instead, a common approach is to evaluate the pressure drop through a clean filter, prior to the collection of particles. In addition, the aggregate flow is assumed to decompose into several simple flows around the fibers.

Based on the Happel-cell model, when the fiber orientation is parallel to flow, the pressure drop over a parallel fibrous assembly can be expressed by:

$$\frac{\Delta P_{\parallel}}{L} = \frac{Q\mu}{Ad_f^2} \frac{32\alpha_f}{(4\alpha_f - \alpha_f^2 - 3 - 2\ln\alpha_f)}. \tag{9.3-13}$$

In the companion case of a perpendicular fibrous assembly with a fiber orientation normal to flow, the pressure drop can be calculated using the expression:

$$\frac{\Delta p_{\perp}}{L} = -\frac{32\,Q\mu\alpha_f}{d_f^2 A}\left(\ln\alpha_f + \frac{1-\alpha_f^2}{1+\alpha_f^2}\right)^{-1}. \tag{9.3-14}$$

In flows through a random labyrinth of fibers, it becomes necessary to modify the expression for the pressure drop such that twice the weight is assigned to the correlation for the case of flow perpendicular to fibrous cylinders as for the scenario of flow parallel to the fibrous cylinders. This is necessary because the model that describes flow perpendicular to the fibrous cylinders does not distinguish between crossed or parallel arrangements of the cylinders (Happel, 1959). Thus, for a random assemblage of fibers with cake formation, assuming the cake is in form of a porous layer, the total pressure drop can be estimated with the expression:

$$\Delta p_m = \frac{1}{3}\Delta p_{||} + \frac{2}{3}\Delta p_\perp + \Delta p_c, \tag{9.3-15}$$

where the pressure drop over the cake is approximated by a similar scenario involving flow over a granular bed and using Ergun's equation, Eq. (2.4-5).

9.4 Separation by External Electric Field

Particles can be separated from a carrying fluid by altering their trajectories with the aid of an external field force. Subjecting a multiphase flow to an electric field is one popular technique that is often employed. In general, the electric field force on a charged particle is given by the expression:

$$\mathbf{F}_E = q_e \mathbf{E} + (\mathbf{p}_d \cdot \nabla)\,\mathbf{E}, \tag{9.4-1}$$

where q_e is the charge, and $\mathbf{p}_d$ is the dipole moment. Equation (9.4-1) indicates that there are two types of electric-field forces: the Coulomb force, which results from the electrostatic charge, and the dielectrophoretic force due to polarization effect of the gradient of electric field.

The dielectrophoretic force can persist without an electrostatic charge on a particle. Quantifying the dielectrophoretic force requires information about the dielectric properties of the carrying fluid in addition to the electric field gradient and particle dielectric properties. The dielectric properties of the fluid and particulate determine whether a dielectric particle in a dielectric fluid is compelled by the dielectrophoretic force toward the maximum or minimum intensity of the administered electric field. An applied electric field can be either direct current (DC) or alternating current (AC), per requirements of the application.

9.4.1 Electrostatic Precipitation

Electrostatic precipitation denotes particle–fluid separation via the aid of electrostatic forces in the presence of an external electric field. In an electrostatic precipitation process, suspended particles are charged, driven by the resulting electrostatic force toward a collection surface, and are removed by a discharging operation. Figure 9.4-1 conceptually illustrates the process. An electrostatic precipitator typically has the advantage of requiring much less pressure-driven power as compared to other separation systems because the separation force is directly applied to the particles without a need to accelerate the fluid phase.

There are two basic types of electrostatic precipitators: the cylinder-type and the plate-type. These are schematically depicted in Figures 9.4-2 (a) and (b), respectively. Cylinder-type electrostatic precipitators, sometimes operated as wet precipitators, can achieve collection efficiencies of up to 99.9%. Plate-type precipitators, commonly used for large flow rates, can attain collection efficiencies of approximately 95% (Ogawa, 1984).

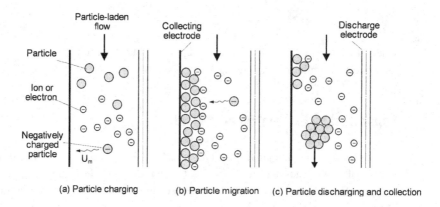

Figure 9.4-1 Mechanisms and process of electrostatic precipitation.

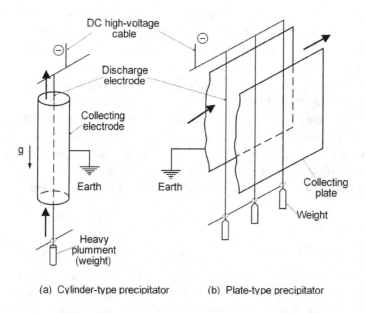

Figure 9.4-2 Type of electrostatic precipitators (after Ogawa, 1984).

9.4.1.1   Charging of Particles

Typical particle charging methods include triboelectric charging via surface contact, field-diffusion charging in a unipolar ion medium, and corona charging in a bipolar ionized fluid medium. An electrostatic charge on particles is often acquired in the natural course of events, such as through triboelectric charging via mutual impacts or via collisions with other solid materials, including pipe walls and valves. However, to achieve highly efficient particle removal in an electrostatic precipitator, the particles are typically charged to a saturated degree by artificial means.

When particles are exposed to unipolar ions in an electric field, ions and particles collide because of two basic mechanisms: field charging and diffusion charging. In a field-charging process, the motion of ions is induced by an applied external electric field. Ion bombardment causes the particles to retain a charge. This charging mechanism dominates when particles are larger than 1 μm in diameter. In a diffusion-charging process, Brownian motion of ions induces a charge on the particles. This mechanism occurs even in the absence of external electric fields and governs when particles are smaller than 0.2 μm in diameter. In the case of an external electric field of strength E_0 charging a particle, the saturated charge acquired by the particle can be estimated by the expression (Fan and Zhu, 1998):

$$q_s = \begin{cases} 3\pi \varepsilon_0 E_0 d_p^2 & \text{conducting particle} \\ \dfrac{9\pi \varepsilon_0 \varepsilon_r}{\varepsilon_r + 2} E_0 d_p^2 & \text{dielectric particle} \end{cases} . \tag{9.4-2}$$

Notably, the saturated charge by field charging is independent of the charging density of ions.

In a diffusion charging process, the saturated charge acquired by a particle can be estimated by the relationship:

$$q_s = \frac{2\pi \varepsilon_0 d_p k_B T}{e} \ln \left(1 + \frac{d_p q_0 e}{8\varepsilon_0 k_B T} \sqrt{\frac{k_B T}{\pi m}} t \right) . \tag{9.4-3}$$

where q_0 is the charge density of the unipolar ions. The saturated charge obtained by diffusion charging is a function of the charging duration and the charging density of ions. In theory, charging by diffusion imposes no limit on the maximum attainable so-called saturation charge, but the breakdown electric field strength of the fluid medium in which the particle is carried provides a realistic constraint.

In corona charging, the ionized fluid medium is typically bipolar; both positive and negative ions exist simultaneously. In this case, particle charging occurs due to the difference in the charging mobility of bipolar ions. The saturated charge limit for a particle is reached when the charging rates from different ions are at equilibrium. The saturated charge per particle can be estimated, implicitly, by the expression:

$$q_s = \frac{q_{io}}{n_p} \left\{ 1 - \sqrt{\frac{m_e}{m_i}} \exp \left(-\frac{q_s e}{\pi \varepsilon_0 d_p k_B T} \right) \right\} . \tag{9.4-4}$$

To avoid interactions of charged particles, the particle number density should be sufficiently dilute; namely, the averaged interparticle distance should be much larger than the Debye shielding distance. This criterion leads to an imposed limit on the particle number density, whereby:

$$n_p < \sqrt[3]{\frac{1}{2\lambda_D}} , \tag{9.4-4a}$$

where λ_D is the Debye shielding distance, as estimated by:

$$\lambda_D = \sqrt{\frac{\varepsilon_0 k_B T}{q_{e0} e}}. \tag{9.4-4b}$$

Particles may not be able to actually attain the saturated charge condition due to the charging limits of the fluid medium. The maximum charge permitted on a particle surface is further limited by the breakdown electric field strength of the fluid medium, beyond which an electric arc forms. For instance, consider an example involving air. At a pressure of 1 atm, the breakdown strength, E_{max}, is approximately 3×10^6 V/m. If a charged particle is suspended in the air, the electric field strength on the particle surface can be estimated from Gauss's law, such that the following is valid:

$$E = \frac{\sum q_{ei}}{S\varepsilon} = \frac{\sigma_e}{\varepsilon}. \tag{9.4-5}$$

Together with the expression for breakdown strength of electric field, the maximum surface charging density of a particle can then be obtained as:

$$\sigma_{max} = E_{max}\varepsilon_0 = 2.6 \times 10^{-5} \frac{C}{m^2}. \tag{9.4-6}$$

Depending upon the size and geometry of the particle, the maximum charge allowed for a particle can thus be estimated.

The charging probability per surface atom is also of interest. For most solids, the atom surface density is approximately $2 \times 10^{19}/m^2$. According to Eq. (9.4-6), the maximum surface electron density in air is predicted as $1.6 \times 10^{14}/m^2$ because each electron carries a charge of 1.6×10^{-19} C. Thus, the probability of surface atoms becoming charged is merely 10^{-5}, a result that indicates less than 10 atoms per million surface atoms can be charged! Such a low charging probability of surface atoms may explain the unpredictable and random nature of particle charging in many industrial applications, especially when particle compositions with a high degree of surface material impurity are involved.

9.4.1.2   Particle Migration in an Electric Field

The size of an electrostatic precipitator depends on the time required to collect charged particles in the presence of an electric field. The duration can be determined from the migratory motion of a charged particle. If it is assumed that particles are only subjected to Stokes drag and the Coulomb force, the equation of motion for a particle in a stagnant fluid is simplified from Eq. (5.2-1) as:

$$m\frac{d\mathbf{U}_p}{dt} = q_e \mathbf{E} - 3\pi \mu d_p \mathbf{U}_p. \tag{9.4-7}$$

With the particles initially at rest, the particle starts to move along the electric field direction, with velocity given by:

$$U_p = \frac{q_e E}{3\pi \mu d_p} \left\{ 1 - \exp\left(-\frac{t}{\tau_S}\right) \right\}. \tag{9.4-8}$$

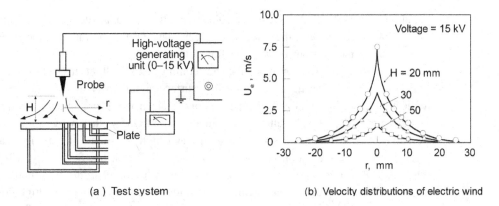

(a) Test system

(b) Velocity distributions of electric wind

Figure 9.4-3 Exemplified electric wind velocity (from Kercher, 1969).

A key measure of charged particle motion is quantified by the migration velocity, or terminal velocity, which is computed from:

$$U_{pt} = \frac{q_e E}{3\pi \mu d_p}. \tag{9.4-9}$$

When ions are present in the fluid medium, the value of U_{pt} may be higher than the theoretical value predicted by the preceding equation due to electric wind from the discharge electrode to the collecting electrode. Electric wind is caused by the motion of ions driven by the electric field. The intensity of the electric wind is a function of ion charge, fluid properties, and electric field intensity. As illustrated in Figure 9.4-3, the electric wind velocity can be on the order of 10 m/s for ionized air at pressure of 1 atm. Hence, particle migration in the electrostatic precipitator, and consequently the collection efficiency, is strongly influenced by the effects of electric wind.

9.4.1.3 Collection Efficiency of Electrostatic Precipitators

Particles are typically discharged on contact with the collecting electrodes in an electrostatic precipitator. Collection of discharged particles is achieved either through particle adhesion to the collection surface or by sliding into collecting hoppers.

Consider a flow of charged particles passing through a gap between two parallel plates that act as collecting electrodes. For this arrangement, it is assumed that (1) the particle concentration at any cross section of the precipitator is uniformly distributed; (2) flow velocity is constant; (3) U_{pt} is constant and relatively small when compared to the flow velocity; and (4) there is no particle re-entrainment. Therefore, the collection efficiency for an electrostatic precipitator can be predicted by:

$$\eta_e = \frac{\alpha_{pi} - \alpha_{p0}}{\alpha_{pi}} = 1 - \exp\left(-\frac{U_{pt}A}{Q}\right), \tag{9.4-10}$$

where α_{pi} and α_{po} are the particle volume fractions at the inlet and the outlet, respectively; and A is the area of the electrode. Equation (9.4-10), known as the Deutsch equation (Deutsch, 1922), reveals the collection efficiency of an electrostatic precipitator that increases with the area of the electrodes and the migration velocity but shares an inverse relationship with gas flow rate.

Similar to any separation methods, shortcomings exist with electrostatic precipitation. Particles adhering to the collecting surface significantly impede the actual collection efficiency realized in practice for an electrostatic precipitator. Particle adhesion may dramatically alter the original electric field by sharply increasing the resistivity of the electrode; severe "coating" and high resistivity of particles may even reverse the ionization or corona discharge (White, 1963; Svarovsky, 1981; Ogawa, 1984). For certain types of particles, electrostatic charging is difficult to accomplish without additional chemical conditioning. To accommodate a high flow rate of particulate-laden gas, the size of the precipitator or total area of collecting electrodes must be large enough to ensure a sufficient potential for collection. The method also requires a regulated high-voltage power supply, which is usually above 10 kV. Additionally, maintenance procedures, such as cleaning the collecting plates, is generally difficult and constantly with environmental concerns. Finally, the overall equipment and operating costs associated with the electrostatic precipitation method are relatively high.

9.4.2   Separation by Polarization of Dielectric Particles

Dielectric particles in the presence of an electric field become polarized, resulting in bound charges induced at the particle surfaces. If the electric field is uniformly applied, the interaction between these bound charges and the electric field produces no net force. However, if the electric field is nonuniform and possesses a spatial gradient across the particles, the interaction between the electric field and the bound charges leads to a net force on the particles that results in particle motion. The magnitude of these forces depends on the magnitude of the electric field gradient and the dielectric constant of the particles relative to the background liquid. This phenomenon is known as dielectrophoresis (DEP). The resultant force can be employed to control dielectric particle motion. Due to individual polarization properties, particles comprised of different materials can be separated via the characteristic difference in speed and direction of their respective dielectrophoretic motion.

For a dielectric material, charges are locally bound but can move short distances if an external electric field is applied. Positive and negative charges move in opposite directions and induce dipoles. There are several mechanisms that can account for such polarization, including polarization of the electron cloud of atoms, small displacement of ions, and alignment of permanent dipole moments in molecules. Polarization also occurs at the interface between two dissimilar materials, where charges are confined due to localized heterogeneity within the medium.

Figure 9.4-4 illustrates the pertinent effects for a single particle. Figure 9.4-4(a) portrays the case where a particle is more readily polarized than the suspending fluid in a uniform electric field; additional charge accumulates on the interior surface of the particle. The net dipole is parallel to the applied field, and the electric

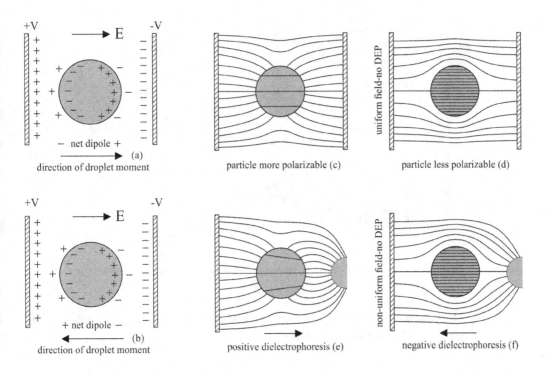

Figure 9.4-4 Suspended dielectric particles in an electric field.

field lines bend toward the particle. The field inside the particle is much weaker than that present in the fluid, which is shown in Figure 9.4-4(c). Conversely, when the particle possesses lower polarization potential than the surrounding fluid medium, more charge accumulates on the exterior surface of the particle. The induced dipole is in the opposite direction to the applied field, and the field lines bend around the particle. If the polarization potential of the particle and the fluid are equivalent, no dipole is induced at the interface, and the field will appear as if the particle did not exist. In a nonuniform field, a particle with higher polarization potential than the surrounding medium will move toward the stronger field region. This is characterized as positive DEP and is depicted in Figure 9.4-4(e). A particle with lower polarization potential than the fluid medium will move toward regions with lower field strength. This is fittingly termed negative DEP and is illustrated in Figure 9.4-4(f).

The DEP force originates from the interaction between the induced dipole and the applied electric field. Thus, an expression for the DEP force can be written as:

$$\mathbf{F}_{DEP} = \left(\mathbf{p}_d \cdot \nabla\right) \mathbf{E}. \tag{9.4-11}$$

For a spherical particle in a DC, or stationary AC, field with magnitude E, the time-averaged DEP force is expressed by the relationship (Morgan and Green, 2002):

$$\mathbf{F}_{DEP} = \frac{\pi}{8}\varepsilon_f d_p^3 Re\left[f_{CM}\right] \nabla|\mathbf{E}|^2. \tag{9.4-12}$$

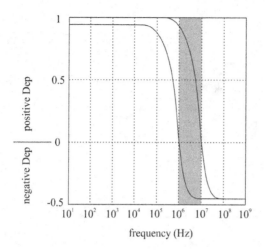

Figure 9.4-5 Normalized DEP force on two different particles versus AC frequency.

Here, Re[f_{CM}] refers to the real component of the complex variable f_{CM}, which is the Clausius–Mossotti factor defined by:

$$f_{CM} = \frac{\varepsilon_p - \varepsilon_f - i\left(\frac{\sigma_p - \sigma_f}{\omega}\right)}{\varepsilon_p + 2\varepsilon_f - i\left(\frac{\sigma_p + 2\sigma_f}{\omega}\right)}, \tag{9.4-12a}$$

where ε and σ stand for the permittivity and electric conductivity, respectively; subscripts p and f represent the particle and suspending fluid, respectively; whereas i is the imagery index. It is revealed in Eq. (9.4-12) that the direction of the DEP force is determined by the real component of the Clausius–Mossotti factor. Positive DEP occurs when Re[f_{CM}] > 0, and the particle moves toward regions of high field strength. Negative DEP occurs when Re[f_{CM}] < 0, and the particle is repelled from regions of high field strength.

Equation (9.4-12a) also indicates that the direction of particle motion is a function of the frequency of the AC field in addition to the properties of the particle and suspending fluid. Thus, the AC field is a viable tool for particle sorting and separation operations. To illustrate, the DEP forces on two solid latex particles with different conductivities suspended in an electrolyte are plotted in Figure 9.4-5.

At low frequencies, charge movement tracks the change of the applied field, and the DEP force is constant. At high frequencies, the DEP force magnitude remains unchanged, but it reverses direction because the dielectric constant of the fluid is much higher than that of latex. A transitional frequency exists for each particle where particle behavior shifts from positive DEP to negative DEP. The transition defines a frequency band in which the two particles move in different directions and thus enables their separation, as exemplified by the shaded frequency window in Figure 9.4-5.

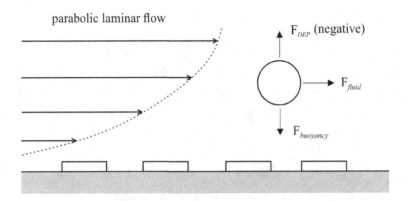

Figure 9.4-6 Schematic illustration of the DEP-FFF separation.

In addition to direction, a difference in the magnitude of the DEP force can also be utilized for particle separation. One such method employing this technique is known as DEP field flow fractionation (FFF) and is illustrated in Figure 9.4-6.

Because the DEP force depends on the gradient of the applied electric field, which is a function of the channel height, the equilibrium levitation height of the particle can be calculated by balancing the DEP force with buoyancy and gravitational forces. Particles with distinct densities will persist at different heights in the channel. In a pressure-driven flow, the parabolic velocity profile of the liquid causes the particles to exit the channel at different times, and thus results in the desired particle separation.

For a general AC field with a spatially dependent phase, the field can be expressed by:

$$\mathbf{E}(\mathbf{r}, t) = Re[\mathbf{E}^*(\mathbf{r})e^{i\omega t}], \tag{9.4-13a}$$

where the vector $\mathbf{E}^*(\mathbf{r})$ is the complex phasor. In this case, an additional term appears in the full DEP force, which is given by the expression (Morgan and Green, 2002):

$$\mathbf{F}_{DEP} = \frac{\pi}{8}\varepsilon_f d_p^3 Re\left[f_{CM}\right] \nabla |\mathbf{E}^*|^2 - \frac{\pi}{4}\varepsilon_f d_p^3 Im\left[f_{CM}\right] \left(\nabla \times (Re[\mathbf{E}^*] \times Im[\mathbf{E}^*])\right). \tag{9.4-13}$$

The first term on the right-hand side of the equation corresponds to the conventional DEP force described above. The square of the complex phasor is obtained as:

$$|\mathbf{E}^*|^2 = |Re[\mathbf{E}^*]|^2 + |Im[\mathbf{E}^*]|^2. \tag{9.4-13b}$$

The second term on the right-hand side depends on the imaginary component of the Clausius–Mossotti factor and the spatially varying phase; this term describes the so-called travelling wave dielectrophoresis (twDEP). In an AC field produced by an interdigitated electrode array, similar to the arrangement shown in Figure 9.4-7, a dielectric particle will simultaneously experience both the conventional DEP force and the twDEP force.

For the twDEP to be effective, the particle must be levitated above the electrodes by a negative DEP force. A nonzero imaginary component in the Clausius–Mossotti

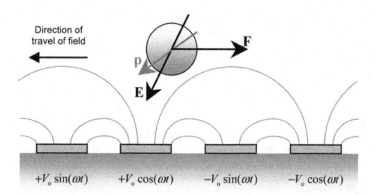

Direction of travel of field

$+V_o \sin(\omega t)$ $+V_o \cos(\omega t)$ $-V_o \sin(\omega t)$ $-V_o \cos(\omega t)$

Figure 9.4-7 Schematic twDEP using an interdigitated electrode array with consecutive phase-shifted signals.

factor indicates the particle will be propelled parallel to the electrode arrays by the twDEP force. Balancing the twDEP force with Stokes drag reveals that the particle velocity depends on several factors, including the particle size, the frequency of the applied field, as well as the electrical properties of both the particle and the suspending fluid. Therefore, particles with different sizes or distinct polarization potentials can be effectively separated by their unique twDEP velocities (Morgan *et al.*, 1997).

In practice, the DEP and twDEP separation techniques are often employed for micron or submicron particles. For example, the techniques are often applied in lab-on-a-chip devices for separating biological particles such as DNA, cells, and bacteria.

9.5   Case Studies

The following case studies serve for further discussion with simple applications or extensions of separation methods discussed in this chapter. The examples include a method to obtain the fractional collection efficiency curve and cutoff size of a cyclone for a polydispersed particulate flow, the overall separation efficiency of a fibrous filter for a fine powder flow with polynomial size distributions of the powder, a CFD modeling of gas–solid flow through a cyclone, and a CFD modeling of charged particle removal by an electrostatic precipitator.

9.5.1   Cyclone Collection Efficiency for a Polydispersed Particulate Flow

Problem Statement: A cyclone separator has an overall collection efficiency of 90% for a dust-laden air flow. The results of the particle size analysis are given in Table 9.5-1 below. Plot the fractional efficiency curve and estimate the cutoff diameter of this cyclone.

Table 9.5-1 Size distribution data for particles at the inlet and outlet of the cyclone.

Size range (μm)	Particle size analysis (wt%)	
	Inlet	Outlet
0–5	8.0	76.0
5–10	1.4	12.9
10–15	1.9	4.5
15–20	2.1	2.1
20–25	2.1	1.5
25–30	2.0	0.7
30–35	2.0	0.5
35–40	2.0	0.4
40–45	2.0	0.3
>45	76.5	1.1

Analysis: Let A and B represent the particle mass flow rate at inlet and outlet, respectively. The overall collection efficiency is given by

$$\eta_c = \frac{A - B}{A} = 0.9. \qquad (9.5\text{-}1)$$

From which $B/A = 0.1$. The fractional collection efficiency, η_{ci}, can be calculated in a similar manner as tabulated below:

d_p (μm)	$d_{p,ave}$ (μm)	Inlet (wt%)	Inlet mass flow	Outlet (wt%)	Outlet mass flow	η_{ci} (%)
0–5	2.5	8.0	0.080A	76.0	0.760B	5.0
5–10	7.5	1.4	0.014A	12.9	0.129B	7.8
10–15	12.5	1.9	0.019A	4.5	0.045B	76.3
15–20	17.5	2.1	0.021A	2.1	0.021B	90.0
20–25	22.5	2.1	0.021A	1.5	0.015B	92.9
25–30	27.5	2.0	0.02A	0.7	0.007B	96.5
30–35	32.5	2.0	0.02A	0.5	0.005B	97.5
35–40	37.5	2.0	0.02A	0.4	0.004B	98.0
40–45	42.5	2.0	0.02A	0.3	0.003B	98.5
> 45	–	76.5	0.765A	1.1	0.011B	99.9

Figure 9.5-1 shows the fractional collection efficiency, η_{ci}, with particle diameter. Based on the curve, the cutoff diameter for this cyclone is about 10 μm.

9.5.2 Inertial Impaction-Dominated Fibrous Filtration of Fine Particles

Problem Statement: Consider a dust-laden gas passing through a loosely, but uniformly, packed fibrous filter, in which the particle size is much less than the fiber

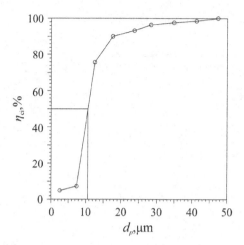

Figure 9.5-1 Fractional collection efficiency with d_p.

diameter. Assume that the flow velocity through the filter is high enough so the collection by Brownian diffusion can be ignored, compared to the impaction and interception. Using the Flagan and Seifeld model of impaction and interception, as represented by Eq. (9.3-3), show that the single-fiber collision efficiency of particle size x can be expressed by a simple 4th-order polynomial in terms of x as:

$$\eta_f(x) = C_2 x^2 + C_3 x^3 + C_4 x^4. \tag{9.5-2}$$

Based on Eq. (9.5-2), estimate the overall collection efficiency if the mass-based particle size distribution follows a Gaussian distribution or a lognormal distribution. Compare the size-averaged collection efficiencies with those of monodispersed particles of the mean size.

Analysis: The following is a simplified model from the original work of Zhu *et al.* (2000).

It is recognized that, for most fibrous filtrations with $d_p/d_f \ll 1$, both $\eta_{imp+int}$ and λ are much smaller than unity. Thus, by using Taylor's expansion, the $\eta_{imp+int}$ equation in Eq. (9.3-3) reduces to

$$\eta_{imp+int} = \frac{1-\alpha_f}{Ku}\lambda^2, \tag{9.5-3}$$

which suggests that $\eta_{imp+int}/\lambda \ll 1$. Thus, the λ-equation in Eq. (9.5-3) becomes

$$\lambda = \frac{d_p}{d_f} + Stk\sqrt{\alpha_f}\left\{1 - \exp\left(-\frac{1}{Stk\sqrt{\alpha_f}}\right)\right\}. \tag{9.5-4}$$

For a loosely packed filter, the volume fraction of fibers is much smaller than unity so that $Stk\alpha_f^{1/2} \ll 1$. Thus, Eq. (9.5-4) can be further simplified to

$$\lambda = \frac{d_p}{d_f} + Stk\sqrt{\alpha_f}. \tag{9.5-5}$$

Substituting Eq. (9.5-5) into Eq. (9.5-3) yields

$$\eta_{imp+int} = \frac{(1-\alpha_f)}{Ku} \left\{ \left(\frac{d_p}{d_f}\right)^2 + 2\sqrt{\alpha_f} Stk \left(\frac{d_p}{d_f}\right) + \alpha_f Stk^2 \right\}. \tag{9.5-6}$$

It is interesting to note that the first term is the collection efficient by interception whereas the last two are due to impaction, and hence Eq. (9.5-6) can be further decomposed into two independent components of Eq. (9.3-4a) and Eq. (9.3-4b). Now, substituting Stk expressed by Eq. (9.3-3a) into Eq. (9.5-6) yields

$$\eta_{imp+int} = \frac{(1-\alpha_f)}{Ku} \left\{ \left(\frac{d_p}{d_f}\right)^2 + 2\sqrt{\alpha_f} \frac{C_c \rho_p d_p^2 U}{18\,\mu d_f} \left(\frac{d_p}{d_f}\right) + \alpha_f \left(\frac{C_c \rho_p d_p^2 U}{18\,\mu d_f}\right)^2 \right\}. \tag{9.5-7}$$

With the neglect of η_d, the total efficiency of a fiber is then equal to $\eta_{imp+int}$, so that Eq. (9.5-7) leads to Eq. (9.5-2), with

$$C_2 = \frac{(1-\alpha_f)}{Ku d_f^2}$$

$$C_3 = \frac{2\sqrt{\alpha_f}\,(1-\alpha_f)}{Ku} \frac{C_c \rho_p U}{18\,\mu d_f^2}. \tag{9.5-2a}$$

$$C_4 = \frac{\alpha_f\,(1-\alpha_f)}{Ku} \left(\frac{C_c \rho_p U}{18\mu d_f}\right)^2$$

For a polydispersed particle-laden flow, the averaged collection efficiency of a fiber can be defined, with a weighing of particle size density distribution, by Eq. (9.3-11) as

$$\eta_{f,ave} = \int_0^\infty f_M(x)\eta_f(x)dx. \tag{9.3-11}$$

When the particle size density distribution follows the Gaussian type, it gives

$$f_M(x) = A_M \exp\left(-\frac{(x-d_0)^2}{2\sigma^2}\right), \tag{9.5-8}$$

where d_0 is the median size, σ is the standard deviation, and A_M is a normalizing coefficient determined by

$$A_M = \frac{1}{\sigma}\sqrt{\frac{2}{\pi}} \left[1 + erf\left(\frac{d_0}{\sqrt{2}\sigma}\right)\right]^{-1}. \tag{9.5-8a}$$

Substituting Eq. (9.5-8) and Eq. (9.5-2) into Eq. (9.3-11) yields

$$\eta_{f,ave} = C_2 d_0^2 + C_3 d_0^3 + C_4 d_0^4 + \left(C_2\sigma^2 + 3C_3\sigma^2 d_0 + 6C_4\sigma^2 d_0^2 + 3C_4\sigma^4\right)$$

$$+ \left(C_2 d_0 + C_3 d_0^2 + 2C_3\sigma^2 + C_4 d_0^3 + 5C_4\sigma^2 d_0\right)\sigma^2 A_M \exp\left(-\frac{d_0^2}{2\sigma^2}\right). \tag{9.5-9}$$

The collection efficiency of monodispersed particles of median size is given by Eq. (9.5-2) as

$$\eta_{f,d_0} = C_2 d_0^2 + C_3 d_0^3 + C_4 d_0^4. \tag{9.5-10}$$

Comparing Eq. (9.5-10) with Eq. (9.5-9) suggests that $\eta_{f,ave} > \eta_{f,d_0}$.

For a lognormal particle size density distribution, it is expressed by

$$f_M(x) = \frac{1}{x\sqrt{2\pi}\,\sigma_{dl}} \exp\left(-\frac{(\ln x - \ln d_0)^2}{2\sigma_{dl}^2}\right), \tag{9.5-11}$$

where σ_{dl} is the natural log of the ratio of the diameter for which the cumulative distribution curve has a value of 0.841 to the mean diameter. Substituting Eq. (9.5-11) and Eq. (9.5-2) into Eq. (9.3-11) gives the averaged fiber collection efficiency of

$$\eta_{f,ave} = C_2 d_0^2 \exp\left(2\sigma_{dl}^2\right) + C_3 d_0^3 \exp\left(\frac{9}{2}\sigma_{dl}^2\right) + C_4 d_0^4 \exp\left(8\sigma_{dl}^2\right), \tag{9.5-12}$$

which once again shows that $\eta_{f,ave} > \eta_{f,d_0}$.

Comments: (1) The overall filter efficiency of polydispersed particles can be estimated from Eq. (9.3-11). Combined with Eq. (9.3-11), it yields

$$\eta_F = 1 - \exp\left(-\int_0^L \frac{4}{\pi d_f} \frac{\alpha_f}{(1-\alpha_f)} \left(\int_0^\infty f_M(x)\eta_f(x)dx\right) dl\right). \tag{9.5-13}$$

It is noted that $f_M(x)$ keeps changing along the filtration path, which is due to the nonuniform removal or capture of different sizes of fibers within the filter. Hence, a coupling relationship among f_M, η_f, and η_F may need to be developed. (2) With the collection of particles by the fibers, the collected particles may also act as granular filtering medium to assist the filtration. Thus, in practice, a fibrous filtration is actually a combined fibrous and granular filtration.

9.5.3 Numerical Modeling of Gas–Solid Flow in a Cyclone Separator

Problem Statement: Establish a modeling approach for the solid particle separation of a turbulent gas–solid flow by a cyclone. The particle-laden flow is not very dilute so that the particle–fluid interaction, interparticle collisions, and particle rotation effect should be included in the modeling, in addition to the turbulent effect.

Analysis: The following is a simplified model based on the original work of Chu *et al.* (2011). A combined Eulerian–Lagrangian–DEM model is adopted for this problem, with the Eulerian modeling for gas flow, the Lagrangian trajectory modeling for the solid particle transport, and the DEM soft-sphere model accounts for the translational and rotational momentum transfer due to interparticle collisions. It is assumed that the solid particles are spherical and monodispersed, without agglomeration, clustering, or breakup during the separation process. The gas flow is regarded as incompressible and isothermal.

For the Eulerian model of the turbulent gas flow, the volume–time-averaged governing equations can be expressed, based on Eq. (6.4-21) and Eq. (6.4-22), as:

$$\frac{\partial}{\partial t}(\alpha\rho) + \nabla \cdot (\alpha\rho\mathbf{U}) = 0 \tag{9.5-14}$$

$$\frac{\partial}{\partial t}(\alpha\rho\mathbf{U}) + \nabla \cdot (\alpha\rho\mathbf{U}\mathbf{U}) = -\alpha\nabla p + \nabla \cdot \left[\alpha\mu\left(\nabla\mathbf{U} + \nabla\mathbf{U}^T\right)\right] + \alpha\rho\mathbf{g} \\ - \mathbf{M}_{gp} + \nabla \cdot (-\rho\overline{\mathbf{u}'\mathbf{u}'}), \tag{9.5-15}$$

where the gas–solid interphase momentum transfer, $\mathbf{M}_{gp}$, is given by Eq. (6.8-22a), and the last term in Eq. (9.5-15) is the turbulent Reynolds stress, which can be further modelled by RSM introduced in Section of 2.3.4.3 or Eq. (2.3-23).

For the deterministic Lagrangian model of solid particles, the equation of translational motion of the ith particle is given, based on Eq. (5.3-30), by

$$m_{pi}\frac{d\mathbf{u}_{pi}}{dt} = \mathbf{F}_{Di} + V_{pi}\nabla p + \sum_{j=1}^{k_i} \mathbf{F}_{C,ij} + m_{pi}\mathbf{g}, \tag{9.5-16}$$

where the drag force can be estimated by

$$\mathbf{F}_{Di} = \frac{\pi}{8}C_{D0i}d_{pi}^2\alpha^{-1.65}\left|\mathbf{U}(\mathbf{r}_{pi}) - \mathbf{u}_{pi}\right|(\mathbf{U}(\mathbf{r}_{pi}) - \mathbf{u}_{pi}). \tag{9.5-16a}$$

The collision force can be estimated from the DEM soft-sphere model introduced in the Section of 5.3.2. Similarly, the equation of rotational motion of the ith particle is given by Eq. (5.3-31):

$$I_{pi}\frac{d\boldsymbol{\omega}_{pi}}{dt} = \sum_{j=1}^{k_i} (\mathbf{T}_{tij} + \mathbf{T}_{rij}), \tag{5.3-31}$$

where the tangential friction torque and the rolling friction torque are estimated by Eq. (5.3-28) and Eq. (5.3-29), respectively.

The equation of trajectory is given by

$$\frac{d\mathbf{r}_{pi}}{dt} = \mathbf{u}_{pi}. \tag{9.5-17}$$

The gas volume fraction is related to the particles within the local cell and is given by:

$$\alpha = 1 - \frac{1}{V_c}\sum_{i}^{k_c} V_{pi}. \tag{9.5-18}$$

Thus, six independent equations, Eqs. (9.5-14)–(9.5-18) and Eq. (5.3-31), are obtained for six unknowns (α, $\mathbf{U}$, p, $\mathbf{u}_p$, $\boldsymbol{\omega}_p$, and $\mathbf{r}_p$). The closure of the modeling is now established. It is noted that the Lagrangian model is applied to each individual particle, respectively, and the gas–solid interaction terms in the Eulerian equation are contributed by all particles found in the local domain of $(\mathbf{r}, t)$. In addition, there are two key sub-models, including RSM for turbulence and DEM soft-sphere model for interparticle collisions.

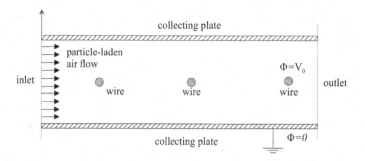

Figure 9.5-2 Configuration of wire-plate electrostatic precipitator.

Comments: (1) The strong rotating flow within the cyclone not only generates a large pressure gradient force or buoyance effect but also makes the turbulence highly anisotropic. Thus, the classical $k-\varepsilon$ model developed for isotropic turbulence may not be directly applicable without proper modifications. In this case, RSM is an appropriate choice among various turbulence models, although it is much more computationally demanding. (2) The boundary layer conditions, including wall functions, can also be significantly affected by the anisotropic turbulence and particle–wall collisions. (3) The above model may be extendable to cyclone separations of gas–solid flow with polydispersed particles, but such an extension may only be limited to very dilute cases where interparticle collisions are less significant. The DEM soft-sphere model for collisions among polydispersed particles is still under development. (4) Case simulation examples can be found in the reference (Chu *et al.*, 2011).

9.5.4   Numerical Modeling of Particulate Removal by Electrostatic Precipitator

Problem Statement: The plate electrostatic precipitator has an array of wires as electrodes, as shown in Figure 9.5-2, which is used for charging and removing the particles suspended in a passing air flow. The plates are grounded and used as particle collectors. Establish a numerical modeling for the characteristics of particle transport, electrostatic charging and removal, and the collection efficiency of the electrostatic precipitator.

Analysis: The following is a simplified model from the original work of Skodras *et al.* (2006), in which an Eulerian–Lagrangian model with a coupled electrostatic charging field is developed. The collection efficiency for the electrostatic precipitator is defined as the ratio of particles that are deposited on the collection plate to the total number of particles present in the inlet flow. In order to determine the likelihood that particles will encounter the collection plate or escape to the outlet along with the gaseous phase of flow, their trajectories must be mapped in Lagrangian coordinates. Recall that particle motion is influenced by both the aerodynamic drag and the electrostatic forces; therefore, the continuous gas phase flow and the electric field need to be interrogated.

Assuming the flow is steady-state and the particle volumetric effect can be ignored in the gas phase governing equation. In addition, the gas flow is incompressible, isothermal, and turbulent with isotropic turbulence. Thus the volume–time-averaged continuity and momentum equations of gas phase can be expressed, based on Eq. (6.4-21) and Eq. (6.4-22), as:

$$\nabla \cdot (\rho \mathbf{U}) = 0 \qquad (9.5\text{-}19)$$

$$\nabla \cdot (\rho \mathbf{U}\mathbf{U}) = -\nabla p + \nabla \cdot \left[(\mu + \mu_t) \left(\nabla \mathbf{U} + \nabla \mathbf{U}^T \right) \right] - \mathbf{M}_{gp} + \rho_{ion} \mathbf{E}. \qquad (9.5\text{-}20)$$

The last two terms on the right-hand side of the gas phase momentum equation account for the momentum transfer due to particle–gas interactions and the electric force on the ions, respectively. The electric force on the ions is produced by corona discharge from the transverse wires that causes macroscopic gas flow toward the plates ("electric wind") via collisions with neutral gas molecules. The turbulent viscosity, μ_t, can be calculated using turbulent models such as the k-ε model introduced in Section 2.3.4.2 or Eqs. (2.3-18), Eq. (2.3-21), and Eq. (2.3-22). The gas–solid interphase momentum transfer, $\mathbf{M}_{gp}$, is assumed to be dominated by drag forces of particles, which can be simplified from Eq. (6.8-22a) as:

$$\mathbf{M}_{gp} = \frac{1}{V_c} \sum_i^{k_c} \frac{3}{4} \frac{C_{Di}}{d_{pi}} \rho \left| \mathbf{U} - \mathbf{u}_{pi} \right| (\mathbf{U} - \mathbf{u}_{pi}). \qquad (9.5\text{-}20a)$$

The trajectory equation and the equation of motion of the ith-particle are governed, respectively, by:

$$\frac{d\mathbf{r}_{pi}}{dt} = \mathbf{u}_{pi} \qquad (9.5\text{-}21)$$

$$m_{pi} \frac{d\mathbf{u}_{pi}}{dt} = \mathbf{F}_{Di} + m_{pi}\mathbf{g} + q_{pi}\mathbf{E}. \qquad (9.5\text{-}22)$$

The charging of the ith-particle can be calculated by:

$$\frac{dq_{pi}}{dt} = \begin{cases} (q_{si} - q_{pi})^2 / \tau q_{si} & \text{if } q_{pi} < q_{si}. \\ 0 & \text{if } q_{pi} \geq q_{si} \end{cases} \qquad (9.5\text{-}23)$$

The saturation charge of a dielectric particle is given by:

$$q_s = \frac{3\pi \varepsilon_0 \varepsilon_r |\mathbf{E}| d_{pi}^2}{\varepsilon_r + 2}. \qquad (9.5\text{-}23a)$$

The relaxation time for the particle to reach half of its saturation charge is given by:

$$\tau = \frac{4\varepsilon_0}{\rho_{ion}\kappa_{ion}}, \qquad (9.5\text{-}23b)$$

where κ_{ion} is the ion mobility.

The electric field is affected by ions as well as the charged particles; thus the electric potential is given, by the modified Eq. (5.4-11), as:

$$\nabla^2 \Phi = -\frac{\rho_{ion} + \rho_{pe}}{\varepsilon_0}. \qquad (9.5\text{-}24)$$

The particle charge density, ρ_{pe}, is contributed by all particles within the local cell, as calculated by

$$\rho_{pe} = \frac{1}{V_c} \sum_i^{k_c} q_{pi}.$$ (9.5-25)

The electric field intensity, $\mathbf{E}$, is calculated by Eq. (5.4-10) as:

$$\mathbf{E} = -\nabla\Phi.$$ (5.4-10)

The transport of the ions is governed by the continuity of the electric current, which is given by:

$$\nabla \cdot \left(\rho_{\text{ion}} \mathbf{U} + \rho_{\text{ion}} \kappa_{ion} \mathbf{E} - D_e \nabla \rho_{\text{ion}} + \rho_{pe} \mathbf{U}_p \right) = 0,$$ (9.5-26)

where $\mathbf{U}_p$ is the local volume-averaged particle velocity, and D_e is the effective diffusivity.

Thus, nine independent equations, Eqs. (9.5-19)–(9.5-26) and Eq. (5.4-10), are obtained for nine unknowns ($\mathbf{U}$, p, $\mathbf{u}_p$, $\mathbf{r}_p$, q_p, $\mathbf{E}$, Φ, ρ_{pe}, and ρ_{ion}). The closure of the modeling is now established. It is noted that the Lagrangian model is applied to each individual particle, respectively, and the gas–solid interaction terms in the Eulerian equation are contributed by all particles found in the local domain of $(\mathbf{r}, t)$. In addition, there are two key sub-models, including the k–ε model for turbulence and the drag force model for particle–gas interactions.

Comments: (1) The overall removal efficiency and collection efficiency characteristics as a function of particle size can be readily calculated from the tracking of particle trajectories. (2) For fine particles, the Brownian force can be important. The randomness from the Brownian motion and turbulent effect may require the use of the stochastic particle trajectory model rather than the current deterministic Lagrangian model. (3) Case simulation examples can be found in the reference (Skodras *et al.*, 2006).

9.6   Summary

The phase separation of a multiphase flow is primarily achieved with an application of a specific mechanism that can lead to a distinctively different dynamic response of each phase in a multiphase medium. Typical applied mechanisms and their effects can be given by:

- gravitational settling (*e.g.*, solids in fluids, droplets in immiscible fluids, bubbles in liquids or slurries);
- flow-induced alternation of phase inertia (*e.g.*, centrifugal acceleration by flow rotation, nozzle jetting and jet dispersion, impaction on a surface);
- selective interception or blockage of phase transport (*e.g.*, sieving, fibrous filtration; granular filtration);
- separation using externally controlled field forces (*e.g.*, external electric field to separate the charged particles from a particle–fluid flow).

A separation system or method can be developed by using one or a combination of more than one of the mechanisms given above. High separation efficiency and low mechanical energy loss (typically characterized by the pressure drop of fluid flow at a given flow rate) are among the most important objectives for system design or selection.

Actual multiphase flows in separation systems are complex as they may involve complicated phase transport, flow regimes, polydispersed size distributions of discrete phases, and system geometries. Thus, aside from numerical modeling, much simplified analytical models with empirical correlations are still popular to account for practical operation of a separation system.

Nomenclature

A	Area
	Constant
Cc	Cunningham slip factor
D	Diffusivity
d	Diameter
E	Electric field strength
e	Charge of an electron
F	Force
f_{CM}	Clausius–Mossotti factor
f_M	Mass density function
g	Gravitational acceleration
H	Height
Ku	Kuwabara factor
k_B	Boltzmann constant
L	Length
$\mathbf{M}$	Momentum transfer
m	Mass
N	Number
n	Number density
Pe	Peclet number
p	Pressure
pd	Dipole moment
Q	Flow rate
q	Charge carried by a particle
R	Radius
Re	Reynold number
r	Radial coordinate
S	Surface
Stk	Stokes number
T	Temperature
t	Time

U	Velocity
	Radial velocity
u	Instant velocity
V	Volume
	Tangential velocity
W	Axial velocity
y	Limiting width of particle stream

Greek Symbols

α	Volume fraction
β	Vortex exponent
Γ	Circulation
Δp	Pressure drop
δ	Stopping distance
ε	Permittivity
η	Collection efficiency
κ	Mobility
λ	Coupling parameter
λ_D	Debye shielding distance
λ_g	Mean free path of gas
μ	Viscosity
ρ	Density
σ	Surface density
	Electric conductivity
τ	Relaxation time
Φ	Electric potential
ω	Angular frequency

Subscript

θ	Center
	Initial
	Particle-free
	Vacuum
C	Collision
c	Cake
	Cyclone
	Computational cell
D	Drag
DEP	Dielectrophoresis
d	Droplet
	Diffuse

E	Electric
e	Electrostatic charge
F	Filter
f	Fiber
g	Gas
i	inlet
imp	Impaction
int	Intercept
ion	Ion
m	Maximum
	Mean
n	Number
o	Outlet
p	Particle
pc	Cutoff
pt	Particle terminal
pu	Pickup
r	Relative
S	Stokes
s	Saturated
	Settling
	Granular particle
t	Turbulent
V	Volume

Problems

P9.1 Consider a clean house bag filter filled with randomly assembled fibers. The porosity of the filter is 0.97, and the fibers have an effective diameter of 10 μm. The length of the filter is 0.1 m. The viscosity of air is 1.8×10^{-5} kg/m·s. If the flow velocity of the dust-laden air is 0.2 m/s, estimate the pressure drop across the filter.

P9.2 Show that the velocity and pressure distributions of Burgers' viscous vortex model, Eqs. (9.2-7a, b and d), satisfy the governing equations of Eq. (9.2-5) and Eq. (9.2-6).

P9.3 According to the Rankine Vortex Model, a rotating flow is incompressible, axisymmetric, absence in body force, and inviscid. Additionally, the flow field is comprised of a forced vortex region and a free vortex region. In the forced vortex region of $0 \leq r \leq r_f$, the tangential velocity and pressure distributions are given by:

$$V = \omega r \qquad p = p_0 + \frac{\rho \Gamma^2}{2 r_f^4} r^2. \qquad \text{(P9.3-1)}$$

In the free vortex region of $r < r_f$, it gives:

$$Vr = \Gamma = \omega r_f^2 \qquad p = p_0 + \frac{\rho \Gamma^2}{2} \left(\frac{2}{r_f^2} - \frac{1}{r^2} \right). \qquad (\text{P9.3-2})$$

Show that the above solutions satisfy the Navier–Stokes equations and the continuity equation.

P9.4 Consider a tangential-inlet cyclone. The flow field follows Eq. (9.2-8a), which is known as Alexander's vortex model. Assume that particles are only subjected to the Stokes drag force and centrifugal force.

(a) Show that the equation of motion can be expressed by:

$$\frac{1}{\tau_S} \frac{dr}{dt} = \frac{V_i^2 R^{2\beta}}{r^{2\beta+1}}, \qquad (\text{P9.4-1})$$

where V_i is the inlet velocity, and the traveling time for particles moving from r_0 at $t = 0$ to r can be written as:

$$t = \frac{1}{2 \tau_S (\beta + 1)} \left(\frac{R}{V_i} \right)^2 \left[\left(\frac{r}{R} \right)^{2\beta+2} - \left(\frac{r_0}{R} \right)^{2\beta+2} \right]. \qquad (\text{P9.4-2})$$

(b) Assume that the initial radial position of the uncollected particles is at the vortex center where $r_0 = 0$. Show that the collection efficiency, η_c, can be expressed by Eq. (9.2-8b).

P9.5 Show that, for a tangential-inlet cyclone separator for particle flows, the cutoff size can be estimated by Eq. (9.2-9).

P9.6 Consider a case where filter fiber orientation is parallel to flow. Using the Happel-cell model, assume that the fluid motion is in the creeping flow regime so that the flow can be modeled in cylindrical coordinates as:

$$\frac{1}{r} \frac{d}{dr} \left(r \frac{dW}{dr} \right) = \frac{1}{\mu} \frac{dp}{dz}, \qquad (\text{P9.6-1})$$

with the boundary conditions of:

$$W \mid_{r=a} = 0; \quad \left(\frac{dW}{dr} \right)_{r=b} = 0. \qquad (\text{P9.6-2})$$

Show that:
(a) Axial fluid velocity inside the annular space is given by:

$$W = -\frac{1}{4\mu} \frac{dp}{dz} \left(a^2 - r^2 + 2b^2 \ln \left(\frac{r}{a} \right) \right). \qquad (\text{P9.6-3})$$

(b) Fluid flow rate through the entire annulus is given by:

$$Q = -\frac{\pi}{8\mu} \frac{dp}{dz} \left(4 a^2 b^2 - a^4 - 3 b^4 + 4 b^4 \ln \frac{b}{a} \right). \qquad (\text{P9.6-4})$$

(c) Pressure drop over the filter thickness, L, is given by:

$$\frac{\Delta p}{L} = \frac{Q\mu}{A\,a^2}\frac{4\,\alpha_f}{(2\,\alpha_f - \alpha_f^2/2 - 3/2 - \ln\,\alpha_f)}. \tag{P9.3-13}$$

P9.7 Consider the case where a liquid is sprayed in the gravity direction in a vertical column, in which a gas flows upward against the droplet settling. Assume that the spraying velocity of all droplets is the same and the droplet concentration is low so that the droplet–droplet interactions can be neglected. Establish a model to estimate the cutoff size of sprayed droplets below which the droplets may be entrained from the column by the gas flow. Derive an expression of the cutoff size as a function of the spraying velocity, gas velocity, and column length.

P9.8 A tangential inlet cyclone separator with a steep cone is operated under the following conditions: $R = 0.12$ m, $Q = 0.20$ m^3/s, $\Gamma_m = 4.2$ m^2/s, $H = 0.87$ m, $d_p = 10$ μm and $\rho_p = 2{,}000$ kg/m^3. The carrier gas is air at ambient conditions. Calculate the collection efficiency. If all conditions except the particle diameter remain unchanged, estimate the cutoff particle size for this cyclone.

P9.9 In a performance test of a cyclone separator, a dust-laden gas with a particle mass flow rate of 600 kg/hr is used. A particle size analysis provides the weight percentage [wt%] at the inlet and outlet of the cyclone, and the results are given in Table P9.9-1.

Table P9.9-1 Results of particle size analysis at inlet and outlet of cyclone

Size range (μm)	Particle size analysis (wt%)	
	Inlet	Outlet
0–5	1.5	28.5
5–10	1.7	20.0
10–15	2.2	13.0
15–20	3.4	10.0
20–30	9.5	14.7
30–40	15.0	8.5
40–50	19.0	3.3
50–60	22.0	1.3
>60	25.7	0.7

Given a dust escape rate of 30 kg/h, what is the overall collection efficiency? Plot the fractional efficiency curve to identify the cutoff particle size for this cyclone.

P9.10 Consider a parallel-plate electrostatic precipitator with a 10 kV applied voltage. The interplate spacing is 1 cm; the particle density is 2,000 kg/m^3; the total surface area is 10 m^2; and the flow rate is 0.5 m^3/s. Use the test results provided in Table P9.10 to estimate the following:
(a) Charge-to-mass distribution versus particle size;
(b) Overall collection efficiency; and
(c) Size distribution (wt%) at the outlet.

Table P9.10 Properties of particles collected in electrostatic precipitator

d_p(μm)	U_m (cm/s)	wt% ($< d_p$)
5	2.2	-
10	5.9	10
20	8.3	50
40	11.2	90
80	13.8	99.6
>80	14.0	-

P9.11 Consider a clean residential bag filter filled with randomly assembled fibers. The porosity of the filter is 0.97; the fibers have an effective diameter of 10 μm; the length of the filter is 0.1 m; the viscosity of air is 1.8×10^{-5} kg/m·s. For a dust-laden air velocity of 0.2 m/s and a collection efficiency of a single fiber equal to 1.50%, estimate the overall efficiency of the filter. To reach a collection efficiency of 99.99%, what is the minimum length required of this type of filter? At this minimum length, what is the overall pressure drop?

P9.12 Consider a flow of charged particles passing through a gap between two parallel plates that act as collecting electrodes. The plates are of length L, height H, and gap δ. For this arrangement, it is assumed that: (1) the particle concentration at any cross section of the precipitator is uniformly distributed; (2) flow velocity is constant; (3) U_m is constant and relatively small when compared to the flow velocity; and (4) there is no particle re-entrainment. Show that the collection efficiency of this electrostatic precipitator can be estimated by the Deutsch equation, Eq. (9.4-11).

P9.13 Consider a dust-laden gas passing through an electrostatic precipitator. Assume that all particles have the same charge-to-mass ratio, q/m, and the electric field is uniform. Use the Deutsch equation to estimate the overall collection efficiency if the mass-based particle size distribution follows: (a) Gaussian distribution; or (b) lognormal distribution. Compare your results with the collection efficiencies of monodispersed particles of the mean size.

P9.14 Consider a dust-laden gas passing through an electrostatic precipitator. Assume that all particles have the same charge-to-mass ratio, q/m, and the electric field is uniform. Use the Deutsch equation to estimate the overall collection efficiency if the particle size distribution, based on weight percentage (wt%), is expressed by the mass density function, f_M, as:

$$f_M = \begin{cases} 0 & 0 < d_p < d_{pm} - \delta \\ \dfrac{3}{4\delta}\left(1 - \dfrac{(d_p - d_{pm})^2}{\delta^2}\right) & d_{pm} - \delta \le d_p \le d_{pm} + \delta \\ 0 & d_{pm} + \delta < d_p < \infty \end{cases} \qquad \text{(P9.14-1)}$$

where d_{pm} and δ are the density function parameters.

P9.15 Determine the optimal diameter, D, and height, L_{ss}, of a vertical gas–liquid separator under the following conditions:

Operating pressure and temperature: 1,000 psi and 60°F, respectively;

Gas flow rate: 0.04 m^3/s, liquid flow rate: 0.00368 m^3/s;

Gas density: 59.5 kg/m^3; liquid density: 825 kg/m^3;

Gas viscosity: 0.013cP; droplet diameter: 140 μm; and

Liquid residence time: 3 min.

Assume the height of the vapor space is 1 m.

References

Alexander, R. M. (1949). Fundamentals of cyclone design and operation. *Proc. Australas. Inst. Min. Met.*, **152**, 203–228.

Beizaic, M. (1977). *Deposition of Particles in a Single Collector*. Ph.D. Dissertation, Syracuse University, Syracuse, New York.

Burgers, J. M. (1948). A mathematical model illustrating the theory of turbulence. *Advances in Applied Mechanics* **1**, ed. von Mises and von Karman. New York: Academic Press.

Chu, K. W., Wang, B., Xu, D. L., Chen, Y. X., and Yu, A. B. (2011). CFD-DEM simulation of the gas-solid flow in a cyclone separator. *Chem. Eng. Sci.*, **66**, 834–847.

Deutsch, W. (1922). Bewegung und Ladung der Electrizitätsträger im Zylinderkondensator. *Ann. Phys.*, **68**, 335–344.

Fan, L.-S., and Zhu, C. (1998). *Principles of Gas-Solid Flows*. Cambridge: Cambridge University Press.

Flagan, R. C., and Seinfeld, J. H. (1988). *Fundamentals of Air Pollution Engineering*. Englewood Cliffs, NJ: Prentice-Hall.

Happel, J. (1959). Viscous flow related to arrays of cylinders. *AIChE J.*, **5**, 174–177.

Kercher, H. (1969). Elektrischer Wind, Rücksprühen und Staubwiderstand als Einflußgrößen im Elektrofilter. *Staub - Reinhalt.* **29**, 314–319.

Leith, D., and Licht, W. (1972). The collection efficiency of cyclone type particle collectors: a new theoretical approach. *AIChE. Symp. Ser.*, **68**(126), 196.

May, K. R. (1945). The cascade impactor: an instrument for sampling coarse aerosols. *J. Sci. Instrum.*, **22**, 187–195.

Mercer, T. T. (1973). *Aerosol Technology in Hazard Evaluation*, New York: Academic Press.

Morgan, H., and Green, N. G. (2002). *AC Electrokinetics: Colloids and Nanoparticles*. Philadelphia, PA: Research Studies Press.

Morgan, H., Green, N. G., Hughes, M. P., Monaghan, W., and Tan, T. C. (1997). Large-area travelling-wave dielectrophoresis particle separator. *J. Micromech. Microeng.*, **7**, 65–70.

Ogawa, A. (1984). *Separation of Particles from Air and Gases, II.* Boca Raton, FL: CRC Press.

Ounis, H., Ahmadi, G., and McLaughlin, J. B. (1991). Brownian diffusion of submicrometer particles in the viscous sublayer. *J. Colloid and Interface Sci.*, **143**, 266–277.

Pfeffer, R., and Happel, J. (1964). An analytical study of heat and mass transfer in multiparticle systems at low Reynolds numbers. *AIChE J.*, **10**, 605–611.

Rajagopalan, R., and Tien, C. (1976). Trajectory analysis of deep-bed filtration with the sphere-in-cell porous media model. *AIChE J.*, **22**, 523–533.

Reist, P. C. (1993). *Aerosol Science and Technology*, 2nd ed. McGraw-Hill, New York.

Skodras, G., Kaldis, S. P., Sofialidis, D., Faltsi, O., Grammelis, P., and Sakellaropoulos, G. P. (2006). Particulate removal via electrostatic precipitators – CFD simulation. *Fuel Process. Technol.*, **87**, 623-631.

Stewart, M., and Arnold, K. (2008). *Gas-Liquid and Liquid-Liquid Separators*. Elsevier.

Svarovsky, L. (1981). *Solid–Gas Separation*. New York: Elsevier Scientific.

Theodore, L., and Buonicore, A. J. (1976). *Industrial Air Pollution Control Equipment for Particulates*. Cleveland, OH: CRC Press.

Tien, C. (1989). *Granular Filtration of Aerosols and Hydrosols*. Boston: Butterworths.

White, H. J. (1963). *Industrial Electrostatic Precipitation*. Reading, MA: Addison-Wesley.

Zenz, F. A., and Othmer, D. F. (1960). *Fluidization and Fluid–Particle Systems*. New York: Reinhold.

Zhu, C., Lin, C.-H., and Cheung, C. S. (2000). Inertial impaction dominated fibrous filtration with rectangular or cylindrical fibers, *Powder Technol.*, **112**, 149–162.

10 Fluidization

10.1 Introduction

Fluidization refers to a state of particle movements in which the particles are suspended and mobilized by a carrying fluid flow with the carrying fluid being a gas, a liquid or a gas–liquid mixture. In a fluid–solid two-phase system, when the fluid velocity is low, the fluid simply percolates through the voids between packed particles while the particles remain motionless. Such a system is noted to be at the fixed bed state. When the fluid velocity sufficiently exceeds the minimum fluidization velocity, the particles move apart and become suspended, and the bed enters the fluidized state. In fluidization, when the fluid velocity is relative low, the bed is characterized by a dense fluid–particle suspension with a small amount of particles entrained from the column by the carrying fluid. At a higher fluid velocity, the dense bed surface becomes indistinguishable with a large amount of particles entrained and transported out of the column. When a significant particle entrainment takes place, the fluidization state can be maintained by external circulation of the entrained particles. The state of fluidization where the majority of particles are present inside the fluidized bed is termed dense-phase fluidization, while the state of fluidization where the operation takes place with the presence of a small amount of the particles inside the fluidized bed and the prevalence of external circulation of particles is termed lean-phase fluidization.

Gas–solid fluidization and liquid–solid fluidization may share some similar hydrodynamic behaviors under low fluid velocity conditions. Liquid–solid fluidization typically results in a smooth expansion of the solid bed when the liquid velocity exceeds the minimum fluidization velocity. Such a condition is known as particulate fluidization or homogeneous fluidization. Gas–solid fluidization, on the other hand, can also be operated as particulate fluidization, but under certain particle property and gas velocity conditions. Aside from particulate fluidization, gas–solid fluidization exhibits more vigorous movements of the solids and less uniform bed expansion, which are the characteristics of the aggregative, or heterogeneous, fluidization. When gas is introduced into a liquid–solid system, gas bubbles are formed and the fluidization is termed gas–liquid–solid (three-phase) fluidization. The salient features of a fluidized bed include simple structure, large particle–fluid contact area, excellent mass and heat transfer, and its suitability for large-scale operations. The exposure of a large contact area is beneficial to any contact-area–based chemical reactions or

Table 9.5-1 Examples of Fluidization Applications.

Types of applications	Gas–solid fluidization	Liquid–solid fluidization	Gas–liquid–solid fluidization
Physical operations	Solids blending, Particle coating and drying, Adsorption, Granulation	Crystallization, Flocculation, Heat exchange	Sand filter cleaning, Granulation, Dust collection, Crystallization, Three-phase transport, Flocculation
Chemical or biochemical syntheses or operations	Phthalic anhydride synthesis, Acrylonitrile synthesis, Ethylene dichloride synthesis, Methanol to gasoline process, Olefin polymerization,	Linear alkylbenzene, Anaerobic fermentation	Methanation, Hydrogenation of heptane, Methanol production, Polymerization of ethylene and propylene, Aerobic waste water treatment, Ethanol fermentation, Cultivation of immobilized animal and plant cell
Metallurgical and mineral processes	Uranium processing, Iron production from iron ore, Pyrolysis of oil shale, Titanium dioxide production, Calcination	Particle size classification, Particle separation, Washing	Air flotation, Leaching of copper ore, Fluidized bed electrodes
Other applications	Coal combustion, Coal gasification, Incineration of solid waste, Fluidized catalytic cracking (FCC)	Wastewater treatment, Leaching	Hydrotreating of petroleum resides, Methanation, Flue gas desulfurization, Fischer–Tropsch synthesis,

physical interactions, whereas the energetic agitation of particles, fluids, and/or bubbles enhances the rate process of the fluid–particle transport. As a result, fluidized beds are effective vessels or reactors used for transport-enhancement or chemical reactions applications. Some examples of applications of fluidized beds are given in Table 9.5-1.

This chapter focuses more on gas–solid fluidization due to its relatively well-developed subject matter and broader industrial applications compared to gas–liquid–solid fluidization. Sections 10.2 and 10.3 discuss, respectively, the dense-phase fluidization characteristics and the lean-phase fluidization behavior. The gas–liquid bubble column and gas–liquid–solid slurry-bubble column are described in Section 10.4. As the operation of a bubble column is considered as the limiting case of that of a slurry-bubble column and a three-phase fluidized bed, the subject of bubble column is discussed in this section as a preamble to the discussion of slurry-bubble columns and three-phase fluidized beds that is introduced in Section 10.5.

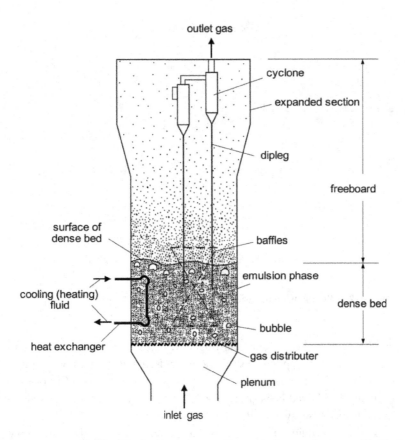

Figure 10.2-1 Fluidized bed with cyclone and internals (heat exchanger and baffles).

10.2   Dense Phase Gas–Solid Fluidized Beds

Common components of a fluidized bed are shown in Figure 10.2-1. One of the most important components is the distributor that provides a desired distribution of the gas flow to the fluidized bed. Typical distributors include sandwiched, stagger perforated, dished, grated, porous, tuyere, and cap types (Kunii and Levenspiel, 1991; Grace *et al.*, 2020). In order to ensure the uniformity of the flow distribution in the bed, a minimum pressure drop across the distributor is required, typically 1/3 of the pressure drop across the bed (excluding the distributor) for gas–solid fluidized beds. A fluidized bed often includes various internals, such as baffles, cyclones, and heat exchangers. The presence of internals in a bed, or variation in geometric configurations, such as tapered geometry, can affect the flow behavior.

While most fluidized beds are designed in a way that particles are fluidized mainly by the balance of hydrodynamic forces and the gravitational force, there are many other fluidized beds conducted in the presence of field forces other than

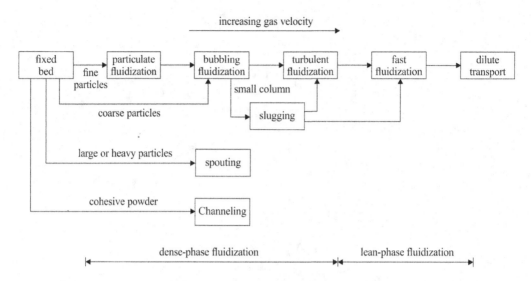

Figure 10.2-2 Relationship of various fluid–particle contact regimes.

gravity. Such fields include centrifugal, vibrational, acoustic, and magnetic fields. Operations with applications of these fields are known as vibro-fluidized beds (Mori *et al.*, 1992), acoustic fluidized beds (Montz *et al.*, 1988; Chirone *et al.*, 1992), centrifugal fluidized beds (Kroger *et al.*, 1979; Qian *et al.*, 2001) and magnetofluidized beds (Rosensweig, 1979; Liu *et al.*, 1991).

Regimes of fluid–solids systems may be classified based on their flow patterns, depending on the particle properties, the fluid velocity, and the geometry of the column given in Figure 10.2-2. In the figure, the interrelationship of these regimes includes, as stated in Section 10.1, the fixed bed state and the dense-phase fluidization state. Depending upon the fluid velocity and properties of fluidized particles, dense-phase fluidization may encompass particulate fluidization, bubbling fluidization, turbulent fluidization, slugging flow, spouting flow, and channeling flow. These flow relationships are graphically shown in Figure 10.2-3. At very high fluid velocities, the system operates in lean-phase fluidization including fast fluidization and dilute transport.

The particulate fluidization refers to the regime where all the gas passes through the interstitial space between the fluidizing particles without forming bubbles. The bed appears to be homogeneous, with fluidized particles more or less uniformly distributed throughout the bed. The bubbling fluidization refers to the regime where bubbles or bubble-shaped voids form and induce vigorous motion of the particles. Bubbles coalesce and break up as they rise in the bed. There are two distinct phases (*i.e.*, the bubble phase and the emulsion phase) that characterize the bed flow structure. The turbulent fluidization refers to the regime where, with the increased gas velocity and enhanced turbulent mixing, the bubble and emulsion phases become less distinguishable in the bed. Further increase in the gas

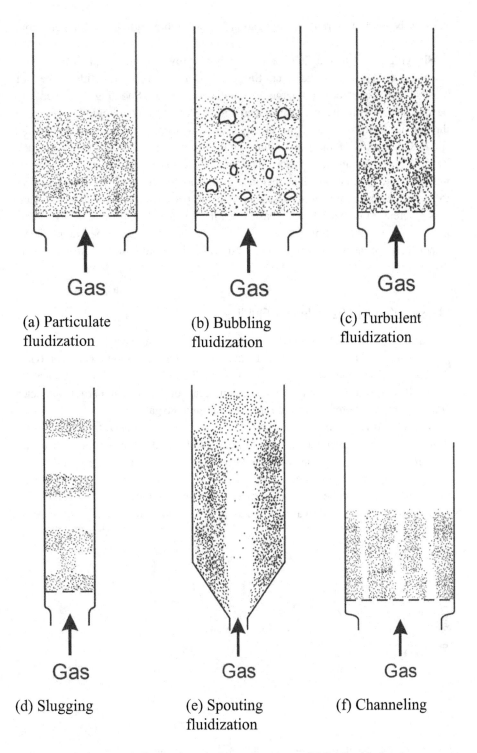

Figure 10.2-3 Flow regimes in dense-phase gas–solid fluidized bed.

velocity beyond the turbulent fluidization leads to more particle entrainment from the bed.

Slugging refers to the fluidization represented by gas slugs with bubble sizes comparable to the bed diameter. Slugging often occurs in a fluidized bed of a small bed diameter with large height/diameter ratio. Spouting in a bed can be conducted with the gas introduced to the center or core region of the bed through a jet nozzle at a high velocity. Particles in the bed are entrained to the spout and ejected at the top of the spout. The ejected particles then fall back by gravity to the annular region. The annular region represents the downward-moving particles that are then recirculated to the core region. Channeling refers to a flow pattern where gas passes through the packed particles via many channels whereas most particles are basically immobilized due to the cohesive inter-particle force. Channeling usually occurs in a bed of fine, cohesive particles. Channeling can also occur when gas distribution is highly nonuniform across the distributor.

10.2.1   Classifications of Particles for Fluidization

Based on experimental data and observation, particles can be classified into four groups (*i.e.*, Groups A, B, C, and D) based on their fluidization behavior (Geldart, 1973). This classification, known as Geldart's classification, is shown in Figure 10.2-4, where particles are grouped in terms of the average particle diameter and the density difference between the particles and the gas.

Both Group A and B particles can easily be fluidized. Fluidization with Group A particles can be operated in both the particulate fluidization regime and in the bubbling fluidization regime, whereas Group B particles can only be fluidized in the bubbling fluidization regime. Group C particles are typically small in size ($d_p < 20$ μm) and cohesive. The particle cohesion is dominated by such interparticle contact forces as the van der Waals force, capillary forces, and electrostatic

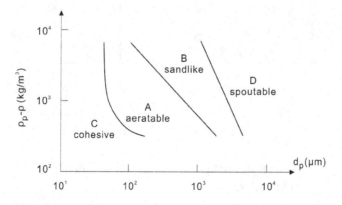

Figure 10.2-4 Geldart classification of particles for gas–solid fluidization.

forces over the hydrodynamic forces. Group C particles are difficult to fluidize and gas channeling is a typical flow pattern. Group D comprises coarse particles ($d_p > 1$ mm) and can readily be fluidized by spouting.

There are empirical or semi-empirical correlations that improve the quantification of the demarcation among Geldart groups. The correlation for demarcation between Group C and A particles is based on the balance between the van der Waals force and the hydrodynamic forces given by (Molerus, 1982):

$$\left(\rho_p - \rho\right) g d_p^3 = 0.001 \, F_H, \tag{10.2-1}$$

where F_H is the cohesive force between particles with a value ranging from 8.8×10^{-8} N (for hard material) to 3.7×10^{-7} N (for soft material). The correlation for the demarcation between Group A and Group B particles is given by (Grace, 1986):

$$Ar = 1.03 \times 10^6 \left(\frac{\rho_p}{\rho} - 1\right)^{-1.275}, \tag{10.2-2}$$

where Ar is the Archimedes number, defined as:

$$Ar = \frac{\rho(\rho_p - \rho) g d_p^3}{\mu^2}. \tag{10.2-2a}$$

The correlation for demarcation between Group B and Group D is given by (Grace, 1986):

$$Ar = 1.45 \times 10^5 \tag{10.2-3}$$

Geldart group classification given in Figure 10.2-4 has been widely used in fundamental and applied research and development of gravity-based gas–solid fluidized beds. However, it should be noted that some fluidization variables other than $(\rho_p - \rho)$ and d_p may alter the demarcation of particle groups. For example, for fluidization using different gases, the cohesive nature of fluidized particles due to the surface or interparticle contact forces can vary significantly because of the difference in the type of gases adsorbed on the particles. Likewise, gas velocity, temperature, and pressure may also dramatically affect the fluidized particle characteristics, which in turn alter the particle group demarcation. It is also important to note that, for fluidized beds influenced by other field forces such as centrifugal, vibrational, acoustic, and magnetic forces, the demarcation of Geldart groups can alter.

10.2.2   Dense Phase Fluidization

As described for Figure 10.2-2, there are several fluidization regimes in the dense-phase fluidization, including the particulate fluidization, bubbling fluidization, turbulent fluidization, slugging flow, spouting flow, and channeling flow. The fluidization characteristics in each regime and regime transition conditions are discussed in the following.

10.2.2.1   Minimum Fluidization Condition and Particulate Fluidization

Define the dynamic pressure drop, Δp_d, as the total pressure drop across the bed, Δp_b corrected for the hydrostatic head of the fluid. The dynamic pressure drop is then independent of the superficial fluid velocity, U, under the fluidization conditions for two-phase (gas–solid or liquid–solid) fluidization. The superficial fluid velocity is given as the volumetric fluid flow rate divided by the cross-sectional area of the bed. When the fluid is gas and the density of the gas is significantly smaller than that of the solid, Δp_d is essentially identical to Δp_b. The dynamic pressure drop variation with the superficial gas velocity can then be expressed in terms of Δp_b versus U, as given in Figure 10.2-5. It is noted that the bed pressure drop excludes the pressure drop over the distributor. As shown in the figure, as U increases in the packed bed, the pressure drop increases, reaches a peak, and then drops to a constant. As U decreases from the fully fluidized state, the pressure drop follows a different path without passing through the peak. Hence, there is a hysteresis loop in the bed pressure variation with increasing and then decreasing gas velocity. The peak under which the bed is operated is denoted as the minimum fluidization condition, and its corresponding superficial gas velocity is defined as the minimum fluidization velocity, U_{mf}. In other words, the common definition of U_{mf} is obtained from the path with increasing fluid velocity. The hysteresis is attributed to the wall friction and cohesive forces between particles (Tsinontides and Jackson, 1993; Srivastava and Sundaresan, 2002).

With an increase in the superficial fluid velocity beyond the fixed bed, the bed expands, as shown in Figure 10.2-6. In gas–solid fluidization, the operational range of the particulate fluidization regime, in which the bed expands uniformly with an

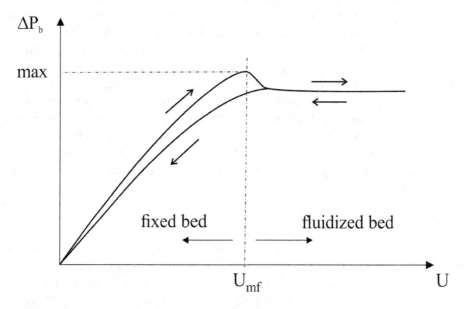

Figure 10.2-5 Characteristics of pressure drop versus superficial velocity.

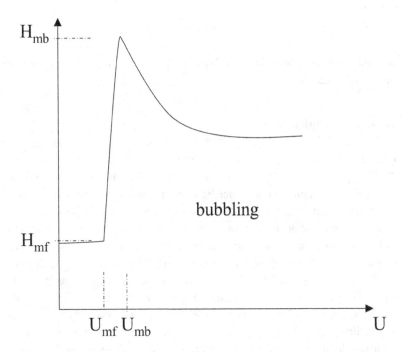

Figure 10.2-6 Schematic characteristics of bed height versus superficial velocity.

increase in the gas velocity, is quite narrow. For $U > U_{mf}$, the bed height monotonically increases with U until the minimum bubbling point U_{mb} is reached. Further, an increase in the gas velocity leads to a decrease in the bed height due to the formation of "bubbles" in the fluidized bed. Thus, the particulate fluidized bed operates between the lower limit of the minimum fluidization velocity U_{mf} and upper limits of the minimum bubbling velocity U_{mb}.

The minimum fluidization velocity can be evaluated based on the incipient condition when the weight of the bed is balanced by the total pressure drop across the bed. The effect of the wall friction is assumed to be negligible. Based on the Ergun equation, for a loosely but uniformly packed granular medium, the minimum fluidization velocity can be linked to the bed voidage at minimum fluidization α_{mf} by:

$$Ar = 150\frac{(1-\alpha_{mf})}{\alpha_{mf}^3\varphi^2}Re_{pmf} + \frac{1.75}{\alpha_{mf}^3\varphi}Re_{pmf}^2, \qquad (10.2\text{-}4)$$

where Re_{pmf} is the particle Reynolds number at minimum fluidization; φ is the sphericity of the particles. Equation (10.2-4) yields a general form of Re_{pmf} in terms of Ar as:

$$Re_{pmf} = \sqrt{m^2 + nAr} - m, \qquad (10.2\text{-}4a)$$

where m and n are empirical constants, depending upon the particle shape, packing modes, and the critical packing density at minimum fluidization (Yang, 2003). For gas–solid fluidized beds, a frequently used semiempirical correlation was given by (Wen and Yu, 1966):

$$\mathrm{Re}_{pmf} = \sqrt{(33.7)^2 + 0.0408\mathrm{Ar}} - 33.7, \qquad (10.2\text{-}4b)$$

which is applicable to both gas–solid and liquid–solid fluidization over a wide range of Reynolds numbers.

10.2.2.2 Bubbling Fluidization

In gas–solid fluidization, a gas "bubble" in suspended solids is a gas cavity containing suspended particles of a much diluted concentration. The formation of such a "bubble" is due to the inherent flow instability of the gas–solid flow. Different from a gas bubble in a liquid, the interface between the cavity and suspended solids is not as sharp. Despite the mechanistic differences between the two types of bubbles, both of them do bear some degrees of similarities in their hydrodynamic characteristics, such as bubble wake, bubble stability, and bubble motion patterns. Thus, the approach in accounting for bubble dynamics in gas–solid fluidization to some degrees is similar to that in gas–liquid systems, such as a large gas slug rising in gas–liquid and gas–solid systems (Clift and Grace, 1985; Fan, 1989; Fan and Tsuchiya, 1990).

The onset of "bubbling" in a gas–solid fluidized bed is believed to be related to the flow instability when the gas passes through fluidized particles beyond a critical velocity. However, there are no reliable mechanistic models that can accurately explain the onset of such instability. An empirical correlation to estimate the minimum bubbling velocity is given by (Abrahamsen and Geldart, 1980):

$$U_{mb} = 2.07 \exp\left(0.716\phi_f\right) \frac{d_p \rho^{0.06}}{\mu^{0.347}}, \qquad (10.2\text{-}5)$$

where ϕ_f is the mass fraction of fine particles smaller than 45 μm, and all units in the equation are S.I. This expression becomes invalid when the calculated U_{mb} is less than U_{mf}.

Based on experimental observation and measurements, there are two basic configurations of bubbles, namely, the clouded bubble and the cloudless bubble. The cloud is characterized by a surrounding dense layer of particles moving around the bubble and confined by circulating gas flow around the bubble, as schematically depicted in Figure 10.2-7. The demarcation of these two types of bubbles depends mainly upon the relative velocity between the bubble rising velocity and the interstitial gas velocity in bed. When the bubble rise velocity is higher than the interstitial gas velocity, the relative velocity may cause a compacting effect on the particles in the frontal region of bubble, leading to the formation of a slightly packed dense layer of particles and a relative circulatory slip around the "bubble," as shown in Figure 10.2-7(a). The degree of compacting effect is also dependent upon the flow properties of the particles. In fluidized beds of Group A or B particles, the bubbles are typically characterized by clouded bubbles. When the bubble rise velocity is less than the interstitial gas velocity, the bubbles are found to be "cloudless," with some emulsion gas directly flowing through the bubbles, as illustrated in Figure 10.2-7(b).

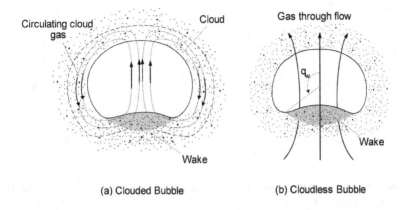

(a) Clouded Bubble   (b) Cloudless Bubble

Figure 10.2-7 Clouded and cloudless "bubbles" in a gas–solid fluidized bed.

The bubble wake is also an important hydrodynamics feature of "bubbles" in fluidized solids. The bubble wake plays an important role in solids movement and solids mixing both within the fluidized bed and in the freeboard region above the bed surface. In a gas–solid fluidized bed, the bubble wake is defined as the region enclosed by a streamline of the emulsion phase behind the bubble base. In the bed, the wake rises with the bubble, thereby providing an essential means for global solids circulation and induces axial solids mixing. Particles carried along in the bubble wake are a source of the particles in the freeboard. The bubble wake fractions in gas–solid fluidized beds are generally much smaller than those in low-viscosity liquid systems. Similar to the wake dynamics of gas bubbles in liquid media, in gas–solid fluidized beds, wake shedding has also been observed. Most of the shed wake fragments appear to be banana-shaped, and the shedding may occur at fairly regular intervals.

10.2.2.3   Turbulent Fluidization

The turbulent regime is the transition regime occurring from bubbling to lean phase fluidization. In this regime, with an increase in gas velocity, the boundary of the bubbles starts to diminish and the gas voids become less distinguishable. Starting from the bubbling regime, as the gas velocity increases, bubble motion becomes increasingly vigorous, resulting in an increase in the amplitude of pressure fluctuations. However, after reaching a maximum, the amplitude of the pressure fluctuation then decreases with an increase in the gas velocity. The onset of the transition to the turbulent regime is commonly defined as the gas velocity, U_c, corresponding to the peak of the pressure fluctuation. Figure 10.2-8 illustrates a schematic relationship between the amplitude of pressure fluctuation and superficial gas velocity for Group A particles, which shows that the pressure fluctuation gradually levels off after a gas velocity U_k (Yerushalmi and Cankurt, 1979). At U_k the bed is usually highly turbulent and may be close to the transition to lean-phase fluidization. For Group B

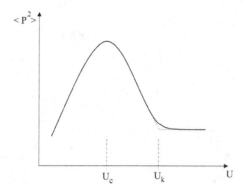

Figure 10.2-8 Intensity of pressure fluctuation with the gas velocity in dense-phase fluidized beds of Group A particles.

and Group D particles, the amplitude of pressure fluctuation at velocities beyond U_c may decrease without having a distinct leveling-off point for U_k. At gas velocities less than U_c the bubble interaction is dominated by bubble coalescence, while at gas velocities greater than U_c it is dominated by bubble breakup (Cai *et al.*, 1989).

The flow and transport characteristics in the turbulent regime are distinctly different from those in bubbling fluidization operated at relatively low gas velocities. They can be described as follows:

(1) The upper surface of the turbulent bed exists yet becomes more diffused with a high particle concentration in the freeboard.
(2) The bubbles are of small sizes with irregular shapes, frequent splitting, and redispersion in the bed.
(3) Bubbles or voids move violently, rendering it difficult to distinguish the emulsion phase from the bubble phase in the bed.
(4) Bubble motion appears to be more random with enhanced interphase exchange and hence intimate gas–solid contact and high heat and mass transfer.

10.2.2.4   Entrainment and Elutriation

The entrainment refers to the ejection of particles from the dense bed into the freeboard region by the fluidizing gas. Particles are ejected into the freeboard via two basic mechanisms: the ejection from the bubble roof and the ejection from the bubble wake. In the roof ejection, as the bubble approaches the bed surface, the particle layer between the bubble roof and the bed surface is thinning up and eventually erupts. Particles on the top of the bubble roof are thrown into the freeboard by the gas originally inside the bubble. In the wake ejection, when the bubble erupts near the bed surface, particles in the bubble wake continue to move into the freeboard due to their inertia and are entrained by the fluidizing gas leaving the bed surface. At high gas velocities, bubble coalescence in the bed due to accelerated trailing bubbles leads to significant ejection of wake particles of the leading bubbles. The wake ejection becomes the dominant mechanism of particle entrainment in this situation,

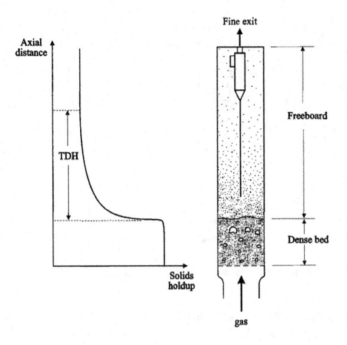

Figure 10.2-9 Solids holdup distribution in the dense bed and freeboard.

and it is much more pronounced in a turbulent regime than in the bubbling regime because of the higher bubble rise velocities.

In the freeboard region of the fluidized bed, coarse particles whose terminal velocity is higher than the fluidizing gas velocity fall back into the dense bed, while fine particles exit from the freeboard. The separation of fine particles from a mixture of particles and their eventual removal from the freeboard is referred to as elutriation. The nonuniform solids holdup distribution in the freeboard region, as shown in Figure 10.2-9, can be related to the nonuniformity of particle properties. Coarse particles are present primarily in the region near the dense bed surface, whereas the upper part of the freeboard mainly consists of fine particles. Design consideration of the fluidized bed requires the freeboard height to be higher than the transport disengagement height (TDH), beyond which the solids holdup and entrainment rate remain nearly constant. The total solids entrainment rate flux above the TDH can be estimated by various correlations (Geldart *et al.*, 1979; Colakyan and Levenspiel, 1984).

10.2.2.5 Slugging

Slugging occurs when bubbles grow to sizes comparable to the bed diameter. This happens most frequently in small-diameter beds with large bed heights, especially with large or heavy particles. A minimum gas velocity and a minimum bed height must be reached in order for slugging to take place (Stewart and Davidson, 1967).

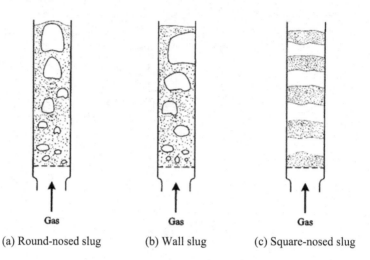

(a) Round-nosed slug   (b) Wall slug   (c) Square-nosed slug

Figure 10.2-10 Schematic forms of various slugs in fluidization.

In a slugging bed, the pressure drop increases with increasing gas velocity, which is due to strong particle–wall friction and momentum dissipation from the gas to the particles. This is in contrast to the particulate or bubbling fluidized bed, in which the pressure drop remains nearly constant when the gas velocity changes. Figure 10.2-10 shows the different forms of slugs in the slugging bed (Kunii and Levenspiel, 1991; Grace *et al.*, 2020). Type (a) is the round-nosed slug, which occurs in systems with fine particles. Type (b) is the wall slug, also called half slug, which appears in beds with rough walls, large particle-to-bed diameter ratios, angular particle shapes, and relatively high gas velocities. Type (c) is the square-nose slug, also known as plug, which is present in coarse particle systems with significant particle bridging effect.

10.2.2.6   Spouted Beds

In a spouted bed, gas coming in through a bottom jet nozzle with diameter D_i forms a spout of diameter D_s in the center of the bed, while particles in the annular region form a downward-moving bed, as shown in Figure 10.2-11.

Particles entrained into the spout are separated from the gas in the solid disengagement fountain (H_F) and return to the bed. Spouted bed operation is typically performed with Group D particles. In some cases a secondary gas is introduced into the bed through the side wall of the conical base to improve the gas–particle contact in the annular region in a fluid-spout or aerated-spout bed operation.

The onset of spouting can be analyzed from the variation of pressure drop as a function of the gas velocity, as shown in Figure 10.2-12. Starting from a fixed bed, as the gas velocity increases, a dilute core region is formed under the bed surface due to the gas penetration and the pressure drop increase along the curve AB. Further increase of the gas velocity leads to a decrease in pressure drop, until point C where the dilute core penetrates the entire bed to form a spout. A slight gas velocity increase beyond point C results in a sharp decrease in pressure drop to point D,

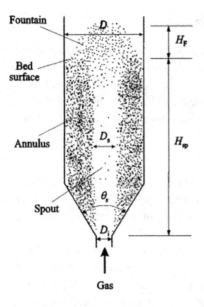

Figure 10.2-11 Schematic diagram of a spouted bed.

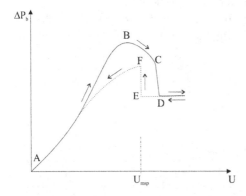

Figure 10.2-12 Bed pressure drop with gas velocity in a spouted bed.

after which the pressure drop levels off. When the gas velocity decreases from point D, the pressure drop variation follows a different route DEFA, forming a hysteresis loop. The minimum spouting velocity U_{msp} is defined as the gas velocity at point E. Unlike the minimum fluidization velocity, which is a function of gas and particle properties, the minimum spouting velocity depends on additional parameters such as the bed geometry and static bed height (Fane and Mitchell, 1984). In order for the spout to form, the bed height cannot exceed the maximum spoutable bed depth H_m, beyond which the spout starts to collapse and the bed starts to change from a spoutable bed to a fluidized bed (Epstein and Grace, 1997; Grace et al., 2020).

10.2.3   External Field Modulated Fluidization

Conventional fluidized beds are operated against gravity and therefore the operating range of the superficial gas velocity is confined to a certain level. To expand, enhance or modulate this level, the external force fields can be applied such as the centrifugal field by flow rotation, the magnetic field by electromagnets, and the acoustic field by loud speakers. Fluidization under external force fields is described below.

10.2.3.1   Centrifugal Fluidization

Centrifugal fluidization is achieved by a rotating fluidized bed with aeration introduced radially inward, as shown in Figure 10.2-13. A cylindrical distributor rotates around its axis of symmetry either vertically or horizontally, allowing gases to enter while preventing particles from leaving. Generally, the centrifugal acceleration is significantly larger than the gravitational acceleration so that the granular particles are driven toward the cylinder wall to form a dense granular bed. The gas flow, moving inward in a counterflow mode to the particle motion, provides the needed hydrodynamic forces for particle fluidization. With sufficiently high gas velocities, the dense granular bed expands toward the central axis. At full fluidization, centrifugal forces on the granules balance the drag forces from the gas. The particulate and bubbling fluidization in the radial direction can be achieved.

Typically flow distributors for centrifugal fluidization include slotted, perforated, or sintered granular or metal cylindrical shells. For distributors made of a slotted or

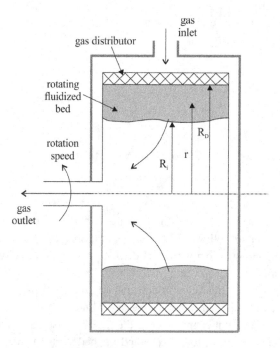

Figure 10.2-13 Rotating fluidized bed.

perforated shell, a thin layer of a fine screen is usually placed between the cylindrical shell and the bed granules to prevent the leakage of granules through the distributor.

The pressure drop over a centrifugal fluidized bed is a function of rotating speed, flow rate, properties of the particles, and the geometry of the distributor. The overall pressure drop for the flow through the rotating fluidized bed can be expressed as the additive combination of three components:

$$\Delta p_T = \Delta p_G + \Delta p_R + \Delta p_S, \tag{10.2-6}$$

where Δp_G is pressure drop due to radial gas flow through the porous granular bed; Δp_R is pressure drop of the gas flowing through the rotating chamber before the distributor; and Δp_S is pressure drop due to viscous dissipation by tangential shear, which mobilizes the tangential movements of the granules. Δp_R can be estimated by using an analogy between flow through a rotating chamber and flow through an array of rotating blades in a centrifugal fan, while Δp_S can be estimated by the change of kinetic energy of the relative tangential-slip velocities.

When the rotating granule bed is in the packed bed regime, the gas pressure gradient can be estimated, based on the Ergun equation, Eq. (2.4-5), by:

$$\frac{dp_G}{dr} = \phi_1 U + \phi_2 U^2, \tag{10.2-7}$$

where U is the superficial velocity, which is a function of radial location and pressure; ϕ_1 and ϕ_2 are the coefficients defined by:

$$\phi_1 = \frac{150(1-\alpha)^2 \mu}{\alpha^3 (\varphi d_p)^2} \qquad \phi_2 = \frac{1.75(1-\alpha)\rho}{\alpha^3 (\varphi d_p)}. \tag{10.2-7a}$$

In the fully fluidized bed regime, the pressure gradient is equal to the centrifugal acceleration of the granule bed:

$$\frac{dp_G}{dr} = (\rho_p - \rho)(1-\alpha) r\omega^2. \tag{10.2-8}$$

The minimum fluidization velocity at a given radial location in a rotating fluidized bed can be obtained, in a similar approach to the gravitational fluidized bed, by equating the pressure drop across the rotating fluidized bed to that across the packed bed. Thus, based on Eqs. (10.2-7) and (10.2-8), it gives:

$$U_{mf} = \frac{\sqrt{\phi_1^2 + 4\phi_2(\rho_p - \rho)(1-\alpha)\omega^2 r} - \phi_1}{2\frac{R_D}{r}\phi_2}. \tag{10.2-9}$$

Equation (10.2-9) indicates that the minimum fluidization velocity increases monotonically with the radial location r, indicating that the initial fluidization starts at the inner surface with a minimum fluidization velocity U_{mfi} (obtained by letting $r = R_i$ in Eq. (10.2-8)). Upon further increases in the flow rate, the bed becomes partially fluidized until the flow rate reaches the critical minimum fluidization velocity U_{mfc} (obtained by letting $r = R_D$ in Eq. (10.2-9)) when granular particles at the outer radius (near the distributor surface) are fluidized. At the velocity beyond U_{mfc}, the bed is fully fluidized.

The pressure drop across the granular bed can be obtained by integrating Eq. (10.2-8) and Eq. (10.2-9), depending on the fluidization regimes, as follows:

(1) When the bed is packed at $U \leq U_{mfc}$,

$$\Delta p_G = \frac{\phi_1 Q}{2\pi L} \ln\left(\frac{R_D}{R_i}\right) + \frac{\phi_2 Q^2}{4\pi^2 L^2}\left(\frac{1}{R_i} - \frac{1}{R_D}\right). \tag{10.2-10a}$$

(2) When the bed is partially fluidized at $U_{mfi} < U < U_{mfc}$,

$$\Delta p_G = (1-\alpha)(\rho_p - \rho)\omega^2 \frac{\left(r_{pf}^2 - R_i^2\right)}{2} + \frac{\phi_1 Q}{2\pi L} \ln\left(\frac{R_D}{r_{pf}}\right) + \frac{\phi_2 Q^2}{4\pi^2 L^2}\left(\frac{1}{r_{pf}} - \frac{1}{R_D}\right). \tag{10.2-10b}$$

(3) When the bed is fully fluidized at $U_{mfc} \leq U$,

$$\Delta p_G = (1-\alpha)(\rho_p - \rho)\omega^2 \frac{\left(R_D^2 - R_i^2\right)}{2}. \tag{10.2-10c}$$

A general expression of the overall pressure drop across the rotating fluidized bed given in Equation (10.2-6) can be further expressed as:

$$\Delta p_T = (c_1 + c_2\omega)\,Q + c_3 Q^2 + c_4\omega^2. \tag{10.2-11}$$

where c's are semiempirical coefficients depending only upon the particle properties, the distributor geometry, and the fluidization status of particles. Equation (10.2-11) shows that the pressure drop can be expressed as a quadratic equation with respect to rotating speed as well as the flow rate (Zhu *et al.*, 2003). It is shown that the rotating fluidization is similar to the conventional fluidization in the pressure drop and gas velocity variation behavior. However, the fluidization characteristics such as the minimum fluidization velocity depend on the rotating speed in addition to the particle-property, as further elaborated in the following section.

10.2.3.2 Modulation of Fluidization by Centrifugal Acceleration

Geldart's particle classification is defined in gravity-driven gas–solid fluidized beds. For fluidization driven by other external forces, such as the centrifugal force, the demarcation of the Geldart's classification needs to be adjusted. For example, by employing a high rotation speed of a centrifugal fluidized bed, Geldart's Group C particles, which are difficult to fluidize in the gravity-driven fluidized bed, can exhibit Group A behavior and thus improve its fluidization quality.

The effect of the field acceleration on the fluidization behavior can be revealed by the transition criteria between different particle groups in Eq. (10.2-1) and Eq. (10.2-2), in which the gravity provides the field acceleration. In a centrifugal field, the gravitational acceleration g is replaced by the centrifugal acceleration. As shown in Figure 10.2-14, when the acceleration is increased, all demarcation curves are shifted so that the powder originally classified as Group C under gravitational acceleration may become Group A/B (Figure 10.2-14(a)) or even Group D

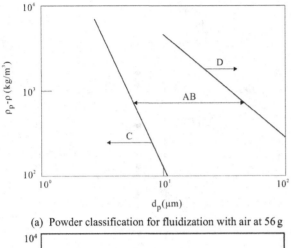

(a) Powder classification for fluidization with air at 56 g

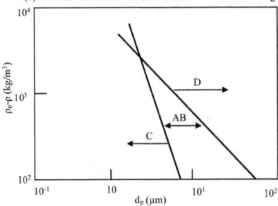

(b) Powder classification for fluidization with air at 278 g

Figure 10.2-14 Effect of centrifugal acceleration on powder classification.

powder (Figure 10.2-14(b)) under centrifugal acceleration. Such behavior is verified experimentally (Qian *et al.*, 2001).

10.2.3.3 Magnetic-Assisted Fluidization

The fluidized beds of magnetizable particles operated under an externally applied magnetic field is referred to as the magnetofluidized beds (MFB). The magnetic field is able to suppress gas bubbling and bypassing in a magnetofluidized bed, and thus the particulate fluidization regime can be extended to superficial gas velocities significantly higher than the minimum fluidization velocity. As the result, the solids are homogeneously fluidized with little back mixing in this magnetically "stabilized" bed under high gas flow rates. A typical magnetic-assisted fluidized bed is sketched in Figure 10.2-15. The magnetic field is produced by wound coils that carry an electric current and surround the fluidization column. The coils produce a magnetic field,

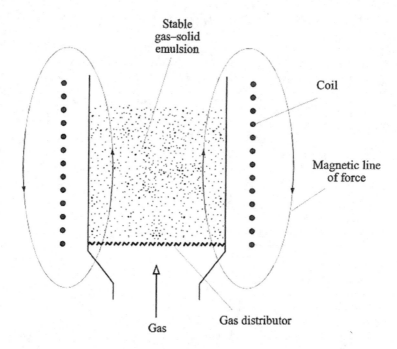

Figure 10.2-15 Schematic diagram of a magnetic-assisted fluidized bed.

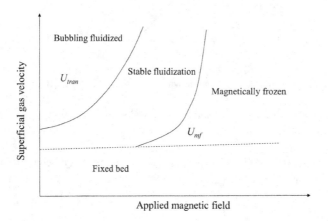

Figure 10.2-16 Regime diagram of magnetofluidized bed.

which is aligned parallel to the flow direction of the fluidizing gas. Often a weak, spatially uniform, and time-invariant magnetic field is employed.

The behavior of the magnetofluidized bed is determined by both the superficial gas velocity and the intensity of the applied magnetic field, as illustrated by the fluidization regime diagram in Figure 10.2-16 (Liu *et al.*, 1991).

When the superficial gas velocity U is below the minimum fluidization velocity U_{mf}, the bed is in the fixed bed mode. For a conventional bubbling bed without magnetic effects, the bed is in aggregative fluidization. However, in a magnetically stabilized fluidized bed, the bed is in homogeneous fluidization without the presence of bubbles at a low U. With an increase in U, bubbling and bed fluctuation will occur. The transition from a homogeneous to a bubbling state occurs at a superficial gas velocity substantially higher than U_{mf}. If the magnetic field is sufficiently strong, the magnetized particles may experience appreciable attractive interparticle forces, and the bed appears to be locked or frozen. The mechanism of bubble suppression in a magnetic field is attributed to the rearrangement of magnetized particles, which form chains or arrays that constrict the size of bubbles. It is also suggested that a gas bubble in a magnetic field represents a nonmagnetic void with decreased magnetic flux density and thus is subjected to an interfacial magnetic force that causes the bubble to collapse.

10.2.3.4   Vibration-Assisted Fluidization

Mechanical vibration has been applied to improve the quality of gas–solid fluidization for both noncohesive and cohesive particles. The fluidization quality is characterized by the fluidization index FI, which is defined by the ratio of the pressure drop Δp over the bed to the bed weight mg per unit cross-sectional area A

$$\text{FI} = \frac{\Delta p A}{mg}. \tag{10.2-12}$$

For good quality fluidization, the pressure drop over the bed is approximately equal to the bed weight per area, which yields a fluidization index close to one. On the other hand, poor-quality fluidization caused by gas bypass and channeling yields a lower pressure drop and the fluidization index is much less than one. For sinusoidal vibration, a vibration number G, defined below, is introduced to evaluate the vibration intensity:

$$G = \frac{A_0 \omega^2}{g}, \tag{10.2-13}$$

where A_0 is the peak amplitude of vibration, ω is the angular frequency, and g is the gravitational acceleration.

For noncohesive Group A particles, the bed voidage is decreased by vibration. The fluidization index slightly increases and is close to unity. In the fluidization of Group C particles, the assistance of vibration energy can effectively reduce the bubbling and channeling behavior through breaking down the cracks and channels in the bed (Marring *et al.*, 1994). When vibration is applied, there exist critical values of the minimum gas velocity and the vibration intensity to achieve good fluidization. In general, the minimum fluidization velocity decreases when the vibration number G increases. Upon vibration at higher frequencies (*e.g.*, higher than 100 Hz), large bubbles tend to form. Thus, typical vibrated fluidized beds operate at lower frequencies. In practice, shallow beds are often used so that a significant vibration energy dissipation in deep beds can be avoided. Besides vertical vibration, horizontal and twist vibration modes are also be used (Mawatari *et al.*, 2001).

10.2.4   Fluidization of Nanoparticles

Nanoparticles refer to the particles whose primary size is in the range of nanometers. In a naturally packed state in a gaseous medium, nanoparticles are found in forms of submillimeter-sized agglomerates rather than in forms of individual primary particles. This may be attributed to the strong adhesive interactions among the nanoparticles (such as van der Waals forces, electrostatic forces, and moisture-induced surface tension forces). Hence the gas fluidization of nanoparticles is actually the fluidization of agglomerates of nanoparticles in gas flows.

The nanoparticles in agglomerates are interlinked as chains by strong interparticle forces. The chain-linked nanoparticles inside agglomerate are not uniformly distributed, typically containing many sub-agglomerates with denser chainlike structures, as schematically depicted in Figure 10.2-17. The agglomerates of nanoparticles are highly porous (and hence of low bulk material density), with the overall porosity of fluidized agglomerates of nanoparticles up to 98% or even higher. For example, in gas fluidization of silicon nanoparticles, the typical size range of a fluidizing agglomerate is from 70 to 700 μm, while the primary particle size ranges from 7 to 500 nm (Morooka *et al.*, 1988; Matsuda *et al.*, 2001). Because of the high porosity of the nanoparticle aggregates, their bulk density is about one to two orders of magnitude smaller than the material density of primary particles.

10.2.4.1   Fluidization Characteristics and Regime Classification

There are two distinctive fluidization regimes for nanoparticles. Fluidization of some nanoparticles results in an extremely high bed expansion without formation of bubbles, while for fluidization of other nanoparticles, only a very limited bed expansion exists with formation of large bubbles rising through the bed. The former type is termed as agglomerate particulate fluidization (APF), and the latter one is termed as agglomerate bubbling fluidization (ABF) (Wang *et al.*, 2002; Zhu *et al.*, 2005). The fluidization characteristics of APF are similar to those of particulate fluidization in gas–solid fluidized bed, and the ABF fluidization behavior mimics that of bubbling

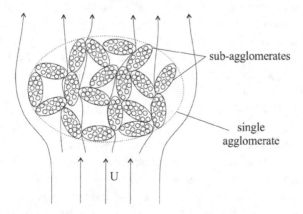

sub-agglomerates

single agglomerate

U

Figure 10.2-17 Schematic structure of aerated agglomerate of nanoparticles.

fluidization. However, the mechanisms for demarcation of APF and ABF are not clear. For silicon-based nanoparticles, the demarcation between APF and ABF may be linked to primary size and the bulk material density of nanoparticles. For example, most silicon-based nanoparticles with bulk density < 100 kg/m^3 and primary size < 20 nm demonstrate APF characteristics, whereas those with bulk density > 100 kg/m^3 and primary size > 20 nm are likely in the ABF regime (Wang *et al.*, 2002; Zhu *et al.*, 2005).

10.2.4.2 Minimum Fluidization Velocity and Agglomerate Size in APF

As indicated, the fluidization behavior of nanoparticles is determined by the fluidization properties of the agglomerates, which are chain-linked assemblies of the primary nanoparticles. The drag force on the nanoparticle agglomerate in a swarm of agglomerates is close to that predicted for a solid sphere of the same size in a swarm of solid spheres as gas permeation through porous aggregates does not impose significant hydrodynamic effects on its fluidization behavior. Thus, existing modeling approaches for fluidization of solid spheres can be applied to that of nanoparticle aggregates when the physical properties of the nanoparticle aggregates are known. As an example, the spherical equivalent agglomerate size can be determined using the Richardson–Zaki equation, a semiempirical hydrodynamics correlation for the solid particles that relates the ratio of the superficial gas velocity, U, to the particle terminal velocity, U_{pt}, to the bed voidage, α. Given the measured superficial gas velocity and bed expansion relationship under the uniform bed expansion, the bed voidage can be expressed by

$$\alpha = 1 - \frac{H_0}{H}(1 - \alpha_0),\tag{10.2-14}$$

which gives rise to the Richardson–Zaki equation as

$$\sqrt[n]{U} = \sqrt[n]{U_{pt}}\left\{1 - \frac{H_0}{H}(1 - \alpha_0)\right\}.\tag{10.2-15}$$

Taking the Richardson–Zaki index, n, of 5 for the creeping flow condition for solid particles (Davis and Birdsell, 1988), particle terminal velocity and original bed voidage can be obtained for the nanoparticle aggregate fluidization from the intercept and slope of a linear plot of $U^{1/n}$ versus bed expansion ratio (H_0/H). The size of nanoparticle agglomerates, d_a, can then be calculated from the Stokes equation for the solid particle terminal velocity as

$$d_a = \sqrt{\frac{18\mu U_{pt}}{(\rho_a - \rho)g}}.\tag{10.2-16}$$

With the agglomerate size determined, the minimum fluidization velocity can be calculated based on the Ergun equation as follows:

$$U_{mf} = \frac{\Delta p}{H}\frac{d_a^2}{150\mu}\frac{\alpha_{mf}^3}{(1 - \alpha_{mf})^2}.\tag{10.2-17}$$

When nanoparticles form relatively large agglomerates, they can be difficult to fluidize. However, their fluidization can be eased with the aid of external excitations such as vibration (Nam *et al.*, 2004) or introducing a magnetic field (Yu *et al.*, 2005) or acoustic field (Zhu *et al.*, 2004), discussed in Section 10.2.3, or introducing one or more microjets above the distributor to produce a high-velocity secondary flow (Quevedo *et al.*, 2010). The external excitation can lead to the fragmentation of the large agglomerates.

10.3   Circulating Fluidized Beds

In dense-phase fluidized beds, the particle entrainment rate is low but increases with increasing gas velocity. At relatively low gas velocities, the bed surface is distinguishable. As gas velocity increases, the bubble/void phase becomes less distinguishable from the emulsion phase and the distinction of the bed surface gradually disappears. At high gas velocities, the solid entrainment rate is high and continuous feeding of the solid particles into the fluidized bed becomes required in order to maintain a steady solids flow in the bed. Fluidization at this state marks the operation of a circulating fluidized bed where solid particles circulate between a riser (where flow moves against gravity) and a downcomer (where flow moves along gravity), as shown in Figure 10.3-1. The circulating fluidized bed is typically operated under the fast fluidization regime, characterized by a heterogeneous flow structure. When the gas velocity further increases, the fast fluidization regime transits to the dilute transport regime, characterized by a homogeneous flow structure. The focus of this section is on the fast fluidization regime and associated flow transition modeling.

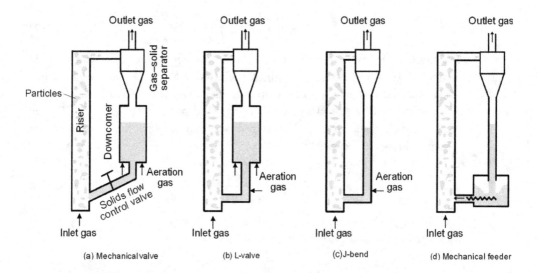

Figure 10.3-1 Circulating fluidization systems with various feeding devices.

10.3.1   Components of a Circulating Fluidized Bed

A circulating fluidized bed (CFB) comprises a riser, a gas–solid separator, a down-comer, and a solid feeding device, as shown in Figure 10.3-1. The figure also indicates various schemes for solid feeding in CFB.

The riser refers to a vertical column in which particles are transported by an upward flow of carrying fluid against gravity. It is the main component of a CFB system. The gas–solid separator is typically in the form of a cyclone, and its efficiency affects the particle size distribution and solid circulation rate in the system. The downcomer provides a holding volume and a static pressure head for particles recycling into the riser. It can be either a large reservoir, as shown in Figure 10.3-1(a) and (b), or simply a standpipe, as shown in Figure 10.3-1(c) and (d). The solid flow control device serves to seal the riser gas flow to the downcomer while controlling the solid circulation rate. Those functions can be performed using either a mechanical valve (such as a rotary screw, or sliding valves) or a nonmechanical valve (such as L-valves or J-bends).

10.3.2   Fast Fluidization Regime

The flow structure of a riser in the fast fluidization regime is represented by a dense flow-developing region at the bottom, with particles accelerating into a dilute region above it. The flow behavior in this regime varies with both the gas velocity and the solid circulation rate. The particle cluster formation characterizes the fast fluidization regime. The cluster is referred to as a lump of solid particles for which their flow properties inside the cluster do not vary significantly. The clusters are formed mainly due to the hydrodynamic effects of the riser flow (Jiang et al., 1993), which are different from agglomerates formed mainly due to the surface attraction effects induced by such forces as van der Waals force (Horio and Clift, 1992). The mapping of the riser flow can be made by the pressure gradient and solid circulation rate relationship, as shown in Figure 10.3-2. The variation of the gas velocity versus the solid circulation rate in relation to the riser flow regimes is shown in Figure 10.3-3 (Bai et al., 1993; Bi and Grace, 1995).

Figure 10.3-2 describes the characteristics of the pressure gradient of riser, $\Delta P/\Delta z$, with respect to the solid circulation rate, J_p. For a low flow velocity of carrying fluid, there exists a value of J_p, corresponding to the saturation carrying capability of solid flows in a riser, beyond which a slight increase in J_p yields a dramatic increase in the pressure gradient. This behavior can be exemplified by curve ABC in the figure. As J_p further increases, the riser transitions to the dense-phase flow as indicated by curve CD with high pressure gradients in the solid transport. The sudden increase in the pressure gradient from B to C signifies the phenomenon of choking in which a sharp increase in the solid concentration takes place in the riser. When the flow velocity of carrying fluid increases beyond a critical value noted as transport velocity, U_{tr}, the choking phenomenon no longer occurs. As the fluid velocity further increases, dilute transport occurs as expressed in curves GH and IJ,

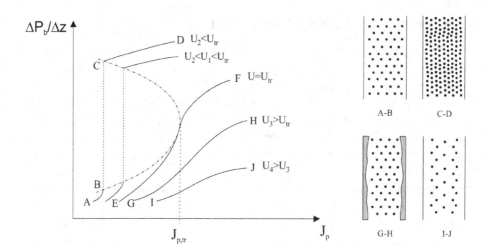

Figure 10.3-2 Mapping of pressure gradient with solid circulation rate.

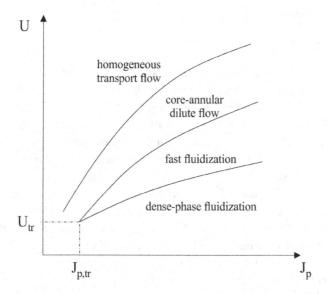

Figure 10.3-3 Mapping of gas velocity with solid circulation rate.

representing a vertically constant core-annular flow structure and a homogeneous flow structure as depicted in the figure.

Figure 10.3-3 shows the map of regimes for different gas velocities and solid circulation rates. The fast fluidization regime is seen to occur when the gas velocity is higher than the transport velocity U_{tr} and the solid circulation rate is higher than $J_{p,tr}$. For a given solid circulation rate J_{p1}, the fast fluidization regime is bounded between the dense-phase flow and the dilute phase flow with a vertically constant core-annular structure. At very high gas velocities, the riser is operated under the homogeneous flow regime.

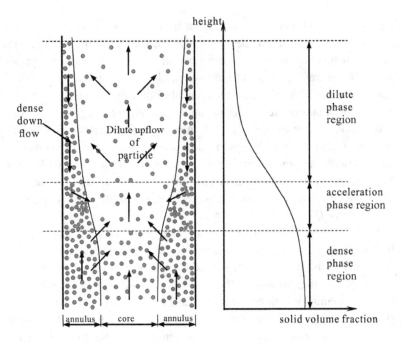

Figure 10.3-4 Heterogeneous flow structure in a riser.

10.3.3 Fast Fluidization Structure and Transition to Choking

As noted, fast fluidization occurs at $U > U_{tr}$ and is characterized by a heterogeneous flow structure with a dense developing region at the bottom, an acceleration region in the middle, and a dilute region at the top, typically with solids concentration variation in both radial and axial direction, as shown in Fig. 10.3-4. The solids concentration in the dense region is often uniform and decreases along the bed height, following a sigmoidal curve, to the dilute region. The inflection of the curve signifies the acceleration region or the transition region in the solids concentration profile. In the transition region, the momentum transfer between the particles and the carrying fluid is significant due to the large velocity difference between the two phases, leading to particle acceleration.

An exchange of solids between the core region and the annulus region or the wall regions, as seen in the figure, occurs along the riser. The radial flow profile in the riser is a result of the interaction between gas–solid momentum transfer, wall effects, and gravity. The wall provides not only the geometric confinement to the radial flow movement but also a surface where the stagnant carrying fluid resides. The wall region, characterized by a dense solids concentration and a low gas velocity, is developed from the riser bottom. Particles migrate into the wall region with their initial upward momentum gained from their interaction with the carrying fluid or by interparticle collision. However, at a certain bed height, in the wall region very close to the wall surface, the velocity of carrying fluid becomes so low that the carrying fluid cannot provide sufficient momentum to sustain the upward movement of the particles. The solids in the wall region then begin to fall down by gravity. Thus, there is a bed height where the averaged solids velocity in the wall region is null. At this bed height, solids from the upper and the lower wall region merge and migrate inwardly toward the center of the riser. Under a high gas velocity and low solids loading, the backflow of solids is low, and the inwardly migrating solids are entrained all upward in the core region. In the upper part of the riser, particles diffuse into the wall region and move downward. A thin layer of the downward flow of particles near the wall along with the upward flow of all phases in the core constitutes a core-annulus two-zone structure in the main part of the riser transport. Such a core-annulus flow structure, where the radial solids concentration gradually decreases toward the centerline of the riser, is common in the upper part of the riser at gas velocities that are far above U_{tr}.

Under another condition (*i.e.*, at low gas velocities with $U < U_{tr}$, high solids loading, and small riser diameter), part of the inwardly migrating solids may reach the central axis of riser. With the axial symmetry of a cylindrical riser, a dense core region is then formed at the center. This solid flow condition is reflected by the core-annulus-wall three-zone structure, as shown in Figure 10.3-5. Specifically, since the axial pressure gradient in all regions at a given bed height is approximately equal, the gas velocity in the relatively dense core tends to be lower than that in the annulus where the solids concentration is relatively low. The slower flowing gas in the core results in a lower acceleration of solids in the core region. In contrast, in the upper part of the riser, more solids depletion takes place in the core than in the annulus, leading to a core-annulus two-zone structure. The core-annulus-wall three-zone structure has been observed experimentally at a high particle flow rate, with a reversal-hump-shaped distribution of solids concentration in the riser (Du *et al.*, 2004). The formation of the peak solids concentration in the core region can trigger the instabilities that lead to the collapse of the stable structure of solids with

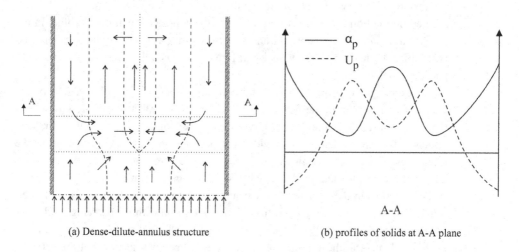

(a) Dense-dilute-annulus structure   (b) profiles of solids at A-A plane

Figure 10.3-5 Heterogeneous flow structures near choking in risers.

a flow transition to choking. Specifically, choking occurs at $U < U_{tr}$ and is a result of the combination of both radial and axial particle transport in the riser, induced by such particle flow phenomena as the back-mixing, and the interparticle collisions and turbulence. The mechanisms of onset of choking are closely related to the effect of the riser wall and its effect is reduced as the riser diameter increases.

10.3.4 Modeling of Flow in Fast Fluidization

The heterogeneous flow structure for solids transport in fast fluidization, shown in Figure 10.3-4, can be modeled by assuming the flow to be steady, isothermal, and axial-symmetric. The transport of solids and gas phases is dominant in the axial direction, with nonuniform radial distributions of phase properties along the riser. The pressure gradient in the radial direction is assumed to be uniform at any cross section of the riser.

The governing equations for the cross-sectional averaged axial distributions of gas–solid phase transport properties can be expressed in terms of differential-integral forms. The mass conservations for the gas and solid particle phase can be written as

$$\frac{d}{dz}\left(\int_0^R 2\pi\alpha\rho\, Urdr\right) = 0 \tag{10.3-1}$$

$$\frac{d}{dz}\left(\int_0^R 2\pi\alpha_p\rho_p U_p\, rdr\right) = 0. \tag{10.3-2}$$

The momentum balance of the gas phase is given by

$$\frac{d}{dz}\left(\int_0^R 2\pi\alpha\rho\, U^2\, rdr\right) = -\int_0^R 2\pi\alpha\left(\frac{dp}{dz}\right)rdr - 2\pi\alpha_w\tau_w R$$
$$-\int_0^R 2\pi\alpha\rho\, grdr - \int_0^R 2\pi f_H\, rdr. \tag{10.3-3}$$

The terms on the right-hand side of Eq. (10.3-3) represent, respectively, the gas acceleration due to pressure gradient, the wall friction, the gravity field force, and the hydrodynamic interactions between gas and solids including drag force, carried mass force, Basset force, and Saffman force. Similarly, the momentum balance of the solid particle phase can be expressed as

$$\frac{d}{dz}\left(\int_0^R 2\pi\alpha_p\rho_p U_p^2\, rdr\right) = -\int_0^R 2\pi\alpha_p\left(\frac{dp}{dz}\right)rdr - 2\pi\alpha_{pw}\tau_{pw}R$$
$$-\int_0^R 2\pi\alpha_p\rho_p\, grdr + \int_0^R 2\pi f_H\, rdr - \int_0^R 2\pi f_C\, rdr. \tag{10.3-4}$$

The terms on the right-hand side of Eq. (10.3-4) represent, respectively, the pressure gradient force, the wall friction, the gravity field force, the hydrodynamic interactions between gas and solids, and the interparticle collision force. It is noted that the interparticle collisions in the acceleration regime are not in an equilibrium or balanced state in the axial direction; rather, they act as a damping force to constrain the solids from free acceleration driven by the hydrodynamic force from the gas phase (Zhu and You, 2007).

The volume fractions of gas and solid phases are constrained by

$$\alpha + \alpha_p = 1. \tag{10.3-5}$$

The gas density obeys the ideal gas law as given by

$$\rho = \frac{p}{R_g T}. \tag{10.3-6}$$

Thus, six independent equations can be established (Eqs. (10.3-1) to (10.3-6)) for six transport properties ($\alpha(r, z)$, $\alpha_p(r, z)$, $\rho(z)$, $p(z)$, $U(r, z)$, and $U_p(r, z)$). The closure requires further modeling on the detailed constitutive relations of radial profiles, especially for the gas velocity, solid concentration, and solid velocity (He *et al.*, 2016).

10.4   Gas–Liquid Bubbling Flows

The behavior of gas–liquid bubbling flows is introduced in this section as a preamble to the macroscopic account of gas–liquid–solid fluidization due to their resemblance in bubbling phenomena. Whenever appropriate for this section, however, the relevant effects of solid particles on the gas–liquid bubbling phenomena are also accounted for. The details of the gas–liquid–solid fluidization is given in Section 10.5. It is noted that the gas–liquid bubbling flow itself is an important multiphase phase flow subject. Figure 10.4-1 illustrates this flow as applied to a configuration of a common reactor configuration in which gas bubbles dispersed in a batched or continuously fed liquid in a vertical column, known as bubble column. The dynamic motions of bubbles agitate the liquid in order to enhance the two-phase contact and interactions.

10.4.1   Bubble Formation and Shape Regime

Bubbles are typically generated when gas is pressured to flow through small orifices in a distributor that is submerged in a liquid medium. When gas directly flows through an orifice when no gas chamber is present, the gas flow rate through the orifice and the supplied flow rate are in equipoise; this is known as the constant flow condition. Conversely, when an orifice is connected to a gas chamber or plenum, bubble formation can present in either constant or variable flow modes. The touchstone for classifying the flow is the capacitance number, Nc, defined as:

$$\text{Nc} = \frac{4Vg\rho_l}{\pi D_o^2 p}, \tag{10.4-1}$$

where V is the chamber volume, D_o is the orifice diameter, and p is the gas pressure. When Nc < 1, the gas flow rate is almost constant; when Nc > 1, the gas flow rate is not constant and the flow rate depends on the difference between the pressure in the chamber and that in the bubble (Yang *et al.*, 2007).

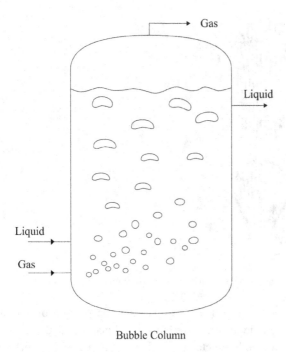

Figure 10.4-1 Schematic illustration of bubble columns.

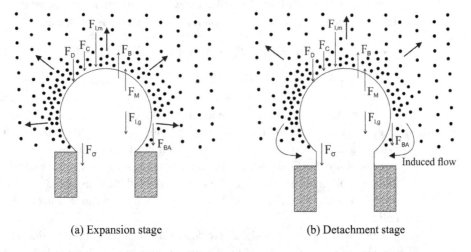

(a) Expansion stage   (b) Detachment stage

Figure 10.4-2 Two-stage model of bubble formation from a submerged orifice.

For a single orifice immersed in a solid–liquid suspension in the constant flow regime, a two-stage mechanistic model that considers the force balance on the bubble may be used to predict the initial bubble size. The basic system and force balance are illustrated in Figure 10.4-2.

In the expansion stage, the bubble enlarges while its base remains attached to the orifice. In the detachment stage, the bubble migrates away from the orifice, although

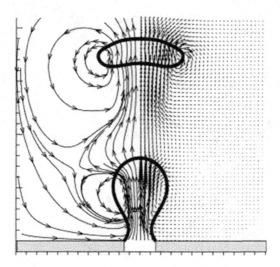

Figure 10.4-3 Streamlines and velocity during bubble formation.

it remains connected with the orifice as a characteristic "neck" forms. The forces acting on the bubble include the effective buoyancy force, F_B, and the gas momentum force, F_M, which are both directed upward. The liquid drag F_D, surface tension force F_σ, bubble inertia $F_{I,g}$, Basset force F_B, bubble–particle collision force F_C, and liquid–particle inertia force $F_{I,m}$ all act downward, functioning to resist detachment. Thus, the total force balance equation for the gas bubble is written as:

$$F_B + F_M = F_D + F_\sigma + F_{BA} + F_{I,g} + F_C + F_{I,m}. \qquad (10.4\text{-}2)$$

The above equation applies for both the expansion and the detachment stages, although the individual forces for each stage may not necessarily be the same. The transition from the expansion stage to the detachment stage occurs when the upward and downward forces in Eq. (10.4-2) reach equipoise, such that the bubble diameter, $d_b(t)$, ceases to increase. Once this condition occurs, the bubble size will remain constant throughout the detachment stage, and its proportions will correspond to the initial bubble sized as formed at the orifice. The bubble is assumed to detach from the orifice when the bubble base has shifted upward a distance equal to the bubble radius.

Note that the above model is predicated upon the assumptions that the bubble remains spherical throughout the formation process, and that the bubble does not interact with other bubbles in the proximate liquid. These conditions are somewhat artificial. For example, the bubble's motion induces circulation currents in the surrounding fluid that contribute to the thinning of the neck and facilitate detachment of the bubble from the orifice. This process is depicted in Figure 10.4-3 (Gerlach *et al.*, 2007).

The cross flow of liquid near the orifice may also develop a shear force that can separate the bubble from the gas stream. As bubbles move away from the orifice, their size may eventually deviate due to breakup and coalescence. The extent of the bubble

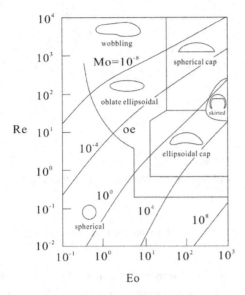

Figure 10.4-4 Shape regime of a single bubble in a liquid (Clift *et al.*, 1978).

size deviations depends on the local velocity field, turbulence energy dissipation, and presence of surfactants. Additionally, phase change, absorption, and pressure variation can also influence deviations in bubble size.

Buoyant ascension of gas bubbles in a liquid medium is a relevant phenomenon for many gas–liquid systems. For this class of flow, interactions between the buoyancy, viscous, and surface tension forces govern the bubble's shape and ascent velocity. Three dimensionless groups that are germane to the gas–liquid systems are the Reynolds number (Re), Morton number (Mo), and Eötvös number (Eo). These quantities are defined, respectively, as:

$$ \text{Re} = \frac{\rho_l U d_b}{\mu_l} \qquad \text{Mo} = \frac{g \mu_l^4}{\rho_l \sigma^3} \qquad \text{Eo} = \frac{g \rho_l d_b^2}{\sigma}. \qquad (10.4\text{-}3) $$

Figure 10.4-4 employs the dimensionless parameters above to present a graphic correlation for the shape and terminal velocity of a single bubble rising freely in an unbounded Newtonian liquid. Note that the figure also applies to droplets as discussed in Section 3.6.1.

In general, small bubbles remain spherical as they rise. Large bubbles, however, are influenced by liquid viscosity: ellipsoidal caps or skirted shapes often manifest in high-viscosity liquids, while spherical cap shapes typically predominate low-viscosity liquids. Figure 10.4-4 can be employed to glean the terminal velocity of a bubble if the Reynolds number (Re) is known. Notably, the Reynolds number (Re) is a function of the Morton number (Mo) and the Eötvös number (Eo). Other parameters, in addition to those appearing in the expressions for Mo and Eo, may also affect

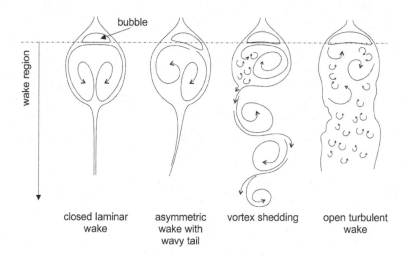

wake region

bubble

| closed laminar wake | asymmetric wake with wavy tail | vortex shedding | open turbulent wake |

Figure 10.4-5 Wake structures behind a spherical cap bubble (Fan and Tsuchiya, 1990).

the bubble shape and rise velocity. For example, contamination on the air bubble surface can restrict the bubble deformation and reduce the ascent velocity in the case of millimeter ellipsoidal air bubbles in water where the surface tension dominates the bubble rise process.

10.4.2   Bubble Wake Dynamics and Interaction

Bubble wake dynamics is key to important transport phenomena occurring in the bubbling systems of a liquid or a liquid–solid suspension (Fan and Tsuchiya, 1990). Figure 10.4-5 provides a schematic representation of general wake structures behind a spherical cap bubble. At low Reynolds numbers, a closed laminar wake manifests with a well-defined boundary that delineates a steady circulation zone with a toroidal internal vortex ring. As the Reynolds number increases, the wake becomes unstable and asymmetric. The streaming tail extending below the toroidal vortex becomes wavy, and vortex shedding becomes identifiable. At high Reynolds numbers, an open turbulent wake forms with pronounced vortical motion in the near-wake region with high turbulence extending far downstream. The effects of a wake on the bubble dynamics can be revealed via synchronization of phenomena between wake vortex shedding and oscillations of the bubble motion. The wake further provides an important mechanism for bubble interactions. For example, the wake of a leading bubble can significantly affect the formation and ascent of trailing bubbles (Chen and Fan, 2004).

In a liquid or a liquid–solid suspension, bubble coalescence typically occurs when a trailing bubble is induced into the wake of a leading bubble and the two subsequently collide. This mechanism, as illustrated in Figure 10.4-6(a), proceeds in three steps: (1) suction of the trailing bubble into the low-pressure region behind the leading bubble with an elongation of the trailing bubble; (2) thinning of the liquid film

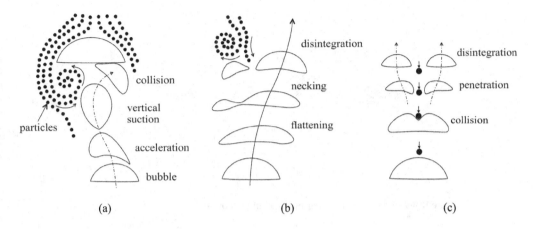

Figure 10.4-6 Mechanisms of coalescence and breakup of bubbles.

between the two bubbles via liquid drainage; and (3) rupture of the film and overall merging of two bubbles into one. An increase in system pressure and the introduction of inhibitors, such as inorganic salt or organic surfactants, can mitigate the effects of bubble coalescence.

Bubble breakup can occur through one of two mechanisms: instability and bubble–particle collision. Breakup due to instability manifests differently depending on pressure. At high pressure, generally over 1.0 MPa, instability breakup is governed by internal circulatory fluid motion within the bubble. Specifically, under high pressures, the bubble breakage is caused mainly by the centrifugal force induced by gas circulation inside the bubble (Luo *et al.*, 1999). At pressures below 1.0 MPa, either Rayleigh–Taylor or Kelvin–Helmholtz instability operate to effectuate breakup (Fan and Yang, 2003). Figure 10.4-6(b) provides a visual representation of bubble breakup resulting from interaction with the wake of a leading bubble. When elongated by shear flow, the bubble experiences flattening, necking, and eventual disintegration. In the process, shear and dynamic pressures overcome the stabilizing surface tension force and cause deformation and rupture of the bubble surface. In the gas–liquid–solid systems, bubble breakup may result from oscillations generated on the bubble surface from bubble–particle collisions, as illustrated in Figure 10.4-6(c). After collision, the particle may either be ejected from the surface or penetrate the bubble. In the latter case, the bubble may or may not break up; relative size between the particle and the deformed bubble determines the outcome. Numerical simulation has revealed that the large pressure oscillation resulting from particle penetration likely contributes to bubble surface instability, leading to eventual rupture (Hong *et al.*, 1999).

10.4.3   Bubble Columns

In a bubble column, gas forms bubbles with liquid as a continuous phase. The liquid in a bubble column can be in a batch or continuous flow condition. Gas bubble motion

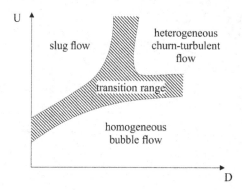

Figure 10.4-7 Schematic flow regime in a bubble column.

agitates the liquid medium that enhances liquid mixing and gas–liquid mass transfer. Some applications of bubble columns contain fine solid particles dispersed in the liquid medium. The bubble column operated with fine solid particles in it is generally referred to as the slurry-bubble column. When the liquid–solid media behave as a pseudo-homogeneous medium, the slurry-bubble columns exhibit similar flow characteristics to the gas–liquid bubble columns.

At low gas velocities, small bubbles are formed in a bubble column that flow upward with mild transverse trajectories and axial oscillations with no bubble coalescence. The bubble size distribution is uniform. This flow condition is noted as the bubbly or homogeneous regime. At high gas velocities, churn-turbulent flow occurs. The flow dynamics is characterized by liquid recirculation and intense bubble coalescence and breakup leading to a wide bubble size distribution. This flow condition is noted as the heterogeneous regime. Bubbles in columns of a small diameter can be in a slug form. In practice, most bubble columns operate in the bubbly or the churn-turbulent regimes with a relative relationship of the superficial gas velocity, U, and column diameter, D, in a flow regime map, as given in Figure 10.4-7 (Shah *et al.*, 1982). The regime map in the figure is presented, based on a given liquid flow rate. Typical liquid flow rates for industrial operation are low. The specific demarcation of regime transitions in the map depends on a number of factors including gas and liquid properties, column configurations, distributor design, and operating conditions such as pressure and temperature (Shaikh and Al-Dahhan, 2007). No universal flow regime map is available to account for a wide range of operating conditions.

A bubble column in the churn-turbulent regime exhibits complex flow patterns that are highly unsteady. On a time-averaged flow pattern, the churn-turbulent regime is characterized by large-scale circulations whereby large bubbles are conveyed upward in the center of the column while small bubbles simultaneously descend with the liquid phase near the wall. These flow patterns are ascertained through measurements using such techniques as particle imaging velocimetry (PIV) (Chen *et al.*, 1994) and electrical capacitance tomography (ECT) (Warsito and Fan, 2005). It reveals that the churn-turbulent regime can be further divided into two flow

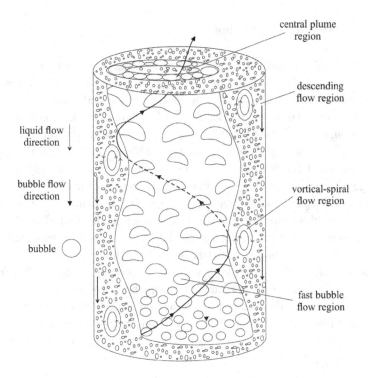

central plume
region

descending
flow region

liquid flow
direction

bubble flow
direction

vortical-spiral
flow region

bubble

fast bubble
flow region

Figure 10.4-8 Vortical-spiral flow structure in gas–liquid bubble column.

sub-regimes: the vortical-spiral sub-regime at intermediate gas velocities and the turbulent sub-regime at high gas velocities. In the vortical-spiral sub-regime, four regions with distinct flow characteristics can be identified. The flow structure of this sub-regime is given in Figure 10.4-8. It is seen in the figure that the descending flow region is located near the wall and is characterized by a downward liquid flow in either a straight or spiral manner. The bubble presence is minimal in this region, which may even be devoid of bubbles at low gas velocities. Following the descending flow region is the vortical-spiral region, which sways laterally, but conveys downward in spiral flow. The fast bubble flow region exists where significant bubble coalescence and breakup occur here. Bubble clusters and large bubbles are conveyed upward in a spiral fashion at high velocity, but the spiral bubble stream can traverse laterally. The central plume region is located at the column center. This region has a relatively uniform bubble size distribution and minimal bubble–bubble interactions. For small-diameter columns, this region may be indistinguishable because of the merging effect of the bubble flow region.

When gas velocity is increased beyond that sufficient to sustain the vortical-spiral sub-regime, bubble coalescence will predominate, and large bubbles can readily form. Bubble wake and drift effects, caused by the chaotic motion of the liquid phase, function to progressively alter the vortical-spiral structure. This occurrence eventually leads to a transition into the turbulence sub-regime.

10.5   Gas–Liquid–Solid Fluidization

Gas–liquid–solid (three-phase) fluidization involves a number of different gas-, liquid-, and solid-phase contact modes and flow regimes (Fan, 1989). A typical three-phase fluidization with concurrent upward flows of gas and liquid phases in contact with a bed of solid particles in suspension is the focus of this section. In this mode of fluidization, the liquid is the continuous phase while the solid particles are fluidized in the bed characterized by an expanded bed with a well-defined bed surface. The gas–liquid–solid fluidized bed can be schematically represented by Figure 10.5-1. In the figure, there are three distinct regions including the distributor region, the bulk fluidized bed region, and the freeboard region. The distributor region is where the initial gas bubbles are formed from the gas distribution manifold. Hydrodynamics in this region are primarily influenced by the design of the gas distributor and the physical properties of the liquid and solid media. The bulk fluidized bed region is the main part of the bed, and its flow behavior varies with operating conditions. The freeboard region contains the particles entrained from the bulk fluidized bed through bubble wakes or bubble ejection in a similar mechanism as described for the gas–solid fluidized bed given in Section 10.2.2.4. The solids concentration decreases with the height above the bed surface. For small or light particulates, the boundary between

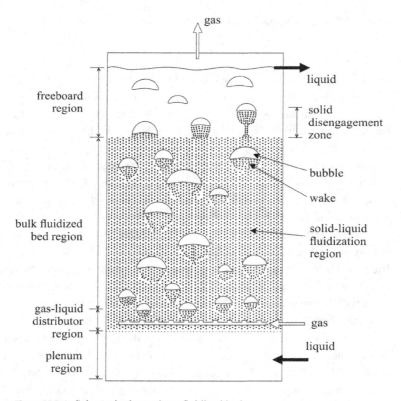

Figure 10.5-1 Schematic three-phase fluidized bed.

the freeboard region and the bulk fluidized bed region is less pronounced than when large or heavy particles are fluidized.

10.5.1 Pressure Drop and Phase Holdup

In a three-phase fluidized bed, the conservation of volume fractions of each phase requires:

$$\alpha_g + \alpha_l + \alpha_s = 1. \tag{10.5-1}$$

For solids with negligibly low entrainment rates, the averaged solids holdup in the bed can be obtained from the measured bed height by:

$$\alpha_s = \frac{4M_s}{\pi \rho_s D^2 H}, \tag{10.5-2}$$

where M_s is the total solid mass; D and H are the column diameter and bed height, respectively. Under steady-state flow conditions, the total pressure drop in the bed can be expressed by:

$$-\frac{dp}{dz} = \left(\rho_g \alpha_g + \rho_l \alpha_l + \rho_s \alpha_s \right) g. \tag{10.5-3}$$

Thus, with the measurement of pressure gradient of the bed, the volume fractions of each phase can be obtained from Eqs. (10.5-1) to (10.5-3).

The behavior of the gas holdup in a three-phase fluidized bed is strongly affected by the flow regime. Depending on whether the particles promote gas bubble coalescence or breakup, the gas holdup can be either higher or lower compared to that of the corresponding bubble column. Liquid holdup increases in direct proportion with liquid velocity and viscosity, and decreases with increasing gas velocity. In the limiting case of no gas flow, the bed porosity reduces to that of a liquid–solid bed where the bed expansion follows the Richardson–Zaki relation. Correlations for gas holdup and liquid holdup in three-phase fluidized beds are available for various operating conditions in the literature (Fan, 1989).

10.5.2 Incipient Fluidization and Flow Regimes

The minimum fluidization velocity, U_{lmf}, of a three-phase fluidized bed refers to the liquid flow velocity at a given gas velocity in which the bed transitions from a fixed state to a fluidized condition. The minimum fluidization velocity is determined from a slope change in the bed pressure drop variation with the fluid velocity in a similar manner as that for two-phase fluidized beds (*i.e.*, gas–solid or liquid–solid), as given in Section 10.2.2.1. Figure 10.5-2 shows the dynamic pressure drop variation with the liquid velocity at a constant gas velocity for a three-phase fluidized bed operation. In the figure, the dynamic pressure drop is used for the fluidized bed section, and its variation with the liquid velocity is not constant but the slope is significantly different to allow the interception of the two lines to be identified. In contrast, for the dynamic

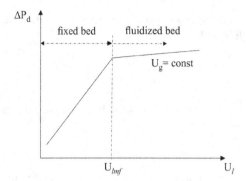

Figure 10.5-2 Dynamic pressure drop versus liquid velocity.

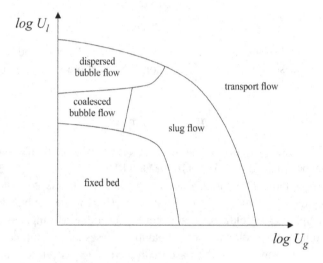

Figure 10.5-3 Flow regime map of gas–liquid–solid fluidized bed.

pressure drop in the two-phase fluidized beds, the dynamic pressure drop is constant in its variation with the fluid velocity.

The minimum liquid velocity required for fluidization also depends on, in addition to the gas velocity, the particle properties. An increase in the gas velocity results in a decrease in U_{lmf}. This effect is significant for large particles and less apparent for small particles. For small or light particles, the transition from a fixed to a fluidized bed is not abrupt. For large or heavy particles, the transition from a fixed to a fluidized bed is clearly marked by an expansion of a distinct bed surface at an increase in either liquid velocity or gas velocity.

The flow regime of a gas–liquid–solid system is determined by both the superficial liquid, U_l, and gas velocities, U_g. In Figure 10.5-3, the fluidized bed regime is bounded by the minimum fluidization velocity and the transport velocity of the liquid phase (Zhang *et al.*, 1997).

The fluidized bed regime encompasses three separate bubble flow patterns including the homogeneous flow pattern with small dispersed bubbles, the heterogeneous flow pattern with large coalesced bubbles, and the slugging flow pattern with slug bubbles. The homogeneous flow pattern predominates at high liquid velocities and low or intermediate gas velocities. The heterogeneous flow pattern persists at low liquid velocities and high gas velocities. The slug flow pattern is limited to small-diameter columns with high gas velocities. The system operating pressure and presence of surfactants also influence the flow patterns. High pressures or surfactants can reduce the bubble size and hence lead to homogeneous flows in the three-phase fluidized bed (Fan and Yang, 2003).

10.5.3 Bed Contraction and Moving Packed Bed

A unique bed expansion characteristic exists when gas is introduced into a liquid–solid fluidized bed composed of small particles. Instead of expanding, the bed actually contracts as gas initially flows into the bed (Epstein, 1981). The bed contraction continues with an increase in the gas velocity until a critical gas velocity is reached when the bed starts to expand. Bed contraction is attributed to the bubble wake phenomenon in which some liquid flows through the bubble wake at a high velocity, thereby bypassing the liquid–solid fluidized region. Thus, the effective liquid flow that is to fluidize the solid particles is reduced. Large bubble systems are prone to bed contraction more than the small bubbles.

The moving packed bed phenomenon occurs during a start-up of a three-phase fluidized bed where the bed is initially at the packed bed condition. When small gas bubbles are first introduced at a start-up, which is followed by the introduction of the liquid flow, a moving packed bed of particles can occur. This is because the small bubbles that are initially introduced can attach to the solid particles and form a bubble blanket so that the subsequent liquid cannot flow through. The introduced liquid then pushes the packed bed forward in a moving bed pattern and elevates the entire bed (Fan, 1999). This phenomenon occurs mainly under conditions of high pressures and/or in the presence of surfactants, yielding the presence of small bubbles in the liquid medium. The moving packed bed is detrimental to the reactor operation. It can be remedied, during the reactor start-up, by first introducing the liquid flow for expanding the bed, and then followed by the introduction of the gas flow.

10.6 Case Studies

The following case studies discuss additional aspects of the subject matter of interest to fundamentals and applications of fluidization. They include a method to estimate the pressure distributions and pressure balance in a CFB operation, a parametric model for energy partition characteristics in a gas–solid riser flow, and an Eulerian–Eulerian modeling of bubbling fluidization in a gas–solid dense fluidized bed.

10.6.1 Pressure Balance in CFB

In a circulating fluidized bed reactor system given in Figure 10.3-1, solids are transported from the fluidized bed through the riser to the cyclone and the standpipe, and recycled back to the fluidized bed through various feeding devices. Hence, the solids feeding rate in a stabilized operation depends largely on the pressure balance of the entire CFB system between the upward transport and downward transport.

Problem Statement: Consider a CFB with the L-valve, as shown in Figure 10.6-1(a). The fluidized bed is operated in the dense-phase bubbling regime with Group B particles, and the fluidizing gas at inlet pressure is controlled at P_4. The cyclone outlet pressure is fixed at P_1. The height of the bubbling fluidized bed is maintained at h_B. The standpipe is operated under the moving bed condition, with an aeration from the L-valve. Develop a model to estimate the relationship between the solids circulation rate and the L-valve aeration rate. Also plot the pressure distribution along the CFB system.

Analysis: The pressure balance in a closed loop of a circulating fluidized bed system can be expressed as

$$\Delta P_{cylone} + \Delta P_{standpipe} = \Delta P_{L-valve} + \Delta P_{fluidized\ bed} + \Delta P_{riser}, \qquad (10.6\text{-}1)$$

where

$$\begin{cases} \Delta P_{cylone} = P_2 - P_1 \\ \Delta P_{standpipe} = P_3 - P_2 \\ \Delta P_{L-valve} = P_3 - P_4 \\ \Delta P_{fluidized\ bed} = P_4 - P_5 \\ \Delta P_{riser} = P_5 - P_1. \end{cases} \qquad (10.6\text{-}2)$$

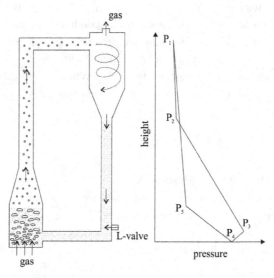

(a) CFB with L-valve   (b) Pressure distribution

Figure 10.6-1 Schematic axial pressure distribution in a CFB.

The pressure drop through the riser and freeboard of the fluidized bed can be estimated, by ignoring hydrodynamic contributions and wall friction, as

$$\Delta P_{riser} = (\alpha_{sr}\rho_s + \alpha_r\rho)\, gH_r, \tag{10.6-3}$$

where H_r is the height of the riser, and the solids volume fraction in the riser, α_{sr}, is very dilute, commonly less than 1%. Similarly, the pressure drop in the fluidized bed can be evaluated by

$$\Delta P_{fluidized\ bed} = (\alpha_{sB}\,\rho_s + \alpha_B\rho)\, gH_B + (\alpha_{sf}\,\rho_s + \alpha_f\rho)\, gH_f, \tag{10.6-4}$$

where H_B and H_f are the heights of dense-phase bubbling bed and freeboard, respectively. It is known that the solids volume fraction in the bubbling bed, α_{sB}, is much higher than that of freeboard, and the solids volume fraction in the freeboard is close to that in the riser. Thus, the pressure gradient in the fluidized bed is typically much larger than that in the riser.

The pressure drop through the standpipe can be estimated using the Ergun equation, Eq. (2.4-5), as given by

$$\Delta P_{standpipe} = \left(\frac{150\alpha_{ss}^2}{(1-\alpha_{ss})^3}\frac{\mu\Delta U}{d_s^2} + \frac{1.75\alpha_s}{(1-\alpha_{ss})^3}\frac{\rho\Delta U^2}{d_s} \right) H_s, \tag{10.6-5}$$

where ΔU is the velocity difference between the gas and solids in the standpipe, and the gas density is a function of local pressure and temperature. Assume that the drag force on the gas induced by the downward moving particles in the standpipe is counterbalanced by the positive downward gradient of pressures so that the velocity of gas is null. The gas–solid relative velocity is identical to the solids velocity. Thus, Eq. (10.6-5) becomes

$$\Delta p_{standpipe} = \left(\frac{150\alpha_{ss}}{(1-\alpha_{ss})^3}\frac{\mu J_{ss}}{\rho_s d_s^2} + \frac{1.75}{\alpha_s(1-\alpha_{ss})^3}\frac{\rho J_{ss}^2}{\rho_s^2 d_s} \right) H_s. \tag{10.6-5a}$$

The pressure drop in the cyclone is normally small compared to that in other sections of the system. The pressure drop through the L-valve is closely related to the design of the L-valve. The pressure at the aeration gas injection point of the L-valve, P_3, is the highest in this circulating fluidized bed system. The pressure profile in the system can be given as shown in Figure 10.6-1(b).

Similar to the flow in the standpipe, as represented by Eq. (10.6-5a), pressure drops in other sections of the CFB system such as bubbling bed, freeboard, riser, and L-valve can all be related to the solid mass fluxes and gas velocities in each section. Thus, substituting these relationships into the system pressure balance equation (i.e., Eq. (10.6-1)), the solid circulation rate in the system can be calculated.

10.6.2 Energy Partitions in Riser Transport

The phase transport properties such as solid holdup and solid velocity are nonuniformly distributed along a riser, typically with a dense-phase transport in the lower part of the riser, a lean-phase transport in the upper part, and an acceleration transport between the two, as shown in Figure 10.3-4. The energy expended for the riser transport is manifested by the pressure drop and the flow rate of the gas and is used for the potential energy for solids lift-up, the kinetic energy for solids transport (including solids acceleration), and energy dissipations for the acceleration and dense-phase transport due to interparticle collisions, in addition to the energy dissipation for wall friction and potential energy gain of the gas phase.

Problem Statement: Assume that the transport energy of solids is represented by the kinetic energy in translational mode with negligible contributions from other modes such as particle rotation and vibration. Provide a modeling analysis for the energy partitions of solids acceleration, solids hold-up, and energy dissipation in the riser transport.

Analysis: The following is a modified model from the original work of Zhu and You (2007). For simplicity and without loss of generality, the cross-section area of riser is considered to be constant, and the gas–solids flow is steady, isothermal, and without any phase changes.

The mass continuity equations of cross-sectional averaged phase parameters of gas and solids can be simplified from Eq. (6.4-21) and the integrated form is expressed, respectively, by

$$J = (1 - \alpha_p) \rho U = \text{constant} \tag{10.6-6}$$

$$J_p = \alpha_p \rho_p U_p = \text{constant}. \tag{10.6-7}$$

The cross-sectional averaged momentum equations of gas phase can be simplified from Eq. (6.4-22) as

$$J \frac{dU}{dz} = -(1 - \alpha_p) \frac{dp}{dz} - \frac{l_w \tau_w}{A} - F_{gp} - (1 - \alpha_p) \rho g, \tag{10.6-8}$$

where F_{gp} is the momentum transfer between gas and solids, which includes drag forces, carried mass force, Basset force, and velocity gradient forces, with the effects of neighboring particle compacting and turbulence. Similarly, the momentum equation of solids is given by

$$J_p \frac{dU_p}{dz} = -\alpha_p \frac{dp}{dz} - \frac{l_{wp} \tau_{wp}}{A} + F_{gp} - \alpha_p \rho_p g - F_C, \tag{10.6-9}$$

where F_C is the cross-sectional averaged interparticle collision force, which is nonzero due to the particle acceleration along the riser. The existence of this axially nonbalanced force by solids collision can be deduced from the following analysis of energy conservations in a riser flow.

Combining Eqs. (10.6-8) and (10.6-9) yields the momentum equation of gas–solids flow in the riser as

$$-\frac{dp}{dz} = J \frac{dU}{dz} + J_p \frac{dU_p}{dz} + ((1 - \alpha_p) \rho + \alpha_p \rho_p) g + \frac{l_w \tau_w + l_{wp} \tau_{wp}}{A} + F_C. \tag{10.6-10}$$

Equation (10.6-10) indicates that the pressure gradient is balanced by the acceleration of gas and solid flow, the phase holdup, the wall fractions, and the net interparticle collision force that is nonzero due to the particle acceleration along the riser. However, a direct evaluation or solving this equation is difficult due to the unavailability of the constitutive model of F_C as well as incomplete measured information on other terms along the riser.

Alternatively, the conservation of total kinetic energy can be examined by combining Eq. (10.6-8) multiplied by U and Eq. (10.6-9) multiplied by U_p, which is then expressed by

$$-\frac{dp}{dz}U = JU\frac{dU}{dz} + J_pU_p\frac{dU_p}{dz} + \left(J + J_p\right)g + \Gamma, \tag{10.6-11}$$

where Γ is the dissipative energy, defined by

$$\Gamma = \alpha_p\left(-\frac{dp}{dz}\right)(U - U_p) + F_{gp}\left(U - U_p\right) + \frac{l_w\tau_w U + l_{wp}\tau_{wp}U_p}{A} + F_C U_p. \tag{10.6-11a}$$

Equation (10.6-11a) shows that Γ is generated via four mechanisms represented by the pressure gradient, gas–solid momentum transfer, wall friction, and net collision force. Based on Eq. (10.6-6) and Eq. (10.6-7), the acceleration of gas can be related to that of solids, and thus Eq. (10.6-11) can also be expressed as

$$\left(-\frac{dp}{dz}\right)U = \left(1 - \frac{\rho}{\rho_p}\left(\frac{U}{U_p}\right)^3\right)J_pU_p\frac{dU_p}{dz} + \left(J + J_p\right)g + \Gamma. \tag{10.6-11b}$$

In order to demonstrate the energy partitions for phase acceleration, lift-up, and energy dissipation, Eq. (10.6-11b) is further transformed into the following form:

$$\beta + \chi + \delta = 1 \tag{10.6-12}$$

with

$$\beta = \frac{\left(J + J_p\right)g}{\left(-\frac{dp}{dz}\right)U}; \quad \chi = \left(1 - \frac{\rho}{\rho_p}\left(\frac{U}{U_p}\right)^3\right)\frac{J_pU_p\frac{dU_p}{dz}}{\left(-\frac{dp}{dz}\right)U}; \quad \delta = \frac{\Gamma}{\left(-\frac{dp}{dz}\right)U}. \tag{10.6-12a}$$

Here β, χ, and δ represent the energy partitions for the phase lift-up, acceleration, and dissipation, respectively. It can be shown (Zhu and You, 2007) that, in the dilute phase transport, $\beta = 1$ with $\chi \approx 0$ and $\delta \approx 0$. In the dense phase transport, however, $\chi \approx 0$ with $\beta + \delta = 1$. A general representation can be schematically shown in Figure 10.6-2.

10.6.3   Kinetic Theory Model for Bubbling Fluidization

Hydrodynamics of fluidization in systems of complex geometries and with internals or complicated phase transport phenomena such as clustering is studied via numerical simulations, in which Eulerian–Eulerian two-fluid modeling is commonly adapted. The dense-phase transport of solids, dominated by interparticle collisions, is typically described via kinetic theory of granular flow that is analogous to the molecular transport of gases, as introduced in Section 2.5.

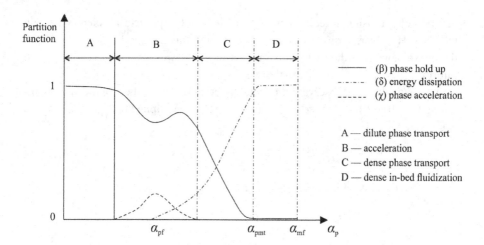

Figure 10.6-2 Energy partitions versus solid volume fractions in riser flow.

Problem Statement: Establish a kinetic theory model to simulate the gas–solid bubbling fluidization in a dense-phase fluidized bed. The particles can be regarded as rough spheres, whose translational and rotational velocity fluctuations due to the interparticle collisions should both be accounted for. The turbulence in gas-phase flow should also be included.

Analysis: The following is a slightly modified model based on the original work of Ding and Gidaspow (1990) and Wang *et al.* (2012). It is assumed that the particles are monodispersed without agglomeration or breakup, and there is no interfacial mass transfer. The turbulent flow of gas phase is assumed to be incompressible and isotropic, and thus can be approximately described using the $k-\varepsilon$ model. The interphase momentum transfer in gas–solid fluidization is assumed to be dominated by the drag force.

The continuity equations of gas and solids are given, based on Eq. (6.4-21), as:

$$\frac{\partial}{\partial t}(\alpha\rho) + \nabla \cdot (\alpha\rho \, \mathbf{U}) = 0 \tag{10.6-13}$$

$$\frac{\partial}{\partial t}(\alpha_p\rho_p) + \nabla \cdot (\alpha_p\rho_p \, \mathbf{U_p}) = 0. \tag{10.6-14}$$

The volume fraction conservation of Eq. (6.3-7a) gives

$$\alpha + \alpha_p = 1. \tag{10.6-15}$$

The momentum equation of gas is given, based on Eq. (6.4-22), by

$$\frac{\partial}{\partial t}(\alpha\rho \, \mathbf{U}) + \nabla \cdot (\alpha\rho \, \mathbf{UU}) = -\alpha\nabla p + \nabla \cdot [\alpha(\mu + \mu_e)2\,\mathbf{S}] + \alpha\rho \, \mathbf{g} - \beta_{gp}(\mathbf{U} - \mathbf{U}_p), \tag{10.6-16}$$

where the stain rate, $\mathbf{S}$, is defined by Eq. (2.2-3), and the turbulent viscosity, μ_e, is approximately determined from k and ε by Eq. (2.3-18), and k and ε equations, with gas–solids modulations, are expressed by

$$\frac{\partial}{\partial t}(\alpha\rho k) + \nabla \cdot (\alpha\rho \mathbf{U}k) = \nabla \cdot \left[\alpha\left(\mu + \frac{\mu_e}{\sigma_k}\right)\nabla k\right] + 2\mu_e(\alpha\mathbf{S}:\mathbf{S}) - \alpha\rho\varepsilon + S_k$$

(10.6-16a)

$$\frac{\partial}{\partial t}(\alpha\rho\varepsilon) + \nabla \cdot (\alpha\rho \mathbf{U}\varepsilon) = \nabla \cdot \left[\alpha\left(\mu + \frac{\mu_e}{\sigma_k}\right)\nabla\varepsilon\right] + 2\mu_e C_1 \frac{\varepsilon}{k}(\alpha\mathbf{S}:\mathbf{S}) - C_{2}\alpha\rho\frac{\varepsilon^2}{k} + S_\varepsilon,$$

(10.6-16b)

where S_k and S_ε are the gas–solid turbulence modulation terms, given by Eq. (6.7-15) and Eq. (6.7-16), respectively. The interphase momentum transfer coefficient for gas–solid fluidization can be formulated by:

$$\beta_{gp} = \begin{cases} 150\dfrac{\alpha_p^2\mu}{\alpha d_p^2} + 1.75\dfrac{\alpha_p\rho\,|\mathbf{U}-\mathbf{U}_p|}{d_p} & \alpha \leq 0.8 \\[3mm] \dfrac{3}{4}\dfrac{C_D}{d_p}\alpha^{-1.65}\alpha_p\rho\,|\mathbf{U}-\mathbf{U}_p| & \alpha > 0.8. \end{cases}$$

(10.6-16c)

The momentum equation of particle phase is given by:

$$\frac{\partial}{\partial t}(\alpha_p\rho_p\mathbf{U}_p) + \nabla \cdot (\alpha_p\rho_p\mathbf{U}_p\mathbf{U}_p) = -\alpha_p\nabla p + \nabla\cdot\mathbf{T}_p + \alpha_p\rho_p\mathbf{g} + \beta_{gp}(\mathbf{U}-\mathbf{U}_p),$$ (10.6-17)

where the total stress tensor of solids, $\mathbf{T}_p$, is due to interparticle collisions and defined by Eq. (2.5-18) as

$$\mathbf{T}_P = (p_p - \lambda_p\nabla\cdot\mathbf{U}_p)\mathbf{I} + 2\mu_p\left[\frac{1}{2}\left(\nabla\mathbf{U}_p + (\nabla\mathbf{U}_p)^T\right) - \frac{1}{3}(\nabla\cdot\mathbf{U}_p)\mathbf{I}\right],$$ (10.6-17a)

in which the particle pressure, p_p, particle bulk viscosity, λ_p, and particle shear viscosity, μ_p, can be expressed, respectively, by Eqs. (2.5-18a), (2.5-18b), and ((2.5-18d)), which are all related to the fluctuating kinetic energy of particles in terms of granular temperature.

The fluctuating kinetic energy equation can be expressed, from Eq. (2.5-15), by

$$\frac{3}{2}\left[\frac{\partial}{\partial t}(\alpha_p\rho_p T_C) + \nabla \cdot (\alpha_p\rho_p T_C\mathbf{U}_p)\right] = -\mathbf{T}_p:\nabla\mathbf{U}_p - \nabla\cdot\mathbf{q}_p + \gamma - 3\beta_{gp}T_C,$$

(10.6-18)

where the total flux of fluctuation energy, $\mathbf{q}_p$, and the rate of energy dissipation, γ, are given by Eq. (2.5-19) and Eq. (2.5-20), respectively.

Thus, ten independent equations can be obtained, Eqs. (10.6-13)–(10.6-18) with Eqs. (2.3-18), (10.6-16a), (10.6-16b), and (10.6-17a), for ten unknowns (α, α_p, $\mathbf{U}$, $\mathbf{U}_p$, p, μ_e, k, ε, $\mathbf{T}_p$, and T_C). The closure of the modeling is now established. It is noted that, for fluidization of rough solids, the contribution of particle rotation can be significant. In that case, the fluctuating kinetic energy equation should be directly expressed in terms of fluctuating kinetic energy, with corresponding modified transport coefficients and constitutive relations (Wang et al., 2012).

Comments: (1) The gas density can be correlated to the local gas-phase pressure, which may vary significantly in a dense-phase fluidized bed or dense-phase riser transport. (2) The current turbulence model is developed for dilute phase transport. A reliable turbulence model in dense-phase fluidization, especially with bubbling, is yet to be developed. (3) More detailed formulations with the inclusion of particle rotation and case simulation examples are available in the literature (Wang *et al.*, 2012).

10.7   Summary

Fluidization represents an important particulate and multiphase operation, featuring dynamic interactions between a continuum fluid (*i.e.*, gas or liquid) and a discrete phase (*e.g.*, solids or solids and bubbles). It is typically realized in a vertical column or pipe. Various fluidization regimes occur, depending on the property of the fluidizing particles, flow rate, and external field force applied. This chapter describes gas–solid fluidization represented by dense-phase fluidized beds and circulating fluidized beds. Fluidization under the gas–liquid–solid flow conditions are also illustrated, which are discussed with the inclusion of its limiting condition of two-phase flows (*i.e.*, gas–liquid bubbling flows).

The following are important fundamental topics to know pertaining to multiphase fluidization:

- classification of fluidization regimes and characteristics of each regime;
- phase-interaction mechanisms (*e.g.*, fluid–particle, particle–particle, particle–wall, and/or bubble–wake–emulsion interactions) during the dense- and dilute-phase fluidization and behavior of nanoparticle fluidization;
- structure and key components of fluidized bed systems;
- multiscaled transport phenomena, such as clustering, agglomeration, breakup, and coalescence of dispersed particles or bubbles.

The Eulerian–Eulerian modeling approach is extensively used for the numerical modeling of fluidization systems. Such continuum modeling approaches are often coupled with the DEM models or kinetic theory models for collision-induced transport in the dispersed phase. Chapter 7 provides details of numerical modeling methods that can directly be applied to account for the fluidization systems.

Nomenclature

A	Area
	Amplitude
Ar	Archimedes number
d	Diameter of particle or bubble
D	Diameter of pipe or column
Eo	Eötvös number
F_H	Cohesive force
FI	Fluidization index

F	Force
G	Vibration number
g	Gravitational acceleration
H	Height
J	Mass flux
L	Length
M	Mass
Mo	Morton number
m	Mass of particle
Nc	Capacitance number
n	Richardson–Zaki exponent
p	Pressure
Q	Volumetric flow rate
R	Radius
	Gas constant
Re	Reynolds number
r	Radial coordinate
T	Temperature
t	Time
U	Velocity
	Superficial gas velocity
V	Volume
z	Axial coordinate

Greek Symbols

α	Volume fraction
Δp	Pressure drop
μ	Dynamic viscosity
ρ	Density
σ	Surface tension
φ	Sphericity
τ	Shear stress
ω	Angular velocity
	Angular frequency

Subscript

0	Initial
a	Agglomerate
b	Bubble
C	Collision
c	Critical

D	Drag
	Distributor
G	Granular
g	Gas
H	Hydrodynamic
i	Inner
l	Liquid
mb	Minimum bubbling
mf	Minimum fluidization
o	Orifice
p	Particle
pf	Partially fluidized
pt	Particle terminal
s	Solid
T	Total
tr	Transport
w	Wall

Problems

P10.1 For the case study of 10.6.1, the pressure distribution along the circulating fluidized bed system is represented schematically by Figure 10.6-1, where P_3 is the inlet pressure of aeration gas at the L-valve and P_5 is the pressure at the height of dense-phase bubbling fluidized bed. How will the pressure distribution be altered if the height of the bubbling fluidized bed is increased from the original height h_{B1} to a new height h_{B2} while the solids circulation rate is maintained the same?

P10.2 The flow distributor of a rotating fluidized bed of granular materials is typically a slotted, perforated, or sintered metal cylindrical shell, which can be simplified as a rotating cylinder containing many curved channels for flow passages, as depicted in Figure P10.2. The pressure drop of the distributor is due to a combination of the effects of flow passing through a rotating channel and tangential shear losses. Assume that there is an analogy between flow through rotating channels and flow

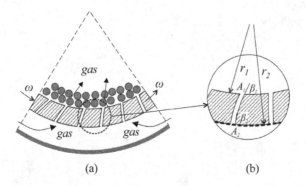

(a)   (b)

Figure P10.2 Model of a rotating fluidized granular bed.

through an array of rotating blades in a centrifugal turbine such as a fan or compressor. Based on the theory of turbine machinery, show that the pressure drop due to flow through rotating channels, Δp_R, can be expressed by

$$\Delta p_R = \rho(r_2^2 - r_1^2)\omega^2 + \omega Q \rho \left(\frac{r_2}{A_2} \cot \beta_2 - \frac{r_1}{A_1} \cot \beta_1 \right), \quad \text{(P10.2-1)}$$

and pressure drop due to viscous dissipation by tangential slip, Δp_S, can be expressed by

$$\Delta p_S = \frac{1}{2}\rho \left\{ \left(\frac{Q}{A_2}(\cot \beta_2 - \cot \beta_1) + \omega r_2 \right)^2 + \left(\frac{Q}{A_1} \cot \beta_1 \right)^2 \right\}, \quad \text{(P10.2-2)}$$

where β is the equivalent blade angle, which is a geometric property of channels.

P10.3 Show that the pressure drop over a rotating fluidized bed of granular materials can be expressed by Eqs. (10.2-10a) – (10.2-10c), respectively, for cases of (a) no fluidization (*i.e.*, rotating packed bed), (b) partial fluidization, and (c) full fluidization.

P10.4 The rise velocity of a single isolated spherical cap bubble in an infinite liquid medium can be expressed as

$$U_{b\infty} = \frac{2}{3}\sqrt{gr_c}, \quad \text{(P10.4)}$$

where r_c is the radius of the frontal curvature of the spherical cap bubble. Derive this expression considering the following assumptions: (1) pressure in the vicinity of the bubble nose is constant, and (2) a potential flow field around a sphere is valid in the vicinity of the bubble nose.

P10.5 As the predominant bubble interaction varies from coalescence to breakup, the prevailing flow regime undergoes a transition from bubbling to turbulent fluidization. Assume that the relationship between the bed voidage, α, and the operating variables is known to be

$$\alpha = f(U, \rho, \mu, d_b, \rho_b). \quad \text{(P10.5)}$$

A criterion for the regime transition can be established according to the variation of the bubble number per unit bed volume, n_b, with the gas velocity. Using the Pi-theorem, provide a general relationship of this criterion in dimensionless groups.

P10.6 For the modeling of a fluidized-bed mixer of two different powders, describe the major advantages and disadvantages between fluid-coupled kinetic theory model and the fluid-coupled DEM model.

P10.7 Describe the flow at different length scales involved in a bubble column and discuss typical numerical models that can be used for each scale.

P10.8 Consider a gas–liquid–solid fluidized bed, in which motions of the liquid and the solid particles are induced mainly by the bubble flow. Describe all the hydrodynamic factors that dictate the motions of the liquid and the solid particles.

P10.9 For a dense bubbling gas–solid fluidized bed with chemical reactions (*e.g.*, solid waste incineration, coal gasification), write a complete set of volume–time-averaged governing equations. Consider that the gas phase is a multicomponent mixture; and the flow is turbulent. List all independent variables and indicate the number of independent equations that are required to ensure a model closure.

P10.10 Survey the literature on modeling of energy loss in gas–solid dense-phase fluidized beds and circulating fluidized beds. Discuss the partition of energy loss between the gas–solids friction and the interparticle collisions.

References

Abrahamsen, A. R., and Geldart, D. (1980). Behavior of gas fluidized beds of fine powders. Part II. Voidage of the dense phase in bubbling beds. *Powder Technol.*, **26**, 47–55.

Bai, D., Jin, Y., and Yu, Z. (1993). Flow regimes in circulating fluidized beds. *Chem. Eng. Tech.*, **16**, 307–313.

Bi, H. T., and Grace, J. R. (1995). Flow regime diagrams for gas-solid fluidization and upward transport. *Int. J. Multiphase Flows*, **21**, 1229–1236.

Cai, P. Chen, S. P., Jin, Y., Yu, Z. Q., and Wang, Z. W. (1989). Effect of operating temperature and pressure on the transition from bubbling to turbulent fluidization. *AIChE Symp. Ser.*, **85** (270), 37–43.

Chen, C., and Fan, L.-S. (2004). Discrete simulation of gas-liquid bubble columns and gas-liquid-solid fluidized beds. *AIChE J.*, **50**(2), 288–301.

Chen, R. C., Reese, J., and Fan, L.-S. (1994). Flow structure in a three-dimensional bubble column and three-phase fluidized bed. *AIChE J.*, **40**(7), 1093–1104.

Chirone, R., Massimilla, L., and Russo, S. (1992). Bubbling fluidization of a cohesive powder in an acoustic field. *Fluidization VII*. Ed. Potter and Nicklin. New York: Engineering Foundation.

Clift, R., Grace, J. R., and Weber, M. E. (1978). *Bubbles, Drops and Particles*. New York: Dover.

Clift, R., and Grace, J. R. (1985). Continuous bubbling and slugging. *Fluidization, 2nd ed.*, Ed. Davidson, Clift, and Harrison. London: Academic Press.

Colakyan, M., and Levenspiel, O. (1984). Elutriation from fluidized beds. *Powder Technol.*, **38**, 223–232.

Ding, J., and Gidaspow, D. (1990). A bubbling fluidization model using kinetic theory of granular flow. *AIChE J.*, **36**, 523–538.

Du, B, Warsito, W., and Fan, L.-S. (2004). ECT studies of the choking phenomenon in a gas-solid circulating fluidized bed. *AIChE J.*, **50**, 1386–1406.

Epstein, N. (1981). Three phase fluidization: some knowledge gaps. *Can. J. Chem. Eng.*, **59**, 649–657.

Epstein, N., and Grace, J. R. (1997). Spouting of particulate solids. *Handbook of Powder Science and Technology, 2nd ed.*, Ed. Fayed and Otten. New York: Chapman & Hall.

Fan, L.-S., (1989). *Gas-Liquid-Solid Fluidization Engineering*. Boston, MA: Butterworths.

Fan, L.-S. and Tsuchiya, K. (1990). *Bubble Wake Dynamics in Liquids and Liquid–Solid Suspensions*. Boston, MA: Butterworths.

Fan, L.-S. (1999). Moving packed bed phenomenon in three-phase fluidization. *Powder Technol.*, **103**, 300–301.

Fan, L.-S., and Yang, G. (2003). Gas-liquid-solid three-phase fluidization. *Handbook of Fluidization and Fluid-Particle Systems*, Ed. W.-C. Yang, Boca Raton, FL: CRC Press.

Fane, A. G., and Mitchell, R. A. (1984). Minimum spouting velocity of scaled-up beds. *Can. J. Chem. Eng.*, **62**, 437–439.

Geldart, D. (1973). Types of gas fluidization. *Powder Technol.*, **7**, 285–292.

Geldart, D., Cullinan, J., Geotghiades, S., Gilvray, D., and Pope, D. J. (1979). The effect of fines on entrainment from gas fluidized beds. *Trans. Inst. Chem. Eng.*, **57**, 269–275.

Gerlach, D., Alleborn, N., Buwa, V., and Durst, F. (2007). Numerical simulation of periodic bubble formation at a submerged orifice with constant gas flow rate. *Chem. Eng. Sci.*, **62**, 2109–2125.

Grace, J. R. (1986). Contacting modes and behavior classification of gas–solid and other two-phase suspensions. *Can. J. Chem. Eng.*, **64**, 353–363.

Grace, J. R., Bi, X. T., and Ellis, N. (2020). Ed., *Essentials of Fluidization Technology*. Weinheim, Germany: Wiley-VCH.

He, P., Wang, D., Patel, R., and Zhu, C. (2016). Modeling of axial-symmetric flow structure in gas-solids risers. *J. Fluid Eng.*, **138**(4), 041302.

Hong, T., Fan, L.-S., and Lee, D. (1999). Force variations on particle induced by bubble-particle collision. *Int. J. Multiphase Flow*, **25**(3), 477–500.

Horio, M., and Clift, R. (1992). A note on terminology: clusters and agglomerates. *Powder Technol.*, **70**, 196.

Jiang, P. J., Cai, P., and Fan, L.-S. (1993). Transient flow behavior in fast fluidization. In *Circulating Fluidized Bed Technology IV*. Ed. A. A. Avidan. New York: AIChE Pub.

Kroger, D. G., Levy, E. K., and Chen, J. C. (1979). Flow characteristics in packed and fluidized rotating beds. *Powder Technol.*, **24**, 9–18.

Kunii, D., and Levenspiel, O. (1991). *Fluidization Engineering*, 2nd ed. Boston: Butterworth-Heinemann.

Liu, Y., Hamby, R. K., and Colberg, R. D. (1991). Fundamental and practical developments of magnetofluidized beds: a review. *Powder Technol.*, **64**, 3–11.

Luo, X., Lee, D. J., Lau, R., Yang, G., and Fan, L.-S. (1999). Maximum stable bubble size and gas holdup in high-pressure slurry bubble columns. *AIChE J.*, **45**(4), 665–680.

Marring, E., Hoffmann, A., and Janssen, L. (1994). The effect of vibration on the fluidization behaviour of some cohesive powders. *Powder Technol.*, **79**(1), 1–10.

Matsuda, S. Hatano, H., Muramota, T., and Tsutsumi, A. (2001). Particle and bubble behavior in ultrafine particle fluidization with high G. *Fluidization X*, 477–484.

Mawatari, Y., Koide, T., Tatemoto, Y., Takeshita, T., and Noda, K. (2001). Comparison of three vibrational modes (twist, vertical and horizontal) for fluidization of fine particles. *Adv. Powder Technol.*, **12**(2), 157–168.

Molerus, O. (1982). Interpretation of Geldart's type A, B, C and D powders by taking into account interparticle cohesion forces. *Powder Technol.*, **33**, 81–87.

Montz, K. W., Beddow, J. K., and Butler, P. B. (1988). Adhesion and removal of particulate contaminants in a high-decibel acoustic field. *Powder Technol.*, **55**, 133–140.

Mori, S., Iwasaki, N., Mizutani, E., and Okada, T. (1992). Vibro-Fluidization of Two-Component Mixtures of Group C Particles. *Fluidization VII*. Ed. Potter and Nicklin. New York: Engineering Foundation.

Morooka, S., Kusakabe, K., Kobata, A., and Kato Y. (1988). Fluidization state of ultrafine powders. *J. Chem. Eng. Jpn.* **21**(1), 41–46.

Nam, C. H., Pfeffer, R., Dave, R. N., and Sundaresan, S. (2004). Aerated vibrofluidization of silica nanoparticles. *AIChE J.*, **50**(8), 1776.

Quevedo, J., Omosebi, A., and Pfeffer, R. (2010). Fluidization enhancement of agglomerates of metal oxide nanopowders by microjets. *AIChE J.*, **56**(6), 1456–1468.

Qian, G. H., Bágyi, I., Burdick, I. W., Pfeffer, R., Shaw, H., and Stevens, J. G. (2001). Gas–solid fluidization in a centrifugal field. *AIChE J.*, **47**(5), 1022–1034.

Rosensweig, R. E. (1979). Fluidization: hydrodynamic stabilization with a magnetic field. *Science*, **204**, 57–60.

Shah, Y., Kelkar, B. G., Godbole, S., and Deckwer, W. D. (1982). Design parameters estimations for bubble column reactors. *AIChE J.*, **28**(3), 353–379.

Shaikh, A., and Al-Dahhan, M. H. (2007). A review on flow regime transition in bubble columns. *Int. J. Chem. Reactor Eng.*, **5**(1).

Srivastava, A., and Sundaresan, S. (2002). Role of wall friction in fluidization and standpipe flow. *Powder Technol.*, **124**(1–2), 45–54.

Stewart, P. S. B., and Davidson, J. F. (1967). Slug flow in fluidized beds. *Powder Technol.*, **1**, 61–80.

Tsinontides, S., and Jackson, R. (1993). The mechanics of gas fluidized beds with an interval of stable fluidization. *J. Fluid Mech.*, **255**, 237–274.

Wang, S., Hao, Z., Lu, H., Liu, G., Wang, J., and Xu, P. (2012). A bubbling fluidization model using kinetic theory of rough spheres. *AIChE J.*, **58**, 440–455.

Wang, Y., Gu, G., Wei, F., and Wu, J. (2002). Fluidization and agglomerate structure of SiO2 nanoparticles. *Powder Technol.*, **124**(1–2), 152–159.

Warsito, W., and Fan, L.-S. (2005). Dynamics of spiral bubble plume motion in the entrance region of bubble columns and three-phase fluidized beds using 3D ECT. *Chem. Eng. Sci.*, **60**, 6073–6084.

Wen, C. Y., and Yu, Y. H. (1966). Mechanics of Fluidization. *Chem. Eng. Prog. Symp. Ser.*, **62**, 100–111.

Yang, G., Du, B., and Fan, L.-S. (2007). Bubble formation and dynamics in gas–liquid–solid fluidization – A review. *Chem. Eng. Sci.*, **62**(1–2), 2–27.

Yang, W.-C. (2003). Ed., *Handbook of Fluidization and Fluid-Particle Systems*. Boca Raton, FL: CRC Press.

Yerushalmi, J., and Cankurt, N. T. (1979). Further studies of the regimes of fluidization. *Powder Technol.*, **24**, 187–205.

Yu, Q., Dave, R. N., Zhu, C., Quevedo, J. A., and Pfeffer, R. (2005). Enhanced fluidization of nanoparticles in an oscillating magnetic field. *AIChE J.*, **51**(7), 1971–1979.

Zhang, J.-P., Grace, J., Epstein, N., and Lim, K. (1997). Flow regime identification in gas–liquid flow and three-phase fluidized beds. *Chem. Eng. Sci.*, **52**(21–22), 3979–3992.

Zhu, C., Lin, C. H., Qian, G. H., and Pfeffer, R. (2003). Modeling of pressure drop and flow field in a rotating fluidized bed. *Chem. Eng. Comm.*, **190**(6), 1–23.

Zhu, C., Liu, G., Yu, Q., Pfeffer, R., Dave, R. N., and Nam, C. H. (2004). Sound assisted fluidization of nanoparticle agglomerates. *Powder Technol.*, **141**, 119–123.

Zhu, C., and You, J. (2007). An energy-based model of gas-solids transport in a riser. *Powder Technol.*, **175**, 33–42.

Zhu, C., Yu, Q., Dave, R. N., and Pfeffer, R. (2005). Gas fluidization characteristics of nanoparticle agglomerates. *AIChE J.*, **51**(2), 426–439.

11 Pipe Flow

11.1 Introduction

Multiphase flows in pipes and ducts commonly occur in material transport processes. For instance, pneumatic conveying employs a gaseous, pressurized airstream to transport solid materials such as flour, granular chemicals, lime, soda ash, plastic chips, coal, gun powder pellets, ore, and grains. In hydraulic conveying, solid materials are transported, or fed, via slurry flows. A vapor–liquid pipe flow is routinely utilized in various heat exchanger applications. In the oil and gas industry, a provision known as flow assurance is regularly included in the design and operation of offshore pipelines whereby multiple gas, liquid, and solid phases are simultaneously transported in an effort to ensure a safe and continuous production process. The advantages of pipeline transport methods include logistical flexibility in routing and space utilization, safer working conditions, and relatively low life-cycle costs. The disadvantages include increased power consumption requirements when compared to other bulk solid transport systems. Another hindrance is the potential for significant wearing and abrasive effects resulting from collisions between transported particles and pipe walls that can cause erosion of the pipeline and reduce service life.

The pipe flows discussed in this chapter focus more on horizontal flows in contrast to vertical flows in a column that portray fluidization described in Chapter 10. A vertical pipe flow occurring in some flow regimes is also given in this chapter. Specifically, Section 11.2 introduces multiphase flow patterns in pipeline transport. Section 11.3 discusses two important characteristic velocities in pipe transport: saltation and pickup velocities. Pressure drop through a pipeline is analyzed in Section 11.4. The fully suspended flows and wavy stratified flows in straight pipes are presented, respectively, in Sections 11.5 and 11.6. Section 11.7 focuses on flows in pipe bends and associated bend erosion. Simple mechanistic models are included in the preceding sections, whereas sophisticated modeling and CFD simulations are exemplified in the case studies in Section 11.8.

11.2 Multiphase Flow Patterns in Pipeline Transport

In multiphase pipe flows, the spatial distribution of different phases can be highly nonuniform. Between the two extremes of fully mixed (*i.e.*, homogeneously dispersed) and fully separated (*i.e.*, stratified) flows, varied spatial structures exist, which can be characterized by the presence of bubbles, slugs, wavy stratified layers, or annulus. Such spatial structures are indicative of characteristic flow patterns and facilitate categorization into distinct regimes. Since the interactions between different phases are highly dependent on their spatial distribution, performance parameters, such as pressure drop and heat transfer capacity, are influenced by the flow patterns. In addition, many mechanistic models and empirical correlations are limited to the specific flow patterns. Therefore, recognizing the characteristics of the flow pattern is of paramount importance in multiphase pipe flow applications. Distinct heterogeneous flow structures are affected by several factors including lateral forces, particle collisions among discrete particles and with pipe wall, or flow instabilities. Typical parameters that influence the flow pattern include physical properties, superficial velocity, flow direction, volume or mass fraction of each phase, pipe size, and pipe orientation with respect to gravity. The flow patterns of gas–solid, liquid–solid, and gas–liquid flows share many similar features, but they can also be differentiated by the density ratio of particle to fluid, deformability of particles, and electrification from interparticle or particle–wall collisions. Flow patterns in each of the gas–solid, liquid–solid, and gas–liquid flows are discussed in this section. The discussion addresses some common flow behaviors while an exhaustive treatment of the subject is not attempted.

11.2.1   Flow Regimes in Horizontal Pneumatic Conveying

Flow regimes for aerated particle transport in horizontal pipe flows are classified either by the particle loading, defined as the mass flow ratio of particles to carrying gas, or by the pressure drop per unit length. The density ratio of solid particle to gas is typically on the order of 10^3 in magnitude, except in cases involving buoyant solid materials such as foam pellets or agglomerates of nanoparticles.

Figure 11.2-1 presents the characteristic flow patterns for the five regimes in pneumatic conveyance of a particle-laden medium in a horizontal pipe at a prescribed flow rate. Enumerated with respect to increasing particle mass loading, the five regimes are (1) fully suspended dilute flow; (2) flow with intermediate occurrence of particle sedimentation, or dune formation; (3) wavy stratified flow; (4) plug flow; and (5) moving bed flow.

Figure 11.2-2 illustrates an alternate method to ascertain the flow regime diagram in a pneumatic conveying process using the pressure drop per unit length versus the superficial gas velocity at a given mass flux of particles. In general, escalating the mass flux of the solid phase results in an increased pressure drop per unit length at a given gas velocity. However, at extremely low solids loadings, the pressure drop can be less than that of a particle-free flow, a phenomenon known as drag reduction discussed in Section 11.4.5. Figure 11.2-2 indicates that the pressure drop per unit length initially decreases and then increases as the superficial gas velocity incrementally intensifies. The point of minimum pressure drop per unit length delineates

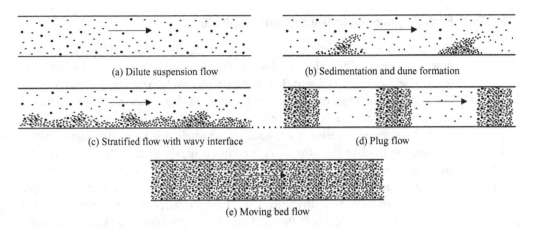

Figure 11.2-1 Flow regimes in pneumatic conveying.

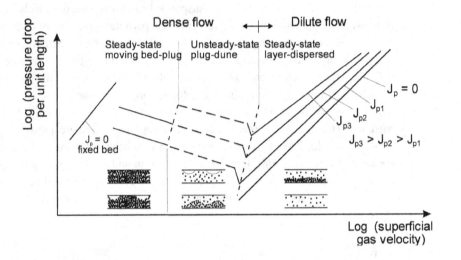

Figure 11.2-2 Flow regime diagram for pneumatic transport in horizontal pipes.

the regime transition from dense flow to dilute flow for a pneumatic conveying system. In horizontal conveying, a saltation phenomenon, where coarse particles begin to settle, occurs at the regime transition between dilute and dense flows, given by the point of minimum pressure drop. In conveyance of fine particles, the saltation manifests at a superficial gas velocity in excess of the minimum pressure drop point (Klinzing *et al.*, 1997). Figure 11.2-2 also reveals a power-law relationship between the pressure gradient and the superficial gas velocity for dilute suspension flow of fine particles, which appears linear on the log–log scale. Visual observation of the flow regimes further concludes that a dense-phase flow often presents as a plug flow or moving bed flow. For reference, the fixed bed flow regime is also marked on the figure.

11.2.2   Flow Regimes in Horizontal Slurry Pipe Flows

Transport of a solid particle suspension in a carrying liquid is known as slurry flow. The densities of solid particles and liquid are typically of the same order of magnitude. For horizontal pipe flow, particles that are heavier than the carrying liquid tend to settle from the slurry due to the vertical gravitational force. However, the sedimentation rate is often undetectable in the time scale of the slurry flow when the suspended particles are small, the particle density approximates that of the carrying fluid, or the suspending medium is highly viscous. In such cases, the slurry maintains a homogeneous state and approximates a single-phase flow with either Newtonian or non-Newtonian rheological properties. Conversely, the degree of sedimentation significantly affects the flow characteristics of a settling slurry. Several distinct characteristic flow regimes have been observed at different slurry velocities (Albunaga, 2002; Pecker and Helvaci, 2008), as schematically mapped in Figure 11.2-3.

At high flow rates, all particles are suspended and distributed uniformly in a homogeneous fashion. However, at low velocities the particles remain fully suspended, but a significant concentration gradient develops, characterizing the heterogeneity. If the slurry velocity is low, liquid turbulence is not sufficiently high to overcome the particle inertia; particles precipitate and accumulate on the bottom of the pipe to form a veritable moving bed. Decreasing the velocity further, the particle bed becomes partially stationary. At an extremely low flow speed, the pipe can be plugged by the solid particles. The transition between these flow regimes is demarcated by prescribed critical velocities that can be calculated using various correlations (Albunaga, 2002; Pecker and Helvaci, 2008). Particle sedimentation and dense region formation contribute to a substantial increase in pressure drop. Thus, care should be exercised to mitigate these undesirable flow patterns for operation of slurry transport in pipes.

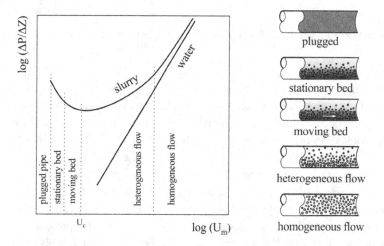

Figure 11.2-3 Flow regime diagram for slurry flows in horizontal pipes.

11.2.3   Gas–Liquid Flow Regimes in Pipes

In gas–liquid pipe flows, the gas or the liquid can be the dominant and continuous phase while the other assumes the role of the discrete phase. The discrete phase is embodied in mist, where the liquid phase is discrete in the gas medium, or in bubbles, with the gas phase suspended in liquid. Frequent deformation, dissolution, and coalescence of the gas–liquid interface modifies the size and shape of the discrete phase in dispersed flow regimes. For vapor–liquid pipe flows with phase changes, such as boiling or condensation, heat flux through the pipe wall may significantly affect the volumetric flow ratio between the two phases and complicate prediction of the flow regime transition. A treatment of phase change in gas–liquid pipe flow is given in Chapter 12.

Figure 11.2-4 illustrates typical gas–liquid flow regimes in a horizontal pipe (Taitel and Dukler, 1976), which is mapped by velocity of liquid versus velocity of gas. The mapped regimes include:

- Dispersed bubble flow: Gas bubbles much smaller than the pipe diameter are dispersed in the continuous liquid flow.
- Elongated bubble flow: Flows are characterized by the formation of elongated, bullet-shaped bubbles, which are sometimes referred to as Taylor bubbles.
- Slug flow: Waves on the liquid surface rise to the top of the pipe, separating the gas slugs. Plug and slug flows are also known as intermittent flows.
- Smooth stratified flow: Gas and liquid are separated into two layers under the effects of gravity, with continuous gas flow at the top and continuous liquid flow at the bottom. A smooth interface exists between the two phases.
- Wavy stratified flow: It is characterized by a stratified flow with formation of waves on the liquid surface.

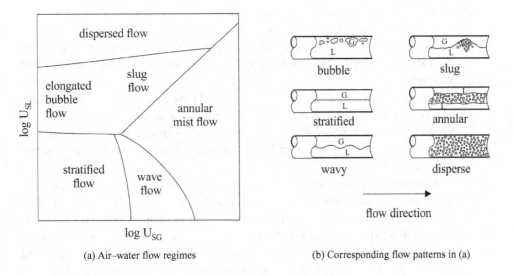

(a) Air–water flow regimes   (b) Corresponding flow patterns in (a)

Figure 11.2-4 Regimes of horizontal gas–liquid pipe flows.

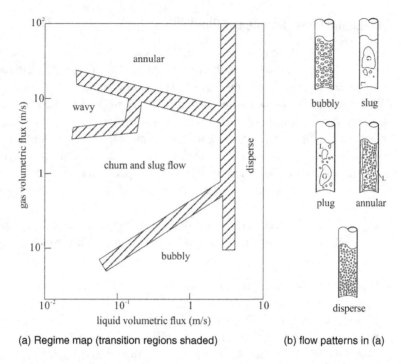

(a) Regime map (transition regions shaded)   (b) flow patterns in (a)

Figure 11.2-5 Illustrative flow regimes for upward gas–liquid pipe flow.

- Annular mist flow: There is a liquid region flow near the inner wall of the pipe, while gas flows in the center. Droplets sheared from the liquid surface are entrained into the center region of the pipe by the flowing gas. Fine liquid droplets are dispersed in the continuous gas flow.

Transition between regimes is mainly induced by Kelvin–Helmholtz instability, which causes the amplitude of waves to increase at the gas–liquid interface. The equilibrium liquid level, controlled by its equivalent volume fraction, also influences the flow regime. Detailed discussion of the governing mechanisms and predictive methods of flow regime transition is given in the literature (Taitel and Dukler, 1976; Weisman, 1983).

The mechanics for gas–liquid flow in a vertical pipe are analogous to the horizontal pipe case in many ways. Figure 11.2-5 illustrates the characteristic flow regimes for upward gas–liquid flow (Weisman, 1983), which is expressed in terms of the volumetric flux of gas versus that of liquid. The shaded areas stand for the transition regions.

At modest superficial gas velocities, small gas bubbles disperse in the continuous liquid flow. Increasing the gas velocity prompts the small bubbles to coalesce into large slugs, known as Taylor bubbles. In high turbulence, persistent deformation and dissolution of large bubbles results in an unstable flow pattern, designated as churn flow. In churn flows, as gas flows upward in the center of the pipe, liquid intermittently flows upward or downward near the pipe wall. If the gas velocity is further increased, a transition to the annular flow regime occurs where a thin layer of liquid flows along the inner pipe wall while gas flows in the centralized core region. The gas velocity considerably exceeds the liquid velocity, and fine liquid droplets are entrained in the gas core. At exceedingly high liquid velocities, the flow is highly dispersed, engendering froth flow at low gas velocities or mist flow at high gas velocities.

11.3 Saltation and Pickup Velocities

In a horizontal gas–solid or liquid–solid flow, gravitational settling of particles can occur if the velocity of carrying fluid is not sufficient. Two critical characteristic velocities can be used to account for the settling phenomenon: minimum transport velocity and pickup velocity. The minimum transport velocity, also known as salta-tion velocity, aims to characterize the onset of settling of suspended particles. Below the minimum transport velocity, particles can form a stationary or sliding layer on the bottom of a horizontal pipeline. That is, if the mean velocity of the carrying fluid is less than the minimum transport velocity, saltation occurs in a horizontal pipeline. Pickup velocity prescribes the minimum flow velocity required to lift settled particles and reincorporate them into the main flow stream in the pipe. The pickup veloc-ity is typically higher than the saltation velocity because upheaving the particles from their resting state on the bottom pipe wall requires the hydrodynamic force to overcome additional resistance due to inertia, interparticle forces, and particle–wall forces. Some resistive particle impetus includes van der Waals forces, liquid-bridge surface forces, and electrostatic forces.

11.3.1 Critical Transport Velocity

Saltation of solids occurs in the turbulent boundary layer where the wall effects on the particle motion must be accounted for. Significant wall effects include lift due to an imposed mean shear (*e.g.*, Saffman lift) and particle rotation (*e.g.*, Magnus effect), and an increase in drag force (*e.g.*, the Faxen effect). Additionally, if the particles are accelerated, the added mass effect and Basset force can be influential. The bounc-ing effect stemming from particle–wall collisions may also forestall particle settling. Thus, a comprehensive model to predict the saltation velocity should include the coupling between the hydrodynamics and the inherent particle–wall collisions. It is common to assume that saltation is primarily dependent on the hydrodynamic behav-ior of particles in the wall-boundary region, with the wall collision effects either ignored or simplified.

The particle settling behavior in the pipe wall region is influenced by factors including the bulk turbulence of pipe flow, wall friction or turbulent flow in the boundary layer, particle properties, and the particle concentration or volumetric fraction. The minimum transport velocity, U_{mt}, may be expressed as a general function of

$$U_{mt} = f\left(\rho, \mu, \rho_p, d_p, D, \alpha_p, \tau_w\right).$$

(11.3-1)

Using the Buckingham Pi-theorem, Eq. (11.3-1) can be expressed as a function of a group of dimensionless parameters, as:

$$\text{Re}_{mt} = f\left(\frac{\rho}{\rho_p}, \frac{d_p}{D}, \frac{U_{pt}}{U_{pf}}, \alpha_p\right),$$ (11.3-1a)

where Re_{mt} is the pipe Reynolds number at the minimum particle transport condition, defined by:

$$\text{Re}_{mt} = \frac{DU_{mt}\rho}{\mu}.$$ (11.3-1b)

U_{pf} is the friction velocity at the particle-free condition (*i.e.*, $\alpha_p = 0$), defined by

$$U_{pf} = \sqrt{\frac{\tau_w}{\rho}}.$$ (11.3-1c)

Notably, the particle terminal velocity, U_{pt}, is not a new independent parameter since it is a function of the fluid and particle properties already accounted for by Eq. (11.3-1). Equation (11.3-1a) indicates that the minimum transport velocity is strongly influenced by the ratio of particle terminal velocity to friction velocity; this is a veritable key to formulate an expression for minimum transport velocity required to prevent settling.

For fully developed turbulent flow of a single-phase mixture in a circular pipe, the mean stream velocity is related to the friction velocity via boundary-layer theory of log-law velocity distributions (Taylor, 1954):

$$\frac{U}{U_f} = 5\log\left(\frac{\rho DU}{\mu}\right) - 3.90,$$ (11.3-2)

where U_f is the friction velocity based on the mixture density, defined by:

$$U_f = \sqrt{\frac{\tau_w}{(1-\alpha_p)\rho + \alpha_p\rho_p}} \approx \sqrt{\frac{D\Delta p}{4L\left((1-\alpha_p)\rho + \alpha_p\rho_p\right)}}.$$ (11.3-2a)

The minimum transport velocity, U_{mt}, may be expressed in an analogous form to Eq. (11.3-2) with a modification factor for the dependence of U_f on α_p, which is given as (Thomas, 1962):

$$\frac{U_{mt}}{U_{pf}} = \frac{U_{mt}}{U_f}\left(\frac{U_f}{U_{pf}}\right) = \left\{5\log\left(\frac{\rho DU_{mt}}{\mu}\right) - 3.90\right\}\left(1 + 2.8(\alpha_p)^{\frac{1}{2}}\left(\frac{U_{pt}}{U_{pf}}\right)^{\frac{1}{3}}\right),$$ (11.3-3)

where U_{pf} can be further related to the particle terminal velocity by:

$$\frac{U_{pt}}{U_{pf}} = \begin{cases} 4.90\left(\frac{d_p}{D}\right)^{0.6}\text{Re}_{pf}^{0.4}\left(\frac{\rho_p}{\rho} - 1\right)^{0.23} & \text{Re}_{pf} > 5 \\ 0.01\text{Re}_{pf}^{2.71} & \text{Re}_{pf} < 5 \end{cases},$$ (11.3-3a)

where Re_{pf} is the particle Reynolds number at U_{pf}. The particle terminal velocity, U_{pt}, can be independently calculated via the balance of the gravitational force and

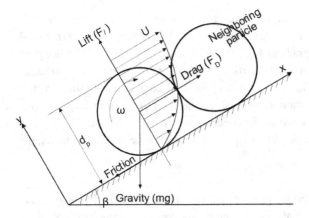

Figure 11.3-1 Forces acting on a particle on an inclined surface.

the particle–fluid drag. For spherical particles, the particle terminal velocity can be expressed, in terms of particle Reynolds number at terminal velocity, by

$$\frac{\pi}{6}d_p^3\left(\rho_p - \rho\right)g = C_D\frac{\pi}{8}\frac{\mu^2}{\rho}\text{Re}_{pt}^2, \tag{11.3-3b}$$

where C_D is a function of Re_{pt} given by Eq. (3.2-3).

As shown in Figure 11.2-3 of horizontal slurry pipe flows, the critical velocity to ensure the solids are fully suspended or maintain the slurry from saltation is marked by the point at which the pressure gradient becomes minimum. This critical or minimum transport velocity is also known as the deposition velocity, which can be estimated by a simple semiempirical correlation (Wasp et al., 1977):

$$U_D = 4(\alpha_p)^{\frac{1}{3}}\left(\frac{d_p}{D}\right)^{\frac{1}{6}}\sqrt{2gD\left(\frac{\rho_p}{\rho} - 1\right)}. \tag{11.3-4}$$

11.3.2 Pickup Velocity

Particle pickup is affected by two competing mechanisms: the shear lifting force acting against gravity and contact resistance (e.g., cohesion) due to particles interacting with the pipe wall and adjacent particles, as illustrated in Figure 11.3-1. The shear lifting force in this context is orthogonal to the inner pipe surface, and it is the predominant mechanism that yields sliding or resting particles at the pipe wall to reenter the main flow stream. In general, the lift force is not necessarily in the opposite direction of gravity; lift is invariably directed opposite to the cohesion forces of particle–wall contact.

The motion of a particle in contact with the pipe wall is governed by:

$$\rho_p V\frac{dU_p}{dt} = \mathbf{F}_D + \mathbf{F}_L + \mathbf{F}_A + \mathbf{F}_B + \mathbf{F}_M + \mathbf{F}_P + \mathbf{F}_C + \mathbf{F}_w + \left(\rho_p - \rho\right)V\mathbf{g}, \tag{11.3-5}$$

where the hydrodynamic forces are represented by the drag, shear lifting, carried mass, Basset history, and Magnus forces. The influence of external fields is captured in the pressure gradient force, the gravitational force, and the buoyancy force. The influence of neighboring particles is accounted for by particle collision force or the contact force. The effects of particle–wall contact are implicit in the wall contact force; both wall friction and wall-contact cohesion are included. Thus, for the main flow to pick up a particle in contact with the pipe wall, superposition of the normal components of all forces in Eq. (11.3-5) must be net positive; namely, the criterion is:

$$F_L > \{\mathbf{F}_M + \mathbf{F}_C + \mathbf{F}_w + (\rho_p - \rho)\,V\mathbf{g}\} \cdot \mathbf{n}. \tag{11.3-5a}$$

The drag force and the pressure gradient force persist in the longitudinal direction of the pipeline such that their normal components are always rendered null; this impacts the above equation. Equation (11.3-5a) can be used as a general criterion to predict the pickup velocity.

For the simplified condition of a small, isolated, and spherical particle at rest within the wall boundary region, flow around the particle is assumed to behave according to the characteristic Stokes regime with viscous shearing. The cohesion force of particle–wall contact is dominated by the Van der Waals force. Thus, the viscous shear lifting force is gleaned by the Saffman force; the Magnus force and interparticle contact force both vanish in the absence of relative rotation and adjacent particles. In this condition, using Eq. (3.2-23a) for the Saffman force and Eq. (4.4-1) for the Van der Waals force, Eq. (11.3-5a) reduces to:

$$\frac{K}{4}\rho U d_p^2 \sqrt{\frac{\mu}{\rho}\left|\frac{\partial U}{\partial y}\right|} > (\rho_p - \rho)\,Vg\cos\beta + \frac{A_H d_p}{12\delta^2}. \tag{11.3-5b}$$

The averaged-shear gradient in the wall boundary layer is crudely estimated using the wall shear stress τ_w, which can be correlated to the friction velocity, U_f, and mean flow velocity, U. With the aid of Eqs. (11.3-2) and (11.3-2a), the pickup velocity is estimated from:

$$U_{pu}^2 > \frac{4}{K}(5\log\mathrm{Re}_D - 3.90)\left\{\frac{\pi(\rho_p - \rho)}{6\rho}d_p g\cos\beta + \frac{A_H}{12\delta^2\rho d_p}\right\}. \tag{11.3-5c}$$

11.4   Pressure Drop

Multiphase flow transport through a pipeline is typically pressure-driven. The pressure loss per pipe length is a function of the flow characteristics including the accelerating or fully developed status, pattern of phase distribution, and laminar or turbulent flow regime. Pipe orientation with respect to gravity, geometry, and system routing are also influential (*e.g.*, horizontal, inclined, or vertical pipe; pipe shapes and variation in cross-sectional areas along the pipe; bends and branches; elevations). For a multiphase flow in a straight pipe of fixed cross-sectional shape and area, the total pressure gradient consists of three components, respectively, due to friction, gravitation, and acceleration, which is expressed as:

$$-\frac{dp}{dz} = \left(-\frac{dp}{dz}\right)_f + \left(-\frac{dp}{dz}\right)_g + \left(-\frac{dp}{dz}\right)_a. \qquad (11.4\text{-}1)$$

For a pipe at an inclination angle of β to the horizontal plane, the gravitational contribution to pressure drop is modified to:

$$\left(-\frac{dp}{dz}\right)_g = (\alpha\rho + \alpha_p\rho_p)g\sin\beta. \qquad (11.4\text{-}1a)$$

For fully developed adiabatic flows, the influence of acceleration is often negligible on pressure drop so that friction loss is the dominant contributor to the pressure drop. However, the acceleration effect can be important in flows with significant phase changes, as discussed later in Chapter 12. This section is mainly focused on fully developed two-phase pipe flows without phase changes.

11.4.1 Pressure Drop of a Fully Developed Suspension Flow

The role of the pressure gradient becomes evident in the momentum equation for steady, fluid–particle two-phase flow in a pipe. For a fully suspended multiphase flow in a straight pipe, the pressure variation at a cross section is generally much less than the pressure gradient along the pipeline axis. Thus, it yields:

$$\nabla p \approx \frac{dp}{dz}\mathbf{n}_z. \qquad (11.4\text{-}2)$$

Interparticle collisions can be neglected if the flow is assumed to be isothermal, turbulent, and devoid of phase change. The momentum equations for the fluid and particle phases, respectively, are obtained using Eq. (6.4-22) as:

$$\nabla \cdot \mathbf{J}_m = -\alpha\nabla p + \nabla \cdot \boldsymbol{\tau}_e + \mathbf{F}_A + \rho\mathbf{g} \qquad (11.4\text{-}2a)$$

and

$$\nabla \cdot \mathbf{J}_{mp} = -\alpha_p\nabla p + \nabla \cdot \mathbf{T}_{ep} - \mathbf{F}_A + \alpha_p(\rho_p - \rho)\mathbf{g}. \qquad (11.4\text{-}2b)$$

The effective shear stress tensor of fluid and the total stress tensor of particles per unit volume are denoted as τ_e and $\mathbf{T}_{ep}$, respectively; $\mathbf{F}_A$ accounts for the interfacial interactions between the fluid and particles; and $\mathbf{J}_m$ and $\mathbf{J}_{mp}$ are the momentum fluxes of fluid and particle phases, respectively. $\mathbf{J}_m$ and $\mathbf{J}_{mp}$ are respectively defined by:

$$\mathbf{J}_m = \alpha\rho\mathbf{U}\mathbf{U} - \rho D_e\{(\nabla\alpha)\mathbf{U} + \mathbf{U}\nabla\alpha\} \qquad (11.4\text{-}2c)$$

and:

$$\mathbf{J}_{mp} = \alpha_p\rho_p\mathbf{U}_p\mathbf{U}_p - \rho_p D_{ep}\{(\nabla\alpha_p)\mathbf{U}_p + \mathbf{U}_p\nabla\alpha_p\}, \qquad (11.4\text{-}2d)$$

where D_e and D_{ep} represent the turbulent eddy diffusivities of fluid and particle phases, respectively.

The equation of pressure gradient in the longitudinal direction of the pipe is derived from the combined momentum equation of the two phases as:

$$-\frac{dp}{dz} = \frac{d}{dz}\left\{\mathbf{n}_z \cdot \left(\mathbf{J}_m + \mathbf{J}_{mp} - \boldsymbol{\tau}_e - \mathbf{T}_{ep}\right) \cdot \mathbf{n}_z\right\} - \left(\alpha\rho + \alpha_p\rho_p\right)\mathbf{g} \cdot \mathbf{n}_z. \quad (11.4\text{-}3)$$

11.4.2 Pressure Drop in Dilute Gas–Solid Flows

In dilute phase conveying, the total frictional pressure drop can be represented via superposition of the proportional contributions from the two phases (Weber, 1981):

$$\frac{\Delta p}{L} = \left(\lambda + \lambda_p\left(\frac{J_p}{J}\right)\right)\frac{\rho U^2}{2D}, \quad (11.4\text{-}4)$$

where the friction factor of gas, λ, is related to the pipe roughness, δ, and pipe Reynold number by (Swamee and Jain, 1976):

$$\lambda = \frac{1.325}{\left[\ln\left(\delta/3.7D\right) + 5.74/\text{Re}^{0.9}\right]^2}. \quad (11.4\text{-}4a)$$

The friction factor of particles, λ_p, can be correlated to the following nondimensional parameters, as (Jones and Williams, 2003):

$$\lambda_p = f\left(\frac{J_p}{J}, \text{Fr}, \frac{U_{pt}}{U}, \frac{d_p}{D}, \frac{\rho_p}{\rho}\right), \quad (11.4\text{-}4b)$$

where Fr is the Froude number, defined as:

$$\text{Fr} = \frac{U^2}{gD}. \quad (11.4\text{-}4c)$$

Various empirical correlations of λ_p are available in the literature (e.g., Weber, 1991; Jones and Williams, 2003).

11.4.3 Pressure Drop in Slurry Flows

When the liquid velocity is high, or particle size is small, the slurry flow is in the homogeneous regime. In the homogeneous flow regime, relative motion between the solid phase and liquid phase is inconsequential, and the liquid–solid mixture essentially behaves as a pseudo single-phase medium with a mixture density and viscosity. Given a low solids concentration, a low particle-to-liquid density ratio, or a high liquid-phase viscosity, the mixture can be modeled as a Newtonian fluid. The mixture viscosity can be represented by the Einstein model (Einstein, 1906), expressed as:

$$\mu_m = \left(1 + 2.5\alpha_p\right)\mu, \quad (11.4\text{-}5a)$$

when the slurry concentration is low at $\alpha_p < 5\%$. At a higher solids concentration, another expression is appropriate (Krieger and Dougherty, 1959):

$$\frac{\mu_m}{\mu} = \left(1 - \frac{\alpha_p}{\alpha_{p\,\text{max}}}\right)^{-\eta\alpha_{p\,\text{max}}}, \quad (11.4\text{-}5b)$$

where η is the intrinsic viscosity of particles, which dispends on particle shape. A rough approximation of $\eta\alpha_{pmax}$ being 2 can be adopted for many slurry flows (Genovese, 2012). At a high solids concentration, the non-Newtonian characteristics prevail, for which most slurries mimic the nature of shear thinning or pseudo-plastic fluids. In cases involving exceedingly fine particles, such as starch, the slurry may present a shear thickening property. The pressure drop of non-Newtonian slurries is calculated via the friction coefficient, defined by

$$\frac{\Delta p_f}{L} = 2f\frac{\rho_m U_m^2}{D}, \tag{11.4-6}$$

where U_m is the averaged velocity of mixture, and ρ_m is the mixture density.

The friction factor, f, is commonly developed using rheological models, such as power-law, Bingham, Casson and Herschel–Bulkley models. The frictional factors for homogeneous slurries in a straight pipe can be formulated under laminar or turbulent flow conditions, as exemplified in Table 9.5-1 (Dodge and Metzner, 1959; Darby and Melson, 1981).

For slurry flows in heterogeneous regimes, it is customary to express the frictional pressure drop by the following relationship:

$$\frac{\Delta p_m}{L} = \left(1 + \phi\alpha_p\right)\frac{\Delta p_l}{L}, \tag{11.4-7}$$

where Δp_m and Δp_l are the frictional pressure drop in slurry and in a pure liquid condition at an equivalent volumetric flow rate, respectively. The factor ϕ is a dimensionless factor for excess pressure loss. In the limiting case of homogeneous slurry flows, ϕ approaches the value of $(\rho_p/\rho_l - 1)$, provided the slurry and the liquid have an identical friction factor for an equivalent pipe diameter and flow rate. In the heterogeneous flow regimes, ϕ is often calculated from various empirical correlations, such as for fully suspended sand–water slurries (Durand and Condolios, 1952) and regime-specific operations (Newitt *et al.*, 1955).

Table 9.5-1 Friction Factors for Homogeneous Slurry Flows

Flow condition	Slurry viscosity	Friction factor	Parameters	
Laminar	Newtonian	$f = \frac{16}{Re_m}$	$Re_m = \frac{\rho_m U_m D}{\mu_m}$	(11.4-6a)
	Power-law $\eta_m = K_p\dot\gamma^{n-1}$	$f = \frac{16}{Re_p}$	$Re_p = \frac{\rho_m U_m D}{K_p}\left(\frac{4n}{3n+1}\right)^n\left(\frac{D}{8U_m}\right)^{n-1}$	(11.4-6b)
	Bingham $\tau_m = \tau_b + \eta_b\dot\gamma$	$f = \frac{16}{Re_p}\left[1 + \frac{1}{6}\frac{He_b}{Re_b} - \frac{1}{3}\frac{He_b^4}{f^3 Re_b^7}\right]$	$Re_b = \frac{\rho_m U_m D}{\eta_b}$ $\quad He_b = \frac{\rho_m \tau_b D^2}{\eta_b^2}$,	(11.4-6c)
Turbulent	Newtonian	$1/\sqrt{f} = 4.0\log(Re_m\sqrt{f}) - 0.4$	Re_m given in Eq. (11.4-6a)	(11.4-6d)
	Power-law	$1/\sqrt{f} = \frac{4.0}{n^{0.75}}\log(Re_p f^{1-n/2}) - \frac{0.4}{n^{1.2}}$	Re_p given in Eq. (11.4-6b)	(11.4-6e)
	Bingham	$f = (f_L^n + f_T^n)^{1/n}$ $f_T = 10^\beta Re_b^{-0.193}$ $\beta = -1.378$ $\left[1 + 0.146\exp\left(-0.29\times10^{-5}Re_b\right)\right]$	f_L given in Eq. (11.4-6c) $n = 1.7 + 40000/Re_b$	(11.4-6f)

11.4.4   Pressure Drop in Gas–Liquid Flows

Frictional pressure drop for gas–liquid flows is a function of the flow pattern and its characteristic laminar or turbulent nature. For dilute dispersed flows where the effect of slip between two phases can be neglected, the homogeneous model is typically employed. The homogeneous model for gas–liquid flows mirrors the form of slurry flows, in which the flow is treated as a single-phase mixture flow. Thus, the equation for the frictional pressure drop in the virtual single-phase flow follows Eq. (11.4-6), with formulations of the Fanning friction factor for the mixture represented by those in Table 9.5-1.

Conversely, for separated flows where the slip is distinct, the frictional pressure gradient of a two-phase flow is correlated to the corresponding single-phase gas and liquid flows by introducing modification multipliers ϕ_g and ϕ_l, as suggested by Lockhart and Martinelli (1949):

$$\left(-\frac{dp}{dz}\right)_f = \left(-\frac{dp}{dz}\right)_{g0}\phi_g^2 = \left(-\frac{dp}{dz}\right)_{l0}\phi_l^2. \tag{11.4-8}$$

The subscripts "$g0$" and "$l0$" stand for limiting cases when the gas or liquid is the only phase in the pipe. The gas and liquid pressure gradients can be calculated, respectively, from the corresponding superficial velocity of each phase:

$$\left(-\frac{dp}{dz}\right)_{g0} = 2f_g\frac{\rho_g U_g^2}{D} \quad \left(-\frac{dp}{dz}\right)_{l0} = 2f_l\frac{\rho_l U_l^2}{D}, \tag{11.4-8a}$$

where the frictional factors can be estimated from Table 9.5-1. The friction multipliers in Eq. (11.4-8) are empirically correlated with the Lockhart–Martinelli parameter X, which is defined as:

$$X^2 = \left(-\frac{dp}{dz}\right)_{l0} \bigg/ \left(-\frac{dp}{dz}\right)_{g0} = \frac{\phi_g^2}{\phi_l^2}. \tag{11.4-8b}$$

A convenient correlation of the friction multipliers is commonly used (Chisholm, 1967):

$$\phi_g^2 = 1 + CX + CX^2. \tag{11.4-8c}$$

The parameter C depends on the flow regime, as prescribed in Table 9.5-2. Thus, given the flow rates of two phases, X, ϕ_g, and ϕ_l can be solved from the coupled Eqs. (11.4-8a)–(11.4-8c).

The Lockhart–Martinelli approach is especially suitable for separated flows or flows with minimal velocity differences between phases, although the technique is

Table 9.5-2 Value of C for the Frictional Multipliers

	Laminar gas	Turbulent gas
Laminar liquid	5	12
Turbulent liquid	10	20

often extended to other regimes. It tends to be less accurate at extremely low or high volume fractions.

11.4.5 Drag Reduction

When particulate is introduced into a flow stream, it is expected that the pressure drop of the new particle-laden flow will be greater than its prior single-phase counterpart. However, experimental data reveal that in some conditions, such as when the mass flux ratio of particles to fluid is within the range of 0.5 to 2, the pressure drop of the two-phase flow in the pipe can be actually much less than that for the particle-free flow. That is, a reduction in pressure drop of up to 30% over a single-phase flow can be observed for multiphase transport of small particles (*e.g.*, less than 200 μm) in a pipe (McCarthy and Olson, 1968; Shimizu *et al.*, 1978). This phenomenon, illustrated in Figure 11.4-1, is known as drag reduction. This result supports the corollary that it is possible to consume less energy transporting a two-phase flow as compared to a single-phase variant.

To explain the drag reduction phenomena, consider a simple case of fully developed horizontal pipe flow with cross-section-averaged flow parameters. In this condition, the pressure drop is solely a function of wall friction, a result that can be deduced from Eq. (11.4-3). It yields:

$$-\frac{dp}{dz} = \frac{4}{D}\left(\tau_{fp} + \tau_{wp}\right),\tag{11.4-9}$$

where τ_{fp} denotes the fluid wall shear stress in the presence of solid particles; and τ_{wp} represents the friction effect due to collisions of particles with the wall. Thus, the ratio of pressure drop when solids are present to the pressure drop of a particle-free flow can be expressed as:

$$\frac{\Delta p}{\Delta p_0} = \frac{\tau_{fp}}{\tau_{f0}} + \frac{\tau_{wp}}{\tau_{f0}}.\tag{11.4-10}$$

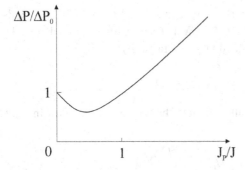

Figure 11.4-1 Pressure drop ratio versus mass flux ratio.

The volume-averaged wall friction of fluid phase, τ_{fp}, is calculated by:

$$\tau_{fp} = \alpha_w \mu_{epw} \left(\frac{\partial U_f}{\partial r} \right)_w. \tag{11.4-11}$$

A sufficiently dilute solids concentration where the averaged distance between particles is at least ten particle diameters is required for the drag reduction phenomenon to manifest; this indicates that interparticle effects have a negligible effect on drag reduction. Empirical measurements further demonstrate that, within this range of particle loading, the velocity profile of the fluid phase is nearly unaltered by the presence of solid particles (Soo et al., 1964). The fluid phase shear friction ratio can be expressed by:

$$\frac{\tau_{fp}}{\tau_{f0}} \approx \alpha_w \frac{\mu_{ep}}{\mu_{e0}}. \tag{11.4-11a}$$

For dilute suspensions of small particles, Eq. (6.6-11) estimates the damping effect of the particles on flow turbulence. Thus, the fluid phase shear friction ratio in Eq. (11.4-11a) can be further simplified to:

$$\frac{\tau_{fp}}{\tau_{f0}} \approx \alpha_w \left(1 + \frac{\alpha_p \rho_p}{\rho} \right)^{-1}. \tag{11.4-11b}$$

The volume-averaged wall friction of the fluid phase can be related to the particle volume fraction, the particle velocity, and the turbulent intensity of particles at the wall (Soo, 1962):

$$\tau_{wp} = \frac{1}{2\sqrt{\pi}} \alpha_{pw} \rho_p U_{pw} \sqrt{\left(u'_{pw} \right)^2}. \tag{11.4-12}$$

The velocity fluctuation for particulate near the pipe wall is gleaned from particle velocity fluctuations in the main flow stream. From the Hinze–Tchen model in Section 6.6.2, the particle velocity fluctuation is further related to the turbulence intensity of the pipe flow through:

$$\sqrt{\left(u'_{pw} \right)^2} \approx \sqrt{\left(u'_p \right)^2} = \sqrt{(u')^2} \left(1 + \frac{\tau_S}{\tau_e} \right)^{-\frac{1}{2}}, \tag{11.4-12a}$$

where τ_S is the Stokes relaxation time defined by Eq. (3.2-22a), and τ_e is the eddy duration estimated by Eq. (6.6-6a). Wall friction for simplified particle-free pipe flows can be related to the Moody friction factor by:

$$\tau_{f0} = \frac{f}{8} \rho U^2. \tag{11.4-12b}$$

Thus, the ratio of particle wall friction to that of particle-free flow in Eq. (11.4-11a) can be refined to:

$$\frac{\tau_{wp}}{\tau_{f0}} = \frac{4\alpha_{pw}}{f\sqrt{\pi}} \left(\frac{\rho_p}{\rho} \right) \left(\frac{U_{pw}}{U} \right) \left\{ \frac{\sqrt{(u')^2}}{U} \right\} \left(1 + \frac{\tau_S}{\tau_e} \right)^{-\frac{1}{2}}. \tag{11.4-12c}$$

Substituting Eq. (11.4-11b) and Eq. (11.4-12c) into Eq. (11.4-10) yields the following:

$$\frac{\Delta p}{\Delta p_0} \approx \alpha_w \left(1 + \frac{\alpha_p \rho_p}{\rho}\right)^{-1} + \frac{4\alpha_{pw}}{f\sqrt{\pi}} \left(\frac{\rho_p}{\rho}\right) \left(\frac{U_{pw}}{U}\right) \left\{\frac{\sqrt{(u')^2}}{U}\right\} \left(1 + \frac{\tau_S}{\tau_e}\right)^{-\frac{1}{2}}.$$

$$(11.4\text{-}13)$$

In the case of pipe flows with suspended small particles, Eq. (11.4-13) demonstrates the presence of the particles will actually reduce the resistive pipe wall friction effect on the fluid because of their damping effect on fluid turbulence, as indicated by the first term; however, solid particulate transport still demands pressure drop be overcome and consumed, as shown by the second term. Thus, the criterion for drag reduction to manifest is that the turbulent damping effect leading to decreased pressure drop must outstrip the pressure drop consumption required for the solids transport.

11.5   Phase Distributions of Suspended Pipe Flows

For suspended multiphase flows in the pipe transport, due to the pipe wall effect and gravity, the phase distributions of velocity and volume fraction in a cross section are unlikely to be uniform, even under fully developed flow conditions. This nonuniformity in phase distributions can also be affected by the electrostatic charges on the transport particles. Using simple modeling approaches, this section provides analyses of phase distributions under three cases of suspended flows in pipes: (1) fully developed and fully suspended dilute phase flows; (2) effect of electrostatic charges on phase transport; and (3) dilute phase transport in a pipe, in which some solids migrate toward pipe wall by turbulent diffusion and consequently slide downward by gravity in a thin boundary layer while the majority of solids are transported upward in the pipe core by the upward turbulent gas flow.

11.5.1   Fully Developed Dilute Pipe Flows

A fully developed dilute flow implies the velocity profiles of both the continuous and the discrete phases remain unchanged along the axial pipe direction. Assume the particles are fully suspended without any wall deposition or stratification. For dilute suspensions, the interparticle collisions are further neglected. Thus, the particle transport is governed by the two mechanisms: hydrodynamic convection of the carrying fluid flow and the particle diffusion.

Consider flow in a circular pipe at an inclination angle of β referenced to the direction of gravitational acceleration $\mathbf{g}$. A fully developed flow requires that the following expressions hold:

$$\frac{\partial}{\partial t} = 0; \quad \frac{\partial}{\partial z} = 0; \quad V = V_p = W = W_p = 0. \tag{11.5-1}$$

Thus, only the axial velocities of phases need to be investigated. The mass continuity of the fluid and particle phases provides:

$$\int_0^{2\pi} \int_0^R \left(1 - \alpha_p\right) \rho U r dr d\ \theta = J \qquad (11.5\text{-}2)$$

and:

$$\int_0^{2\pi} \int_0^R \alpha_p \rho_p U_p r dr d\ \theta = J_p, \qquad (11.5\text{-}3)$$

where J and J_p are the specified total mass flow rates for the fluid and particle phases, respectively.

For the gas phase flow at high Re and Fr, the gravity has an insignificant effect on the gas velocity distribution. With this stipulation, the gas velocity becomes a function of the radial coordinate only. Thus the momentum equation for the gas phase is simplified from Eq. (6.4-22) and expressed as:

$$- \left(1 - \alpha_p\right) \frac{dp}{dz} - \rho g \cos \beta + \frac{1}{r} \frac{d}{dr} \left(r \mu_e \frac{dU}{dr}\right) - \frac{\alpha_p \rho_p}{\tau_{rp}} \left(U - U_p\right) = 0, \quad (11.5\text{-}4)$$

where μ_e is the effective viscosity of turbulence, and the last term in Eq. (11.5-4) is the drag force formulated from Eq. (6.5-9c) that also defines the particle relaxation time τ_{rp} as:

$$\tau_{rp} = \left(C_D \frac{3}{4} \frac{\rho}{\rho_p} \frac{\mathbf{U} - \mathbf{U}_p}{d_p}\right)^{-1} = \left(\frac{3}{4} \frac{\mu}{d_p^2 \rho_p} C_D \mathrm{Re}_p\right)^{-1}. \qquad (11.5\text{-}4a)$$

Similarly, the momentum equation of particles is given as:

$$- \alpha_p \frac{dp}{dz} - \alpha_p(\rho_p - \rho)g \cos \beta + \frac{1}{r} \frac{\partial}{\partial r} \left(r \mu_{ep} \frac{\partial U_p}{\partial r}\right)$$
$$+ \frac{1}{r} \frac{\partial}{\partial \theta} \left(\mu_{ep} \frac{\partial U_p}{r \partial \theta}\right) + \frac{\alpha_p \rho_p}{\tau_{rp}} \left(U - U_p\right) = 0, \qquad (11.5\text{-}5)$$

where the particle viscosity μ_p can be related to the particle diffusivity D_{ep} through:

$$\mu_p = \alpha_p \rho_p D_{ep} \mathrm{Pr}_p, \qquad (11.5\text{-}5a)$$

where Pr_p is the Prandtl number of the particle phase. For a dilute suspension, the diffusion of particles is primarily due to the particle–gas interactions via turbulence of the gas phase, which leads to Pr_p of approximately unity.

The above four independent equations, Eqs. (11.5-2) to (11.5-5), can be solved jointly for the four independent variables: α_p, U, U_p, and p. Assuming the pertinent correlations for transport coefficients can be obtained, model closure has thus been achieved.

11.5.2   Effect of Electrostatic Charge on Phase Transport

In multiphase pipeline transport, the effects of electrostatic charge on solids, primarily generated via triboelectric charging from particle–wall contact, can be remarkable. The following is an analysis of the particle volume fraction distribution in a fully developed and dilute suspension flow containing monodispersed and charged spherical particles (Soo and Tung, 1971). For simplification, the motion of a single particle is subjected to the drag force, gravitational and buoyancy force, and the electrostatic force, while other forces such as the pressure gradient force are ignored. Based on Eq. (3.5-1) with $d\mathbf{U}_p/dt$ as zero for a fully developed flow, the equation of motion is given by:

$$\frac{\mathbf{U} - \mathbf{U}_p}{\tau_{rp}} + \left(1 - \frac{\rho}{\rho_p}\right)\mathbf{g} - \frac{q_e}{m}\nabla\Phi_E = 0, \tag{11.5-6}$$

where the electric potential, Φ_E is governed by Eq. (4.2-15) and expressed as:

$$\nabla \cdot (\varepsilon\nabla\Phi_E) = -\frac{6\alpha_p q_e}{\pi d_p^3}. \tag{11.5-6a}$$

In a fully developed and fully suspended transport, there should be no depositing in solid particle phase (*i.e.*, zero deposition rate), which means that the flux due to diffusion must equate to the drift flux. Thus, it yields:

$$-D_{ep}\nabla\left(\alpha_p\rho_p\right) + \alpha_p\rho_p\left(\mathbf{U}_p - \mathbf{U}\right) = 0. \tag{11.5-7}$$

Substituting Eq. (11.5-6) into the above expression gives the governing equation for the volume fraction distribution of particles as:

$$\frac{\nabla\left(\alpha_p\rho_p\right)}{\alpha_p\rho_p} = \frac{\tau_{rp}}{D_{ep}}\left\{\left(1 - \frac{\rho}{\rho_p}\right)\mathbf{g} - \frac{q_e}{m}\nabla\Phi_E\right\}. \tag{11.5-8}$$

Assuming that the ratio of relaxation time to particle diffusivity can be approximated as a constant across a pipe cross section, Eq. (11.5-8) may be directly integrated to yield the expression:

$$\frac{D_{ep}}{\tau_{rp}}\ln\alpha_p = C_1 - \left(1 - \frac{\rho}{\rho_p}\right)gr\sin\beta\cos\theta - \frac{q_e}{m}\Phi_E, \tag{11.5-8a}$$

where C_1 is the constant of integration determined from the imposed boundary conditions for particle concentration. Equation (11.5-8a) reveals that the volume fraction of particles, when acted upon by gravity and electrostatic charges, approximately follows an exponential distribution in the radial direction.

11.5.3   Dilute Transport in a Vertical Pipe

In a vertical pipe, the particles are transported upwards against gravity by an upward-carrying fluid flow in a similar manner as the riser flow discussed in Section 10.3. For both pipe and riser flows, the mixing and acceleration of particles near the bottom of the pipe or riser strongly depend on the methods or devices employed for particle feeding, such as bends from horizontal transport (*e.g.*, Figure 11.7-2) or nonmechanical valves from CFB applications (*e.g.*, L-valve or J-bends in Figure 10.3-1). For a long riser or pipe at a high fluid velocity, the particle transport can reach a state of dilute-phase transport given in Figures 10.3-2 and 10.3-3. At a moderate particle loading, the particle transport can be roughly represented by a core-annulus two-zone structure, with an upward dilute flow of particles in the core zone and a downward dense flow of particles in the annulus or a thin wall zone. There is a net particle migration from core to wall that sustains the downward moving particles in the annulus zone. It is noted that, due to this net particle migration, the particle concentration in the core keeps depleting so that the particle–fluid flow cannot be fully developed along the pipe or riser. At a very low particle loading, however, there is no core-annulus flow structure since the particles migrating to the wall region by turbulent diffusion will have a complete re-entrainment back to the mainstream, a condition described by the model in Section 11.5.1.

A simple model of the core-annular pipe flow can be extended from the flow transport model in risers given in Section 10.3.4, with neglecting the effects of phase acceleration and interparticle collisions. It is further assumed that the radial profiles within the core-zone are approximately uniform, and the gas in the annulus zone is stagnant (which may be partially justified by the counterbalancing effects of axial gradient in pressure and the downward moving of particles). Thus, the mass continuity equations of gas and particles in the core region can be simplified from Eq. (10.3-1) and Eq. (10.3-2), respectively, by

$$\frac{d}{dz}\left(\alpha_c \rho U_c A_c\right) = 0 \qquad (11.5\text{-}9)$$

$$\frac{d}{dz}\left(\alpha_{pc} \rho_p U_{pc} A_c\right) = -\Gamma_{pc}, \qquad (11.5\text{-}10)$$

where A_c is the cross-sectional area of core region, Γ_{pc} is the net mass transfer rate of solids across the core boundary from the core to the annulus wall zone. The momentum equation of gas phase can be obtained from the simplified Eq. (10.3-3) as

$$\frac{d}{dz}\left(\alpha_c \rho U_c^2 A_c\right) = -\alpha_c \frac{dp}{dz} A_c - \tau_c 2\pi R_c - \alpha_c \rho g A_c - f_D A_c, \qquad (11.5\text{-}11)$$

where the hydrodynamic force is dominated by the drag force, τ_c is the gas shearing force at the artificial core boundary, and R_c is the radius of the core. Similarly, the momentum equation of the solids phase in the core region is given from the simplified Eq. (10.3-4) as

$$\frac{d}{dz}\left(\alpha_{pc} \rho_p U_{pc}^2 A_c\right) = -\alpha_{pc} \frac{dp}{dz} A_c - \tau_{pc} 2\pi R_c - \alpha_{pc} \rho_p g A_c + f_D A_c - \Gamma_{pc} U_{sw}, \qquad (11.5\text{-}12)$$

where τ_{pc} is the solids shearing force at the artificial core boundary, and U_{sw} is the averaged solids velocity in the annulus wall zone, which is most likely negative (downward by gravity).

The area of the core region, A_c, in the above governing equations could be related to the wall boundary thickness, δ, as

$$A_c = \pi R_c^2 = \pi (R - \delta)^2. \qquad (11.5\text{-}13)$$

The wall film thickness is generally correlated to the geometric parameters as well as some local flow characteristics, including pipe diameter, solids mass flux, superficial gas velocity, and the relative measurement height against the total height of pipe, such as the one proposed by Harris *et al.* (2002):

$$\frac{\delta}{R} = 1 - 0.4014\alpha_p^{-0.0247}\mathrm{Re}^{0.0585}\left(1 - \frac{z}{H}\right)^{-0.0663}. \tag{11.5-14}$$

The mass transfer of solids across the core-annulus regional boundary can be formulated based on turbulent diffusion by (Bolton and Davidson, 1988):

$$\Gamma_{pc} = 2\pi\kappa_d R_c \rho_p \left(\alpha_{pce} - \alpha_{pc}\right), \tag{11.5-15}$$

where α_{pce} is an equilibrium solids concentration in the core region where deposition and entrainment occur at the same rate; and κ_d is the deposition coefficient accounting for the net turbulent diffusion of particles in the core region to the wall region. The deposition coefficient can be related to the turbulent fluctuation velocity of carrying fluid by

$$\kappa_d = \frac{1.2\sqrt{\pi}\,|u'|}{12 + \mathrm{St}}, \tag{11.5-16a}$$

where the amplitude of turbulent velocity fluctuation is estimated by

$$|u'| = U_c\left(1 - \frac{2.8}{\sqrt[8]{\mathrm{Re}}}\right), \tag{11.5-16b}$$

Therefore, with the available constitutive relations for the drag force and boundary frictional forces, the dilute transport model can be closed and solved (Wang *et al.*, 2010).

11.6   Stratified Flows in Pipes and Ducts

Two-phase stratified flow indicates that two phases flow in distinct parallel layers separated by a common interfacial boundary. The phase interactions through the interface cause the stratified flow to move in a wavy pattern, as shown in Figure 11.6-1. The analysis of stratified pipe flow is typically predicated on the conservation equations of regional-averaged phase properties (Delhaye, 1981; de Crecy, 1986; Zhu, 1991; Fan and Zhu, 1998).

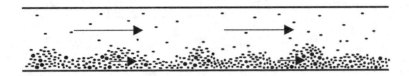

Figure 11.6-1 Two-phase stratified flow.

11.6.1   Regional-Averaged Theories of Stratified Flows

To simplify the analysis of two-phase stratified flow in a pipe or duct, the following assumptions are adopted:

(1) The stratified pipe flow is composed of two immiscible moving fluids, separated by an infinitesimal thickness at the interface;
(2) Each phase is continuous at any cross section;
(3) Regionally averaged flows are unsteady and one-dimensional; and
(4) The area and shape of the cross section of the pipe are constant, and the pipe wall is impermeable.

Regional-averaged modeling is considered as a limiting case of the volume-averaged approach described in Chapter 6. In an analogous process used to define volume-averages, an instantaneous regional average of phase j in region k is defined as:

$$\langle \phi_{jk} \rangle = \frac{1}{A_k} \int_{A_k(z,t)} \phi_{jk} dA. \tag{11.6-1}$$

Figure 11.6-2 conceptually illustrates the relevant derivation configuration.

The instantaneous regional-averaged continuity and momentum equations of phase j in region k can be expressed, respectively, as (Delhaye, 1981):

$$\frac{\partial}{\partial t} \left(A_k \langle \rho_{jk} \rangle \right) + \frac{\partial}{\partial z} \left(A_k \langle \rho_{jk} U_{jk} \rangle \right) = - \int_{C(z,t)} \left(\frac{dm_{jk}}{dt} \right) \frac{dC}{\mathbf{n}_k \cdot \mathbf{n}_{kc}} \tag{11.6-2}$$

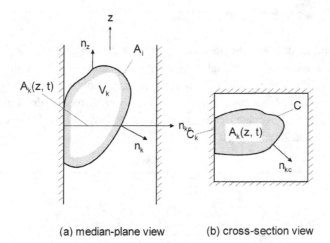

(a) median-plane view   (b) cross-section view

Figure 11.6-2 Configuration for regional averages of region k.

and

$$
\frac{\partial}{\partial t}\left(A_k\langle\rho_{jk}U_{jk}\rangle\right) + \frac{\partial}{\partial z}\left(A_k\langle\rho_{jk}U_{jk}^2\rangle\right) = \frac{\partial}{\partial z}\left(A_k\langle p_{jk}\rangle\right) + \frac{\partial}{\partial z}\left(A_k\langle\mathbf{n}_z\cdot\boldsymbol{\tau}_{jk}\cdot\mathbf{n}_z\rangle\right)
$$

$$
+ A_k\langle\rho_{jk}f_{zj}\rangle - \int_{C(z,t)}\left(\frac{dm_{jk}}{dt}\right)\frac{U_{jk}dC}{\mathbf{n}_k\cdot\mathbf{n}_{kc}} - \int_{C\cup Ck(z,t)}\mathbf{n}_z\cdot\left(\mathbf{n}_k\cdot(p_{jk}\mathbf{I}-\boldsymbol{\tau}_{jk})\right)\frac{dC}{\mathbf{n}_k\cdot\mathbf{n}_{kc}}.
$$

$$(11.6\text{-}3)$$

The quantities A_k, A_i, C_k, C, $\mathbf{n}_z$, $\mathbf{n}_{kc}$, and $\mathbf{n}_k$ are illuminated in Fig. 11.6-2. Specifically, $C(z,t)$ represents the common boundary of interface, A_i, within a cross section plane; $C_k(z,t) + C(z,t)$ defines an enclosed curve in the cross section with C_k on the wall of the pipe; and $\mathbf{n}_{kc}$ is the unit vector normal to C, located in a cross-sectional plane directed away from the region of interest, k.

The preceding equations can be further simplified using the limiting form of Gauss theorem and intrinsic phase properties of volume fraction, velocity, and tensor (Fan and Zhu, 1998). Accordingly, Eq. (11.6-2) becomes:

$$
\frac{\partial}{\partial t}\left(A_k\alpha_{jk}\rho_{jk}\right) + \frac{\partial}{\partial z}\left(A_k\alpha_{jk}\rho_{jk}U_{jk}\right) = \sum_{l=1}^{N-1}\left(\dot{m}_{jlk} - \dot{m}_{jkl}\right)C, \qquad (11.6\text{-}4)
$$

where $\dot{m}_{jkl}$ is the mass transfer from phase l to phase j in region k, and $\dot{m}+jkl$ is the mass transfer from phase j to phase l in region k. Eq. (11.6-3) can be expressed by:

$$
\frac{\partial}{\partial t}\left(A_k\alpha_{jk}\rho_{jk}U_{jk}\right) + \frac{\partial}{\partial z}\left(A_k\alpha_{jk}\rho_{jk}U_{jk}^2\right) = A_k\alpha_{jk}\rho_{jk}\langle f_z\rangle - \frac{\partial}{\partial z}\left(A_k\alpha_{jk}p_k\right) - p_{jki}\frac{\partial A_k}{\partial z}
$$

$$
+ \frac{\partial}{\partial z}\left(A_k\alpha_{jk}^i\langle\mathbf{n}_z\cdot\boldsymbol{\tau}_{jk}\cdot\mathbf{n}_z\rangle\right) + \sum_{l=1}^{N-1}\left(\dot{m}_{jlk}U_{jl} - \dot{m}_{jkl}U_{jk}\right)C - \varepsilon_k\tau_{jki}C - C_k\chi_{jf}\tau_{jkw},
$$

$$(11.6\text{-}5)$$

where p_{jki} and τ_{jki} are the interfacial pressure and stress of phase j in region k, respectively; ε_k is the sign function; χ_{jf} is the fraction of the frictional perimeter of phase j; and τ_{jkw} is the wall shear stress of phase j in region k.

11.6.2   Stratified Gas–Liquid Flows

For stratified gas–liquid horizontal pipe flows, the region of interest, k, is entirely occupied by either the liquid or the vapor phase; thus, the subscript j in the equations of the previous section is superfluous if the volume fraction $\alpha_{jk} = 1$. When the interfacial mass transfer is negligible, the continuity equations of the liquid and vapor phases or Eq. (11.6-4) simplify to (Barnea, 1991):

$$
\frac{\partial}{\partial t}\left(\rho_l A_l\right) + \frac{\partial}{\partial z}\left(\rho_l A_l U_l\right) = 0 \qquad (11.6\text{-}6a)
$$

$$
\frac{\partial}{\partial t}\left(\rho_g A_g\right) + \frac{\partial}{\partial z}\left(\rho_g A_g U_g\right) = 0. \qquad (11.6\text{-}6b)
$$

In the momentum equations of Eq. (11.6-5), the field force term vanishes because the direction of the gravitational force is perpendicular to the pipe-axis. The pressure, p_k, is then obtained from:

$$p_k = p_{ik} - \rho_k g \Delta h_k, \qquad (11.6\text{-}7)$$

where p_{ik} is the pressure of phase k at the interface, and Δh_k is the height of the center of gravity of phase k above the interfacial boundary. If the surface tension effect is neglected, the interface condition becomes $p_{il} = p_{ig} = p_i$. With further neglecting the viscous stress gradient, the momentum equations of liquid and gas phases can be expressed from Eq. (11.6-5), respectively, as:

$$\frac{\partial}{\partial t}(\rho_l A_l U_l) + \frac{\partial}{\partial z}\left(\rho_l A_l U_l^2\right) + A_l \frac{\partial p_i}{\partial z} + \rho_l g A_l \frac{\partial h}{\partial z} = -\tau_{wl} C_l - \tau_i C \qquad (11.6\text{-}8a)$$

$$\frac{\partial}{\partial t}\left(\rho_g A_g U_g\right) + \frac{\partial}{\partial z}\left(\rho_g A_g U_g^2\right) + A_g \frac{\partial p_i}{\partial z} - \rho_g g A_g \frac{\partial h}{\partial z} = -\tau_{wg} C_g - \tau_i C. \qquad (11.6\text{-}8b)$$

Equating the pressure drops of the above two momentum equations yields:

$$\rho_l \frac{\partial U_l}{\partial t} - \rho_g \frac{\partial U_g}{\partial t} + \rho_l U_l \frac{\partial U_l}{\partial z} - \rho_g U_g \frac{\partial U_g}{\partial z} + (\rho_l - \rho_g)g \frac{A}{(dA_l/dh)} \frac{\partial \alpha_l}{\partial z} = F, \qquad (11.6\text{-}9)$$

where:

$$\alpha_l = \frac{A_l}{A} \qquad \alpha_g = \frac{A_g}{A} \qquad (11.6\text{-}9a)$$

and

$$F = \frac{\tau_i C_i}{A}\left(\frac{1}{\alpha_g} + \frac{1}{\alpha_l}\right) - \frac{1}{A}\left(\frac{\tau_{wl} C_l}{\alpha_l} - \frac{\tau_{wg} C_g}{\alpha_g}\right). \qquad (11.6\text{-}9b)$$

Rearranging the continuity equations to the following expressions for the liquid and gas phases, respectively:

$$\frac{\partial \alpha_l}{\partial t} + \alpha_l \frac{\partial U_l}{\partial z} + U_l \frac{\partial \alpha_l}{\partial z} = 0 \qquad (11.6\text{-}10a)$$

$$\frac{\partial \alpha_g}{\partial t} + \alpha_g \frac{\partial U_g}{\partial z} + U_g \frac{\partial \alpha_g}{\partial z} = 0. \qquad (11.6\text{-}10b)$$

The linear stability of the set of differential equations prescribed by Eqs. (11.6-9) and (11.6-10) can be examined via a Kelvin–Helmholtz analysis, which yields a stability criterion for a gas–liquid stratified horizontal pipe flow where the viscous shearing is ignored (Barnea, 1991):

$$(U_g - U_l)^2 < (\alpha_l \rho_g + \alpha_g \rho_l)\frac{\rho_l - \rho_g}{\rho_l \rho_g}\frac{A}{(dA_l/dh)}g. \qquad (11.6\text{-}11)$$

Equation (11.6-11) can be used to estimate the regime transition from stratified flow to other patterns, such as slug or annular flows. If the stability condition is violated, a disturbance of finite amplitude will propagate and eventually destroy the stratified flow pattern. When the liquid level is high, a wave of increasing amplitude may forestall gas flow and facilitate plug or slug flow. Conversely, when there is insufficient

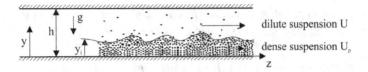

Figure 11.6-3 Stratified gas–solids flow in a horizontal pipe with selected coordinates.

liquid to completely bridge the cross section of the pipe, the onset of annulus flow is possible (Taitel and Dukler, 1976).

11.6.3 Stratified Gas–Solids Flow

The stratified gas–solid flow in a horizontal pipe comprises a dense layer of solids sliding along the pipe bottom while a dilute solids suspension flows atop at an increased velocity; Figure 11.6-3 provides a conceptual illustration of the process. The interface between the two flow layers is often unsteady and wavy, yielding pronounced mass and momentum transfer across the boundary. The concentration profile of solids is dominated by the effects of gravity settling, and it can be represented by a function of the position component in the direction of gravity.

In addition to the assumptions enumerated in Section 11.6.1, other simplifications are proffered for the efficient analysis of stratified flow in a horizontal pipe:

(1) The pressure distribution over the cross section is induced by gravity;
(2) Mass transfer between phases is neglected;
(3) The presence of particles in the dilute region is inconsequential, and particles in the dense region are considered to act as a moving bed;
(4) The fluid expansion rate in the axial direction of pipe for the gas phase ($j = $ gas) may be ignored. This is written as:

$$\frac{\partial}{\partial z}\left(A_k \alpha_k \left(\mathbf{n}_z \cdot \boldsymbol{\tau}_k \cdot \mathbf{n}_z\right)\right) \approx 0. \tag{11.6-12a}$$

(5) The solids expansion rate along the pipe axis ($j = $ particles) can be expressed as:

$$\frac{\partial}{\partial z}\left(A_k \alpha_{pk}\left(\mathbf{n}_z \cdot \mathbf{T}_{pk} \cdot \mathbf{n}_z\right)\right) = -\frac{\partial}{\partial z}\left(A_k \alpha_{pk} C_{pp}\frac{\partial U_{pk}}{\partial z}\right) + \beta_{pk}\left(U_k - U_{pk}\right),$$
$$\tag{11.6-12b}$$

where C_{pp} is the momentum transfer coefficient due to phase expansion from interparticle collisions; β_{pk} is the momentum transfer coefficient between gas and particle phases in region k.

For brevity, the case of stratified horizontal flow in a rectangular duct is only considered. In the dilute region, denoted by subscript "1," $\alpha_1 \approx 1$ and $U_1 \approx U$; the regional cross-sectional area is given by:

$$\frac{A_1}{A} = 1 - \frac{y_i}{h}, \tag{11.6-13}$$

where y_i is the depth of interface, and h is the height of the duct. From simplification (1) given above, the regional-averaged pressure can be estimated by:

$$p_1 = p_i + \frac{1}{h - y_i} \int_{y_i}^{h} g \left(\int_{y}^{y_i} \rho dy \right) dy = p_i - \frac{1}{2} \rho g \, (h - y_i), \qquad (11.6\text{-}14)$$

where p_i is the pressure at the interface. Based on Eqs. (11.6-4) and (11.6-5), it gives:

$$\frac{\partial A_1}{\partial t} + \frac{\partial}{\partial z} (A_1 U) = 0 \qquad (11.6\text{-}15)$$

and:

$$\frac{\partial}{\partial t} (A_1 U) + \frac{\partial}{\partial z} \left(A_1 U^2 \right) - \frac{g}{2} \frac{\partial}{\partial z} (A_1 (h - y_i)) = -\frac{A_1}{\rho} \frac{\partial p_i}{\partial z} + \frac{\tau_i C + \tau_{1w} C_1}{\rho}, \qquad (11.6\text{-}16)$$

where τ_{1w} is the wall shear stress in the dilute region projected on the axial direction of pipe.

In the dense region denoted by subscript "2," the solid particles are taken to behave as a contiguous moving bed. Assuming no relative slip in motion between the gas and solids, the mixture can be modeled as pseudo single-phase flow (i.e., $\alpha_m = 1$ and $U_2 = U_p$). The continuity and momentum equations of the mixture can be obtained from Eqs. (11.6-4) and (11.6-5), respectively, as:

$$\frac{\partial A_2}{\partial t} + \frac{\partial}{\partial z} \left(A_2 U_p \right) = 0 \qquad (11.6\text{-}17)$$

and:

$$\frac{\partial}{\partial t} \left(A_2 U_p \right) - \frac{A}{h} \left(2 U_p \frac{\partial y_i}{\partial t} + U_p^2 \frac{\partial y_i}{\partial z} \right) + g A \frac{y_i}{h} \frac{\partial y_i}{\partial z} = -\frac{A_2}{\rho_m} \frac{\partial p_i}{\partial z} + \frac{\tau_{2w} C_2 - \tau_i C}{\rho_m},$$
$$(11.6\text{-}18)$$

where τ_{mw} is the wall shear stress of the mixture in the dense region.

For internal unsteady wave motion, it is assumed that the total mass flow is constant. Thus, it yields:

$$\rho A_1 U + \rho_m A_2 U_p = const. \qquad (11.6\text{-}19)$$

Combining Eqs. (11.6-16), (11.6-18), and (11.6-19) yields a wave equation at the interface:

$$C_t \frac{\partial y_i}{\partial t} + C_z \frac{\partial y_i}{\partial z} = C_c, \qquad (11.6\text{-}20)$$

where:

$$C_t = \frac{2 \left(\rho U - \rho_m U_p \right)}{h} \qquad (11.6\text{-}20a)$$

$$C_z = \frac{\left(\rho U^2 - \rho_m U_p^2 \right)}{h} + \rho g + (\rho_m - \rho) \frac{y_i}{h} \qquad (11.6\text{-}20b)$$

$$C_c = -\frac{\partial p_i}{\partial z} - \frac{(C_1 \tau_{1w} + C_2 \tau_{mw})}{A}. \qquad (11.6\text{-}20c)$$

According to the preceding expressions, unsteady flow persists because $C_t \neq 0$. The wave speed of the interface is obtained from:

$$U_i = \frac{C_z}{C_t} = \frac{\left(\rho U^2 - \rho_m U_p^2\right) + \rho g h + (\rho_m - \rho) y_i}{2\left(\rho U - \rho_m U_p\right)}. \qquad (11.6\text{-}20d)$$

A steady wave motion, evidenced when $C_c = 0$, facilitates a balance of pressure drop and wall friction of the form:

$$-\frac{\partial p_i}{\partial z} = \frac{C_1 \tau_{1w} + C_2 \tau_{mw}}{A}. \qquad (11.6\text{-}20e)$$

11.7 Flows in Bends

Curved pipes, or bends, contribute to flexibility in flow transport systems by facilitating routing and distribution. Particle-laden multiphase flows through pipe bends commonly occur. Understanding their flow characteristics is practically important as it affects the pressure drop along the pipeline and may manifest erosion resulting from the directional impacts of the solids on interior pipe wall.

11.7.1 Single-Phase Flow in a Pipe Bend

When a flow passes through a curved pipe, a pressure difference develops between the inner radius and outer radius of the bend. The pressure difference induces a pair of circulatory fluid movements near the wall region of the bend cross section; this is commonly referred to as secondary flow. The flow characteristics in a bend are schematically illustrated in Figure 11.7-1, with the secondary flow occurring in the shedding layer near the pipe wall.

A simple flow model is used to demonstrate the parametric effect on flow structure and pressure drop in a pipe bend. It is assumed that variations of the flow pattern along the axis of the bend can be neglected. Thus, the momentum equations governing the shedding layer, expressed in integral form for the peripheral and axial pipe directions, can be written, respectively, as (Ito, 1959):

$$\rho \frac{\partial}{\partial \theta} \int_0^\delta W^2 d\xi = -\tau_\theta R_d + \rho \frac{\delta R_d}{R_B} U_1^2 - \rho \frac{R_d}{R_B} \sin\theta \int_0^\delta U^2 d\xi \qquad (11.7\text{-}1a)$$

and:

$$\rho \frac{\partial}{\partial \theta} \int_0^\delta (U - U_1) W d\xi = -\tau_\phi R_d - \rho \frac{R_d \delta}{R_B} \frac{\partial p}{\partial \phi}, \qquad (11.7\text{-}1b)$$

where R_B is the radius of curvature along the pipe axis; R_d is the radius of the circular cross section of the pipe; δ is the thickness of the shedding layer; and U_1 denotes the axial velocity at the edge of the shedding layer. The pressure gradient along the

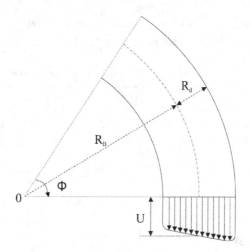

(a) Axial velocity distribution

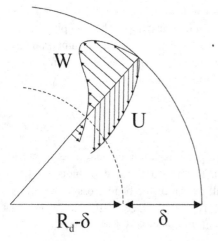

(b) Velocity distribution in the shedding layer

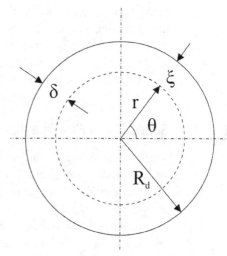

(c) Notation of pipe cross section

Figure 11.7-1 Schematic representation of flow in a bend.

peripheral direction is balanced by the second-flow induced shear friction at the wall as given by:

$$\frac{\partial p}{\partial \phi} = -\frac{2}{\pi} \frac{R_B}{R_d} \int_0^\pi \tau_\phi d\theta. \tag{11.7-1c}$$

The axial and peripheral flow velocities within the shedding layer are estimated, respectively, by:

$$U = U_1 \left(\frac{\xi}{\delta}\right)^{1/7} ; \qquad W = B U_1 \left(\frac{\xi}{\delta}\right)^{1/7} \left(1 - \frac{\xi}{\delta}\right), \tag{11.7-2}$$

where B is a factor addressed subsequently. The shear stress at the wall is given as a function of the tangential velocity near the pipe wall and ξ as (Schlichting, 1979):

$$\tau_w = 0.0225 \left(U^2 + W^2 \right)^{7/8} \rho^{3/4} \left(\frac{\mu}{\xi} \right)^{1/4}. \tag{11.7-3a}$$

Considering the relationships $\tau_\theta / \tau_\varphi = W/U$ and $\tau_\theta^2 + \tau_\varphi^2 = \tau_w^2$, it gives:

$$\tau_\varphi = \frac{\tau_w}{\sqrt{1 + B^2}} \qquad \tau_\theta = \frac{B\tau_w}{\sqrt{1 + B^2}}. \tag{11.7-3b}$$

The factor B and shedding layer thickness, δ, are determined by numerically solving the integral momentum equations of Eq. (11.7-1), together with the velocity and wall shear stress profiles of Eq. (11.7-2) to Eq. (11.7-3). The axial component in the central core of the fluid flow is assumed to be of the form:

$$U = U_m + Ar \cos \theta, \tag{11.7-4a}$$

where the constant A is determined by the condition of continuity imposed on the secondary flow. Prescribing the condition $R_d \ll R_B$, B and δ can be approximated from:

$$B^2 \approx \frac{R_d}{R_B} \qquad \frac{\delta}{R_d} \approx \left(\frac{\mathrm{Re}\, R_d^2}{R_B^2} \right)^{-1/5}. \tag{11.7-4b}$$

Ultimately, an expression for the friction factor of a pipe bend can be gleaned from its fundamental definition as:

$$f_B = -\frac{R_d}{R_B} \frac{\partial p}{\partial \phi} \frac{4}{\rho U_m^2} \approx 0.3 \left(\frac{1}{Re^2} \frac{R_d}{R_B} \right)^{1/10}. \tag{11.7-4c}$$

11.7.2 Particulate Flow in a Pipe Bend

Gravitational and centrifugal settling are relevant considerations for particle-laden suspension flows passing through a curved pipe. The effect of gravity depends on the orientations of both the flow and pipe bend. For horizontal pipe flow entering a bend, a dense layer of particles typically undergoes a swirling motion as it progresses through the bend, a phenomenon known as particle roping; the process is schematically represented in Figure 11.7-2.

Particle roping is primarily manifested via gravity-induced settling of a nonuniform distribution of particle concentration at the bend inlet. It is generally with much denser and slower moving particles remaining near the bottom of the horizontal pipe. The combined effects of gravity, the Coriolis force, the centrifugal force, the hydrodynamic forces from both the main and secondary flow streams, and the collision forces with the bend wall cause particles to separate from the main fluid stream and tend toward the wall in a swirling or "roping" fashion. Due to the frictional collisions with the bend wall, particles reduce momentum as they decrease the overall transport velocity while redirecting the flow. Increased drag forces result from the

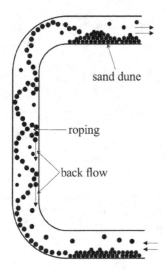

Figure 11.7-2 Schematic representation of roping and back flow in bend.

increased velocity slip, which accelerates slowed particles. A severe particle deceleration may lead to reverse flow by gravity in a vertical section or dune formation in the pipe section downstream of the bend, either of which contributed to flow instability. These general dual-phase flow phenomena are shown in Figure 11.7-2. Because the particles are invariably subjected to fluctuating acceleration as they progress through the bend, hydrodynamic forces that control particle motion should include any acceleration-dependent forces, such as the carried mass force and Basset force as appropriate.

For a dilute suspension, motion of the particle phase may be modeled by tracking a single particle in a bend, as illustrated in Figure 11.7-3 where the pipe bend is located in the vertical plane with vertical approaching flow. Consider a particle sliding along the outer surface of the bend via centrifugal force and by the inertia effects, with a negligible rebounding effect. For simplicity, the secondary-flow effects and the acceleration-dependent hydrodynamic forces are neglected.

The equation of motion for a particle in the axial direction of the pipe is given by:

$$\frac{U_p}{R_d + R_B}\frac{dU_p}{d\phi} = \frac{3\rho}{4\rho_p}\frac{C_D}{d_p}(U - U_p)^2 - g\cos\phi - f_p\left(\frac{U_p^2}{R_d + R_B} - g\sin\phi\right),\quad (11.7\text{-}5)$$

where f_p is the sliding friction coefficient between the particles and the pipe wall. The last term on the right-hand side of the above expression requires the normal force against the wall to be positive and within the range of ϕ considered. Thus, the criterion can be given as:

$$U_p^2 \geq (R_d + R_B)\, g \sin\phi. \qquad (11.7\text{-}5a)$$

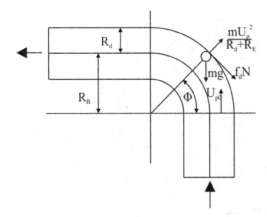

Figure 11.7-3 Particle-laden flow through a bend with vertical flow inlet.

Hence, once the local fluid velocity and drag coefficient are defined, the distribution of the particle velocity along the bend can be obtained via numerical integration of Eq. (11.7-5).

To estimate the decrease in particle velocity due to pipe curvature, it is first assumed that the drag force is inconsequential when compared with the sliding friction force between the particle and the bend surface. With this consideration, Eq. (11.7-5) reduces to

$$\frac{dU_p^2}{d\phi} + 2f_p U_p^2 = -2\,(R_d + R_B)\,g\,(\cos\phi - f_p\sin\phi). \tag{11.7-5b}$$

Imposing the boundary condition of $U_p = U_{p0}$ at $\phi = 0$ allows the preceding equation to yield a particle velocity distribution along the bend or ϕ, expressed as:

$$U_p^2 = \left\{ U_{p0}^2 + \frac{6f_p\,(R_d + R_B)\,g}{1 + 4f_p^2} \right\} e^{-2f_p\phi} - \frac{2\,(R_d + R_B)\,g}{1 + 4f_p^2}\left(3f_p\cos\phi + \left(1 - 2f_p^2\right)\sin\phi\right). \tag{11.7-5c}$$

The pressure drop for the particle-laden flow through the pipe bend is correlated to the particle deceleration and wall friction resulting from both particle–wall collisions and fluid flow wall shearing. These effects may also be influenced by particle roping, particle accumulation, reverse flow, reacceleration, or turbulence modulation.

11.7.3 Bend Erosion by Particle Collision

Abrasive erosion by particle impacts with the pipe wall in a bend is usually much more pronounced than the wear in the straight pipe sections. The location of the most severe erosion is typically near the extrados of a pipe bend, but attrition of the pipe wall primarily depends on material properties and particle trajectories inside the bend section. For ductile materials, the typical impact angles where maximum erosion occurs range from 20° to 45°, with the smaller angles being associated with pointed eroding particles. For brittle materials, however, maximum erosion occurs

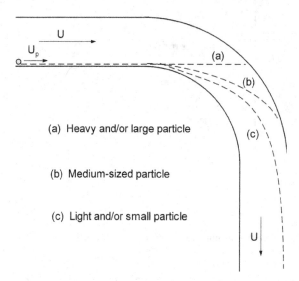

Figure 11.7-4 Trajectories of particles at different inertia in a bend.

due to brittle fractures caused by particle impingement at angles close to 90°. Particle trajectories in a pipe bend are governed by particle inertia and hydrodynamic forces. Figure 11.7-4 conceptually illustrates the likely paths of particles with different inertias. For large or heavy particles, trajectories follow nearly straight paths, resulting in oblique impingements on the bend wall. For very small or light particles, trajectories closely follow the primary fluid motion, with particle–wall collisions occurring sparsely.

Parametric effects on abrasive erosion of a pipe bend may be analyzed via Hertzian contact and collision theory in Section 4.3. (Soo, 1977). In this model, ductile or brittle erosion is assumed to be caused by impingements of granular materials suspended in a gas flow at a moderate speed. The effects of gliding, scattering, and lifting on the particles by the gas stream are neglected.

Abrasive erosion is quantified by the mass of material removed per unit mass of colliding particles. For ductile erosions, the volume loss per impact is equivalent to the work of ductile wear via the ratio of tangential force to the energy required to remove a unit volume of material, which can be expressed by:

$$w_D = \frac{A_C t_C}{\varepsilon_D} \left(\frac{F_t}{A_C} - Y_D \right) U_C \cos \alpha_i, \qquad (11.7\text{-}6)$$

where w_D represents the wear in volume loss per impact; ε_D denotes the energy required to remove a unit volume of material in the ductile mode; U_C is the collision velocity, α_i is the impact angle, whereas A_C and t_C are contact area and collision time, respectively; F_t is the tangential force and Y_D is the yield stress in ductile mode. In an approximation of Hertzian contact and assuming sliding friction for F_t, Eq. (11.7-6) can be expressed in dimensionless form as:

$$E_D \equiv \frac{w_D \varepsilon_D}{F_t t_C U_C} = \frac{w_D \varepsilon_D}{f F_n t_C U_C} = \cos \alpha_i \left\{ 1 - K_D (\sin \alpha_i)^{-1/5} \right\}, \quad (11.7\text{-}6a)$$

where E_D is a ductile erosion parameter, and the normal impulsion during collision is given by:

$$F_n t_C = 2.94 \frac{5}{16} m U_C (1 + e), \quad (11.7\text{-}6b)$$

where F_n is the normal force, f is the sliding friction coefficient, and the restitution coefficient, e, is defined as the ratio of rebounding velocity to collision velocity. K_D is a ductile resistance parameter, defined by:

$$K_D = 3\pi 2^{3/5} \frac{Y_D}{f} \sqrt{k_1 k_2} (1 + e)^{-1/5} N_{Im}^{-1/5}, \quad (11.7\text{-}6c)$$

where $k = (1 - v)/\pi E$; subscripts "1" and "2" denote each of the colliding pairs; and N_{Im} is the impact number, defined as:

$$N_{Im} = \frac{5\pi^2}{2} \rho_p U_C^2 \sqrt{k_1 k_2} \left(\sqrt{\frac{k_1}{k_2}} + \sqrt{\frac{k_2}{k_1}} \right)^4. \quad (11.7\text{-}6d)$$

The above correlates with the dynamic and material properties of the impacting system. Equation (11.7-6c) indicates that, because $(1 + e)^{-1/5}$ nearly equates to unity, K_D only depends on material properties and the collisional speed. Thus, with K_D treated as a parametric constant, Eq. (11.7-6a) embodies the dimensionless ductile wear at various angles of impact. Hence, the impact angle at which the maximum wear occurs can be determined by:

$$K_D \left(1 + 4(\sin \alpha_m)^2 \right) = 5(\sin \alpha_m)^{11/5}. \quad (11.7\text{-}6e)$$

Similarly, for brittle erosions, the brittle erosion parameter, E_B, can be defined as:

$$E_B \equiv \frac{w_B \varepsilon_B}{F_n t_C U_C} = \sin \alpha_i \left\{ 1 - K_B (\sin \alpha_i)^{-1/5} \right\}, \quad (11.7\text{-}7)$$

where the brittle resistance parameter, K_B, is defined by:

$$K_B = 3\pi 2^{3/5} Y_B \sqrt{k_1 k_2} (1 + e)^{-1/5} N_{Im}^{-1/5}. \quad (11.7\text{-}7a)$$

Equation (11.7-7) reveals the dimensionless brittle erosion as a function of impact angle, which suggests that the maximum brittle wear occurs at $\alpha_i = 90°$, (*i.e.*, at normal collision).

The above delineation of ductile and brittle modes through directional impacts represents the most fundamental surface wear model by granular particles. In actuality, surface wear often results from the simultaneous effects of the ductile and brittle modes. Effects of particle gliding, scattering, and lifting by the gas stream may also contribute to particle impacts. For example, the Saffman force arising from the boundary layer motion of the gas may induce gliding (*i.e.*, barely touching) or lifting (*i.e.*, noncontact) of particles; conversely, the deflective motion of fluid flow causes a centrifugal force to act on the particles that augments the compressive stresses associated with impacts.

11.8   Case Studies

The following case studies provide additional discussion on multiphase pipe flows involving simple applications or direct extensions. The studies include a Lagrangian–Eulerian modeling of a turbulent gas–solid flow over a bend and associated bend erosion, an Eulerian–Eulerian modeling of a dense slurry flow over a bend, a simple mechanistic model to estimate the flow conditions for the transition of a stratified flow into a nonstratified one in horizontal pipes, and an Eulerian–Eulerian modeling of a fully developed slurry pipe flow.

11.8.1   Particle–Laden Gas Flow and Erosion in a Bend

As schematically described in Figure 11.7-2, a particle-laden flow over a bend is very complex, with nonuniform phase distributions complicated by the gravity setting, peripheral flow, particle rebounding and sliding from particle–wall impaction, and frequently a turbulent flow that is further modulated by fluid–particle interactions. The analytical flow modeling given in Section 11.7 provides a simple analysis of basic characteristics for such complex flows. However, with rapid development in computational capability, numerical modeling becomes popular, which can lead to more descriptive and reliable field information of the transport phenomena.

Problem Statement: Consider a turbulent and dilute gas–solid suspension flow over a bend. Establish a CFD model that can be solved for the nonuniform phase distributions as well as the pipe erosion rate and distributions over the bend.

Analysis: As introduced in Section 11.7.3, the determination of pipe erosion rate and location depends on the flow-related properties concerning the particle impact such as particle impact velocity and impact angle. Such properties can be obtained by tracking the particle trajectories in Lagrangian coordinates. Since the gas phase flow is turbulent, the turbulence dispersion of particles can be accounted for using stochastic particle trajectory tracking.

The following is a simplified model over the work of Njobuenwu *et al.* (2013) and Duarte *et al.* (2015). An Eulerian–Lagrangian model is used with the Eulerian modeling for gas flow and the Lagrangian trajectory modeling for the solid particle transport. The particle phase is sufficiently dilute so that the Eulerian–Lagrangian model of the gas–solid flow is considered as two-way coupling, with the interparticle collisions ignored in the governing equations of phase transport. It is assumed that the solid particles are spherical and monodispersed, without agglomeration, clustering, or breakup during the transport process. The gas flow is regarded as incompressible and isothermal. For simplicity, only the drag force and the pressure gradient force are considered for the gas–solid interphase momentum transfer, while all the other forces such as added mass, Basset, and Saffman forces are neglected. Due to the very large density ratio of particle to gas in this case, such neglect is appropriate.

For the Eulerian model of the turbulent gas flow, the volume–time-averaged governing equations can be expressed, based on Eq. (6.4-21) and Eq. (6.4-22), as:

$$\frac{\partial}{\partial t}(\alpha\rho) + \nabla \cdot (\alpha\rho\mathbf{U}) = 0 \tag{11.8-1}$$

$$\frac{\partial}{\partial t}(\alpha\rho\mathbf{U}) + \nabla \cdot (\alpha\rho\mathbf{U}\mathbf{U}) = -\alpha\nabla p + \nabla \cdot \left[\alpha\mu\left(\nabla\mathbf{U} + \nabla\mathbf{U}^T\right)\right] + \alpha\rho\mathbf{g} - \mathbf{F}_D + \nabla \cdot (-\rho\overline{\mathbf{u'u'}}), \tag{11.8-2}$$

where the gas–solid drag force, $\mathbf{F}_D$, is given by Eq. (6.8-4a), and the last term in Eq. (11.8-2) is the turbulent Reynolds stress, which can be further modelled either by $k - \varepsilon$ model with turbulence modulation (such as Eqs. (6.7-7), (6.7-8), (6.7-15), and (6.7-16)) or by RSM introduced in Section 2.3.4.3 or Eq. (2.3-23).

For the Lagrangian trajectory model of solid particles, the equation of translational motion of the ith particle is given, based on Eq. (5.3-30), by

$$m_{pi}\frac{d\mathbf{u}_{pi}}{dt} = \beta_i(\mathbf{u} - \mathbf{u}_{pi}) + V_{pi}\nabla p + m_{pi}\mathbf{g}, \tag{11.8-3}$$

where β is the momentum transfer coefficient; $\mathbf{u}$ is the instant gas velocity at the particle position, calculated by

$$\mathbf{u} = \mathbf{U} + \mathbf{u'}, \tag{11.8-3a}$$

where $\mathbf{U}$ is the averaged velocity determined from the Eulerian model whereas the random fluctuating component $\mathbf{u'}$ needs to be obtained from a statistical turbulent dispersion model such as using Eq. (5.2-6a) or Langevin dispersion model. Similarly, the equation of rotational motion of the ith particle is given by Eq. (5.3-31):

$$I_{pi}\frac{d\omega_{pi}}{dt} = \mathbf{T}_t + \mathbf{T}_r, \tag{11.8-4}$$

where the tangential friction torque $\mathbf{T}_t$ and the rolling friction torque $\mathbf{T}_r$ are due to particle–wall collisions, estimated by Eq. (5.3-28) and Eq. (5.3-29), respectively. The equation of trajectory is given by

$$\frac{d\mathbf{r}_{pi}}{dt} = \mathbf{u}_{pi}. \tag{11.8-5}$$

The gas volume fraction is related to the particles within the local cell, and is given by:

$$\alpha = 1 - \frac{1}{V_c}\sum_i^{k_c} V_{pi}. \tag{11.8-6}$$

Thus, six independent equations, Eqs. (11.8-1)–(11.8-6), are obtained for six unknowns (α, $\mathbf{U}$, p, $\mathbf{u}_p$, ω_p, and $\mathbf{r}_p$). With the assistance of the turbulent model of Reynolds stress and the statistical model of turbulent fluctuating velocity, the modeling closure is now established.

To solve the equations, boundary conditions should be imposed. One particular boundary condition involves the detailed modeling of particle–wall collisions. One popular approach is based on the restitution coefficients, defined as ratios of rebounding velocities to the incoming collision velocities. For impact of hard particles on wall surfaces of ductile materials, the restitution coefficients can be empirically correlated to the impact angle by (Grant and Tabakoff, 1975)

$$e_n = 0.993 - 1.76\alpha_i - 1.56\alpha_i^2 + 0.49\alpha_i^3$$
$$e_t = 0.988 - 1.66\alpha_i - 2.11\alpha_i^2 + 0.67\alpha_i^3,$$

(11.8-7)

where the subscripts n and t stand for the normal and tangential directions respectively, and α_i is the impact angle in radians.

Once the particle trajectories and impact properties such as velocity and impact angle are solved from the above CFD model, the wall erosion rate can be further evaluated from the erosion model of impact. There are various erosion models, including Finnie's model (Finnie and McFadden, 1978), Soo's model such as Eq. (11.7-6), and empirical-correlation models (Oka *et al.*, 2005). The selection of these models depends on the collision modules such as ductile or brittle modes as well as particle shapes and material properties of colliding parties.

Comments: (1) Even if the overall flow is dilute, the gravity sedimentation and roping may still yield a dense layer of particles and can form stagnant piles of particles when the transport velocity is below the saltation velocity during the particle transport over a bend. In such a case the interparticle collusions can be as important as the particle–wall collision to the phase transport. Hence, a coupled DEM model may be added to the preceding equations. (2) The boundary conditions of gas phase, including wall functions, can also be significantly affected by the anisotropic turbulence and particle–wall collisions. (3) Case simulation examples can be found in the references (Njobuenwu *et al.*, 2013; Duarte *et al.*, 2015), which also reveal the roping phenomena in particle-laden flows over bends.

11.8.2   Modeling of Slurry Flow over a Bend

When the particle loading is high, the particle-laden flow over a bend can be conveniently described for its transport phenomena using a multifluid model.

Problem Statement: Consider a turbulent and dense slurry suspension flow over a bend. Establish a CFD model for the evaluation of the slurry loading effect on the nonuniform phase distributions and pressure over the bend. The model should include the effect of interparticle collisions as well as particle–wall collisions.

Analysis: The following is a model simplified over the work of Kaushal *et al.* (2013). An Eulerian–Eulerian model is used for this problem, coupled with a modified k-ε model for the Reynolds stress and a kinetic theory modeling for the solid phase stresses from interparticle collisions. The solid particles are assumed to be spherical, monodispersed without agglomeration or breakup.

For the Eulerian–Eulerian model of liquid–solid flow, the continuity equation for the liquid and solid phases can be expressed, based on Eq. (6.4-21), respectively as:

$$\frac{\partial}{\partial t}(\alpha\rho) + \nabla \cdot (\alpha\rho\mathbf{U}) = 0 \tag{11.8-8}$$

$$\frac{\partial}{\partial t}(\alpha_p\rho_p) + \nabla \cdot (\alpha_p\rho_p\mathbf{U}_p) = 0, \tag{11.8-9}$$

where the volume fraction conservation gives

$$\alpha + \alpha_p = 1. \tag{11.8-10}$$

Considering the interphase momentum transfer due to pressure gradient force, drag force, carried mass force, and lift forces, the momentum equation of the liquid phase is given, based on Eq. (6.4-22), by:

$$\frac{\partial}{\partial t}(\alpha\rho\mathbf{U}) + \nabla \cdot (\alpha\rho\mathbf{U}\mathbf{U}) = -\alpha\nabla p + \nabla \cdot \left[\alpha\left(\mu + \mu_e\right)\left(\nabla\mathbf{U} + \nabla\mathbf{U}^T\right)\right] + \alpha\rho\mathbf{g}$$
$$+ \beta(\mathbf{U}_p - \mathbf{U}) + C_A\alpha_p\rho\left(\mathbf{U}_p \cdot \nabla\mathbf{U}_p - \mathbf{U} \cdot \nabla\mathbf{U}\right) + C_L\alpha_p\rho\left(\mathbf{U} - \mathbf{U}_p\right) \times \left(\nabla \times \mathbf{U}\right), \tag{11.8-11}$$

where the momentum transfer coefficient, β, is given by Eq. (6.5-9c), and the turbulent viscosity, μ_e, is determined by Eq. (6.8-6c) and k and ε equations of a homogeneous slurry by Eq. (6.8-6) and Eq. (6.8-7), respectively. The momentum equation of particle phase is given, based on Eq. (6.4-22), by:

$$\frac{\partial}{\partial t}(\alpha_p\rho_p\mathbf{U}_p) + \nabla \cdot (\alpha_p\rho_p\mathbf{U}_p\mathbf{U}_p) = -\alpha_p\nabla p + \nabla \cdot \left[\mathbf{T}_p\right] + \alpha_p\rho_p\mathbf{g}$$
$$- \beta(\mathbf{U}_p - \mathbf{U}) - C_A\alpha_p\rho\left(\mathbf{U}_p \cdot \nabla\mathbf{U}_p - \mathbf{U} \cdot \nabla\mathbf{U}\right) - C_L\alpha_p\rho\left(\mathbf{U} - \mathbf{U}_p\right) \times \left(\nabla \times \mathbf{U}\right), \tag{11.8-12}$$

where the total stress of the particle phase, $\mathbf{T}_p$, is given by Eq. (6.6-1) and determined from the kinetic theory of granular flows in Section 2.5.

For nine unknowns (α, α_p, $\mathbf{U}$, $\mathbf{U}_p$, p, μ_e, k, ε, and $\mathbf{T}_p$), nine independent equations, Eqs. (11.8-8)–(11.8-12) along with Eq. (6.8-6c), Eq. (6.8-6), Eq. (6.8-7), and Eq. (6.6-1), are obtained. The closure of the modeling is now established.

Comments: (1) The particle–wall interactions are included in the boundary conditions of the particle phase as well as in the wall functions of the turbulent boundary layer of the fluid phase. The local information on the particle phase stresses on the pipe wall obtained may be used for the evaluation of pipe erosion. However, a thorough erosion model that is based on two-fluid modeling is yet to be developed. (2) Case simulation examples can be found in the reference (Kaushal et al., 2013).

11.8.3 Modeling of Transition of Stratified to Nonstratified Flow

Problem Statement: The regime map of horizontal gas-liquid pipe flow is depicted in Figure 11.2-4. For gas–liquid flows in horizontal pipes, develop a model to account for the regime transition from stratified to nonstratified flow.

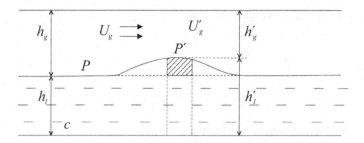

Figure 11.8-1 Instability in stratified gas–liquid flow.

Analysis: The following model is based on the work of Taitel and Dukler (1976). For brevity, consider a two-dimensional, gas–liquid stratified duct flow between parallel plates with an initially stratified gas layer of height, h_g, and a liquid layer of height, h_l, as schematically illustrated in Figure 11.8-1.

When a small disturbance occurs at the interface, a corresponding variation of the gas flow cross-sectional area elicits a change in the bulk gas velocity. Using Bernoulli's principle, the pressure change in the gas phase is given by:

$$p - p' = \frac{1}{2}\rho_g U_g^2 \left(\frac{h_g^2}{h_g'^2} - 1 \right).$$
(11.8-13)

In a similar fashion, the liquid pressure at elevation h_l under the now-raised liquid level in the presence of the disturbance is obtained as:

$$p_l = p' + (\rho_l - \rho_g) g \left(h_g - h'_g \right).$$
(11.8-14)

For the wave to propagate, the pressure at the undisturbed liquid surface, which is presently equal to the gas phase pressure p, must have the capacity to overcome the increase in liquid pressure under the wave. The required condition is therefore $p > p_l$, which yields:

$$U_g > C_1 \sqrt{\frac{(\rho_l - \rho_g)g h_g}{\rho_g}}.$$
(11.8-15)

The coefficient, C_1, in the above expression is given by:

$$C_1 = \sqrt{\frac{2h_g'^2}{h_g \left(h_g + h'_g \right)}}.$$
(11.8-15a)

For infinitesimal waves, h'_g approaches h_g, and C_1 approaches 1. Thus, for a two-dimensional duct, a small disturbance of the stratified flow encourages interfacial instability when the gas velocity is higher than the minimum requirement predicted by Eq. (11.8-15). The wave will ultimately cause the local liquid level to rise. At low gas velocities, the liquid can completely impinge the cross section and form plug or result in slug flow. At high gas velocities, insufficient liquid remains to sustain the plug flow, and the liquid wave sweeps around the walls, resulting in an annular flow.

The criteria of Eq. (11.8-15) can be extended to circular pipes, and the transition from stratified to nonstratified flow regimes occurs when:

$$U_g > C_2 \sqrt{\frac{(\rho_l - \rho_g)gA_g}{\rho_g\,(dA_l/dh_l)}}.$$ (11.8-16)

The coefficient, C_2, is approximated by:

$$C_2 = 1 - \frac{h_l}{D}.$$ (11.8-16a)

Comments: (1) Equation (11.8-15) or Eq. (11.8-16) provides the transition criterion for horizontal gas–liquid pipe flows from stratified to nonstratified flows. This model is proven to work well for small-diameter pipes. It is even suitable for pipes with small inclination angles β (*e.g.*, less than $10°$) if g is replaced by $g \cos \beta$ in the corresponding equations. (2) All geometric information, such as A_g and dA_l/dh_l, can be calculated from dimensions of the pipe cross-section, provided the equilibrium liquid level, h_l, is determined. (3) Equation (11.8-16) is equivalent to the stability condition of Eq. (11.8-11), if it is sensible to assume that $U_l \ll U_g$ and $\alpha_g \rho_l \ll \alpha_l \rho_g$.

11.8.4 Modeling of Fully Suspended Slurry Pipe Flow

As shown in Figure 11.2-3, when the transport velocity at a given solids loading is higher than the corresponding deposition velocity predicted by Eq. (11.3-4), the slurry flow can be maintained to be fully suspended. While the pressure gradients of homogeneous slurry pipe flows can be estimated by Eq. (11.3-4) with friction coefficients calculated using the formula given in Table 9.5-1, details of the pressure gradients and flow nonhomogeneity of slurries of various rheological properties from different liquids, particle sizes, and solids loadings may still need to be resolved from numerical modeling.

Problem Statement: Consider a turbulent and dense slurry suspension flow in a horizontal pipe. Establish a CFD model for evaluating the slurry loading effect on the nonuniform phase distributions and pressure gradients.

Analysis: The following is a simplified model based on the work of Messa *et al.* (2014). An Eulerian–Eulerian model is adopted for this problem, coupled with a modified k-ε model for the Reynolds stress and turbulent diffusion and a mixture viscosity model of solid stresses from the particle packing effect. The solid particles are assumed to be spherical, monodispersed without agglomeration or breakup. The flow is assumed to be steady-state and isothermal.

The volume–time-averaged continuity equation of the liquid can be expressed, based on Eq. (6.4-21), as:

$$\nabla \cdot (\alpha \rho \mathbf{U}) = \nabla \cdot (\rho D_e \nabla \alpha),$$ (11.8-17)

where the eddy diffusion coefficient can be related to the eddy viscosity by

$$D_e = \frac{\mu_e}{\rho \sigma_\alpha}.$$ (11.8-17a)

Similarly, the volume–time-averaged continuity equation of solids is given by

$$\nabla \cdot (\alpha_p \rho_p \mathbf{U}_p) = \nabla \cdot (\rho_p D_{ep} \nabla \alpha_p), \tag{11.8-18}$$

where the solids eddy diffusion coefficient, D_{ep}, can be related to D_e by the Hinze–Tchen model or Eq. (6.6-9a).

The momentum equation of liquid phase can be obtained, based on Eq. (6.4-22), as:

$$\nabla \cdot (\alpha \rho \mathbf{U} \mathbf{U}) = -\alpha \nabla p + \nabla \cdot \left[\alpha \left(\mu + \mu_e \right) \left(\nabla \mathbf{U} + \nabla \mathbf{U}^T \right) \right] + \alpha \rho \mathbf{g} + \nabla \cdot (\rho D_e \mathbf{U} \nabla \alpha)$$
$$+ \beta(\mathbf{U}_p - \mathbf{U}) + C_A \alpha_p \rho \left(\mathbf{U}_p \cdot \nabla \mathbf{U}_p - \mathbf{U} \cdot \nabla \mathbf{U} \right) + C_L \alpha_p \rho \left(\mathbf{U} - \mathbf{U}_p \right) \times (\nabla \times \mathbf{U}), \tag{11.8-19}$$

where the momentum transfer coefficient, β, is given by Eq. (6.5-9c). The turbulent viscosity, μ_e, is related to k and ε by Eq. (2.3-18), whereas k and ε are governed by the modified k-ε equations to include the effect of the dispersed phase and turbulent diffusion (Elghobashi and Abou-Arab, 1983), as expressed, respectively, in a simplified form:

$$\nabla \cdot (\alpha \rho \mathbf{U} k) = \nabla \cdot \left(\alpha \left(\mu + \frac{\mu_e}{\sigma_k} \right) \nabla k \right) + \alpha G - \alpha \rho \varepsilon + (1 - \alpha)\rho S_k + \nabla \cdot \left(\frac{\mu_e}{\sigma_\alpha} k \nabla \alpha \right) \tag{11.8-20}$$

$$\nabla \cdot (\alpha \rho \mathbf{U} \varepsilon) = \nabla \cdot \left(\alpha \frac{\mu_{eff}}{\sigma_\varepsilon} \nabla \varepsilon \right) + \alpha \frac{\varepsilon}{k} \left(C_{1\varepsilon} G - C_{2\varepsilon} \rho \varepsilon \right) + (1 - \alpha)\rho S_\varepsilon$$
$$+ \nabla \cdot \left(\frac{\mu_e \varepsilon}{\sigma_\alpha} \nabla \alpha \right). \tag{11.8-21}$$

Similarly, the momentum equation of solids is given by:

$$\nabla \cdot (\alpha_p \rho_p \mathbf{U}_p \mathbf{U}_p) = -\alpha_p \nabla p + \nabla \cdot \left[\mathbf{T}_p \right] + \alpha_p \rho_p \mathbf{g} + \nabla \cdot (\rho_p D_{ep} \mathbf{U}_p \nabla \alpha_p)$$
$$- \beta(\mathbf{U}_p - \mathbf{U}) - C_A \alpha_p \rho \left(\mathbf{U}_p \cdot \nabla \mathbf{U}_p - \mathbf{U} \cdot \nabla \mathbf{U} \right) - C_L \alpha_p \rho \left(\mathbf{U} - \mathbf{U}_p \right) \times (\nabla \times \mathbf{U}). \tag{11.8-22}$$

The total stress of particle phase, $\mathbf{T}_p$, is defined by Eq. (6.6-1) and consists of a pressure component and a shear stress term of solids. In this example, the determination of $\mathbf{T}_p$ is based on the empirical correlation approach, as discussed in Section 6.6.1. Specifically, depending on the solids volume fraction of the slurry, the mixture viscosity is firstly calculated using Eq. (6.6-2a) or Eq. (6.6-2b). Thus, the shear viscosity of solids can be determined from Eq. (6.6-2), which allows the calculation of shear stress of solids by Eq. (6.6-1a). The pressure of solids can be estimated using Eq. (6.6-3).

Thus, with the volume fraction conservation of Eq. (11.8-10), there are nine independent equations, including Eqs. (11.8-17)–(11.8-22) along with Eq. (2.3-18) and Eq. (6.6-1), for the nine unknowns (α, α_p, $\mathbf{U}$, $\mathbf{U}_p$, p, μ_e, k, ε, and $\mathbf{T}_p$), The closure of the modeling is now established.

Comments: (1) The use of empirical correlations for $\mathbf{T}_p$ may significantly simplify the modeling of solid stresses, compared to the kinetic theory modeling as an alternative option. The inclusion of the maximum volume fraction in the empirical correlation such as Eq. (6.6-2b) can easily prevent the solids from overpacking, which is beneficial to the numerical stability. (2) Case simulation examples can be found in the reference (Messa *et al.*, 2014).

11.9   Summary

This chapter introduces the multiphase pipe flows represented by gas–solid pneumatic transport in horizontal pipes, solid–liquid slurry transport in horizontal pipes, and gas–liquid pipe flows with either the gas or the liquid as the continuous phase. For fully developed multiphase flows in straight pipes, the flows can be characterized by a number of different transport regimes with distinctively different flow patterns and phase interactions dominated by such factors as mass flow ratio of phases, density ratio of phases, pipe orientation relative to gravity direction, transport velocity of continuum phase, and sizes of pipe and particles. This chapter discusses the behavior of the multiphase flow through bends that are also important to the design and operation of a pipeline system. The following topics are important to understand:

- regime classification and flow characteristics in each regime for gas–solid pneumatic transport, solid–liquid slurry transport, and gas–liquid pipe flows;
- critical transport conditions, such as saltation and pickup velocities;
- mechanisms dominating the pressure drop (*e.g.*, drag reduction or phase-interactions with pipe wall);
- suspended flow characteristics in straight pipes and effects of particle loading, electrostatic charges and pipe orientation;
- characteristics of flow over a bend, such as roping phenomena and bend erosion;
- stratified multiphase pipe flow with wavy interfaces.

The multiphase flow transport phenomena in actual pipeline systems is currently still much relied on simplified models of analytical, simple differential equations, or empirical correlations to account for, coupled with measurements of flow properties such as pressure drops. More advanced approaches in progress consider regime-dependent modeling with improved numerical simulation capability for enhancing flow predictability.

Nomenclature

$< >$	Volume or area average
A	Area
A_H	Hamaker constant
C_D	Drag coefficient

C	Perimeter
D	Diameter of pipe
	Diffusivity
d	Diameter of particle
E	Young's modulus
	Erosion parameter
e	Restitution coefficient
Fr	Froude number
$\boldsymbol{F}$	Force
f	Friction factor
g	Gravitational acceleration
h	Height
$\mathbf{I}$	Unit tensor
I	Moment of inertia
J	Mass flux
$\boldsymbol{J}_\mathrm{m}$	Momentum flux
K	Coefficient in Saffman force
	Resistance parameter
k	Turbulence kinetic energy
L	Length
m	Mass of a particle
N_Im	Impact number
$\mathbf{n}$	Normal unit vector
Pr	Prandtl number
p	Pressure
q_e	Charges carried by a particle
R_B	Bend radius
Re	Reynolds number
R_d	Pipe radius
R_g	Gas constant
R	Spherical coordinate
$\mathbf{r}$	Position vector
r	Cylindrical coordinate
$\mathbf{T}$	Total stress tensor
T	Temperature
t	Time
U	Velocity vector
U	Averaged velocity
	Radial velocity
$\mathbf{u}$	Instant velocity vector
V	Tangential or transverse velocity
	Volume
W	Peripheral velocity
	Weight

w	Wear in volume loss
X	Lockhart–Martinelli parameter
Y	Yield strength
y_i	Depth of interface
z	Axial coordinate

Greek Symbols

α	Volume fraction
	Impact angle
β	Inclined angle of pipe
	Momentum transfer coefficient
Γ	Mass transfer rate
Δp	Pressure drop
δ	Interparticle distance
	Pipe roughness
	Shedding layer thickness
ε	Energy per unit volume wear
	Permittivity
	Rate of dissipation of k
η	Intrinsic viscosity
θ	Cylindrical coordinate
λ	Friction factor
μ	Viscosity
ξ	Distance normal to wall
ρ	Density
σ	Normal stress
τ	Shear stress tensor
τ	Shear stress
τ_e	Eddy duration
τ_{rp}	Particle relaxation time
τ_S	Stokes relaxation time
Φ_E	Electric potential
ϕ	Multiplier for pressure gradient
ω	Angular velocity

Subscript

0	Particle-free
$'$	Fluctuating component
A	Added mass
	Interfacial area
a	Acceleration
B	Basset
	Brittle

C	Collision
c	Core
D	Drag
	Deposition
	Ductile
e	Turbulent eddy
	Effective
f	Friction
g	Gas
	Gravitation
i	Interface
L	Lift
l	Liquid
M	Magnus
m	Mean
	Mixture
max	Maximum
mb	Moving bed
mt	Minimum transport
n	Normal
P	Pressure
p	Particle
pf	Particle-free friction
pu	Pickup
t	Tangential
w	Wall
z	Axial

Problems

P11.1 (a) Based on governing equations for fully developed dilute suspension flows in a pipeline, derive the corresponding dimensionless equations, with the following dimensionless variables:

$$r^* = \frac{r}{R_d}, z^* = \frac{z}{R_d}, \quad \alpha_p{}^* = \frac{1-\alpha}{1-\alpha_0}, \quad V_E{}^* = \frac{q\, V_E\, \tau_{rp}}{m\, D_p}$$

$$U^* = \frac{U}{U_0}, \quad U_p{}^* = \frac{U_p}{U_0}, \quad \mu^* = \frac{\mu_e}{\mu}, \quad \rho^* = \frac{\rho_p}{\rho}, \quad p^* = \frac{p}{\rho U_0^2} \qquad \text{(P11-1)}$$

where subscript "0" represents $r = 0$.

(b) Based on boundary conditions for fully developed dilute suspension flows in a pipeline, derive the corresponding dimensionless boundary conditions.

(c) Show that there are five dimensionless numbers in the dimensionless equations and boundary conditions:

- Reynolds number Re

$$Re = \frac{U_0 R_d \rho}{\mu} \tag{P11-1a}$$

- Froude number Fr

$$Fr = \frac{U_0}{\sqrt{2 g R_d}} \tag{P11-1b}$$

- Electrodiffusion number N_{ED}

$$N_{ED} = \frac{q}{m} \frac{R_d^2}{D_{ep}} \sqrt{\frac{(1 - \alpha_0)\rho_p}{\varepsilon_0}} \tag{P11-1c}$$

- Diffusion response number N_{DF}

$$N_{DF} = \frac{D_{ep} \tau_{rp}}{R_d^2} \tag{P11-1d}$$

- Momentum transfer number N_m

$$N_m = \frac{U_0 \tau_{rp}}{R_d} \tag{P11-1e}$$

(d) Describe the meaning and the significance of these dimensionless numbers.

P11.2 The minimum transport velocity may be expressed as a function of

$$U_{mt} = f(\rho, \mu, \rho_p, d_p, D, \alpha_p, \tau_w) \tag{P11.2-1}$$

(a) Based on the Pi-theorem, show that the above equation can be expressed in the following dimensionless form of

$$Re_{mt} = f\left(\frac{\rho}{\rho_p}, \frac{d_p}{D}, Re_{pf}, \alpha_p\right) \tag{P11.2-2}$$

where

$$Re_{mt} = \frac{D U_{mt} \rho}{\mu}; \quad Re_{pf} = \frac{d_p U_{f0} \rho}{\mu} = \frac{d_p \rho}{\mu} \sqrt{\frac{\tau_w}{\rho}} \tag{P11.2-2a}$$

(b) Show that Eq. (11.3-3) is a special form of the above general dimensionless equation.

P11.3 Perform a literature survey on modeling of multiphase flow (such as slurry flow or bubbling flow or gas–solids flow) in a bend and on the effect of roping, concentration profile, or pressure drop over the bend.

P11.4 In a gas–solid horizontal pipe flow, the particle used is 100 μm glass beads with a density of 2,500 kg/m³. The pipe diameter is 50 mm. The average particle volume fraction in the pipe is 1%. The gas density and kinematic viscosity are 1.2 kg/m³ and 1.5×10^{-5} m²/s, respectively. Estimate the minimum transport velocity and power consumption per unit length.

P11.5 For a single-phase flow in a bend, the momentum integral equations for the shedding layer of the secondary flow can be expressed by Eqs. (11.7-1a)–(11.7-1c) (Ito, 1959). On the basis of these equations and the approximations given in Eq. (11.7-2) and Eq. (11.7-3), show that the thickness of the shedding layer can be estimated by Eq. (11.7-4b).

P11.6 The pressure drop of a gas flow over a 90° bend may be estimated by Eq. (11.7-4c). For the ratios of the pipe radius to the bending radius, R_d/R_B, of 0.1, 0.05, and 0.01, calculate the pressure drop predictions for the Reynolds numbers ranging from 4,000 to 10^5.

P11.7 Consider a pneumatic transport of grains in a sheet metal duct. The sliding frictional coefficient f_p is about 0.36 (Haag, 1967). To ensure a smooth operation over a 90° bend, the grain particles should always be sliding against the wall during the turn. Estimate the minimum particle velocity required for such a smooth operation. The pipe diameter is 0.1 m. The bend radius is 0.5 m.

P11.8 Consider a dilute gas–solid flow in a horizontal rectangular pipe made of electrically conducting materials. The pipe is well grounded. The flow is fully developed. It is assumed that the particle volume fraction distribution in the vertical direction is the same as that in a circular pipe flow. Find the cross-section averaged particle volume fraction in terms of the particle volume fraction at the centerline.

P11.9 Consider a fully developed dilute gas–solid flow in a vertical pipe in which solid particles carry significant electrostatic charges. The particle charges vary radially. Assume that the flow and the electrostatic field are axisymmetric and the radial charge distribution $q(r)$ is known. Derive a governing equation for the radial volume fraction distribution of the particles.

P11.10 For a fully developed turbulent pipe flow of dilute particle-laden flow, the eddy duration in the wall region can be approximated by

$$\tau_e = \frac{0.4d_p}{\sqrt{u'^2}} \tag{P11.10}$$

Estimate the ratios of τ_{wp} to τ_{fp} for the following particle–air pipe flows with pipe diameter of 50 mm: (1) for 100 μm glass beads in air with a mass flux ratio of 1, ρ_p/ρ of 2,000, and U of 15 m/s; and (2) for 10 μm alumina in air with a mass flux ratio of 1, ρ_p/ρ of 2,400, and U of 30 m/s. The turbulence intensity is assumed to be 5%.

P11.11 Consider a dilute gas–solid flow in a horizontal pipe made of electrically conducting materials. The pipe is well grounded. The flow is fully developed. Show that (1) the particle concentration is exponentially distributed and is a function of the vertical distance only and (2) the ratio of the particle volume fraction at the top to that at the bottom depends on the density ratio of gas to particle, particle diffusivity, and pipe diameter.

P11.12 For ductile erosion, the maximum value of E_D, E_{Dm}, occurs at an angle of impact α_m, whereas α_m can be determined from Eq. (11.7-6a) for a given K_D. Define the relative ductile erosion, E_D^*, as the ratio of E_D to E_{Dm}. Plot E_D^* as a function of impact angle, α_i, over a range from 0 to 90° for K_D of 0.1, 0.5, and 0.9.

P11.13 For brittle erosion, the maximum value of E_B, E_{Bm}, occurs at an angle of impact α_m, whereas α_m can be determined from Eq. (11.7-7a) for a given K_B. Define the relative ductile erosion, E_B^*, as the ratio of E_B to E_{Dm}. Plot E_B^* as a function of impact angle, α_i, over a range from 0 to 90° for K_B of 0.1, 0.5, and 0.9.

P11.14 In a survey of turbine bucket erosion, the friction coefficient can be related to the impact angle and restitution coefficient by (Soo, 1977)

$$f = \left(\frac{(1-e)\cos\alpha_i}{(1+e)\sin\alpha_i} \right)^{3/5} \tag{P11.14}$$

Assuming that the erosion is purely ductile, derive an expression for the impact angle at which the maximum wear occurs.

References

Albunaga, B. E. (2002). *Slurry Systems Handbook*, New York: McGraw-Hill.

Barnea, D. (1991). On the effect of viscosity on stability of stratified gas-liquid flow-application to flow pattern transition at various pipe inclinations. *Chem. Eng. Sci.*, **46**, 2123–2131.

Bolton, L. H., and Davidson, J. F. (1988). Recirculation of particles in fast fluidized risers. *Circulating Fluidized Bed Technology II*, eds. Basu and Large. Oxford: Pergamon Press.

Chisholm, D. (1967). Pressure gradients during the flow of incompressible two-phase mixtures through pipes, venturis and orifice plates. *Br. Chem. Eng.*, **12**, 1368–1371.

de Crecy, F. (1986). Modeling of stratified two-phase flow in pipes, pumps, and other devices. *Int. J. Multiphase Flow*, **12**, 307–323.

Darby, R., and Melson, J. (1981). How to predict the friction factor for the flow of Bingham plastics. *Chem. Eng.*, **88**, 59–61.

Delhaye, J. M. (1981). Basic equations for two-phase flow modeling. *Two-Phase Flow and Heat Transfer in the Power and Process Industries*, eds. Bergles, Collier, Delhaye, Hewitt, and Mayinger. Washington: Hemisphere.

Dodge D. W., and Metzner A. B. (1959). Turbulent flow of non-Newtonian systems. *AIChE J.*, **5**, 189–204.

Duarte, C. A. R., de Souza, F. J., and dos Santos, V. F. (2015). Numerical investigation of mass loading effects on elbow erosion. *Powder Technol.*, **283**, 593–606.

Durand, R., and Condolios, E. (1952). Experimental investigation of the transport of solids in pipes. *Deuxieme Journée de l'hydraulique*, Societé Hydrotechnique de France.

Einstein, A. (1906). Zur theorie der brownschen bewegung. *Annalen der Physik*, **19**, 248–258.

Elghobashi, S. E., and Abou-Arab, T. W. (1983). A two-equation turbulence model for two-phase flows. *Phys. Fluids*, **26**, 931–938.

Fan, L.-S., and Zhu, C. (1998). *Principles of Gas-Solid Flows*, Cambridge: Cambridge University Press.

Finnie, I., and McFadden, D. H. (1978). On the velocity dependence of the erosion of ductile metals by solid particles at low angles of incidence. *Wear*, **48**, 181–190.

Genovese, D. B. (2012). Shear rheology of hard sphere, dispersed, and aggregated suspensions, and filler-matrix composites. *Adv. Colloid Interface Sci.*, **171**, 1–16.

Grant, G., and Tabakoff, W. (1975). Erosion prediction in turbomachinery resulting from environmental solid particles. *J. Aircraft*, **12**, 471–478.

Haag, A. (1967). Velocity losses in pneumatic conveyer pipe bends. *Br. Chem. Eng.*, **12**, 65.

Harris, A. T., Thorpe, R. B., and Davidson, J. F. (2002). Characterization of the annular film thickness in circulating fluidized-bed risers. *Chem. Eng. Sci.*, **57**, 2579–2587.

Ito, H. (1959). Friction factors for turbulent flow in curved pipes. *ASME, J. Basic Eng.*, **81**, 123–134.

Jones, M. G., and Williams, K. C. (2003). Solids friction factors for fluidized dense-phase conveying. *Particulate Sci. Tech.*, **21**, 45–56.

Kaushal, D. R., Kumar, A., Tomita, Y., Kuchii, S., and Tsukamoto, H. (2013). Flow of mono-dispersed particles through horizontal bend. *Int. J. Multiphase Flow*, **52**, 71–91.

Klinzing, G. E., Marcus, R. D., Rizk, F., and Leung, L. S. (1997). *Pneumatic Conveying of Solids: A Theoretical and Practical Approach*. New York: Chapman & Hall.

Krieger, I. M., and Dougherty, T. J. (1959). A mechanism for non-Newtonian flow in suspensions of rigid spheres. *J. Rheol.*, **3**, 137–152.

Lockhart, R. W., and Martinelli, R. C. (1949). Proposed correlation of data for isothermal, two-phase, two-component flow in pipes. *Chem. Prog.*, **45**, 39–48.

McCarthy, H. E., and Olson, J. H. (1968). Turbulent Flow of Gas–Solids Suspensions. *I&EC Fund.*, **7**, 471–483.

Messa, G. V., Malin, M., and Malavasi, S. (2014). Numerical prediction of fully-suspended slurry flow in horizontal pipes. *Powder Technol.*, **256**, 61–70.

Newitt, D. M., Richardson, J. F., Abbott, M., and Turtle, R. B. (1955). Hydraulic conveying of solids in horizontal pipes. *Trans. Inst. Chem. Eng.*, **33**, 93–113.

Njobuenwu, D. O., Fairweather, M., and Yao, J. (2013). Coupled RANS-LPT modeling of dilute, particle-laden flow in a duct with a 90° bend. *Int. J. Multiphase Flow*, **50**, 71–88.

Oka, Y. I., Okamura, K., and Yoshida, T. (2005). Particle estimation of erosion damage caused by solid particle impact Part 1: Effects of impact parameters on a predictive equation. *Wear*, **259**, 95–101.

Pecker, S. M., and Helvaci, S. S. (2008). *Solid-Liquid Two-Phase Flow*, Amsterdam: Elsevier.

Schlichting, H. (1979). *Boundary Layer Theory*, 7th ed. New York: McGraw-Hill.

Shimizu, A., Echigo, R., Hasegawa, S., and Hishida, M. (1978). Experimental study on the pressure drop and the entry length of the gas–solid suspension flow in a circular tube. *Int. J. Multiphase Flow*, **4**, 53–64.

Soo, S. L. (1962). Boundary layer motion of a gas–solid suspension. *Proc. of Symposium on Interaction between Fluids and Particles*. Inst. Chem. Engineers, London.

Soo, S. L. (1977). A note on erosion by moving dust particles. *Powder Tech.*, **17**, 259–263.

Soo, S. L., Trezek, G. L., Dimick, R. C., and Hohnstreiter, G. F. (1964). Concentration and mass flow distributions in a gas–solid suspension. *I&EC Fund.*, **3**, 98.

Soo, S. L., and Tung, S. K. (1971). Pipe flow of suspensions in turbulent fluid: electrostatic and gravity effects. *Appl. Sci. Res.*, **24**, 83–97.

Swamee, P. K., and Jain, A. K. (1976). Explicit equations for pipe flow problems. *J. Hydrau. Eng., ASCE*, **102**, 657–664.

Taitel, Y., and Dukler, A. E. (1976). A model for predicting flow regime transitions in horizontal and near horizontal gas-liquid flow. *AIChE J*, **22**, 47–55.

Taylor, G. I. (1954). The dispersion of matter in turbulent flow through a pipe. *Proc. R. Soc. London*, **A223**, 446–468.

Thomas, D. G. (1962). Transport characteristics of suspensions. Part IV. Minimum transport velocity for large particle size suspension in round horizontal pipes. *AIChE J.*, **8**, 373–378.

Wang, D., You, J., and Zhu, C. (2010). Modeling of core flow in a gas-solids riser. *Powder Technol.*, **199**(1), 13–22.

Wasp, E. J., Kenny, J. P., and Gandhi, R. L. (1977). *Solid-Liquid Flow Slurry Pipeline Transportation*. Clausthal, Germany: Trans Tech Pub.

Weber, M. (1981). Principles of hydraulic and pneumatic conveying in pipes. *Bulk Solids Handling*, **1**, 57–63.

Weber, M. (1991). Friction of the air and the air/solid mixture in pneumatic conveying. *Bulk Solids Handling*, **11**, 99–102.

Weisman, J. (1983). Two-phase flow patterns. *Handbook of Fluids in Motion*, eds. Cheremisinoff and Gupta, Ann Arbor, MI: Ann Arbor Science Pub.

Zhu, C. (1991). *Dynamic Behavior of Unsteady Turbulent Motion in Pipe Flows of Dense Gas-Solid Suspensions*. Ph.D. Dissertation. Univ. of Illinois at Urbana-Champaign.

12 Flows with Phase Changes and/or Reactions

12.1 Introduction

The hydrodynamics of multiphase flows can be significantly altered by mass transfer between different phases, which can result from either physical phase changes such as evaporation, condensation, cavitation, dissolution, and crystallization, or chemical reactions such as solid or liquid fuel combustion, chemical absorption or adsorption, and gas phase polymerization. From the viewpoint of multiphase fluid dynamics, a phase change can also occur between phases defined by hydrodynamic properties, such as particles with different sizes. Examples of such changes include the breakup and coalescence of bubbles and droplets, and the agglomeration of particles. This chapter is focused on the physical and chemical phase changes where the phase coupling is dominantly influenced by the latent and reaction heats.

Phase change and reactions may affect the multiphase flow dynamics in a number of ways. They may change the phase volume fraction and coupled transport velocity via the change of species concentration in each phase, which is particularly important to the gas phase. The latent heat in phase transitions and the chemical reaction heat often have a significant impact on energy transport and resulted phase temperatures, which may lead to changes in the material properties. Phase change and reactions also introduce additional rate-competing processes into the multiphase flow. If the phase change or reaction time scale is much shorter than the flow time scale, the system is close to phase or chemical equilibrium, and thus equilibrium conditions that are derived from thermodynamics can be applied. If the phase change or reaction is much slower than the flow time scale, they can be assumed to have a negligible impact on the flow dynamics. When the time scale of a phase change or a chemical reaction is comparable to the flow time scale, the phase change or reaction is strongly coupled with the flow, and it is thus necessary to consider their rates, which can be controlled by the availability of either species or thermal energy.

This chapter introduces the basic phenomena and modeling approaches of the multiphase flow dynamics that are coupled with phase change or reactions. Representative transport processes and relevant background in industrial applications are also introduced. Section 12.2 discusses the physical phase change occurring between vapor and liquid during boiling and vapor–liquid flow in a heated pipe. Section 12.3 discusses dynamic transport of evaporating atomized sprays, and its applications in spray cooling and spray drying. Then, Section 12.4 introduces the flow and interfacial mass transfer in bubbling reactors in gas–liquid and gas–liquid–solid systems, including sparged stirred tank, gas–liquid bubble column, and slurry–bubble column. Section 12.5 is focused on reactions in gas–solid systems, with examples of fluid catalytic cracking in risers and gas phase polymerization in fluidized beds. Section 12.6 introduces the basic mechanisms and modeling of dispersed particles in combustion, represented by combustion of a fuel droplet and combustion of a solid fuel particle.

12.2   Boiling in Vapor–Liquid Flows

Boiling refers to the rapid vaporization of a liquid due to heat addition. The liquid can be in continuous contact with a heating surface, such as in a boiler, or in an intermitted contact mode, such as droplet–surface collisions. The presence of a solid surface is usually necessary in order to provide the latent heat required for the phase change. For boiling to occur, the solid surface temperature, T_w, must exceed the saturated temperature, T_{sat}, of the fluid under the operating pressure. The temperature difference between the two, also known as the wall superheat, is a quantity often used in the analysis of boiling heat transfer. The initial temperature of the bulk liquid, denoted as T_b, may be lower than or equal to T_{sat}. The condition occurring when $T_b = T_{sat}$ is known as saturated boiling, whereas the situation where $T_b < T_{sat}$ is known as subcooled boiling. Depending on the state of the bulk liquid, the boiling process can also be divided into two categories, including pool boiling where the bulk liquid is stagnant, and flow boiling where the liquid is subjected to convective flow.

It is noted that another rapid vaporization phenomenon known as flashing or cavitation, which is caused by depressurization, is different from the boiling noted above. Technically, the boiling is a heterogeneous vaporization by solid-contact heating or external heat transfer in a nearly constant pressure environment, whereas the flashing is normally a homogeneous vaporization by heat transfer mostly from depressurization-induced oversaturation of liquid. In general, the flashing is much more violent than the boiling in bubble formation and transport dynamics (Berry *et al.*, 1997). The introduction on mechanistic modeling of flashing is beyond the scope of this book, and hence in the following only the surface-heating induced boiling is discussed.

12.2.1   Boiling in Stagnant Liquid

For a stagnant fluid over a heated surface, a pool boiling curve is typically plotted using the heat flux from the solid surface against the wall superheat using log-log scale, as exemplified in Figure 12.2-1 (Nukiyama, 1934). Along the boiling curve, different boiling regimes can be identified with different contact modes between the solid surface and the fluid. Heat addition can be realized by controlling either the heat

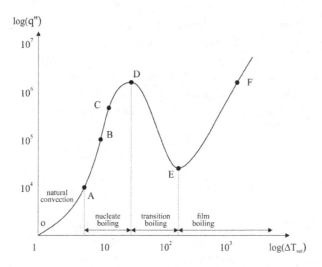

Figure 12.2-1 Pool boiling characteristics and regime definitions.

flux or the wall temperature, and these two controlling mechanisms lead to slightly different paths along the boiling curve.

At low wall superheat, within the range of O-A line in Figure 12.2-1, the liquid remains in a single phase and the heat transfer is due to natural convection. At point A, bubble nucleation is initiated and the slope of the heat flux curve increases rapidly compared to the constant slope in the natural convection region, and hence the condition of point A is defined as the onset of nucleate boiling (ONB). A further increase in the wall superheat, as marked by the B-C-D line, leads to an increase in the bubbling frequency and bubble coalescence, which produces irregular-shaped and unsteady vapor voids with periodic dry patches. The maximum heat flux that can be sustained under nucleate boiling is called the critical heat flux (CHF) as marked by point D. If the heating is power-controlled, such as that of using electrically heated resistive wire, a further increase in the heat flux causes the surface temperature to abruptly jump to point F, where the surface is fully covered by a thin layer of vapor film. Such a condition is thus referred to as film boiling.

In the reverse direction, starting from film boiling, a cooling curve can be obtained by gradually reducing the heat flux. The film boiling condition is maintained until point E, known as the Leidenfrost condition, where a minimum heat flux is reached. Further decreasing of the heat flux leads to a direct contact of liquid to the heat surface, causing a sudden quenching and hence a drastic temperature reduction of the surface. This brings the boiling condition directly back to the regime of nucleate boiling.

The heating or cooling process can also be controlled by the wall temperature. For example, heat can be supplied from an imbedded pipe with internal high pressure steam flow. In this case, the entire boiling curve, including the transition boiling regime (D-E), can be accessed. The transition regime is characterized by intermittent wetting and vapor blanket formation at the heating surface.

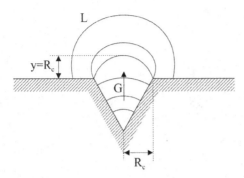

Figure 12.2-2 Growth of a vapor bubble at a cavity on heating surface.

The onset of nucleate boiling marks the start of vapor–liquid two-phase flow phenomena, with the initiation of the bubble growth at the cavities present on the heating surface. Expansion of the residual vapor or gas volume in the cavity is driven by the pressure difference between the vapor and the liquid, while the surface tension force acts as a resistance for bubble growth. As a result of the surface heating, the vapor–liquid interface moves outward, starting from within the cavity, as schematically depicted by a simple model in Figure 12.2-2. When the interface moves to the mouth of the cavity, the process of adjusting the contact angle from the inner cavity surface to the heating surface makes the interface anchored at the cavity's mouth. The minimum bubble radius, and hence the largest resistance due to surface tension, corresponds to the instant when the bubble cap has a semispherical shape, with the cap radius equal to the cavity's mouth radius. In order to initiate nucleate boiling, the liquid temperature at the top of the hemispherical bubble cap must exceed the saturation temperature at the pressure inside the bubble (Hsu and Graham, 1961; Hsu, 1962).

According to the Laplace equation, the pressure inside the bubble can be expressed as

$$p_v = p_l + \frac{2\sigma}{R_c}. \tag{12.2-1}$$

The slope of the p-T curve of a liquid–vapor mixture at equilibrium is given by the Clausius–Clapeyron relation, as:

$$\frac{dp}{dT} = \frac{L}{T_{sat}(v_v - v_l)} \approx \frac{L}{T_{sat}v_v}. \tag{12.2-2}$$

Combining Eq. (12.2-2) with Eq. (12.2-1) leads to

$$T_{sat,v} - T_{sat} \approx \frac{2\sigma T_{sat}v_v}{LR_c}, \tag{12.2-3}$$

where $T_{sat,v}$ and T_{sat} are the saturation temperature at p_v and p_l, respectively. In the vicinity of the heating surface, heat transfer in the liquid is due to conduction. With the linear approximation of the liquid temperature over the boundary layer and the

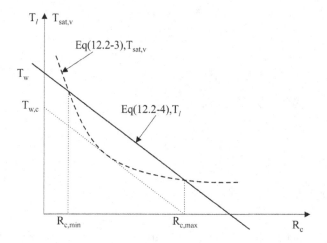

Figure 12.2-3 Cavity size and wall temperature for onset of nucleate boiling.

boundary layer thickness roughly represented by the cavity radius, it can be shown that

$$T_w - T_l \approx \frac{q''}{K_l} R_c. \tag{12.2-4}$$

The criterion for nucleate boiling (*i.e.*, $T_l > T_{sat,v}$), can be graphically explained by Figure 12.2-3.

Based on Eq. (12.2-3) and Eq. (12.2-4), for a given wall superheat $\Delta T_{sat} = T_w - T_{sat}$, and heat flux q'', the region between $R_{c,min}$ and $R_{c,max}$ where the line T_l is above the $T_{sat,v}$ curve corresponds to a range of cavity sizes for nucleate boiling. The critical cavity sizes can thus be calculated by

$$R_{c,max}, R_{c,min} = \frac{\Delta T_{sat} K_l}{q''} \pm \sqrt{\left(\frac{\Delta T_{sat} K_l}{q''}\right)^2 - \frac{8\sigma T_{sat} v_v K_l}{L q''}}. \tag{12.2-5}$$

In addition, the minimum wall temperature for nucleate boiling, T_{wc}, can be found from the tangent line to the $T_{sat,v}$ curve.

$$T_{wc} = T_{sat} + \sqrt{\frac{8\sigma T_{sat} v_v q''}{L K_l}}. \tag{12.2-6}$$

The critical heat flux (CHF) refers to the maximum heat flux that can be sustained during nucleate boiling. A further increase in the heat flux would cause a drastic increase in the bubbling intensity and dryout of the heat surface, leading to the transition to the film boiling condition. The CHF and transition to film boiling can be modeled by analyzing the instability conditions (Zuber, 1959), which is further extended to include the effect of receding contact angle θ_r and inclination angle

β of the heating surface with respect to the horizontal plane (Kandlikar, 2001):

$$q''_{CHF} = \left(\frac{1 + \cos\theta_r}{16}\right) \left[\frac{2}{\pi} + \frac{\pi}{4}(1 + \cos\theta_r)\cos\beta\right]^{1/2} \rho_v L \left[\frac{\sigma g(\rho_l - \rho_v)}{\rho_v^2}\right]^{1/4}. \tag{12.2-7}$$

12.2.2 Boiling in Liquid Pipe Flow

A liquid flowing in a heated confined passage may experience a variety of boiling conditions. To avoid the flow asymmetry from the gravity effect, flow regime characterization is typically obtained via simplified studies on vertical pipe flows, as illustrated in Figure 12.2-4 (Collier and Thome, 1994).

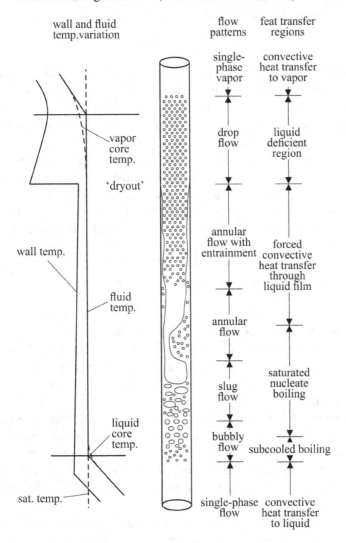

Figure 12.2-4 Boiling of liquid flow in a heated vertical pipe.

For simplicity, consider a subcooled liquid entering a very long, uniformly heated circular pipe. The flow regime may transit from a single-phase liquid flow to a vapor–liquid two-phase flow, and eventually to a single-phase vapor flow. At the same time, the heat transfer regime may go through changes from liquid phase convection to flow boiling, post dryout, and finally vapor phase convection. The state of boiling is often characterized using the flow quality χ, which is defined as the ratio of the vapor mass flow rate to the total mass flow rate.

$$\chi = \frac{\rho_v Q_v}{\rho_v Q_v + \rho_l Q_l}. \tag{12.2-8a}$$

The thermodynamic equilibrium quality χ_e, on the other hand, is defined based on the ratio of the specific enthalpy difference between fluid and saturated liquid to the latent heat of phase change, as expressed by

$$\chi_e = \frac{h - h_{l,sat}}{L}. \tag{12.2-8b}$$

Note that while the actual flow quality ranges from 0 to 1, the thermodynamic equilibrium quality can be negative (subcooled liquid), positive and less than 1 (vapor–liquid mixture), or greater than 1 (superheated vapor).

When the subcooled liquid is heated through single-phase convective heat transfer, both the wall temperature and the liquid temperature increases while the heat transfer coefficient remains almost constant. Similar to the pool boiling, the wall temperature must reach a certain superheat above the saturation temperature before nucleate boiling can occur. The condition when bubbles begin to appear on the wall is defined as the onset of nucleate boiling (ONB). The wall temperature starts to level off, while the heat transfer coefficient increases. Although a thin layer of bubbles is present in the vicinity of the wall, the net vapor-phase flow rate is zero since the bubbles generated at the wall are condensed by the subcooled liquid core. Therefore this condition is called subcooled boiling, and the equilibrium quality χ_e remains negative.

With more bubbles formed at the wall and an increase in bubble size in the flow direction, bubbles start to detach from the wall and enter the flow. This point is known as the net vapor generation (NVG) condition. The flow after NVG becomes two-phase bubbly flow in which motions of bubbles deviate from that of liquid. In this case, some bubbles are still condensed along the flow, and the negative equilibrium quality may still remain.

At the point where the liquid reaches the saturation temperature, $\chi_e = 0$. Beyond this point, the flow enters the saturated flow boiling region. The two-phase flow pattern starts to evolve from bubbly flow, to slug flow, to annular flow where the vapor phase switches into the continuous phase. The transition from annular flow to mist flow is referred to as the "dryout" condition. The heat flux just before dryout is the critical heat flux (CHF), which is the upper limit of the flow boiling heat transfer. Under the dryout condition, the absence of liquid film on the wall dramatically reduces the heat transfer coefficient and increases the wall temperature. After dryout, the temperature of the vapor core rises beyond the saturation temperature, while the temperature of the liquid droplets remains at T_{sat}. The fraction of liquid droplets in the flow decreases as a result of continuous heating, and eventually all of the liquid evaporates and the flow becomes a single-phase vapor flow. At some point before all of the liquid evaporates, the equilibrium quality χ_e reaches unity.

The simplest model for vapor–liquid flows in heated pipes is a one-dimensional flow model with local homogeneous equilibrium (Wallis, 1980), which neglects the relative motion between the two phases, and thus assumes a pseudo-single-phase transport of homogeneous fluid. The model further assumes that the two phases are in thermodynamic equilibrium, giving rise to the vapor–liquid flow quality, χ, that is equal to the thermodynamic equilibrium equality, χ_e. The resulted conservation equations of mass, momentum, and energy are thus given, respectively, by

$$\frac{\partial \rho_m}{\partial t} + \frac{\partial}{\partial z}(\rho_m U_m) = 0 \tag{12.2-9a}$$

$$\frac{\partial U_m}{\partial t} + U_m \frac{\partial U_m}{\partial z} = -\frac{1}{\rho_m}\frac{\partial p}{\partial z} - f_w U_m |U_m| - g\sin\beta \tag{12.2-9b}$$

$$\frac{\partial \rho_m h_m}{\partial t} - \frac{\partial p}{\partial t} + \frac{\partial}{\partial z}(\rho_m h_m U_m) - U_m \frac{\partial p}{\partial z} = q''. \tag{12.2-9c}$$

In Eq. (12.2-9b), f_w is the two-phase friction factor of the pipe wall, and β is the inclined angle of pipe with respect to the horizontal plane. To closure the model of the four variables χ, U_m, h_m, and p, a fourth equation must be introduced, which is often given in the form of the equation of state of the mixture:

$$\rho_m = \rho_m(p, h_m, \chi). \tag{12.2-9d}$$

To account for the phase interaction and nonequilibrium flow characteristics of vapor and liquid phases, a vapor–liquid flow in a boiling pipe (with surface heating) can be, in principle, described by the governing equations based on either the Eulerian–Lagrangian approach in Chapter 5 or the Eulerian–Eulerian approach in Chapter 6. However, the bubble generation from nucleation boiling leads to a wide range of bubble size distribution that may be dependent on the wall roughness as well as heat flux or temperature coupling of the pipe wall. In addition, the complicated bubble–bubble interactions and hydrodynamic coupling with carrying turbulent fluid flow, particularly in the wall boundary layer region, may pose challenges for the problem closing.

An alternative modeling approach, known as the drift flux model, is popularly adopted for vapor–liquid flows with phase changes. The drift flux model is also based on a volume averaging treatment to the governing equations but introducing a simplified concept of "drift velocity" to describe the dynamic phase interactions between vapor and liquid phases. The drift velocity is defined as the relative velocity of the vapor phase with respect to the volumetric center of the vapor–liquid mixture, which is different from the actual relative velocity of vapor and liquid.

12.3   Liquid Spray Dispersion and Evaporation

A liquid spray is typically generated by pressuring liquid through a spray nozzle. Since the droplet formation originates from liquid disintegration via competing mechanisms of viscous forces, inertia forces, and surface tension forces, the regime classification of spray disintegration is typically correlated to the liquid Reynolds number and Weber number or Ohnesorge number (Oh $= \sqrt{\text{We}}/\text{Re}$), as exemplified in Figure 12.3-1 for pressurized atomization by a circular nozzle (Reitz and Bracco, 1982). At low Re, the breakup is dominated by surface tension, and this action forms identical droplet sizes. This is known as the Rayleigh regime. At intermediate Re, the breakup is due to the growth of interfacial wave under the influence of aerodynamic forces, and is thus referred as the aerodynamic regime. The jet shape can be either symmetric (the first wind-induced mode) or asymmetric (the second wind-induced mode). At high Re, the jet breakup is in the atomization mode in which the disintegration is almost spontaneous at the nozzle exit.

The spray can be in a form of intermitted liquid segments, known as ligaments, or in a form of droplets. The latter one, known as atomized spray, provides a much larger total surface area of liquid phase than a ligament spray of the same flow rate. The finer the droplets, the larger surface area of the atomized spray. The evaporation rate, however, depends not only on the available surface area for evaporation but also on the ambient environment in which the spray disperses as well as on the thermodynamic properties for phase change of the evaporating liquid. Most importantly, the

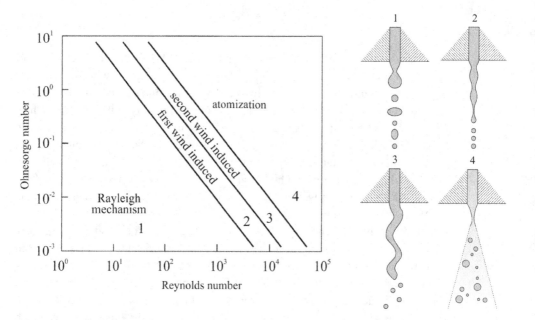

Figure 12.3-1 Regimes of a circular spray in pressurized atomization.

evaporation rate is also governed by the heat transfer rate to the spray, as required for the phase change.

The spray dispersion can be strongly altered by the spray vaporization, via its phase interactions with the surrounding fluid as well as the changes in mass and thermal energy of the spray itself, in particular when the vaporization rate is at a similar time scale of hydrodynamic transport of the spray. Examples of such rapid spray vaporization include sprays under sudden depressurization (known as spray flashing), sprays with direct impacts onto the hot solid surfaces or suspended solids, and spray drying under enhanced heat transfer such as by thermal radiation or in the extremely hot and dry gases. It should be noted that, when the time scale of spray evaporation is much longer than the flow time scale, the two-way coupling between the evaporation and hydrodynamic transport in spray dispersion may be ignored.

12.3.1 Spray Atomization

Atomization is a process in which a continuous stream of liquid flow is transformed into a spray system of highly dispersed droplets. It significantly increases the contact area between the liquid and the surrounding gas, and thus greatly enhances the interfacial heat and mass transfer, and facilitates processes such as evaporative cooling and spray absorption. The disintegration of the continuous liquid is typically caused by the interaction between the liquid and its jetting environment via some externally delivered mechanical energy. Spray atomization can be achieved by atomizers with different energizing principles, including: (a) pressure atomizer that uses the potential energy in pressurized fluid to form small droplets at high speed; (b) pneumatic atomizer that uses a pressurized gas to disintegrate the liquid; (c) rotary atomizer that uses the centrifugal force to disintegrate liquid into droplets; (d) ultrasonic atomizer that makes liquid disintegrate with the assistance of acoustic vibration; (e) electrospray atomizer that accelerates an electrically charged liquid by the electric field to break up the liquid into fine droplets. The choice of an atomizer is based on considerations of the liquid material properties, the liquid mass flow rate, and the spray characteristics such as droplet size distribution and spray shape. Atomized droplets are generated either from the primary atomization or from the secondary atomization. The primary atomization refers to the direct disintegration of a continuous liquid jet or sheet into dispersed droplets. The secondary atomization refers to the further breakup of a large droplet or ligament into smaller droplets, with large relative velocities or shearing between the injected primary droplet or ligament and the ambient fluid.

12.3.2 Evaporating Spray Jets

When the evaporation time scale is comparable to the flow time scale during a spray jetting, the rapid evaporation of droplets generates a quite complicated coupling effect on the flow of gaseous components of the spray jet, and in turn on the droplet

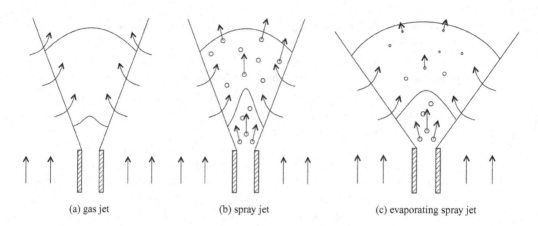

(a) gas jet   (b) spray jet   (c) evaporating spray jet

Figure 12.3-2 Schematic characteristics of cocurrent gas and spray jets.

dynamic transport and the evaporation themselves. For simplicity, consider only the aerated and atomized spray jets in the following discussion.

Figure 12.3-2 illustrates a spray jet into a stagnant gaseous medium or a cocurrent gaseous flow, with arc curves representing the gas velocity profiles. To reveal the effects of phase interaction and evaporation, the evaporating spray jet is compared against a gaseous jet in the absence of droplets as well as a nonevaporating spray under similar flow conditions. For the case of a gaseous jet, as shown in Figure 12.3-2(a), the injected gas mixes with the ambient gaseous medium by diffusion and convection, with the jet entrainment and even jet similarities in velocity, concentration, and temperature profiles (Abramovich, 1963). The addition of nonevaporating droplets leads to likely enhanced momentum of gas phase and hence more entrainment and deeper jet penetration via inertia-led droplet dispersion and droplet–gas interactions, as illustrated in Figure 12.3-2(b). The impact of droplet evaporation further alters the gaseous jet structure and entrainment, with the complications of mass and heat transfer to the gas phase, as illustrated in Figure 12.3-2(c). In this case, an excessive vapor generation speeds up the local gas velocity and expands the jet boundary, which tends to enhance spray penetration and entrainment; however, the excessive vapor may reduce jet-induced pressure difference, which then leads to a reduction of the entrainment (Qureshi and Zhu, 2006). The excessive vapor may also increase the vapor concentration in the immediate vicinity of the evaporating droplets, which in turn restricts the vaporization rate (from the reduced concentration gradient across the boundary layers of droplets) but may promote a further penetration of droplets (from enhancement of the gas velocity).

Figure 12.3-3 depicts similar but more complicated cases of a spray jet into a cross gaseous flow, with comparisons among a gaseous jet, a nonevaporating spray jet, and an evaporating spray jet under similar flow conditions. As shown in Figure 12.3-3(a), the convective mixing of two gaseous streams not only leads to the bending of the jet but also significantly alters the jet structure, with paired vortices

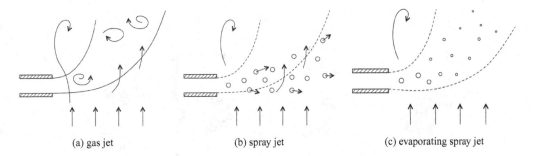

(a) gas jet (b) spray jet (c) evaporating spray jet

Figure 12.3-3 Schematic characteristics of cross-flow gas and spray jets.

induced by secondary-flow within cross sections of the bending gaseous jet, distorted and asymmetric jet boundaries as well as the disappearance of jet similarities (Abramovich, 1963). In the spray cases, as shown in Figures 12.3-3(b) and (c), due to the inertia of droplets, the droplet trajectories can severely depart from the cross-flow mixing region of gaseous component of spray. In cases of rapidly evaporating sprays, the instant and excessive vapor generated along the droplet penetration path may greatly extend the gaseous mixing region, as indicated in Figure 12.3-3(c).

12.3.3 Spray Drying

Spray drying is a popular technology in food, chemical, and pharmaceutical industries for manufacturing powders such as milk powders, coffee, penicillin, amino acids, and catalysts. In spray drying, a liquid solution or a liquid–solid suspension, known as slurry, is atomized into small droplets that are dried in a hot and drying gas medium to form solid particles. After the liquid from slurry droplets evaporates, the dried particles may be nonuniform, exhibiting various morphologies and sizes that depend on the physical properties of the slurry droplets and the drying conditions.

Generally, two stages can be identified during the drying of a droplet, namely the constant-rate stage and the falling-rate stage. These are schematically illustrated in Figure 12.3-4(a). The corresponding changes in temperature, liquid residue, and evaporation rate are exemplified in Figure 12.3-4(b) (Mezhericher et al., 2011). The first stage commences when the droplet is sprayed into the gas medium, which is usually at a higher temperature. An initial heating and evaporation process commences, causing the droplet temperature to quickly rise to the wet bulb temperature. The pure liquid from the slurry droplet ultimately evaporates because the primary resistance to evaporation of gas-phase diffusion is overcome as liquid is continuously brought to the surface. The volume of the droplet decreases as liquid evaporates, but the solid concentration near the surface increases. The evaporation rate and droplet temperature remains almost constant until the solid fraction at the surface reaches a critical value and a porous crust starts to form, which signals the beginning of the second

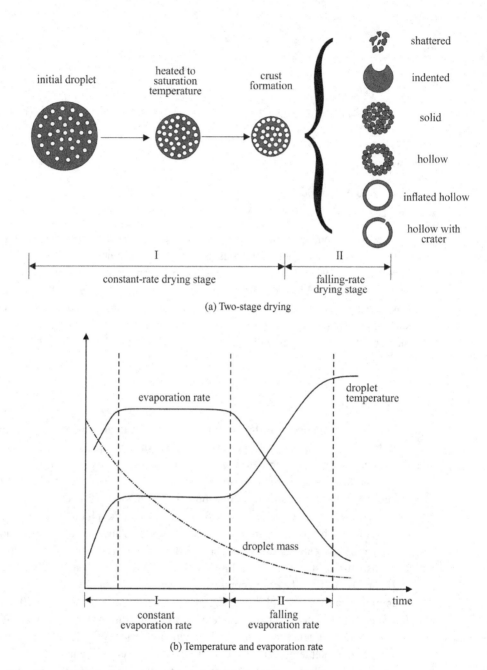

(a) Two-stage drying

(b) Temperature and evaporation rate

Figure 12.3-4 Stages and characteristics of drying a slurry droplet.

stage. From this point onward, resistance to evaporation increases because the liquid must be transported through the resistive porous shell as it makes its way to the droplet's surface, resulting in a proportional decrease in evaporation rate. While the first stage is essentially the same for different materials and conditions, the second

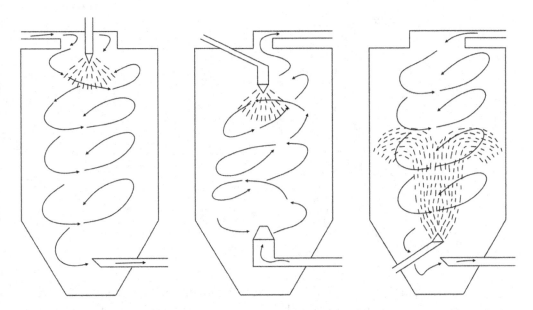

Figure 12.3-5 Configurations of gas–droplet contact mode in drying chamber.

stage is associated with a number of different mechanisms and particle morphologies, as illustrated in Figure 12.3-4(a). Depending on the property of the crust, the droplet may exhibit varied evaporation patterns. If the solid crust is sufficiently mobile and the droplet front is subjected to substantial drag, a dented region may appear in the rear surface; this is due to the faster evaporation near the front of the droplet, inducing an internal liquid flow from rear to front that appears as a void on the droplet surface (Sloth *et al.*, 2009). Under some conditions, the particle may even be shattered, inflated, or develop surface craters. However, if the crust is sufficiently rigid, the drying droplet remains spherical with its outer diameter fixed. Further removal of moisture requires the expansion of the vapor-saturated space take place within the droplet, which can occur through two distinct mechanisms. In the first scenario, a receding of the gas–liquid interface into the droplet causes the wet droplet core to shrink, resulting in a packed solid particle when the drying process completes. In the second scenario, an internal vapor bubble forms and expands at the center of the droplet, leaving a hollow particle when the process completes.

Among the many stages in a spray drying process, the atomization is considered the most critical operation, since it determines the droplet size distribution and the initial velocity, which directly influences the final product size and drying rate. The design of the drying chamber is also affected by the choice of atomizer type and its characteristics. There are three types of gas–droplet contact patterns typically used in the drying chamber, namely the cocurrent flow, the countercurrent flow, and the mixed flow, as shown in Figure 12.3-5.

In the cocurrent flow configuration, droplets and air move in the same direction, and the product temperature is lower than the inlet air temperature. It can be

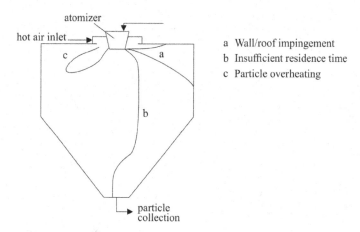

atomizer

hot air inlet

c

a

b

particle
collection

a Wall/roof impingement
b Insufficient residence time
c Particle overheating

Figure 12.3-6 Effects of air flow pattern on droplet trajectory.

used to produce fine particles of heat sensitive material when combined with rotary
atomizers, or coarse particles of heat-sensitive material when combined with nozzle
atomizers. The countercurrent flow pattern is typically used with nozzle atomizers to
produce coarse particles with special porosity or high bulk density for heat-sensitive
materials. In this case, the product temperature is higher than the exit air temperature. The mixed flow pattern is used for coarse particles, which can withstand high
temperatures. It is the most economical design for coarse particles when the chamber
dimension is limited.

The air flow pattern in the drying chamber has considerable impact on the drying operation. Improper air flow will result in poor drying due to the variation in
the droplet trajectory, as shown in Figure 12.3-6 (Oakley, 2004). Droplets may collide with the chamber wall or roof and deposit to form a dry crust, which hardens
under high temperature. The crust can cause local hot spots on the dryer wall and
thus could damage the dryer. The product quality may also deteriorate due to debris
from the deposit. Undesired droplet trajectories may also affect the product quality
by changing the droplet residence time. Droplets bypass the chamber with significantly reduced residence time, thus leading to incomplete drying, while an excessive
residence time of the droplets within the chamber causes overheating.

12.3.4   Spray on a Heated Surface

A droplet that is gently deposited onto a heated surface will undergo a heating process similar to boiling characteristics in Figure 12.2-1. When a droplet collides with
a heated solid surface, inherent coupling between heat transfer and hydrodynamics
introduces additional complexity into the analysis. Three coupled hydrodynamics-
heat transfer behaviors can be identified (Naber and Farrell, 1993). In the case of a
low surface temperature, the droplet will be in complete contact with the wall; this
contact type is associated with both the evaporation and nucleate boiling heat transfer

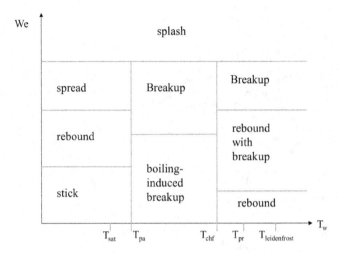

Figure 12.3-7 Impact regimes of a droplet on a heated dry wall.

regimes. If the wall temperature is maintained above the Leidenfrost temperature, the droplet will be levitated by the vapor film; this noncontact type corresponds to the film-boiling heat transfer regime. A transitional contact type exists between the contact and noncontact categories where the vapor layer is only formed intermittently, corresponding to the transitional thermodynamic regime.

A categorization matrix for droplet impingement on a heated surface reveals a general dependency on the wall temperature and Weber number, as shown in Figure 12.3-7 (Bai and Gosman, 1996) in which T_{pa} and T_{pr} denote the pure adhesion and pure rebound temperature. Examples of spray on heated surfaces include the spray coating and spray quenching. In a spray coating, the atomized droplets of liquid solutions or slurries are firstly impinged and spread over the heated surface and then form a thin layer of coating via evaporation. The spray quenching is to use the latent heat from collision-induced evaporation of sprayed liquid to cool down the heated surface, such as to quench a forest fire or burning objects.

12.3.5 Evaporating Spray in Gas–Solid Flows

A rapid spray evaporation requires the timely and sufficient heat transfer to the atomized droplets. Such heat transfer can be greatly enhanced by droplet–particle collisions when a spray is injected into a hot gas–solid flow, such as in the fluidized catalytic cracking and gas-phase polymerization. Transport phenomena of this kind of three-phase flow are much more complicated than that of a liquid spray jet without suspended solid particles. The complication is caused by the droplet–particle collisions as well as additional gas–solid interactions that couple with the droplet dynamics. The droplet–particle collisions can lead to many different consequences for the transport of droplets and particles, including liquid coating onto the particles, liquid-bridged particle agglomeration, and momentum and mass repartition of

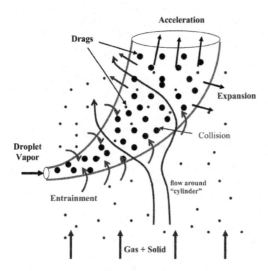

Figure 12.3-8 Schematic illustrations of a liquid spray into a gas–solid flow.

the droplets and particles. When colliding with hot solid particles, rapid vaporization may result, which can significantly alter the flow characteristics of gas jet and species mixing.

Consider an atomized liquid spray injected into a gas–solid flow at an inclined angle of injection with respect to the direction of gas–solid flow. The phase flow patterns and phase interactions can be schematically illustrated in Figure 12.3-8. The spray domain is defined by the liquid phase of droplets. The dynamics of atomized droplets are governed by the evaporation-affected hydrodynamic forces (such as the drag forces from a gas–vapor mixture flowing around the droplets), the particle-droplet collision forces, and the field forces (such as gravity). Thus the spray is deflected, while penetrating, by drag and collisions. The gas phase is in general a mixture of ambient gas and evaporated vapor, since the ambient gas composition can be different from the vapor generated by the droplet evaporation. The flow of ambient gas is mixed with the jetting vapor via jet entrainment, convection, and diffusion. The flow of suspended particles follows a combination of three different paths: (1) some solid particles penetrate through the spray flow without any collisions; (2) some particles directly collide with the liquid sprays; and (3) the rest of the particles follow the ambient gas flow that goes around or bypasses the evaporating spray jet. The governing mechanisms that promote the particles entering the spray region include the particle convection and jet entrainment.

The particle–droplet collisions constantly leads to liquid attachment onto the colliding particles or to particle wetting. The dynamic behavior of wetted particles can be quite different from that of dry particles. The interparticle collisions of wetted particles can cause the liquid-bridged particle agglomeration. To study the characteristics of the spray jet (such as penetration length and expansion), the interactive relations among four phases (*e.g.*, gas phase, droplet phase, dry particle phase, and wetted particle phase) shall be considered in the physical model. The interphase transfers can have significant impact on the spray expansion, deflection, penetration, solids wetting, and vapor dissipation.

12.3.6   Modeling of Spray Transport and Phase Interactions

Modeling of the flow dynamics of liquid sprays includes the modeling of spray atomization or droplet formation, modeling of dispersion and transport of the injected droplets or atomized spray, and modeling of interactions with surrounding media, including neighboring droplets or particles, as well as impact of external fields such as thermal radiation and electric field.

12.3.6.1   Modeling of Spray Atomization

A key to the modeling of spray systems is the treatment of the atomization process where the initial droplets are formed. The length scale covering the entire range of spray atomization is very wide, from a few millimeters to micron-sized droplets. Hence, a complete spray atomization process is normally divided into two sub-processes: primary breakup (for primary formation of ligaments and droplets) and secondary breakup (for the generation of daughter droplets, typically from the further breakup of ligaments and droplets). Since there are relatively few fragments or droplets in the primary breakup, an Eulerian–Eulerian approach is typically adopted for the interface dynamics. The direct numerical simulation (DNS) approaches, such as the volume of fluid model (VOF), have been employed to explicitly capture the primary breakup process in the vicinity of the atomizer (Gorokhovski and Herrmann, 2008; Shinjo and Umemura, 2010). However, with the use of an extremely large number of computational cells (about 10^9), the simulation domain is still less than the nozzle diameter while the atomization process typically lasts over hundreds of nozzle diameters, which renders the direct applications of DNS too prohibitively expensive to be applied for practical purposes. A cost-effective alternative is to use the Large Eddy Simulation (LES) and VOF that mainly resolves the interface at the LES grid level in the near nozzle region (de Villier et al., 2004). In the secondary breakup region, the spray is regarded as sufficiently dilute so that droplets are treated as point sources having interphase transfer with the ambient fluid flow. The Eulerian approach with LES is applied to describe the ambient flow that is typically turbulent. The dynamics of droplets are described by the Lagrangian approach, with a stochastic sub-grid model to account for the secondary breakup (Jones and Lettieri, 2010).

The flow in the vicinity of a spray nozzle involves complicated interactions among different flow features. When the atomization is driven by a back pressure drop much higher than the ambient pressure, the pressure gradient would extend to a zone immediately outside the nozzle. The pressure gradient beyond the nozzle tip, as perhaps evidenced by the existence of the iso-core region of a single-phase jet, imposes a driving force to further accelerate the atomized droplets. In addition, the local volume fraction and number density of the droplets could still be high in this region, so that the coalescence of the already atomized droplets may not be negligible. Besides the possible coalescence, the interactions of the droplets are expected to alter the gas flow immediately surrounding each individual droplet; thus, the hydrodynamic forces upon each droplet may not be deterministic. Moreover, the primary atomization process may originally generate a highly nonspherical liquid fragment, which then either recoils to a spherical droplet due to surface tension or further deforms and breaks up into smaller droplets. Thus, in order to avoid the complexities of the flow physics and interactions mentioned above, modeling of spray dispersion is typically limited to the flow beyond the immediate vicinity from the nozzle tip, without dealing with the internal flow inside the nozzle or the detailed flow across the nozzle tip.

Once beyond the initial jet region, the droplets are usually quickly dispersed, with the dispersion enhancement by the jet entrainment. Thus, in the spray dispersion domain, it is typically assumed that the transport of droplets is in the dilute transport regime so that the interactions among droplets (especially interdroplet collisions) can be neglected. The dynamic transport of atomized spray, as a discrete phase, is primarily based on Lagrangian trajectory models of the droplets, which is coupled with the Eulerian model of the surrounding gas flow.

In practical spray applications, it is popular to exclude the immediate vicinity of the nozzle from the modeling domain of spray dispersion, which leads to a significant reduction in the model complexity by cutting off the linkage between the internal flow inside the nozzle and the external flow in the vicinity of nozzle tip as well as the linkage between spray atomization and spray dispersion. However, this treatment bypasses the key atomization process, where the results are necessary to form the injection boundary conditions of droplets for their dispersion and transport. Instead, now it requires the use of empirical correlations or sub-modeling to formulate the "boundary conditions" of the droplets that include the droplet size distributions, the number density distributions, and the injection velocity distributions of droplets at the artificially defined spray inlet (Lefebvre, 1989).

12.3.6.2   Modeling of Evaporating Spray

For non–highly evaporating liquid sprays, the effect of evaporation of droplets during the spray jet dispersion process is generally negligible. This is because the droplet Reynolds number, defined by the slip velocity and droplet size, is normally at a moderate level so that the evaporation of atomized droplets is mostly dominated by vapor diffusion. In this case, the characteristic time scale in the diffusive evaporation of a droplet is typically much longer than that of the surrounding jet flow. It is noted that the evaporation rate of droplets depends not only upon the driving power of mass transfer (characterized by the concentration gradient and diffusivity) but also, at the same time, the heat transfer rate to the droplets (as an energy requirement for the phase change of liquid vaporization). Hence, the characteristic time scale for droplet evaporation is determined from a combined effect of mass and heat transfer of the droplets. The sprays of highly evaporative liquids or rapid-evaporating liquid sprays can thus be referred to as that where the characteristic time scales of evaporation are on the same (or shorter than) orders of magnitude as those of jet flows. For an evaporating spray jet, there is a strong coupling effect between a spray jetting and the droplet vaporization that may have a significant influence on the transport dynamics of both the jet and the droplets.

Mechanistically, droplet evaporation occurs when the vapor pressure at droplet surface exceeds the vapor pressure in the surrounding environment. Such pressure difference can be induced by the temperature difference or directly caused by the sudden pressure reduction of surrounding environment. The heat transfer modes for droplet evaporation are typically in forms of the heat convection with surrounding fluid media, heat conduction either from the droplet itself or via collision with hot solids or solid surfaces, field heating such as thermal radiation, and a combination of the aforementioned.

There are several simplest models of droplet evaporation. The most classical one is based on the diffusive mode of a spherical and saturated droplet in the stagnant fluid, in which the latent heat for evaporation is provided through heat conduction from the surrounding media. This diffusive evaporation model, also known as d^2-law as introduced in Section 3.4.3, can be further modified with empirical correlations of convective heat and mass transfer coefficients (such as the Ranz and Marshall equation) to account for the effect of convection. For flash evaporation of the droplet, a simple approach is based on the lumped-heat capacity of the droplet, with heat transfer from both the droplet itself and surrounding environment, while the evaporation rate is either modeled as diffusion-controlled (Shin *et al.*, 2000) or governed by the modified Hertz–Knudsen equation (Marek and Straub, 2001; Persad and Ward, 2016).

For spray drying, the evaporation of a solution droplet is driven by the difference between the vapor pressure of the solvents and their partial pressure in the surrounding gaseous medium, which is strongly coupled with the heat and mass transfer within and outside the evaporating droplet as well as the energy balance at the droplet surface where the phase change occurs. During the evaporation, the receding droplet surface results in an increase in solute concentration near the surface, which leads to a diffusive flux of solutes toward the center of the droplets. Consider a multicomponent solution: Assuming a radially symmetric droplet evaporation without internal convection, the diffusion equation of solute i is governed by Fick's second law of diffusion, expressed as (Vehring *et al.*, 2007):

$$\frac{\partial (\rho Y_i)}{\partial t} = D_i \left(\frac{\partial^2 (\rho Y_i)}{\partial r^2} + \frac{2}{r} \frac{\partial (\rho Y_i)}{\partial r} \right) + \frac{1}{R_d} \frac{\partial (\rho Y_i)}{\partial r} \frac{\partial R_d}{\partial t} \qquad (12.3\text{-}1)$$

where Y_i is the mass fraction of the solute i, D_i is the diffusion coefficient of solute i in the liquid phase, and R_d is the radius of the droplet. At the beginning of the droplet drying, the evaporation is nonstationary with the droplet temperature rapidly changing. Then it may reach a nearly constant evaporation rate until the reduction in evaporation rate immediately prior to solidification of the particle. In general, the diffusion coefficient is a strong function of the solute concentration, especially when the drying process leads to supersaturation of the solute or the liquid viscosity increases prior to solidification. In a steady-state evaporation, the coupled heat and mass transfer equations outside the droplet are similar to those introduced in Section 3.4.3.

12.3.6.3   Modeling of Evaporating Spray in Gas–Solids Suspension

There are no well-established hydrodynamic models available for atomized sprays injected into gas–solid flows. This is mainly due to the difficulties in dealing with two interacting discrete phases (particles and droplets) in a continuum fluid medium (vapor–gas mixture). The issue becomes more difficult if the droplet breakup and particle wetting result from the particle–droplet interactions, in addition to the collision-induced heat and mass transfer. One approximate approach is to treat both gas and particle phases as interacting continuum fluids, whereas the atomized droplets are regarded as a discrete phase (Wang *et al.*, 2004). Thus, the dynamics of gas and particles are described by the Eulerian field coordinates, and the droplet dynamics are monitored along their Lagrangian trajectory coordinates. Such a Lagrangian–Eulerian modeling approximation not only needs sub-models for particle–droplet collisions (that results in droplet evaporation, droplet breakup and scattering, and particle wetting) but also requires formulation of the corresponding interaction terms in Lagrangian and Eulerian governing equations.

The following discussion is focused on a representative case where a round nozzle injects an atomized spray at a high speed into a cross-flow of very hot solids, as shown in Figure 12.3-9. The interactions between the high-momentum spray jet and the convective hot solids generate substantial momentum and heat transfer, prompting a collision-dominated rapid vaporization of droplets and coupled three-phase transport.

Assuming the temperature of hot solids far exceeds the boiling point of the injected droplets, the spray immediately vaporizes when the droplets contact with the particles. The rapid vaporization of feed droplets invokes a droplet–particle–vapor three-phase hydrodynamic transport. The vapor is both carried by the penetrating spray and simultaneously permitted to drift from the plume toward the downstream region of the gas–solids ambient flow. These trajectories are sketched in Fig. 12.3-9(a). The solids within the spray region either originate from vapor-jet entrainment or from convection of the constituent solids within the flow. Notably, solid flow convection is characterized by both a direct pass of solids through the spray region and an indirect trajectory whereby some solids flow around the spray region due to droplet vaporization, as shown schematically in Figure 12.3-9(b).

The phase trajectory and mixing characteristics within the jet region can be estimated from a simplified analytical model, using a deterministic Lagrangian trajectory approach in a coordinate system situated along the centerline of the spray jet. Under the steady-state condition, differential-integral governing equations can be derived from the mass, momentum, and energy balance over a control volume in the (ξ, η) coordinates for each phase (Zhu *et al.*, 2002).

Spray of fast-evaporating liquids typically evaporates within the jet mixing region so that the axis of the spray jet is conveniently approximated to be the same as the jet flow of gas–vapor mixture. For simplicity, it is assumed that vaporization

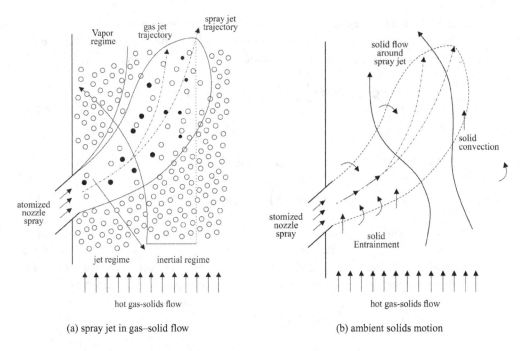

(a) spray jet in gas–solid flow

(b) ambient solids motion

Figure 12.3-9 Schematics of a spray injected into a cross-flow of hot solids.

is heat-transfer controlled and only droplets on the outer layer of the spray can be evaporated. The continuity equation of the droplets is expressed by

$$\frac{d}{d\xi}\left(\int_{A_d}\alpha_d\rho_d U_d dA\right) = -\int_{A_d} n_d \dot{m}_e dA, \qquad (12.3\text{-}2)$$

where A_d is the cross-sectional area of droplets jet; n_d denotes the droplet number density; and $\dot{m}_e$ stands for the evaporation rate per droplet. The ξ-component momentum equation is generated from a ξ-component momentum balance among the increase rate of droplet momentum flow, interfacial forces between droplets and the gaseous mixture, solids–droplets collision, and the momentum transfer due to droplet evaporation, which leads to

$$\frac{d}{d\xi}\left(\int_A\alpha_d\rho_d U_d U_d dA\right) = -\int_{A_d} n_d\dot{m}_e U_d dA - \int_A n_d C_{Dd}\frac{\pi}{8}d_d^2\rho_m(U_d - U)^2 dA$$
$$+ \int_A f_{dp}\frac{\pi}{6}d_p^3\rho_p(U_p - U_d)dA, \qquad (12.3\text{-}3)$$

where C_{Dd} is the drag coefficient of a droplet; and f_{dp} is the solids–droplets collision frequency. The characteristic bending of a spray jet is due to the solids entrainment, which gives the η-component momentum equation:

$$\frac{d\theta}{d\xi}\left(\int_A\alpha_d\rho_d U_d U_d dA\right) = -(1 - C_m)(1 - \alpha_\infty)\rho_p U_e U_{p\infty} l \sin\theta, \qquad (12.3\text{-}4)$$

where θ is the spray angle, defined between ξ and the riser axis; C_m is a partition factor for momentum transfer by solids entrainment to the gas mixture; U_e is the gas entrainment velocity. The energy equation of droplets is obtained by

$$\frac{d}{d\xi}\left(\int_A \alpha_d \rho_d U_d c_{pd} T_d dA\right) = \int_{A_d} n_d \pi d_d^2 h_d (T - T_d) dA - \int_{A_d} n_d \dot{m}_e L dA$$
$$+ \int_A f_{dp} \frac{\pi}{6} d_p^3 \rho_p c_{pp}(T_p - T_d) dA. \tag{12.3-5}$$

Similarly, for the gas–vapor mixture phase, the continuity equation is given by

$$\frac{d}{d\xi}\left(\int_A \rho_m U dA\right) = \alpha_\infty \rho_\infty U_{el} + \int_{A_d} n_d \dot{m}_e dA, \tag{12.3-6}$$

where ρ_m and ρ_∞ are the density of gas–vapor mixture phase and density of ambient gas, respectively; A is the cross-sectional area of the gas–vapor component of the spray jet, which can be wider than A_d; α_∞ is the volume fraction of ambient gas; l is the vapor perimeter. The ξ-component momentum equation is governed by

$$\frac{d}{d\xi}\left(\int_A \rho_m U U dA\right) = \alpha_\infty \rho_\infty U_\infty U_{el} \cos\theta + \int_{A_d} n_d \dot{m}_e U_d dA$$
$$+ \int_{A_d} n_d C_{Dd} \frac{\pi}{8} d_d^2 \rho_m (U_d - U)^2 dA + \int_A n_p C_{Dp} \frac{\pi}{8} d_p^2 \rho_m (U_p - U)^2 dA, \tag{12.3-7}$$

where n_p is the particle number density; C_{Dp} represents the drag coefficient of a particle. The η-component momentum gas–vapor mixture is expressed as:

$$\frac{d\theta}{d\xi}\left(\int_A \rho_m U U dA\right) = -\alpha_\infty \rho_\infty U_\infty U_{el} \sin\theta - C_g C_D \sqrt{\frac{A}{\pi}} \alpha_\infty \rho_\infty U_\infty^2 \sin^2\theta$$
$$- C_m (1 - \alpha_\infty) \rho_p U_{el} U_{p\infty} \sin\theta, \tag{12.3-8}$$

where C_D and C_g denote, respectively, the drag coefficient and a permeability correction factor of drag force for flow around the bending jet tube. It is noted that Eq. (12.3-4) and Eq. (12.3-8) are not independent from each other due to the previous assumption on the same centerline of droplets and vapor phase of spray. Nevertheless, one of these two equations can be regarded as an equation to estimate the partition factor C_m. The energy equation of the gas–vapor mixture is given by:

$$\frac{d}{d\xi}\left(\int_A \rho_m U c_p T dA\right) = \alpha_\infty \rho_\infty U_e c_p T_\infty l + \int_{A_d} n_d \dot{m}_e L dA$$
$$+ \int_{A_d} n_d \pi d_d^2 h_d (T_d - T) dA + \int_A (n_p - f_{dp}) \pi d_p^2 h_p (T_p - T) dA, \tag{12.3-9}$$

where h_d and h_p stand for the heat transfer coefficients of a single droplet and a single particle, respectively, and h_p is greatly influenced by droplet–particle collision.

For the solids in spray, it is assumed that solids enter the mixing region only by jet entrainment and all entrained solids flow along the ξ-direction only. Once solids collide with a droplet, they remain trapped until the host droplet is completely evaporated. The effect of the solids reoccurrence from evaporated droplets is neglected. Thus the mass balance equations is obtained by

$$\frac{d}{d\xi}\left(\int_A \alpha_p \rho_p U_p dA\right) = (1 - \alpha_\infty)\rho_p U_{ep}l. \tag{12.3-10}$$

The momentum equation is given by

$$\frac{d}{d\xi}\left(\int_A \alpha_p \rho_p U_p U_p dA\right) = (1 - \alpha_\infty)\rho_p U_{p\infty} U_{ep}l \cos\theta$$
$$- \int_{A_d} f_{dp}\frac{\pi}{6}d_p^3\rho_p(U_p - U_d)dA - \int_A n_p C_{Dp}\frac{\pi}{8}d_p^2\rho_m(U_p - U)^2 dA. \tag{12.3-11}$$

The energy equation is governed by

$$\frac{d}{d\xi}\left(\int_A \alpha_p \rho_p U_p c_{pp} T_p dA\right) = \alpha_{p\infty}\rho_p U_{ep} c_{pp} T_{p\infty}l$$
$$- \int_{A_d} f_{dp}\frac{\pi}{6}d_p^3\rho_p c_{pp}(T_p - T_d)dA - \int_A (n_p - f_{dp})\pi d_p^2 h_p(T_p - T)dA. \tag{12.3-12}$$

For problem closure, the gas density of the vapor mixture is applied using the ideal gas law, which yields:

$$\rho_m = (1 - \alpha_p - \alpha_d)\frac{p}{R_g T}. \tag{12.3-13}$$

Equations (12.3-2) to (12.3-13) explicate twelve governing equations of hydrodynamics for three-phase flow in the spray zone, for the twelve independent variables of: $A, A_d, \theta, U_d, U, U_p, \alpha_d, \alpha_p, T, T_p, \rho_m$, and C_m. A closed form for the model is therefore achieved. In order to solve these equations, additional intrinsic correlations on the flow entrainment velocity, drag coefficients, heat transfer coefficient, droplet evaporation, particle collision frequency, and collision efficiency must be provided. In addition, the phase distributions over the jet cross section also need to be formulated.

12.4   Bubbling Reactors in Liquid and Liquid–Solid Media

Gas–liquid and gas–liquid–solid contactors discussed in Sections 1.4 and 10.5 are often employed in process systems. The key feature of these contactors is gas bubbles dispersed in a continuous liquid phase or liquid–solid suspension. The industrial embodiments that leverage this bubbling phenomenon are represented by bubble columns or slurry-bubble columns. Other systems such as sparged stirred tanks and static in-line mixers are also popular. The overall rate process in the bubbling systems encompasses various steps of the rate processes involved in the interfacial mass transfer and the reaction kinetics. To accurately account for the reactor performance, both the overall rate process and transport phenomena of the multiphase flows need to be considered.

12.4.1   Mass Transfer in Gas–Liquid Media

For gas–liquid mass transfer in a gas–liquid bubbling reactive system, a gaseous species participating in a liquid phase reaction will be transported from the bulk gas phase to the gas–liquid interface, diffuse from the interface to the bulk liquid phase, and then react with another reactant in the liquid. This type of process is typically analyzed via the film model, which assumes a thin film exists between the interfacial boundaries of each bulk phase. Mass transfer within the gas and liquid films occurs by diffusion, and the total mass transfer resistance is the sum of the mass transfer resistances in the gas and liquid films with the gas and liquid concentrations on the interface being in equilibrium. For bubbling reactors, it is commonly assumed that the predominant resistance for the film diffusion is on the liquid side instead of the gas side of the film. The relative rates of the reaction and the mass transfer process influence the concentration profile in the film and, hence, the interfacial mass transfer rate (Levenspiel, 1999).

For simplicity, consider a reactant, A, diffusing from gas bubbles into the liquid and participating in the liquid-phase reaction. If the liquid phase is well mixed, such that the concentration of A in the bulk liquid is uniform, the macroscopic (*i.e.*, reactor scale) mass balance of species A in the reactor can be written as:

$$\frac{dC_A}{dt} = \frac{Q_l}{V}(C_{A0} - C_A) + k_l a' \alpha (C_A^* - C_A) - \gamma \qquad (12.4\text{-}1)$$

Equation (12.4-1) indicates the accumulation of A within the reactor volume, V, manifests as a consequence of A being transported in and out of the reactor by liquid flow Q_l, interfacial mass transfer from gas to liquid, and consumption due to the volumetric reaction rate, γ. The mass transfer coefficient, k_l, varies with the reactant diffusivity in the liquid phase, thickness of the liquid film, and the liquid phase reaction rate. The interfacial area per unit volume of gas bubbles, a', strongly depends on the bubble size and bubble shape. The volume fraction of gas bubbles in the reactor, α, is a function of the gas velocity. C_A^* is the concentration of species A in the liquid phase, which is in equilibrium with the partial pressure of species A in the gas phase. The equilibrium relationship is dictated by the Henry's law. C_A^* is practically applied to the gas–liquid interface. Equation (12.4-1) indicates that bubble size and volume fraction of gas bubbles play an important role in the interfacial mass transfer performance of a gas–liquid reactor.

12.4.2   Sparged Stirred Tank

Mechanically stirred, sparged tanks are commonly used for mixing applications in industry. Two key components are in the tank (*i.e.*, sparger and impeller). A sparger introduces gas bubbles. The impeller creates either a radial or an axial liquid flow pattern that disperses the gas bubbles into the liquid volume as they pass through the impeller region. Different impeller types create distinct different liquid flow characteristics. Typical radial impellers include flat and concave blade disk turbines. The axial category of impellers comprises pitched blade turbines and hydrofoil blades.

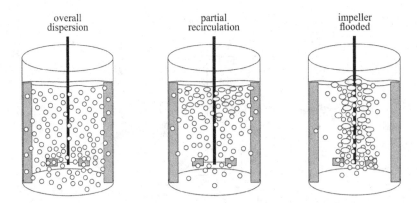

Figure 12.4-1 Flow regimes in gas–liquid stirred tank.

Axial impellers can operate either in an up-pumping or down-pumping mode. In a tall tank, multiple impeller stages are often employed to redisperse large bubbles and improve liquid mixing effectiveness. Wall baffles are generally necessary to be used that function to restrict circumferential liquid motion, encourage vertical flow, and enhance recirculation of the liquid and gas phases. Typical configuration and flow regimes of a sparged stirred tank for gas–liquid mixing are shown in Figure 12.4-1. The flow pattern of the gas phase is a function of gas–impeller interactions.

At a low gas flow rate and high impeller speed, gas bubbles are distributed throughout the domain of the tank, and significant gas recirculation is realized; this condition is known as complete dispersion. At a moderate gas flow rate and impeller speed, the gas dispersion is concentrated in the upper region of the tank; this mode is known as partial recirculation. At a high gas flow rate, bubbles accumulate in the impeller region, and the mixing effect becomes significantly attenuated; this phenomenon is referred to as flooding, and it should be avoided. Flooding results due to the tendency of the gas bubbles to be attracted by the low-pressure region developed behind the impeller blades. When a large volume of gas accrues near the impeller, it forms large "cavities" that obstruct liquid discharge from the impeller. Flooding also significantly reduces the power output of the impeller, resulting in impaired mixing with a corresponding reduction in heat and mass transfer effectiveness. Flooding is imminent when the gas flow rate exceeds a critical value. The critical value for a given system is typically a function of impeller type and impeller speed.

12.4.3   Fischer–Tropsch Synthesis in Slurry Bubble Column

The Fischer–Tropsch (F-T) synthesis is an important process used to produce hydrocarbon from syngas, which mainly consists of hydrogen and carbon monoxide. It is a crucial step in the production of liquid fuels and chemicals from sources such as natural gas, coal, and biomass. The process can be conducted at high temperature (*e.g.*, 300–350°C) in fluidized beds or circulating fluidized beds to produce gasoline and linear low-molecular-weight olefins. Alternatively, the process can

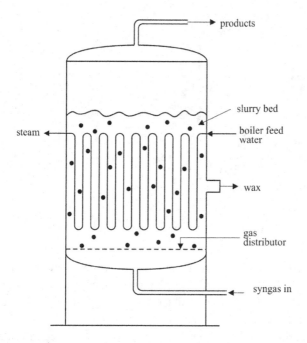

Figure 12.4-2 Simplified structure of a slurry-bubble column for F-T synthesis.

be conducted at low temperatures (*e.g.*, 200–240°C) in slurry-bubble columns to yield high-molecular-weight waxes. The high-temperature process is limited to iron-based catalysts, while the low-temperature alternate relies on either iron-based or cobalt-based catalysts. The reaction for the F-T synthesis can be expressed as:

$$CO + 2H_2 = (\text{-}CH_2\text{-}) + H_2O \tag{12.4-2}$$

The reaction is highly exothermic (*e.g.*, $\Delta H(227°C) = -165$ KJ/mol); therefore, efficient heat removal is required. Various collateral reactions can proceed in the process, including the water-gas shift (WGS) reaction that is moderately exothermic (*e.g.*, $\Delta H(227°C) = -39.8$ KJ/mol). The WGS reaction is given by:

$$CO + H_2O = H_2 + CO_2 \tag{12.4-3}$$

The slurry-bubble column offers excellent heat transfer characteristics and inherent temperature uniformity and is commonly used for low-temperature F-T synthesis. Suspending small catalyst particles (about 50 µm) in the slurry phase eliminates undesirable intra-particle diffusion limitation. A typical slurry-bubble column reactor application for F-T synthesis is shown in Figure 12.4-2 (Dry, 2002). For large-scale industrial operations, the column diameter typically ranges from 6 to 10 m, and column height usually varies from 30 to 40 m. The reactor is operated at a temperature of approximately 240°C and at a pressure of 3 to 4 MPa. Because the reaction is exceedingly exothermic, it is common to rely on a quantity of 5,000 to 8,000 vertically cooling tubes, with a typical diameter of 50 mm, for heat removal. The catalyst particles are suspended in a pseudo-homogeneous slurry phase in a batch or semi-batch mode of operation. The syngas is introduced into the reactor from the distributor at the bottom of the reactor. It is common for the superficial gas velocity to exceed the superficial liquid velocity by at least one order of magnitude. Thus, the hydrodynamics of the column is dominated by the gas phase dynamics (van Baten and Krishna, 2004).

At a low superficial gas velocity, the slurry-bubble column is operated in the homogeneous regime with small bubble sizes and a small size variation, typically between 1 and 7 mm. Conversely, the slurry-bubble column is operated in the heterogeneous regime with large bubble sizes and a large size variation, typically between 20 to 70 mm, when the superficial gas velocity is high. The overall gas holdup and bubble size distribution parameters are functions of solid concentration, column diameter, degree of the F-T reaction, and operating pressure (Krishna and Sie, 2000).

12.5 Reactive Flows in Gas–Solid Fluidized Beds

Fluidized bed (FB) reactors discussed in Section 10.2 and circulating fluidized bed (CFB) riser reactors described in Section 10.3 are extensively used in industry for gas phase reactions with solid catalyst particles, such as fluid catalytic cracking (FCC) and gas phase polymerization, and the hydrodynamics associated with these systems are strongly coupled with concurrent reactions that proceed along the reactor. Key transport phenomena include dramatic changes in phase compositions, volume fractions, and velocities; variations in temperature stemming from reaction heats and pressure changes from the production and/or consumption of gas species in the reaction are also relevant. In FCC and polymerization reactors, the reactants are introduced via liquid spray injection such that a zone of spray vaporization and phase mixing develops in the entrance region. The phase transport in this zone, which may also be coupled with catalytic reactions, can have a considerable impact on the downstream reactions.

12.5.1 Fluid Catalytic Cracking

Fluid catalytic cracking (FCC) is a well-established conversion process in the oil refining industry. A typical FCC riser reactor, as schematically shown in Figure 12.5-1, is designed to use zeolite-containing catalysts to crack the liquid feed of high-boiling hydrocarbons, such as vacuum gas oil (VGO), into valuable lighter hydrocarbons and byproducts such as coke. Both the vaporization of liquid feed and the endothermic FCC reaction rely on the heat supply from preheated fresh catalysts entering the riser. Hence, an FCC process involves not only the transition of a gas–liquid-solids three-phase flow into a gas–solid two phase flow but also the coupled heat and mass transfer among the phases with dynamic interactions such as droplet–solid collisions and catalytic reactions between the gas and solids.

As shown in Figure 12.5-1, the liquid feed enters the riser reactor through multiple feed atomizing nozzles, contacts hot regenerated catalyst, and vaporizes. The

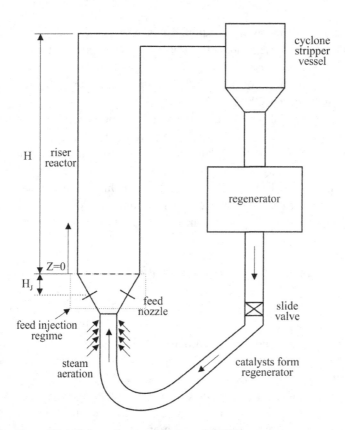

Figure 12.5-1 Schematic representation of a FCC system.

resulting vapor cracks as it is conveyed upward along with the catalyst against the gravitational field in the riser. Cracking and depressurization cause the vapor to expand, thereby increasing the velocities of both vapor and catalyst through the riser. Coke accumulation on the catalyst surface is an inherent consequence of the VGO cracking process, and the fouling effect reduces the catalyst's cracking activity. Spent catalyst is extracted from the hydrocarbon stream using a cyclone separator at the riser exit. Regeneration occurs by burning coke in a high-temperature regenerator, and the catalyst is then fed back to the riser to complete the cycle. The endothermic nature of cracking reactions yields a monotonic decrease of the local equilibrium temperature between the vapor and catalyst along the riser.

The FCC process performance is dictated by the riser hydrodynamics and catalytic reaction kinetics. Specifically, the yield rate of a product depends upon the cracking rate and the reaction time duration, both of which are interdependent functions of local transport properties such as the temperature, catalyst concentration, catalyst deactivation, and transport velocities. The FCC reaction has its genesis in the main transport section of the riser with the vapor phase (VGO) and the solid phase (*e.g.*, catalyst-containing particles) initially participating. However, the liquid spray and its interaction with feed particles does have a subsequent influence on downstream FCC

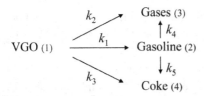

Figure 12.5-2 Simple reactions for FCC cracking of VGO.

reactions, such as vaporization and partial cracking in the spray region, as shown in Section 12.5.2.

The petroleum FCC process involves a large number of reacting species. For simplicity, the FCC cracking chemistry has been represented by a simple four-lump kinetic model for gasoline production (Lee *et al.*, 1989) with the reaction scheme given in Figure 12.5-2, where the rates of reactions are proportional to the rate coefficients (k's).

It is noted that the gasoline, the main product of the process, is an intermediate product that may overcrack down to gases if the crack reaction is not terminated in time. Since the reaction time and conditions are controlled via a riser transport, the hydrodynamics and its control are critical to the cracking reaction rates and gasoline yield.

12.5.2 Vaporization and Reaction in a Riser

In the FCC process, liquid feed enters the riser reactor through a multitude of feed atomizing nozzles that are circumferentially disposed on the reactor wall. To make an effective contact between the atomized spray and hot catalysts, the nozzles are typically of rectangular shape with large aspect ratios and wide fan angles. Spray droplets vaporize on contact with the hot catalyst. Since cracking reactions start almost instantly upon the contact of vapor and catalysts, a significant portion of the cracking process may progress in the feed injection region where the temperature, concentration, and activity of the catalyst exceed other sections of the riser. The cracking reactions that proceed inside the feed injection region undoubtedly influence vapor composition and volume fraction; this effect may subsequently alter the follow-up riser transport and reaction characteristics. Thus, the spray transport, vaporization, and cracking reaction in the feed injection region, which is patently coupled together, dictate the overall performance of a riser reactor.

In the FCC riser reactor, steam is introduced upstream of the feed injection region to facilitate catalyst dispersion. While the steam should be considered in the hydrodynamics of spray, its influence on kinetics can be ignored. Thus, steam is reasonably regarded as an inert gaseous component, whereas the cracking reaction scheme is described using a lumped approach accounting for the effects of reacting gaseous components. Modeling of the reacting spray transport follows the basic approach expounded in Section 12.3.5, albeit with modification to the reaction and vapor phase convection transport to the downstream region external to the spray region. The modeling results are then used as the inlet conditions for the continued vapor cracking in the riser transport.

Without losing generality, the interactions of flow and FCC reaction in the main transport regime of a riser can be revealed by a simple one-dimensional two-phase flow model (Zhu *et al.*, 2011). The riser hydrodynamics are captured by a cross-section-averaged model in which the heterogeneous transport is governed by gas–solids interfacial momentum transfer. The modeling domain starts right after the feed oil completely vaporizes from the spray injection. The transport of catalyst particles and gases in the riser is governed by convection with negligible diffusion. The reaction is approximated by a four-lump kinetic model.

The overall mass balance of the gas phase is given by:

$$\frac{d}{dz}(\alpha\rho U) = -(\gamma_3 + \gamma_5). \tag{12.5-1}$$

The two terms on the right-hand side account for mass lost due to coke formation. The average gas density can be determined from the ideal gas law as:

$$\rho = \frac{p \sum\limits_{i=1}^{3}(C_i M_i)}{R_g T \sum\limits_{i=1}^{3} C_i}. \tag{12.5-1a}$$

Because the component mass balance presupposes chemical reactions, the species equations can be conveniently expressed in terms of molar concentration fractions. Based on the reaction scheme depicted in Fig. 12.5-2, the variations of the molar fractions of concentrations ($C_i, i = 1 - 4$) in the reactor for VGO, gasoline, light gases, and coke can be given, respectively, by:

$$\frac{d}{dz}(\alpha\rho C_1 U) = -\alpha\rho\,\Phi_s\,(k_1 + k_2 + k_3)\,C_1{}^2 \rho_p \alpha_p \tag{12.5-2a}$$

$$\frac{d}{dz}(\alpha\rho C_2 U) = \alpha\rho\,\Phi_s\left(\frac{M_1}{M_2}k_1 C_1{}^2 - (k_4 + k_5)C_2\right)\rho_p \alpha_p \tag{12.5-2b}$$

$$\frac{d}{dz}(\alpha\rho C_3 U) = \alpha\rho\,\Phi_s\left(\frac{M_1}{M_3}k_2 C_1{}^2 + \frac{M_2}{M_3}k_4 C_2\right)\rho_p \alpha_p \tag{12.5-2c}$$

$$\frac{d}{dz}(\alpha\rho C_4 U) = \alpha\rho\,\Phi_s\left(\frac{M_1}{M_4}k_3 C_1{}^2 + \frac{M_2}{M_4}k_5 C_2\right)\rho_p \alpha_p \tag{12.5-2d}$$

$$\sum_{i=1}^{4} C_i = 1, \tag{12.5-2e}$$

where Φ_s is the decay function of the catalyst activity resulting from coke deposits on the catalyst surface (Pitault *et al.*, 1994). Equation (12.5-1) is functionally equivalent to the summation of the individual species equations: Eqs. (12.5-2a) to (12.5-2d). Notably, coke is considered as a component of the gas phase in the reaction model, but its mass is added to the solid phase because of its physical deposition onto the catalyst particle surfaces. The solid phase mass balance is therefore given by:

$$\frac{d}{dz}(\alpha_p \rho_p U_p) = (\gamma_3 + \gamma_5),$$ (12.5-3)

where γ_3 and γ_5 denote the production rate coke from VGO and from gasoline, respectively. The gas phase momentum balance is then written as:

$$\frac{d}{dz}\left(\alpha \rho U^2\right) = -\alpha \frac{dp}{dz} - F_D - \alpha \rho g - (\gamma_3 + \gamma_5)\, U,$$ (12.5-4)

where F_D represents the total momentum transfer between the gas and solid phases. The momentum balance for the solid phase is likewise expressed as:

$$\frac{d}{dz}\left(\alpha_p \rho_p U_p{}^2\right) = -\alpha_p \frac{dp}{dz} + F_D - \alpha_p \rho_p g - F_C + (\gamma_3 + \gamma_5)\, U,$$ (12.5-5)

where F_C is a collision force that modulates the axial acceleration of solids in the dense phase and acceleration regions (Zhu and You, 2007).

Assuming local thermal equilibrium between solid and gas phases yields the overall energy balance equation of:

$$\left(\alpha_p \rho_p U_p c_{pp} + \alpha \rho U c_p\right)\frac{dT}{dz} + (\gamma_3 + \gamma_5)\left(c_{pp} - c_p\right) T = -\sum_{i=1}^{5} \gamma_i . \Delta H_i,$$ (12.5-6)

where ΔH_i is the heat of reaction for the ith endothermic cracking reaction. Thus, it gives ten independent equations, Eq. (12.5-1)–Eq. (12.5-6), for ten independent variables, α_p, U_p, U, C_1, C_2, C_3, C_4, ρ, p, and T. In principle, the condition that the number of independent equations equates to the number of independent results in model closure, or the closed form. However, detailed constitutive relationships that describe gas–solid momentum transfer (F_D), collision-induced momentum transfer (F_C), and the catalyst-influenced reaction coefficients (k's) must be formulated as closed functions of the above independent variables to actualize the closed analytical form of the model.

12.5.3   Gas Phase Polymerization

Gas phase polymerization is exemplified by the polyethylene production process. Polyethylene (PE) is a ubiquitous thermoplastic material and the largest synthetic commodity polymer as measured in terms of annual production. PE is commonly produced by gas phase polymerization in a gas–solid fluidized bed reactor represented commercially by the UNIPOL^{TM} process, as schematically given in Figure 12.5-3 (Jazayeri, 2003).

The vertical reactor comprises a lower reaction zone and an upper disengagement region. A gas mixture containing the monomer (C_2H_4), comonomer, hydrogen, and an inert gas such as nitrogen are introduced through a distributor into the reactor. Catalyst particles are supplied above the distributor, and they are encapsulated by the newly formed polymer as the reaction proceeds. Particle size increases as the

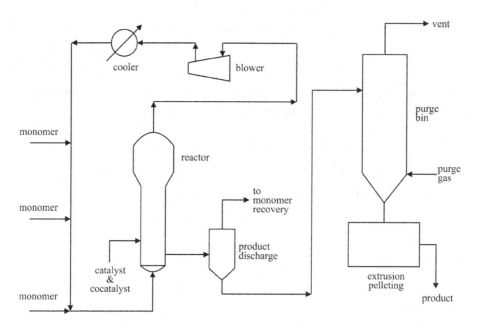

Figure 12.5-3 Simplified diagram for the UNIPOL^{TM} process.

result of the polymerization reaction, and the bed height increases as large parti-cles settle to the bottom of the bed. When the bed height reaches a specified level, mature polymer particles are intermittently discharged from the bed bottom. As the gas progresses through an expanded disengagement zone in the top of the reactor, its velocity decreases and causes the entrained particles to recede to the reaction zone. The effluent gas is compressed, cooled, and subsequently recycled to the reactor. The reactor typically operates at a low temperature (about 75 to 105°C) but at an elevated pressure (*e.g.*, 20 to 25 atm).

The polymerization reaction can be envisioned to occur at the interface between the catalyst particle and the polymer matrix. During this process, gas phase monomers are converted to a solid phase polymer, which influences particle mor-phology. The fresh porous catalyst particles comprise fine primary particles. As the reaction proceeds, monomers initially diffuse into the particle and are adsorbed onto the surface of the primary particles by the active centers. As the monomers con-tinue to convert into polymer, primary particles are encapsulated by the formed polymer to produce aggregates. Key elementary reactions include the initialization of active sites, polymer chain propagation, chain transfer, and catalyst deactivation (Xie *et al.*, 1994).

Details of the reaction pathway depend on the catalyst used. For example, with metallocene catalyst, the reaction pathway can be written as:

$$\text{Initiation:} \qquad c \xrightarrow{k_i} c^* \qquad\qquad (12.5\text{-}7a)$$

$$\text{Propagation:} \qquad P_n^*(c^*) + M \xrightarrow{k_p} P_{n+1}^* \qquad\qquad (12.5\text{-}7b)$$

Decay: $\qquad P_n^*(c^*) \xrightarrow{k_d} P_n + c^0,$ $\qquad$ (12.5-7c)

where c is a potential catalyst active site, c^* is an active catalyst site, and c^0 is a deactivated catalyst site; P_n^* and P_n are active and dead polymer with chain length n, and M is a monomer. The reactions are first order with respect to the reactant concentrations, and the rate constants, k_q ($q = i, p$, and d, respectively representing initial, propagation, and decay), obey the Arrhenius equation of:

$$k_q = k_{q0} \exp \left(-\frac{E_q}{R_g T} \right).$$ $\qquad$ (12.5-7d)

The polymerization reaction is highly exothermic, and therefore the reaction heat must be effectively removed to avoid melting the polymer.

In commercial reactors employing the UNIPOL process, the superficial gas velocity is between 2 and 6 times the minimum fluidization velocity; the fluidized beds therefore operate in or near the turbulent fluidization regime. Fluidization is achieved at a high gas recycle rate, generally exceeding the feed rate of makeup gas by two orders of magnitude. Because of attenuated single-pass conversion, gaseous monomers consumed in the reaction do not have significant effects on the hydrodynamics.

The hydrodynamics of a fluidized bed polymerization reactor can be described using the Eulerian multiphase flow model; N solid phases are each characterized by a specific particle size. The momentum equations of the gas and solid phases are then given by:

$$\frac{\partial}{\partial t}(\alpha \rho \mathbf{U}) + \nabla \cdot (\alpha \rho \mathbf{U}\mathbf{U}) = \nabla \cdot \mathbf{T} + \alpha \rho \mathbf{g} + \sum_{i=1}^{n} \mathbf{F}_i$$ $\qquad$ (12.5-8a)

$$\frac{\partial}{\partial t}\left(\alpha_{pi}\rho_p \mathbf{U}_{pi}\right) + \nabla \cdot \left(\alpha_{pi}\rho_p \mathbf{U}_{pi}\mathbf{U}_{pi}\right) = \nabla \cdot \mathbf{T}_{pi} + \alpha_{pi}\rho_p \mathbf{g} - \mathbf{F}_i + \sum_{j=1, j \neq i}^{n} \mathbf{F}_{pji} + \mathbf{F}_{Ei},$$

$\qquad$ (12.5-8b)

where $\mathbf{T}$ and $\mathbf{T}_{pi}$ are the total stress tensors of gas and the ith solid phase, respectively; $\mathbf{F}_i$ is the interaction force between the gas and the ith solid phase, which typically includes the pressure gradient force and drag force; $\mathbf{F}_{pji}$ is the interaction between the jth and ith solid phases; $\mathbf{T}_{pi}$ and $\mathbf{F}_{pji}$ can be closed via the kinetic theory for granular flow; $\mathbf{F}_{Ei}$ is the electrostatic force acting on the ith solid phase (Fan, 2006; Rokkam et al., 2010).

Because the polymerization reaction is highly exothermic and the reactor operates at a temperature near the melting point of the polymer, temperature control is critical. Maldistribution of temperature can cause polymer sheet formation, bed collapse, or even complete solidification of the reactor in severe cases. Thus, the production rate of the polymer in a reactor of a given size is necessarily constrained by the heat removal rate. Internal cooling coils are impractical because they would quickly be rendered ineffective by the fouling caused by the formation of polymer films.

Instead, temperature control is achieved via cooling of the recycled stream. Specifically, a supercondensed mode is employed whereby the recycle stream is cooled to a temperature below its dew point, forming a two-phase gas–liquid mixture (Jenkins *et al.*, 1986; Burdett, 2008). As the recycled stream reenters the reactor beneath the gas distributor, the liquid phase in the partially condensed mixture persists as fine droplets entrained in the gas phase until evaporation. Rapid vaporization of the liquid dramatically increases the heat removal capacity and consequently leads to a substantial increase in production rate. The extent of condensation is limited by the stipulations that liquid droplets remain suspended in the recycled stream and sufficient superficial gas velocity maintained to sustain bed fluidization.

Particle size growth is an essential characteristic in the gas phase polymerization reactor, and it can be modeled with the population balance approach. Because a change in the spatial and temporal particle size distributions is attributed to chemical reactions, particle transport, and interactions, the population balance model necessitates input parameters from both the reaction kinetics model and the hydrodynamic model. At the same time, particle size is an important criterion in the interphase transport coefficients for the gas and solid phases, and it also appears in the constitutive equation of the solid phase. Thus, the effects of two-way coupling between the population balance model and the transport and reaction models must be accommodated.

A general form of the population balance equation can be expressed as:

$$\frac{\partial n(d_p)}{\partial t} + \nabla \cdot \left[n(d_p) \langle \mathbf{U}_p \mid d_p \rangle \right] = -\frac{\partial}{\partial d_p} \left[G(d_p) n(d_p) \right] + S_{agg}(d_p) + S_{br}(d_p). \quad (12.5\text{-}9)$$

The above equation reveals that the number density of particles, $n(d_p)$, evolves as the result of convection and source terms because of chemical reaction, aggregation, and breakage. The solid phase velocity, $\langle \mathbf{U}_p \mid d_p \rangle$, is the mean value, which is conditioned on the particle size, d_p. Thus, the following equation can be written:

$$\langle \mathbf{U}_p \mid d_p \rangle = \frac{1}{n(d_p)} \int_{-\infty}^{+\infty} \mathbf{U}_s n(d_p, \mathbf{U}_s) d\mathbf{U}_s. \quad (12.5\text{-}9a)$$

In the above expression, $G(d_p) n(d_p)$ represents the particle growth due to the polymerization reaction, and it is calculated from the reaction rate model; the quantities $S_{agg}(d_p)$ and $S_{br}(d_p)$ account for number density distribution influence resulting from particle aggregation and breakage, respectively (Fan, 2006; Chen *et al.*, 2011).

Electrostatics have a significant effect on the minimum fluidization velocity, bubble size, and particle entrainment behavior of a gas–solid fluidized bed. Electric charge generation can be caused by particle–particle and particle–wall collisions as well as gas–solid friction (Fotovat *et al.*, 2017). Small polymer particles are often positively charged, while medium and large polymer particles are typically negatively charged. Charge accumulation is also affected by the physical properties of the polymer material. Polymers with a high molecular weight are predisposed to developing considerable levels of electrostatic charge, and low-molecular-weight polymers are less susceptible. The electric potential of a fluidized bed can often exceed 20 kV (Gajewski, 1985). Electrostatic effects are primarily concentrated in the vicinity of the distributor, near the expansion region, and at the dome section of the reactor. The top of the reactor has a positive potential while the bottom of the bed and the distributor have a negative potential. This type of charge distribution has been attributed to a particle segregation effect along the axial direction in the reactor.

The influence of electrostatics on a polymerization reactor can be modeled by calculating the electric field in the reactor and the electrostatic forces acting on the polymer particles. The electric potential in a gas–solid fluidized bed is governed by the equation:

$$\nabla \cdot (\varepsilon_0 \varepsilon_m \nabla \Phi) = -\rho_e. \tag{12.5-10}$$

The relative permittivity of a gas–solid mixture can be gleaned by solving the Bruggeman equation, represented by:

$$\alpha = \left(\frac{\varepsilon_p - \varepsilon_m}{\varepsilon_p - \varepsilon} \right) \left(\frac{\varepsilon}{\varepsilon_m} \right)^{\frac{1}{3}}. \tag{12.5-10a}$$

The total charge density is the sum of the contributions from polymer particles of all sizes. Thus, it gives:

$$\rho_e = \sum_{i=1}^{N} \rho_{ei} \alpha_{pi}. \tag{12.5-10b}$$

The electrostatic force on the ith particle phase can then be calculated using the local charge density and the electric field as:

$$\mathbf{F}_{Ei} = -\rho_{ei} \alpha_{pi} \nabla \Phi. \tag{12.5-11}$$

Strong electrostatic forces will engender agglomeration of polymer particles and encourage particles to adhere to the reactor wall. These consequences function to impair heat transfer efficiency; as a result, the temperature of polymer particles may increase beyond the melting point. Prolific melting and fusing of the polymer can plug the distributor or product discharge system and cause defluidization. Thus, various techniques are routinely employed to mitigate electrostatic charge in a fluidized bed reactor. Some notable approaches are grounding of the reactor wall to reduce charge generation, injecting an antistatic agent to increase particle surface conductivity, injecting charge particles of opposite polarity to neutralize the charges in the bed, and careful specification of the approved start-up procedure (Hendrickson, 2006).

12.6 Dispersed Fuel Combustion

Dispersed fuel combustion represents an important branch in applications of multiphase reacting flows. The fuel particles in the dispersed phase can be fuel droplets or solid fuel particles. This section covers basic mechanisms and modeling of combustions of a fuel droplet and a solid fuel particle. These mechanistic models are used to describe the reaction nature of the fuel particles, which can be strongly coupled

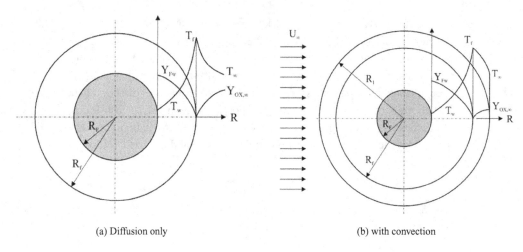

(a) Diffusion only   (b) with convection

Figure 12.6-1 Simple mode of droplet combustion.

with the transport equations of multiphase flows. Further introduction on such modeling approaches is beyond the basic scope of this book, and interested readers may continue to references such as Zhou (2018).

12.6.1   Combustion of a Fuel Droplet

There are many multiphase flow applications involving fuel droplet combustion. The diffusive evaporation model of a nonreacting droplet has been previously introduced in Section 3.4.3. Now, consider a simple quasi-steady model for the gaseous combustion of a fuel droplet in which only fuel vapor, oxygen, inert gas, and combustion product are included, as shown in Figure 12.6-1. In the classic Spalding double film model, the space surrounding the droplet is divided into the fuel film and the oxidizer film separated by a flame front at R_f, as shown in Figure 12.6-1(a). The fuel vapor diffuses from the droplet surface, R_F, to the flame front, while the oxidizer diffuses from the surrounding ($R = \infty$) to the flame front. The combustion is assumed to be stoichiometric, one-step, and complete; therefore, neither fuel nor oxidizer can penetrate the flame zone. Consequently, the fuel mass fraction, Y_F, is zero in the far field when $R > R_f$, and oxidizer mass fraction, Y_{Ox}, equals zero at the droplet surface. This does not necessarily require an infinitely thin flame front, although in practice, the fractions of the fuel and the oxidizer are often found to quickly drop to zero in a thin flame zone. The heat released by the combustion produces a high temperature of T_f at the flame front, and the rate of heat transfer from the flame front to the droplet surface is generally considered to determine the evaporation and combustion rate of the fuel droplet.

The continuity equation of the gaseous mixture is given by

$$\frac{d}{dR}\left(\rho U R^2\right) = 0. \tag{12.6-1}$$

The equation of vapor-species conservation is given by:

$$\rho U \frac{dY_s}{dR} = \frac{1}{R^2} \frac{d}{dR}\left(R^2 D\rho \frac{dY_s}{dR}\right) - \gamma_s, \tag{12.6-2}$$

where γ_s is the reaction rate of s-species, which is assumed to follow the Arrhenius law as:

$$\gamma_s = k_{s0} \prod_{s=1}^{z} \rho_s{}^{v_s} \exp\left(-\frac{E}{R_g T}\right). \tag{12.6-2a}$$

For the inert gas, $\gamma_s = 0$. The boundary conditions, both at the fuel droplet surface and at the ambient at $R \to \infty$, are prescribed for all the species and gas temperature (assuming that the reaction occurs in the whole space domain). The mass balance at the droplet interface also requires:

$$D\rho\left(\frac{dY_s}{dR}\right)_b = (\delta_{sf} - Y_{sb})(\rho U)_b, \tag{12.6-2b}$$

For all species, it gives:

$$\sum Y_s = Y_F + Y_{Ox} + Y_{Pr} + Y_{In} = 1. \tag{12.6-3}$$

The energy equation is given by:

$$\rho U c_p \frac{dT}{dR} = \frac{1}{R^2} \frac{d}{dR}\left(R^2 K \frac{dT}{dR}\right) + \gamma_s \Delta H_s, \tag{12.6-4}$$

where ΔH_s stands for the reaction heat, determined by:

$$\Delta H_R = \sum_p n_p (h_f^0 + h - h^0)_p - \sum_r n_r (h_f^0 + h - h^0)_r, \tag{12.6-4a}$$

where n is the molar concentration, and h with superscript "0" means the enthalpy of formation. Hence, the first summation represents the combined enthalpy of products, whereas the second is the combined enthalpy of reactants. For a positive reaction heat, the reaction is termed endothermic, meaning heat absorption. Otherwise, it is exothermic, meaning heat generation.

The heat balance at the droplet interface requires:

$$K\left(\frac{dT}{dR}\right)_b = (\rho U)_b L. \tag{12.6-5}$$

Assuming a Lewis number of unity and constant transport properties, the evaporation rate of a fuel droplet may be estimated using the Zeldovich transformation as (Zhou, 1993):

$$\dot{m}_e = 4\pi R_d \frac{K}{c_p} \ln\left(1 + \frac{c_p(T_\infty - T_{db}) + (Y_{Ox,\infty} - Y_{Ox,b})\Delta H_{Ox}}{L + Y_{Ox,b}\Delta H_{Ox}}\right). \tag{12.6-6}$$

The evaporation (or burning) rate can be found without knowing the exact location or temperature of the flame front, as indicated by Equation (12.6-4). If the above integral approach is employed and the boundary conditions at the droplet surface and the outer boundary are fixed, the evaporation rate is irrelevant to the location of the flame front, provided that it is within the boundaries of integration. However, the position of the flame can be calculated using an equivalent form of the above evaporation rate expressed in terms of the mass fractions of the fuel and the oxidizer from:

$$\dot{m}_e = R\rho D \ln \left(\frac{1 + Y_{Ox,\infty}}{1 - Y_F + (F/O)_{st} Y_o} \right),$$ (12.6-6a)

where $(F/O)_{st}$ is the stoichiometric proportion of the fuel and the oxidizer. Assuming that the flame front exists at the position where the fuel and the oxidizer are at their stoichiometric ratio, or $Y_F = Y_O = 0$, the location of the flame front can be found from Eq. (12.6-6a), using the evaporation rate first obtained using Eq. (12.6-6).

In reality, many other factors affect the evaporation and reaction rate of a droplet. For example, relative motion between the droplet and ambient gas flow will not only change the spherical symmetry but also lead to an internal liquid circulation within the droplet. The thermal properties are often variable and strongly dependent on the local temperature and species concentrations. It is also possible for the Lewis number to exceed unity. In addition, most of reactions are multistep, nonstoichiometric, and incomplete (with various reversible effects). Sirignano (1999) and Zhou (2018) offer an expanded presentation of these effects.

12.6.2   Combustion of a Solid Fuel Particle

Solid fuels include solid hydrocarbons (such as coal, wood, rubber, paper, plastic, and solid waste), metals (such as Al, Mg, Mo, Ti, K, Na, and Zr), solid propellants, and nonmetals (such as S). Most solid hydrocarbon fuels consist of volatile matter, moisture, carbon, combustible minerals, and noncombustible minerals (ash). The typical combustion process of a solid fuel particle is schematically shown in Figure 12.6-2. In the initial stage of preheating and devolatilization, a coal particle releases constituent volatile gases and water vapor. Such a release process opens many micro-channels or micropores and enlarges the particle size as volatile gases and water vapor are forced out. Once the ambient temperature is above the ignition point, the gaseous combustion of volatiles occurs. The dried and porous charcoal can be ignited to start the combustion process. The residuals from charcoal combustion form ash particles or molten ash lava.

The mechanisms and modeling of preheating, drying, and devolatilization processes are similar to the evaporation of a droplet without reaction, whereas the mechanisms and modeling of the volatile combustion process are similar to those of the evaporation of a droplet with reaction. For the char (*i.e.*, dry and pure solid fuel) combustion, there are two simple models. One is the shrinking sphere (or shrinking core) model, as shown in Figure 12.6-2, that assumes noncombustible minerals, or ash, always remains at the core of the solid fuel particle so that only oxides (*e.g.*, oxygen) need to reach to the surface to react; the size of the solid particle reduces during the combustion process while density remains constant. The other model assumes

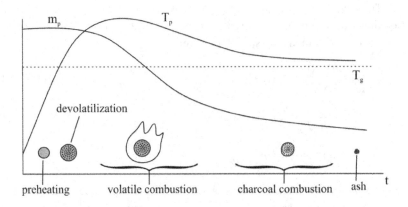

Figure 12.6-2 Schematic combustion process of a solid fuel particle.

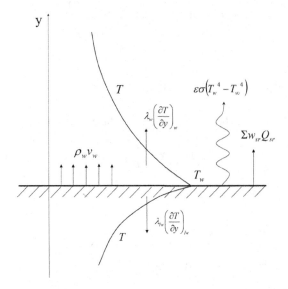

Figure 12.6-3 Boundary condition with flame on the particle surface.

noncombustible minerals, or ash, are uniformly distributed throughout the solid particle and oxides (such as oxygen) must penetrate the porous ash shell via diffusion in order to reach the solid fuel for a reaction; the size of the solid particle in the combustion process remains unchanged.

The shrinking sphere model with a single-step, stoichiometric, and complete reaction on a solid surface (and without any gas phase combustion) represents the most fundamental model of solid fuel combustion. Consider a combusting solid fuel sphere suspended in a binary mixture of oxygen and inert gas. The combustion process is quasi-steady and spherically symmetric. The combustion flame occurs on the solid surface, as shown in Figure 12.6-3.

The governing equations for the gas phase are the same as those for droplet evaporation with reactions, as presented in Section 12.6.1. The boundary conditions on the surface of combusting sphere are given by:

$$-D\rho\left(\frac{dY_s}{dR}\right)_b + Y_{sb}(\rho U)_b = \rho Y_{sb}k_{s0}\exp\left(-\frac{E}{R_g T_p}\right) \qquad s \neq inert \qquad (12.6\text{-}7a)$$

$$-D\rho\left(\frac{dY_{In}}{dR}\right)_b + Y_{In,b}(\rho U)_b = 0 \qquad (12.6\text{-}7b)$$

$$\dot{m}_p = (\rho U)_b = \frac{1}{\beta}\left(D\rho\left(\frac{dY_{Ox}}{dR}\right)_b - Y_{Ox,b}(\rho U)_b\right), \qquad (12.6\text{-}7c)$$

where $\dot{m}_p$ is the reaction rate (mass flux) of solid particle, and β is a constant that depends on the stoichiometric relations and the molar weights of reactants. By solving the species equation and energy equation of gas mixture, it gives:

$$\dot{m}_p = \frac{4\pi}{\beta}R_p D\rho \ln\left(1 + \frac{(Y_{Ox,\infty} - Y_{Ox,b})}{\beta + Y_{Ox,b}}\right) \qquad (12.6\text{-}8)$$

and:

$$\dot{m}_p = \frac{4\pi}{\beta}R_p\frac{K}{c_p}\ln\left(1 + \frac{c_p(T_\infty - T_p)}{q_b}\right), \qquad (12.6\text{-}9)$$

where q_b stands for the specific heat transfer at the surface, defined by:

$$q_b\dot{m}_p = 4\pi R_p^2 K\left(\frac{dT}{dR}\right)_b. \qquad (12.6\text{-}9a)$$

From Eq. (12.6-7), it gives:

$$\dot{m}_p = \frac{1}{\beta}\rho Y_{Ox,b}k_{Ox,0}\exp\left(-\frac{E}{R_g T_p}\right). \qquad (12.6\text{-}10)$$

The energy equation of the particle yields:

$$\frac{4\pi R_p^3}{3}\rho_p c_{pp}\frac{dT_p}{dt} = 4\pi R_p^2\sigma_b\varepsilon_p(T_\infty^4 - T_p^4) - \dot{m}_p q_p + \dot{m}_p\Delta H_p, \qquad (12.6\text{-}11)$$

where ε_p is the particle surface emissivity; ΔH_p is the reaction heat. The mass balance of the particle yields:

$$\rho_p\frac{d}{dt}\left(\frac{4\pi R_p^3}{3}\right) = -\dot{m}_p. \qquad (12.6\text{-}12)$$

Therefore, the six unknowns, $\dot{m}_p$, $Y_{Ox,b}$, q_b, T_p, $(dT/dR)_b$, and R_p, can be solved via six independent equations, Eqs. (12.6-8)–(12.6-12). This model also indicates that, in general, the solid fuel combustion is a kinetics-diffusion controlled process; namely, the rate of reaction depends on the chemical kinetics as well as the heat and mass transfer of both gas phase and the solid particle.

12.7 Case Studies

The following case studies serve for further discussion, with simple applications or extensions, of flows with phase changes or reactions in this chapter. The examples include a Lagrangian modeling of condensing bubbles during their transport in a solution, an Eulerian–Lagrangian modeling of FCC reacting flow in a riser, an Eulerian–Eulerian modeling of bubble-slurry reacting flows in a Fisher–Tropsch reactor, and an Eulerian–Lagrangian modeling of dense-phase gas–solid (*i.e.*, steam-coal) reacting flows in a coal gasification reactor. All these examples are focused on the modeling establishment and closure.

12.7.1 Motion of a Condensing Bubble in a Solution

Absorption heat transformers and absorption chillers involve the condensation of saturated bubbles in hot aqueous solutions. Understanding of the dynamic motion of such bubble condensing, with coupled changes in bubble size and shape, is important to the proper design and operation control of the absorption systems.

Problem Statement: Consider a saturated bubble condensing in an infinite body of quiescent liquid. Establish a model to characterize the bubble motion with condensing in the liquid. The bubble is assumed to be an oblate spheroid, and during the condensing process, both the semimajor axis and the aspect ratio vary linearly with time.

Analysis: The following model is based on the original work of Donnellan *et al.* (2015), with minor modifications. A Lagrangian trajectory model is adopted for the bubble dynamics.

Assume that the bubble rises up against gravity in a linear motion, without rotation and vibration as well as in the absence of internal motion within the bubble. It is further assumed that the forces governing the motion of bubble include only the gravitational, buoyancy, drag, added mass, and Basset historic forces. The bubble motion is thus given by the general B.B.O. equation, Eq. (3.5-3), expressed as:

$$\rho_b V_b \frac{dU_b}{dt} = -\frac{1}{2} C_D A_p \rho_l U_b^2 + F_A + F_B + (\rho_l - \rho_b) V_b g. \tag{12.7-1}$$

For an oblate spheroid, the bubble volume and projected area can be related to the semimajor axis, a, and the aspect ratio, λ, by

$$\lambda = \frac{b}{a} \qquad V_b = \frac{4}{3} \pi a b \qquad A_p = \pi a^2. \tag{12.7-1a}$$

In Eq. (12.7-1a), the semiminor axis, b, is regarded as a dependent parameter to λ. The added mass force, with the bubble volume changing, can be expressed by

$$F_A = -\frac{d}{dt} (C_A \rho_l V_b U_b) = -C_A \rho_l \left(V_b \frac{dU_b}{dt} + U_b \frac{dV_b}{dt} \right). \tag{12.7-1b}$$

The Basset force, with a modification factor of higher Re and nonspherical effect by using the semimajor axis as the characteristic length, is based on Eq. (3.2-21) as:

$$F_B = -6C_B \sqrt{\frac{\pi \mu_l}{\rho_l}} \int_0^t \left(\frac{a^2}{\sqrt{t - \tau}} \frac{\partial U_b}{\partial \tau} \right) d\tau. \tag{12.7-1c}$$

The bubble size reduction in the condensing process depends on the interfacial mass and heat transfer, whose formulation can be complex. A simple alternative approach is to use empirical relationships of the semimajor axis and aspect ratio, as approximated by linear equations with time:

$$a = a_0 - \beta_a t \qquad \lambda = \lambda_0 - \beta_\lambda t. \tag{12.7-2}$$

Thus, with β_a and β_λ determined empirically, the time-dependent velocity of bubble can be solved from the integral-differential equation, Eq. (12.7-1).

Comments: (1) The detailed solution of Eq. (12.7-1) also depends on the selection of constitutive equations of added mass coefficient, drag coefficient, and Basset force coefficient as functions of the aspect ratio. (2) The above model relies on the empirical determination of β_a and β_λ, which should be replaced by mechanism-based sub-models of interfacial heat and mass transfer of bubble condensing. (3) Case simulation examples can be found in the reference (Donnellan et al., 2015).

12.7.2 Modeling of FCC Reacting Flow

As introduced in Section 12.5, the reacting flow in a FCC riser is strongly coupled with both hydrodynamics and chemical kinetics. Since the flow is typically turbulent and complex, CFD-based modeling of FCC reactions in riser flows has been rapidly developed for the analysis of reaction performance.

Problem Statement: Establish a model for the FCC reactions in a gas–solid turbulent riser flow. Since the flow is not very dilute, the interparticle collisions should be included. For FCC reactions, a four-lump reaction kinetics model can be used.

Analysis: The following model is modified from the original work of Wu et al. (2010). An Eulerian–Lagrangian modeling is adopted, with an Eulerian approach for the gas phase and the discrete element method (DEM) for the particle phase. The FCC particles are assumed to be monodispersed, without agglomeration or breakup. For simplicity, the turbulent diffusion is ignored.

For the gas phase, the continuity equation is simplified from Eq. (6.4-21) to

$$\frac{\partial}{\partial t} (\alpha\rho) + \nabla \cdot (\alpha\rho\mathbf{U}) = 0. \tag{12.7-3}$$

The momentum equation is given by Eq. (6.4-22):

$$\frac{\partial}{\partial t} (\alpha\rho\mathbf{U}) + \nabla \cdot (\alpha\rho\mathbf{U}\mathbf{U}) = -\alpha\nabla p + \nabla \cdot \left(\alpha(\mu + \mu_e)(\nabla\mathbf{U} + \nabla\mathbf{U}^T)\right) + \alpha\rho\mathbf{g} - \mathbf{F}_D, \tag{12.7-4}$$

where μ_e is the turbulent eddy viscosity, which can be solved from additional turbulence models, such as modified k-ε model with interphase turbulent modulation, introduced in Section 6.7.3. $\mathbf{F}_D$ is the gas–solid interaction force per unit volume, represented by the drag force and expressed as the sum of all drag forces of particles within the computing cell:

$$\mathbf{F}_D = \frac{1}{V_{cell}} \sum_i^n \mathbf{F}_{Di}. \tag{12.7-4a}$$

The energy equation is governed by Eq. (6.4-23):

$$\frac{\partial}{\partial t} (\alpha \rho c_p T) + \nabla \cdot (\alpha \rho c_p \mathbf{U} T) = \nabla \cdot (\alpha (K + K_e) \nabla T) + Q_A, \tag{12.7-5}$$

where Q_A is the interfacial heat transfer per unit volume, and the sum of all particles in a computing cell is thus expressed by

$$Q_A = \frac{1}{V_{cell}} \sum_i^n Q_{Ai}. \tag{12.7-5a}$$

It is assumed that Q_A includes only the convective heat transfer over the particle surface whereas the reaction heat from surface-based catalytic cracking is all added to the particle phase. The equation of jth-species is given by

$$\frac{\partial}{\partial t} (\alpha \rho Y_j) + \nabla \cdot (\alpha \rho \mathbf{U} Y_j) = \nabla \cdot \left(\alpha (\rho D_j + \frac{\mu_e}{Sc_e}) \nabla Y_j \right) + S_j. \tag{12.7-6}$$

According to the four-lump model in Fig. 12.5-2, the rate law of the reaction for each species (VGO, gasoline, gases, and coke denoted, respectively, as species 1 to 4) gives:

$$S_1 = -\alpha \Phi_s (k_1 + k_2 + k_3) \rho^2 Y_1{}^2 \frac{\rho_p(1-\alpha)}{M_1} \tag{12.7-6a}$$

$$S_2 = \alpha \Phi_s \rho \left(\frac{\rho}{M_1} k_1 Y_1{}^2 - (k_4 + k_5) Y_2 \right) \rho_p(1-\alpha) \tag{12.7-6b}$$

$$S_3 = \alpha \rho \Phi_s \left(\frac{\rho}{M_1} k_2 Y_1{}^2 + k_4 Y_2 \right) \rho_p(1-\alpha) \tag{12.7-6c}$$

$$S_4 = \alpha \rho \Phi_s \left(\frac{\rho}{M_1} k_3 Y_1{}^2 + k_5 Y_2 \right) \rho_p(1-\alpha). \tag{12.7-6d}$$

The reaction coefficients follow the Arrhenius equation, which depends on the temperature of the catalyst particle:

$$k_i = k_{i0} \exp \left(-\frac{E_i}{R_g T_p} \right). \tag{12.7-6e}$$

For the particle motion of the ith particle, the governing equations for the translational and rotational motions are given, respectively, by Eqs. (5.3-30) and (5.3-31), expressed as:

$$m_i \frac{d\mathbf{U}_{pi}}{dt} = \mathbf{F}_{Di} + m_i \mathbf{g} + \sum_j^{n_i} (\mathbf{F}_{nij} + \mathbf{F}_{tij}) \tag{12.7-7a}$$

$$I_i \frac{d\omega_{\text{pi}}}{dt} = \sum_j^{n_i} \left(\mathbf{T}_{tij} + \mathbf{T}_{rij}\right), \tag{12.7-7b}$$

where $\mathbf{F}_{nij}$, $\mathbf{F}_{tij}$, $\mathbf{T}_{tij}$, and $\mathbf{T}_{rij}$ are defined, respectively, by Eqs. (5.3-22a), (5.3-22b), (5.3-28), and (5.3-29). The drag force, $\mathbf{F}_{Di}$, is formulated by (Ding and Gidaspow, 1990):

$$\mathbf{F}_{Di} = \begin{cases} 150 \dfrac{\alpha_p \mu}{\alpha d_p^2} + 1.75 \dfrac{\rho \left|\mathbf{U} - \mathbf{U}_{pi}\right|}{d_p} \left(\mathbf{U} - \mathbf{U}_{pi}\right) & \alpha \leq 0.8 \\[3mm] \dfrac{3}{4} \dfrac{C_{Di}}{d_p} \alpha^{-1.65} \rho \left|\mathbf{U} - \mathbf{U}_{pi}\right| \left(\mathbf{U} - \mathbf{U}_{pi}\right) & \alpha > 0.8 \end{cases}. \tag{12.7-7c}$$

The particle temperature, which dictates the reaction rate as shown in Eq. (12.7-6e), is governed by the particle energy equation, expressed by:

$$m_i c_i \frac{dT_{\text{pi}}}{dt} = \frac{\pi}{4} d_p^2 h_i (T - T_{pi}) + \sum_j^{n_i} q_{ij} + \sum_{j=1}^5 \frac{S_j \Delta H_j V_{cell}}{n}, \tag{12.7-8}$$

where q_{ij} is the heat transfer by collision between a pair of particles, as formulated by Eq. (4.3-16a).

Thus, a CFD-DEM model is established for the FCC reacting species (Y_i), with the coupling of hydrodynamic transport and heat transfer of reacting gas and catalytic particles.

Comments: (1) The selection of turbulence models may affect the FCC modeling predictability. The $k - \varepsilon$ turbulence model in this example is developed for incompressible, isothermal, and isotropic turbulent flows. However, a FCC reacting riser flows is typically compressible, non-isotropic turbulent, and strongly non-isothermal. Unfortunately, no specific and reliable turbulent models have been available on that regards. (2) Backflow mixing with spent catalyst may be accounted for, indirectly, in a proper selection of empirical correlation of deactivation factor Φ_s. (3) Case simulation examples can be found in the reference (Wu et al., 2010).

12.7.3   Modeling of Fisher–Tropsch Slurry Bubble Reactor

Fisher–Tropsch (F-T) slurry-bubble column reactors involve three-phase reacting flow. An F-T synthesis system typically consists of compartmentalized arrangements that induce complex internal flows such as large-scaled flow recirculation as well as turbulent transport. Since the reacting characteristics of F-T synthesis in terms of heat and mass transfer depend largely on the hydrodynamic transport of bubble-solids-liquid flows, an accurate account of hydrodynamics coupled with reaction kinetics can be made from numerical modeling.

Problem Statement: Establish a model for the F-T reactions in a turbulent slurry-bubble column reactor. The slurry is considered to be an ideal liquid–solid mixture. While the F-T reaction is exothermic, there is an internal tubular heat exchanger that ensures an isothermal operation with heat removal.

Analysis: The following model is modified from the original work of Troshko and Zdravistch (2009). An Eulerian–Eulerian two-fluid modeling approach is adopted, with one fluid representing the slurry phase and the other the bubble phase. For simplicity, in this example, the bubbles are approximated to be monodispersed, without coalescence or break-up, and the turbulence is also assumed to be isotropic. The F-T synthesis follows the reaction given by Eq. (12.4-2).

The mass conservation equation for both phases can be written, based on Eq. (6.4-21), as:

$$\frac{\partial}{\partial t}(\alpha\rho) + \nabla \cdot (\alpha\rho\mathbf{U}) = \dot{m}_b \tag{12.7-9a}$$

$$\frac{\partial}{\partial t}(\alpha_b\rho_b) + \nabla \cdot (\alpha_b\rho_b\mathbf{U_b}) = -\dot{m}_b, \tag{12.7-9b}$$

where $\dot{m}_b$ is the mass transfer rate of gas absorption, diffused from the bubble interface into the slurry. The mass flow rate per unit volume is given by

$$\dot{m}_b = \alpha_b(\gamma_{CO}M_{CO} + \gamma_{H_2}M_{H_2}). \tag{12.7-9c}$$

The volume fraction conservation requires

$$\alpha + \alpha_b = 1. \tag{12.7-9d}$$

The momentum equation of slurry phase is given, based on Eq. (6.4-22), by:

$$\frac{\partial}{\partial t}(\alpha\rho\mathbf{U}) + \nabla \cdot (\alpha\rho\mathbf{UU}) = -\alpha\nabla p + \nabla \cdot [\alpha\mu_{eff}(\nabla\mathbf{U} + \nabla\mathbf{U}^T)] + \alpha\rho\mathbf{g} + \mathbf{M}, \tag{12.7-10}$$

where the effective viscosity of liquid phase has three additive contributors: the molecular viscosity, the shear-induced turbulent viscosity, and the bubble-induced turbulent dispersion term, expressed by

$$\mu_{eff} = \mu + \mu_e + \mu_B. \tag{12.7-10a}$$

The turbulent viscosity, μ_e, is determined from k and ε by Eq. (2.3-18). The bubble-induced turbulence viscosity is given by (Sato and Sekoguchi, 1975):

$$\mu_B = \rho\alpha_b C_{\mu B}d_b |\mathbf{U}_b - \mathbf{U}|. \tag{12.7-10b}$$

The interphase momentum transfer, $\mathbf{M}$, includes both hydrodynamic forces and the reaction-induced momentum transfer, as expressed by:

$$\mathbf{M} = \frac{3}{4}\frac{C_D}{d_b}\alpha_b\rho |\mathbf{U}_b - \mathbf{U}|(\mathbf{U}_b - \mathbf{U}) + \alpha_b\rho C_L(\mathbf{U}_b - \mathbf{U}) \times \nabla \times \mathbf{U}$$
$$+ \alpha_b\rho C_A\left(\frac{d\mathbf{U}_b}{dt} - \frac{d\mathbf{U}}{dt}\right) + \dot{m}_b(\mathbf{U}_b - \mathbf{U}), \tag{12.7-10c}$$

whereas C_D, C_L, and C_A stand for the drag coefficient, lift coefficient, and added mass coefficient, respectively.

The momentum equation of bubble phase is given, based on Eq. (6.4-22), by:

$$\frac{\partial}{\partial t}(\alpha_b\rho_b\mathbf{U}_b) + \nabla \cdot (\alpha_b\rho_b\mathbf{U}_b\mathbf{U}_b) = -\alpha_b\nabla p$$
$$+ \nabla \cdot [\alpha_b(\mu_b + \mu_{be})(\nabla\mathbf{U}_b + \nabla\mathbf{U}_b^T)] + \alpha_b\rho_b\mathbf{g} - \mathbf{M}, \tag{12.7-11}$$

where the turbulent viscosity of bubble phase is related to the turbulent viscosity of liquid phase, as suggested by Jakobsen *et al.* (1997):

$$\mu_{be} = \frac{\rho_b}{\rho} \mu_e. \tag{12.7-11a}$$

The k and ε equations are, respectively, given by

$$\frac{\partial}{\partial t}(\alpha \rho k) + \nabla \cdot (\alpha \rho U k) = \nabla \cdot \left[\alpha \left(\mu + \frac{\mu_e}{\sigma_k}\right) \nabla k\right] + 2\mu_e(\alpha S : S) - \alpha \rho \varepsilon + S_k \tag{12.7-12}$$

$$\frac{\partial}{\partial t}(\alpha \rho \varepsilon) + \nabla \cdot (\alpha \rho U \varepsilon) = \nabla \cdot \left[\alpha \left(\mu + \frac{\mu_e}{\sigma_k}\right) \nabla \varepsilon\right] + 2\mu_e C_1 \frac{\varepsilon}{k}(\alpha S : S) - C_2 \alpha \rho \frac{\varepsilon^2}{k} + S_\varepsilon, \tag{12.7-13}$$

where S is the stain rate, defined by Eq. (2.2-3), S_k and S_ε are bubble-induced turbulence terms, respectively, defined by Eq. (6.7-17) and Eq. (6.7-16).

The species mass conservation equation of jth-species is given by

$$\frac{\partial}{\partial t}(\alpha \rho Y_j) + \nabla \cdot (\alpha \rho U Y_j) = \nabla \cdot \left(\alpha(\rho D_j + \frac{\mu_e}{Sc_e})\nabla Y_j\right) + \gamma_j - \gamma_{jb}, \tag{12.7-14}$$

where γ_j is the F-T reaction rate, and γ_{jb} is the rate of absorption reactions.

Thus, eleven independent equations, Eqs. (12.7-9)–(12.7-14) with Eqs. (2.3-18), (6.8-11b), and (6.8-12a), are obtained for ten unknowns (α, α_b, U, U_b, p, μ_e, μ_{be}, μ_B, k, ε, and Y_j). The closure of the modeling is now established. In order to solve these equations, the transport coefficients (such as drag coefficient and diffusivity) and reaction rates need to be further formulated from corresponding sub-models.

Comments: (1) Bubble size in F-T reactors actually may change greatly due to the interfacial mass transfer and bubble coalescence or breakup. Hence, a population balance model needs to be added into above modeling, as done in the original paper of Troshko and Zdravistch (2009). In addition, due to their distinctively different dynamic transport of different-sized bubbles, multifluid modeling approach for polydispersed bubbles may require further subgrouping of "fluid" phases within the bubbles. (2) Selection of proper boundary conditions can be very challenging in F-T reactor simulation, due to the presence of bubbles, solid particles, and internal cooling tubes. (3) Case simulation examples can be found in the reference (Troshko and Zdravistch, 2009).

12.7.4   Modeling of Reacting Flow in Coal Gasifier

Problem Statement: Assuming the use of dry-ash-free (DAF) charcoal (in the absence of moisture and volatile gases) in a coal gasifier, the simplest mode of heterogeneous steam gasification reaction can be expressed by

$$C + H_2O \rightarrow CO + H_2. \tag{12.7-15}$$

Based on this reaction, establish a CFD model of dense-phase gas–solid reacting flow for the steam gasification in a coal gasifier.

Analysis: The following model is modified from the original work of Snider et al. (2011). An Eulerian–Lagrangian modeling approach is adopted, with an Eulerian model for the gas phase and a stochastic Lagrangian model for the particle phase. The probability density function of particle distribution, f, is assumed to be given, and is a function of particle spatial location, velocity, mass, temperature, and time. Thus, $f(\mathbf{r}_p, \mathbf{u}_p, m_p, T_p, t) d\mathbf{u}_p dm_p dT_p$ is the average number of particles per unit volume, within the intervals of $(\mathbf{u}_p, \mathbf{u}_p + d\mathbf{u}_p)$, $(m_p, m_p + dm_p)$, and $(T_p, T_p + dT_p)$.

The equation of motion of a particle is given by the general B.B.O. equation, Eq. (3.5-3). Assuming the forces acting on the particle include only the gravitational, drag, and pressure-gradient forces, the equation reduces to:

$$\frac{d\mathbf{u}_p}{dt} = \beta_p(\mathbf{U} - \mathbf{u}_p) - \frac{1}{\rho_p}\nabla p + \mathbf{g}, \tag{12.7-16}$$

where β_p is the momentum transfer coefficient. The equation of particle trajectory is given by:

$$\frac{d\mathbf{r}_p}{dt} = \mathbf{u}_p. \tag{12.7-17}$$

Assume that the heat from surface reactions is all released to the gas phase. Thus the particle energy equation is governed only by the convective heat transfer:

$$\frac{dT_p}{dt} = \frac{h_p A_p}{m_p c}(T - T_p). \tag{12.7-18}$$

The rate of mass change of a particle can be related to the rate of molar concentration change, as expressed by

$$\frac{dm_p}{dt} = \frac{\alpha M_C}{\rho_p \alpha_p} m_p \frac{dC(s)}{dt}, \tag{12.7-19}$$

where $C(s)$ stands for the molar concentration of solid carbon, and M_C is the molecular weight of carbon. The rate of change of $C(s)$ for the steam gasification reaction is related to the molar concentration of steam, C_{H2O}, and other properties by

$$\frac{dC(s)}{dt} = -k_0 \rho_p \alpha_p C_{H_2O} T \exp\left(-\frac{E}{T}\right). \tag{12.7-20}$$

The mass conservation equation of gas phase is based on Eq. (6.4-21), expressed by:

$$\frac{\partial}{\partial t}(\alpha\rho) + \nabla \cdot (\alpha\rho\mathbf{U}) = \dot{m}_g, \tag{12.7-21}$$

where $\dot{m}_g$ is the mass generation rate from gasification, which can be calculated by:

$$\dot{m}_g = \iiint f \frac{dm_p}{dt} - dm_p d\mathbf{u}_p dT_p. \tag{12.7-21a}$$

The gas-phase volume fraction is related to the solid volume fraction by

$$\alpha = 1 - \iiint f \frac{m_p}{\rho_p} dm_p d\mathbf{u}_p dT_p. \tag{12.7-21b}$$

The momentum equation of gas phase is based on Eq. (6.4-22), given by:

$$\frac{\partial}{\partial t}(\alpha\rho\mathbf{U}) + \nabla\cdot(\alpha\rho\mathbf{U}\mathbf{U}) = -\alpha\nabla p + \nabla\cdot\left[\alpha(\mu+\mu_e)\left(\nabla\mathbf{U}+\nabla\mathbf{U}^T\right)\right] + \alpha\rho\mathbf{g} + \mathbf{M},$$

(12.7-22)

where the turbulent viscosity, μ_e, is determined from a simple homogeneous $k-\varepsilon$ model of gas–solid mixture without interphase turbulent modulation, as described by Eq. (6.8-6) and Eq. (6.8-7). The interfacial momentum transfer per volume is given by

$$\mathbf{M} = -\iiint f\left\{m_p\left(\beta_p(\mathbf{U}-\mathbf{u}_p) - \frac{\nabla p}{\rho_p}\right) + \mathbf{u}_p\frac{dm_p}{dt}\right\}dm_pd\mathbf{u}_pdT_p.$$

(12.7-22a)

The equation of jth-species of gas phase is given by

$$\frac{\partial}{\partial t}(\alpha\rho Y_j) + \nabla\cdot(\alpha\rho\mathbf{U}Y_j) = \nabla\cdot\left(\alpha(\rho D_j + \frac{\mu_e}{Sc_e})\nabla Y_j\right) + \alpha\rho M_j\frac{dC_j}{dt}.$$

(12.7-23)

The energy equation is governed by Eq. (6.4-23). By ignoring the pressure work and viscous dissipation but including the enthalpy diffusion, the equation is given by:

$$\frac{\partial}{\partial t}(\alpha\rho c_p T) + \nabla\cdot(\alpha\rho c_p \mathbf{U}T) = \nabla\cdot(\alpha(K+K_e)\nabla T) + Q_A + \sum_{i=1}^{N}\nabla\cdot(h_i\alpha\rho D\nabla Y_i),$$

(12.7-24)

where h_i is the enthalpy of ith-species, and Q_A is the interfacial heat transfer per volume that is calculated by:

$$Q_A = \iiint f\left\{m_p\left(\beta_p(\mathbf{U}-\mathbf{u}_p)^2 - c\frac{dT_p}{dt}\right) - \frac{dm_p}{dt}h_p\right\}dm_pd\mathbf{u}_pdT_p,$$

(12.7-24a)

where h_p is the particle enthalpy, including reaction heat.

Thus, in principle, a basic CFD model is established for dense-phase reacting gas–solid flows. The model include ten independent equations, Eqs. (12.7-16)–(12.7-24) and Eqs. (12.7-21a), for ten unknowns (α, $\mathbf{U}$, $\mathbf{u}_p$, p, $\mathbf{r}_p$, m_p, $C(s)$, T, T_p, and Y_j). To determine the transport coefficients (such as turbulent transport properties, drag coefficient, and diffusivity) and reaction rates, additional constitutive equations (such as $k-\varepsilon$ equations) need to be implemented or further formulated.

Comments: (1) There are many reactions taking place in gasifiers or gasification processes, including steam gasification, CO_2 gasification, methanation, combustion, and water-gas shift. The reaction model discussed above considers only one key reaction associated with steam gasification. (2) Effects of ash, moisture, and volatile matter as well as coal porosity and morphology may be important in modeling of a real system. (3) The particle distribution function, f, can be further considered via its own transport equation (Andrew and O'Rourke, 1996). (4) Case simulation examples can be found in the reference (Snider et al., 2011).

12.8　Summary

This chapter focused on multiphase flows with phase changes (*e.g.*, boiling, evaporation of liquid or slurry droplets, catalytic cracking reactions, polymerization, and fuel combustion), where the phase interactions are further complicated by the mass transfer and mass-transfer-induced momentum and energy transfer. The following topics are important to understand:

- regime classification and flow characteristics in each regime for boiling and spray dispersion;
- phase interaction mechanisms involving phase changes;
- flow characteristics of atomized spray jetting with vaporization;
- characteristics of gas–solid reacting flows in risers;
- characteristics of bubbling flow in sparged stirred tanks, and their impact on reactions;
- combustion characteristics and coupling of transport mechanisms of dispersed fuel particles.

Multiphase flow modeling with phase changes is quite incomplete compared to that without phase changes. More physical understanding is necessary and hence the mathematical descriptions of flow-influence phase changes (especially in turbulent flow transport), multiscaled (on both spatial and time scales) phase changes, and transport phenomena of multiphase flows under highly pressurized and high-temperature operation conditions. Progress in both modeling and numerical solution techniques is evidenced in the case studies as given for the applications of multiphase flows with phase changes.

Nomenclature

A	Area
$a\prime$	Interfacial area per unit volume of bubbles
C_D	Drag coefficient
C	Concentration
c_p	Specific heat at constant pressure
D	Diffusivity
d_p	Diameter of particle
E	Activated energy
F	Force
f	Friction factor
f_{dp}	Droplet–particle collision frequency
g	Gravitational acceleration
h	Heat transfer coefficient
	Specific enthalpy
K	Heat conductivity
k	Mass transfer coefficient
	Reaction rate constant

L Latent heat
l Perimeter
M Molecular weight
m   Mass
$\dot{m}_e$ Evaporation rate per droplet
$\dot{m}_p$ Reaction rate of a solid particle
n   Number density
p Pressure
Q Flow rate
Q_A Interfacial heat transfer per unit volume
q'' Heat flux
R Radius
        Spherical coordinate
R_g Gas constant
$\mathbf{T}$ Total stress tensor
T Temperature
U Averaged velocity
U Axial or radial velocity
V Volume
Y   Mass fraction
z   Axial coordinate

Greek Symbols

α Volume fraction
β Inclined angle of pipe or surface
        Momentum transfer coefficient
γ Reaction rate
ΔH Reaction heat
ε Permittivity
η Curved coordinate
θ Spray angle
θ_r Receding contact angle
μ Viscosity
ξ Coordinate along spray axis
ρ Density
σ Surface tension
Φ Electric potential
χ Quality

Subscript

∞ Ambient
A   Added mass
B   Basset

b	Bubble
	Boundary
CHF	Critical heat flux
c	Cavity
D	Drag
d	Droplet
E	Electrostatic
e	Electron
	Entrainment
	Equilibrium
	Evaporation
F	Fuel
f	Flame
In	Inert gas
i	ith species or component
l	Liquid
m	Mixture
max	Maximum
min	Minimum
Ox	Oxygen
Pr	Product
p	Particle
s	Species
sat	Saturated
v	Vapor
w	Wall

Problems

P12.1 A circular pipe ($D = 1$cm) is uniformly heated along pipe wall. Initially saturated liquid water with mass flow rate of 0.15 kg/s flows through the pipe. Use homogenous equilibrium model, find the axial position where the void fraction is 5%. The flow properties include $\rho_l = 848$ kg/m^3, $\rho_v = 10$ kg/m^3, $L = 1,888$ kJ/kg, $q' = 20$ kW/m.

P12.2 Consider a vapor–water mixture flow. Assume the volume flux of each phase is the same and is equal to 2.0 m^3/s-m^2, with a void fraction of 0.45. The vapor density is 1 kg/m^3, and water density is 1,000 kg/m^3. Estimate the mass flux and velocity of each phase, and the overall quality of the mixture.

P12.3 Prove that, in general vapor–liquid flow, relationship between the void fraction and flow quality can be expressed as

$$\alpha = \frac{1}{1 + \frac{1-\chi}{\chi}\frac{\rho_v}{\rho_l}\frac{U_v}{U_l}} \tag{P12.3-1}$$

P12.4 Show that, in the homogenous equilibrium model, (a) the density of a vapor–water mixture can be related to quality by

$$\rho_m = (1/\rho_l + \chi/\rho_v)^{-1} \qquad\qquad \text{(P12.4-1)}$$

(b) $\chi = \chi_e$ in the saturated region ($0 < \chi_e < 1$).

P12.5 Evaporating sprays in a cooling tower are employed for thermal energy recovery and for the wet scrubbing of airborne particulates, as shown in Fig. 1.1-7. A relatively dry and hot exhaust gas flow (also contains suspended fine particulates such as flying ash from coal-fired boilers) passes through the cooling tower from bottom to the top, whereas the atomized droplets from sprays settle down in a counterflow mode against the upward moving gas flow. Establish a CFD model for energy recovery and wet scrubbing in a cooling tower application.

P12.6 Plasma (ionized hot gas) jet is used to melt fine metallic particles to weld a joint of two metal pieces. The particles are introduced to the jetting region either by direct injection or by the induction of the gas jet. Assume that the particles become molten only after they are injected into the gas jet. Develop a CFD model for the particle-laden jet transport and surface coating and heating (for welding).

P12.7 Flash spray is caused by a sudden depressurization in which the liquid spray is injected. Consider an atomized spray is injected into a vacuum chamber whose low pressure is kept by continuously vacuuming the vapor generated from the spray. Assume that the injected droplets have a sufficiently high temperature so that the saturated vapor pressure of the inlet droplets is much higher than the vacuuming pressure in the chamber. It is also assumed that chamber height is designed in a way that the spray cooling ends before icing. Establish a model to estimate the temperature decrease in the flash spray cooling.

P12.8 For fuel atomization in a gaseous combustion chamber (*e.g.*, fuel-injection engine), develop a model of the phase transport that can be used to estimate the reaction-induced pressure and temperature distributions inside the combustion chamber. Note that the gas phase is a multicomponent mixture; the flow is turbulent, and the droplet phase is dilute. List all independent variables and the number of independent equations to ensure model closure.

P12.9 A couple of high-speed air jets discharged cocurrently from the expansion wall of a sudden expansion chamber, as shown in 12.9, produce a large recirculation zone induced by the high-velocity jets. Such a flow recirculation zone, connecting the expansion corner and central flow region, can be used as a flame stabilizer in the coal-fired furnace. Develop a model to characterize the trajectories of coal particles and distributions of gas velocity and gas temperature.

P12.10 Air and pulverized coal are fed into a tangentially fired furnace via four corner combustors (such as the one described in P12.9). The main chamber of furnace can be simplified as a rectangular chamber, and the four combustors discharging the reacting air–coal flow at an injection angle that generates a swirling combustion

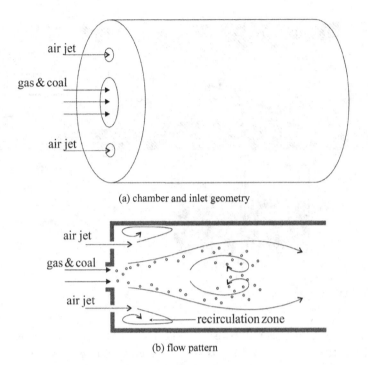

(a) chamber and inlet geometry

(b) flow pattern

Figure P12.9 Jets into sudden-expansion chamber

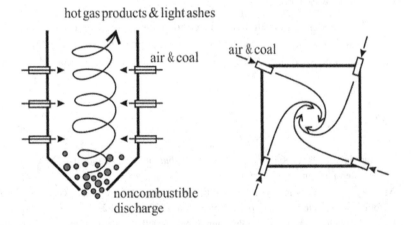

Figure P12.10 Tangentially fired furnace.

zone inside the furnace, as shown in P12.10. The noncombustible part of coal (such as ash) may be collected and discharged from the bottom of the furnace, whereas the hot gas products and light fly ashes are discharged from the top of the furnace. Develop a model to characterize the trajectories of coal particles and distributions of gas velocity and gas temperature in the furnace.

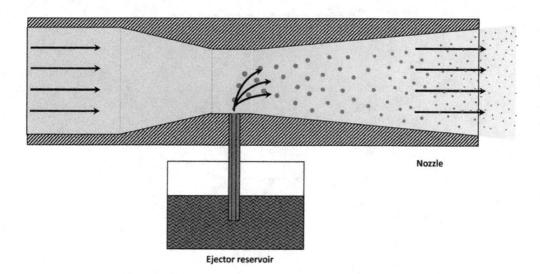

Figure P12.11 Spray ejector.

P12.11 A spray atomizer is based on the induction of an ejector, as shown in Fig. P12.11, in which the droplets are evaporating during their transport by air dispersion. Develop a Lagrangian–Eulerian model for the evaporating spray dispersion from the atomizer nozzle exit, assuming the initial droplet size and velocity distributions are given.

P12.12 Consider a saturated bubble condensing in a turbulent subcooled boiling flow. Establish a model to characterize the dynamic transport of the condensing bubble in the subcooled liquid flow. Use VOF method for the interfacial heat and mass transfer.

References

Abramovich, G. N. (1963). *The Theory of Turbulent Jets.*, Cambridge, MA: Massachusetts Institute of Technology Press.

Andrew, M. J., and O'Rourke, P. J. (1996). The multiphase particle-in-cell (MP-PIC) method for dense particle flow. *Int. J. Multiphase Flow*, **22**, 379–402.

Bai, C., and Gosman, A. (1996). Mathematical modelling of wall films formed by impinging sprays. *SAE Technical Paper* 960626.

Berry, R. A., Zou, L., Zhao, H., Zhang, H., Peterson, J. W., Martineau, R. C., Kadioglu, S. Y., and Andrs, D. (2016). *RELAP-7 Theory Manual*, Idaho National Laboratory report INL/EXT-14-31366 (Revision 2).

Burdett, I. (2008). New innovations drive gas phase PE technology. *Hydrocarbon Eng.*, **13**, 67–76.

Chen, X.-Z., Luo, Z.-H., Yan, W.-C., Lu, Y.-H., and Ng, I.-S. (2011). Three-dimensional CFD-PBM coupled model of the temperature fields in fluidized-bed polymerization reactors. *AIChE J.*, **57**, 3351–3366.

Collier, J. G., and Thome, J. R. (1994). *Convective Boiling and Condensation*. Oxford: Clarendon Press.

de Villiers, E., Gosman, A., and Weller, H. (2004). Large eddy simulation of primary diesel spray atomization. *SAE Technical Paper* 2004-01-0100.

Ding, J., and Gidaspow, D. (1990). A bubbling fluidization model using kinetic theory of granular flow. *AIChE J.*, **36**, 523–538.

Donnellan, P., Byrne, E., and Cronin, K. (2015). Analysis of the velocity and displacement of a condensing bubble in a liquid solution. *Chem. Eng. Sci.*, **130**, 56–67.

Dry, M. E. (2002). The Fischer–Tropsch process: 1950–2000. *Catalysis Today*, **71**, 227–241.

Fan, R. (2006). *Computational Fluid Dynamics Simulation of Fluidized Bed Polymerization Reactors*. Ph.D. dissertation, Iowa State University, Iowa.

Fotovat, F., Bi, X., and Grace, J. R. (2017). Electrostatics in gas-solid fluidized beds: a review. *Chem. Eng. Sci.*, **173**, 303–334.

Gajewski, A. (1985). Investigation of the electrification of polypropylene particles during the fluidization process. *J. Electrostatics*, **17**, 289–298.

Gorokhovski, M., and Herrmann, M. (2008). Modeling primary atomization. *Annu. Rev. Fluid Mech.*, **40**, 343–366.

Hendrickson, G. (2006). Electrostatics and gas phase fluidized bed polymerization reactor wall sheeting. *Chem. Eng. Sci.*, **61**, 1041–1064.

Hsu, Y. Y. (1962). On the size range of active nucleation cavities in a heating surface. *Trans. ASME, J. Heat Transfer*, **84**, 207–216.

Hsu, Y. Y., and Graham, R. W. (1961). An analytical and experimental study of the thermal boundary layer and ebullition cycle in nucleate boiling. *NASA TND-594*, NASA Lewis Research Center, Cleveland, OH.

Jakobsen, H. A., Sannæs, B. H., Grevskott, S., and Svendsen, H. F. (1997). Modeling of vertical bubble-driven flows. *Ind. Eng. Chem. Res.*, **36**, 4052–4074.

Jazayeri, B. (2003). Applications for chemical production and processing. *Fluidization and Fluid-Particle Systems*, ed. W. C. Yang, New York: Marcel Dekker Inc.

Jenkins, III, J. M., Jones, R. L., Jones, T. M., and Beret, S. (1986). *Method for Fluidized Bed Polymerization*. US Patent 4588790.

Jones, W. P., and Lettieri, C. (2010). Large eddy simulation of spray atomization with stochastic modeling of breakup. *Phys. Fluids*, **22**, 115106.

Kandlikar, S. G. (2001). A theoretical model to predict pool boiling CHF incorporating effects of contact angle and orientation. *J. Heat Transfer*, **123**, 1071–1079.

Krishna, R., and Sie, S. T. (2000). Design and scale-up of the Fischer–Tropsch bubble column slurry reactor. *Fuel Process. Technol.*, **64**, 73–105.

Lee, L., Chen, Y., Huang, T., and Pan, W. (1989). Four lump kinetic model for fluid catalytic cracking process. *Can J. Chem. Eng.*, **67**, 615–619.

Lefebvre, A. H. (1989). *Atomization and Sprays*, New York: Hemisphere.

Levenspiel, O. (1999). *Chemical Reaction Engineering*, 3rd ed., New York: John Wiley & Sons.

Marek, R., and Straub, J. (2001). Analysis of the evaporation coefficient and the condensation coefficient of water, *Int. J. Heat Mass Transfer*, **44**, 39–53.

Mezhericher, M., Levy, A., and Borde, I. (2011). Modelling the morphological evolution of nanosuspension droplet in constant-rate drying stage. *Chem. Eng. Sci.*, **66**, 884–896.

Naber, J., and Farrel, P. (1993). Hydrodynamics of droplet impingement on a heated surface. *SAE Technical Paper* 930919.

Nukiyama, S. (1934). The maximum and minimum values of the heat A transmitted from metal to boiling water under atmospheric pressure. *J. Jpn. Soc. Mech. Engrs.*, **27**, 367–374.

Oakley, D. E. (2004). Spray dryer modeling in theory and practice. *Drying Technol.*, **22**, 1371–1402.

Persad, A. H., and Ward, C. A. (2016). Expressions for the evaporation and condensation coefficients in the Hertz-Knudsen relation, *Chem. Rev.*, **116**, 7727–7767.

Pitault, I., Nevicato, D., Blasetti, A. P., and Delasa, H. I. (1994). Fluid catalytic cracking catalyst for reformulated gasolines-kinetic modeling. *Ind. Eng. Chem. Res.*, **33**, 3053–3062.

Qureshi, M. M. R., and Zhu, C. (2006). Gas entrainment in an evaporating spray jet. *Int. J. Heat Mass Transfer*, **49**, 3417–3428.

Reitz, R. D., and Bracco, F. V. (1982). Mechanism of atomization of a liquid jet. *Phys. Fluids*, **25**, 1730–1742.

Rokkam, R. G., Fox, R. O., and Muhle, M. E. (2010). Computational fluid dynamics and electrostatic modeling of polymerization fluidized-bed reactors. *Powder Technol.*, **203**, 109–124.

Sato, Y., and Sekoguchi, K. (1975). Liquid velocity distribution in two-phase bubble flow. *Int. J. Multiphase Flow*, **2**, 79–95.

Shin, H. T., Lee, Y. P., and Jurng, J. (2000). Spherical-shaped ice particle production by spraying water in a vacuum chamber, *Appl. Therm. Eng.*, **20**, 439–454.

Shinjo, J., and Umemura, A. (2010). Simulation of liquid jet primary breakup: Dynamics of ligament and droplet formation. *Int. J. Multiphase Flow*, **36**, 513–532.

Sirognano, W. A. (1999). *Fluid Dynamics and Transport of Droplets and Sprays*. Cambridge: Cambridge University Press.

Sloth, J., Jorgensen, K., Bach, P., Jensen, A. D., Kiil, S., and Dam-Johansen, K. (2009). Spray drying of suspensions for pharma and bio products: drying kinetics and morphology. *Ind. Eng. Chem. Res.*, **48**, 3657–3664.

Snider, D. M., Clark, S. M., and O'Rourke, P. J. (2011). Eulerian-Lagrangian method for three-dimensional thermal reacting flow with application to coal gasifiers. *Chem. Eng. Sci.*, **66**, 1285–1295.

Troshko, A. A., and Zdravistch, F. (2009). CFD modeling of slurry bubble column reactors for Fisher–Tropsch synthesis. *Chem. Eng. Sci.*, **64**, 892–903.

van Baten, J. M., and Krishna, R. (2004). Eulerian simulation strategy for scaling up a bubble column slurry reactor for Fischer-Tropsch synthesis. *Ind. Eng. Chem. Res.*, **43**, 4483–4493.

Vehring, R., Foss, W. R., and Lechuga-Ballesteros, D. (2007). Particle formation in spray drying. *Aerosol Sci.*, **38**, 728–746.

Wallis, G. B. (1980). Critical two-phase flow. *Int. J. Multiphase Flow*, **6**, 97–112.

Wang, X., Zhu, C., and Ahluwalia, R. (2004). Numerical simulation of evaporating spray jets in concurrent gas-solids pipe flows. *Powder Technol.*, **140**, 56–67.

Wu, C., Cheng, Y., Ding, Y., and Jin, Y. (2010). CFD-DEM simulation of gas-solid reacting flows in fluid catalytic cracking (FCC) process. *Chem. Eng. Sci.*, **65**, 542–549.

Xie, T., McAuley, K. B., Hsu, J. C. C., and Bacon, D. W. (1994). Gas phase ethylene polymerization: production processes, polymer properties, and reactor modeling, *Ind. Eng. Chem. Res.*, **33**, 449–479.

Zhou, L. (1993). *Theory and Numerical Modeling of Turbulent Gas-Particle Flows and Combustion*. Beijing: Science Press and CRC Press.

Zhou, L. (2018). *Theory and Modeling of Dispersed Multiphase Turbulent Reacting Flows*. Butterworth-Heinemann.

Zhu, C., Liu, G. L., Wang, X., and Fan, L.-S. (2002). A parametric model for evaporating liquid jets in dilute gas-solid flows. *Int. J. Multiphase Flows*, **28**, 1479–1495.

Zhu, C., and You, J. (2007). An energy-based model of gas-solids transport in a riser. *Powder Technol.*, **175**, 33–42.

Zhu, C., You, J., Patel, R., Wang, D., and Ho, T. C. (2011). Interactions of flow and reaction in fluid catalytic cracking risers. *AIChE J.*, **57**(11), 3122–3131.

Zuber, N. (1959). *Hydrodynamic Aspects of Boiling Heat Transfer*. Ph.D. dissertation, UCLA.

Index

Printed in the United States
by Baker & Taylor Publisher Services